Recent Advances in Broadband Integrated Network Operations and Services Management

Varadharajan Sridhar
Sasken Communication Technologies, India

Debashis Saha
Indian Institute of Management Calcutta, India

Information Science
REFERENCE

Senior Editorial Director: Kristin Klinger
Director of Book Publications: Julia Mosemann
Editorial Director: Lindsay Johnston
Acquisitions Editor: Erika Carter
Development Editor: Myla Harty
Production Editor: Sean Woznicki
Typesetters: Mike Brehm, Keith Glazewski, Natalie Pronio, Jennifer Romanchak, Milan Vracarich Jr.
Print Coordinator: Jamie Snavely
Cover Design: Nick Newcomer

Published in the United States of America by
Information Science Reference (an imprint of IGI Global)
701 E. Chocolate Avenue
Hershey PA 17033
Tel: 717-533-8845
Fax: 717-533-8661
E-mail: cust@igi-global.com
Web site: http://www.igi-global.com/reference

Library of Congress Cataloging-in-Publication Data

Recent advances in broadband integrated network operations and services management / Varadharajan Sridhar and Debashis Saha, editors.
p. cm.
Includes bibliographical references and index.
Summary: "This book covers the principles of both wired and wireless communications of voice, data, images, and video and the impact of their business values on the organizations in which they are used"--Provided by publisher.
ISBN 978-1-60960-589-6 (hbk.) -- ISBN 978-1-60960-590-2 (ebook) 1. Broadband communication systems. 2. Wireless communication systems. 3. Multimedia communications. 4. Information technology--Management. 5. Telecommunication. I. Sridhar, Varadharajan. II. Saha, Debashis, 1965-
TK5103.4.R43 2011
004.6--dc22

2011009462

British Cataloguing in Publication Data
A Cataloguing in Publication record for this book is available from the British Library.

Table of Contents

Detailed Table of Contents

An ad hoc network comprises mobile devices with limited computing and energy resources together with wireless communication, which have to cooperate to provide networking services. This communication scenario presents many specific challenges that make ad hoc networks very different from traditional wired and wireless data networks. It makes classical approaches for network analysis insufficient. To deal with the design, implementation and test of this innovative communication paradigm, simulation techniques are of primary importance, since they allow to specify the level of detail of the simulated model. At the same time, the complex interaction among different entities make the performance evaluation of real ad hoc systems through simulation very hard. This chapter discusses traditional simulation strategies for ad hoc networks, highlighting their limits, drawbacks and possible overcoming. It presents efforts of the research community in improving the quality of simulation analysis according to different aspects, such as metrics definition, model design and simulation tools extensions. Then, the chapter focuses its attention on the benefits that the Discrete Fourier Transform analysis can produce if it is applied on simulation data. It describes a detailed methodology to gather and elaborate simulation measurements in order to avoid loss of information on rare events that occur in simulations. The presented methodology gets advantages (such as simplicity and flexibility) from simulative investigation approaches and, at the same time, offers a new analysis tool suitable for both protocol debugging and system performances evaluation. In fact, it transfers time-dependent measurements into the frequency domain, allowing to point out the occurrence of events which take place only under particular conditions and to detect occasional misbehaviors of the system.

The broadcast diffusion of messages in Delay Tolerant Networks (DTNs) is heavily dependent on nodes mobility, since protocols must rely on contact opportunities among devices to diffuse data. This work is a first effort to study how the dynamics of nodes affect both the effectiveness of the broadcast protocols in diffusing data, and their efficiency in using the network resources. The paper describes three control mechanisms. The mechanisms characterize a family of protocols able to achieve some awareness about the surrounding environment, and to use this knowledge in order to keep the broadcast overhead low, while ensuring high node coverage. Simulation results allow to identify the winning mechanisms to diffuse messages in DTNs under different conditions.

A very promising approach to discovering services and context information in ad-hoc networks is based on the use of Attenuated Bloom filters. In this paper we analyze the impact of changes in the connectivity of an ad-hoc network on this approach. We evaluate the performance of the dis-covery protocol while nodes appear, disappear, and move, through analytical and simulative analysis. The analytical results are shown to be accurate when node density is high. We show that an almost linear relation exists between the density of the network and the number of update messages to be exchanged. Further, in case of nodes moving, the number of messages exchanged does not increase with the speed of movement.

This paper presents a model-based framework to support the automated and adaptive deployment of communication services for QoS. The application domain targets cooperative group activities applied to military emergency operation management systems. Various models are introduced to represent the different levels of cooperation (applicative / middleware / transport). The adaptation decision process relies on structural model transformations while its enforcement is based on the dynamic composition of micro-protocols and software components. Automated deployment is performed both at the transport (i.e. UDP-TCP level) and middleware level. The architecture to support automated network management based on these models is introduced and its performance is evaluated through the use of a Java prototype.

Multimedia and real-time applications require bandwidth guarantees, which may be achieved by resource reservation. Several researches were done to propose efficient reservation MAC protocols for ad-hoc networks. In these schemes, channel is segmented into super-frames composed of fixed number of slots. They allocate slots to each traffic source, and make sure that neighbor nodes record the reservation in order to ensure consistency of reservations between neighbor nodes. However, resource reservation in ad-hoc networks remain a very challenging task due to the instability of radio channels, node

This study demonstrates that an integrated operator can benefit from cost savings, customer retention and prevention of revenue erosion by having a fixed-mobile convergence (FMC) migration strategy including introduction of advanced service packages. This development is driven by increasing importance of mobile network capabilities and services, as well as the decreasing gap between fixed and mobile systems, in terms of technological models and prices, making FMC both requested by the market and commercially feasible to provide. FMC is expected to offer benefits for network and service operators as well as businesses and consumers. We have analyzed the operator's dilemma on proper migration strategy in exploiting the benefits of cost savings and generating new revenues, but exposing oneself to the risk of substitution effects among its fixed and mobile products. We provide quantitative comparison of some strategic scenarios utilizing techno-economic case study methodology in modeling an integrated operator business in a Western European context.

The Policy and Charging Control (PCC) architecture was firstly introduced in the 3GPP's Release 7. However, the PCC has its problems. The main problems include the incapability of performing policy control with consideration of subscriber profiles and missing specification on how to organize and express the policy information. In addition, no policy control at application session establishment stage also contributes to its imperfectness. In this paper, the authors propose a subscription-based policy control framework that implements a subscription-centered approach for policy control and to enable flexible policy definitions based on the subscriber's profile at the application level. The framework also provides functionalities of organizing the subscription data, identifying the policy, regulating the policy control process, interpreting, managing and enforcing the corresponding policies. The main objective is to qualify the subscribers and thus, enhance the network customization through defining flexible policies based on policy control requirements for different subscribers.

we introduce two basic mobile query processing, namely: (i) mobile device side processing (MDSP), and (ii) server side processing (SSP). Our focus in this paper is on performing relational division and multiple division operations using both MDSP and SSP techniques. In addition, walkthrough examples are also illustrated. A series of performance evaluation of our proposed techniques are presented and analyzed.

Research on traffic characterization and analysis of cellular networks has been very active in the past decades. However, it is difficult for networks planners to incorporate new results to network engineering. On the one hand, some models are very complex and need advanced programming and skills to be properly computed. On the other hand, reliability on those models is poor because there is a general lack of published field studies to corroborate them. This paper proposes simple well-known teletraffic models for cellular networks compared to the conventional Erlang-B frequently applied when a first estimate is needed. To this purpose, the latest results on the characterization of the arrival process and service time in cellular systems are reviewed. According to the arrival and service characteristics, three models are proposed to obtain a first approach to the performance characteristics of the cellular system. The first two models are extremely simple, allowing direct computation of performance through closed formulas, while the third requires simple programming.

Air pollution is an important environmental issue that has a direct effect on human health and ecological balance. Factories, power plants, vehicles, windblown dust and wildfires are some of the contributors to pollution. Reasonable simulation tools exist for evaluating large scale sensor networks; however, they fail to capture significant details of node operation or practical aspects of wireless communication. Real life testbeds capture the realism and bring out important aspects for further research. In this article, we present an implementation of a wireless sensor network testbed for automatic and real-time monitoring of environmental pollution for the protection of public spaces. The article describes the physical setup, the sensor node hardware and software architecture for "anytime, anywhere" monitoring and management of pollution data through a single, Web-based graphical user interface. The article presents practical issues in the integration of sensors, actual power consumption rates and develops a practical hierarchical routing methodology.

The scarcity of the radio channel is the main bottleneck toward maintaining the quality of service (QoS) in a mobile cellular network. As channel allocation schemes become more complex and computationally demanding, alternative computational models that include knowledge-based algorithms and provide the means for faster processing are becoming a topic of research interest. An efficient deterministic technique, capable of handling channel allocation problems, is introduced as an alternative. The proposed model utilizes the Global Positioning System (GPS) data for tracing the hosts' likely movements within and across the cells and allocates the channels to the mobile devices accordingly. The allocation of the channels to the mobile hosts is deterministic in the sense that the decision of the channel allocation is based on the realistic data received from the GPS about the hosts' movements. The performance of the proposed technique has been evaluated by conducting the simulation experiments for the two parameters—call blocking and handoff failures. Also, a comparison of the proposed model with an earlier model has been carried out to estimate the effectiveness of the proposed technique. Experimental results reveal that the proposed technique performs better and is more realistic as well.

The instant deployment without relying on an existing infrastructure makes mobile ad hoc networks (MANET) a striking choice for many dynamic situations. An efficient MANET protocol may be applied to other important emerging wireless technologies, such as wireless mesh and sensor networks. This paper proposes a hierarchical routing scheme that is scalable, energy efficient, and self-organizing. The new algorithm that is discussed in this paper is the Dynamic Leader Set Generation (DLSG). This algorithm dynamically selects leader nodes based on traffic demand, locality, and residual energy level, and de-selects them based on residual energy. Therefore, energy consumption and traffic load are balanced throughout the network, and the network reorganizes itself around the dynamically selected leader nodes. Time, space, and message complexities are formally analyzed and implementation issues are also addressed. Incorporating the IEEE 802.11 medium access control mechanism and including the power saving mode, performance evaluation is carried out by simulating DLSG and four existing hierarchical routing algorithms. It shows that DLSG successfully extends network lifetime by 20 to 50% while achieving a comparable level of network performance.

Preface

1 INTRODUCTION

As we witness global macro economic downturn in many parts of the world, an evolution of some sorts is slowly happening in multimedia telecommunication and networking, much as the same we witnessed during the pre-Dot-com era. There has been a huge paradigm shift in the emerging networking landscape from an orderly, predictable, moderate-but-steady growth industry to a chaotic marketplace of disruptive technologies, rapidly changing regulations, complex mergers and acquisitions, and inorganic growth. These are dramatically changing the networking products and services market year-by-year. Currently, the existing networking infrastructure in most parts of the world is adequate only to deliver voice and text applications, but the demand for broadband-based multimedia services, such as streaming video and 3D telepresence is accelerating everyday. The explosion in cellular telephony and the Internet usage (both wired and wireless) has already created a huge demand for bandwidth/capacity and speed. Demand for integrating all data, voice, video and multimedia networks (both wired and wireless), which have matured individually, is gaining momentum to create an all-pervasive platform for user. A key driver behind this unprecedented level of mutation is certainly the rapid fall of the cost/gigabit-kilometer, resulting in sharp rise in the net market value of network products/services.

In today's networked world, two major trends have become prominent- first, the Internet is growing stronger everyday as the backbone service network to house myriads of multimedia applications; second, mobile wireless networks are offering a plethora of choices to the users to access those applications. In fact, wireless mobile networks revolutionized the scenario since late nineties with its concept of tetherless communication in short range (e.g., Bluetooth), middle range (e.g., IEEE 802.11, GSM) and long range (e.g., Satellite). Consequently, as mobile wireless handheld devices, such as laptops, PDAs and smartphones are becoming widely available, customers' needs for portable Internet services are growing commensurately. The Internet backbone, being mostly wired, is traditionally broadband, and now the bandwidth is leaping forward with optical fiber (supporting DWDM) providing the core level speed and capacity. Wireless networks, on the other hand, started in narrowband fashion but quickly elevated themselves to a level where most of them promise to be 'broadband'. Thus, as a whole, the global network scenario is pacing towards a high-bandwidth high-speed era where applications will be available at a lightning speed with a simple key-stroke or mouse-click.

The advantage of high-speed networking is changing the way businesses around the globe operate. The workforce, especially in the Information and Communications Technology industry is more geographically dispersed than ever before. Enabled by technologies, businesses have also started implementing "telecommuting" to reduce office space costs and employee commuting time. The productivity of such a workforce depends on the ability to work together on common projects as well as the ability to share knowledge and build consensus as quickly as possible. Businesses have started harnessing the collaborative models that first blossomed in non-business settings (from online games to open-source software projects to the so-called wiki encyclopedias and blogs) to speed up innovation. Apart from traditional collaboration tools such as Voice over IP (VoIP), Web 2.0 and application Mashups along with bandwidth intensive "telepresence" technologies are set to bring in the much-needed rich features to global collaborative work force.

Now comes the important poser, how the wired and wireless siblings operating in the back end and front end respectively, and will gel seamlessly to offer to users an unobtrusive experience for which they will not hesitate to shell out a considerable portion of their earnings. The bright point is that though the access networks are heterogeneous in nature, they all talk to the Internet and the common language is the IP. Naturally, when we talk about 'integration' today, we usually refer to IP-based integration as if the job will be done by IP alone. Easy to say, but hard to conceive!

Consider, for example, a train embedded in the global '*Broadband Integrated*' Network (BIN). Each compartment of the train may be a separate mobile WLAN, with each WLAN separately connected to a central Gigabit Ethernet switch which is homed at a mobile router (placed, say, on top of the driver's cabin) via an optical backbone. The mobile router again may be connected to the global Internet through 3G cellular networks (or, satellite networks). Individual users may have PDAs (or, laptops), which connect to the nearest access point via WLAN cards; finally, additional Bluetooth-enabled appliances may connect to WLAN using the PDA/laptop as an intermediary. It is a multi-tier network involving numerous devices, multiple technologies, and different service providers. It requires complex routing, intricate interworking, unmatched application provisioning, and finally complicated revenue sharing. But users want simple seamless experience at an affordable cost!

Take, another example, the case of telecom operators. Today, telecom operators of every mature nation are facing serious challenges introduced by the ever-increasing demand for provisioning diversified and digital convergent services in a fast-changing multi-technology network environment. It is extremely essential that they quickly respond to market and technology changes, satisfy customers' needs, while keeping total cost of ownership (TCO) under control. They need to rationalize their organizations to present a more nimble and lean competitive profile so that the growing user demand for voice, video, and data services and the ability to modify services on demand is met on the fly. Similar situation exists for data network operators too. They need solutions, which provides capabilities on a single platform for converging legacy broadband and IP networks, offer service providers an operational and competitive advantage. To address these issues, the operator's Operations and Services Management System (OSMS) (TM Forum, 2009a; TM Forum, 2009b; TM Forum, 2009c) must meet the requirements such as automated service provisioning for fast service fulfillment, proactive and reactive monitoring for end-to-end quality assurance, efficient and effective trouble handling, and high flexibility of customization and adjustment for offering new products to market in time. Again, easy to spell out, but hard to realize in reality!

While the underlying technologies continue to advance at a rapid pace, the fundamental principles governing the design, deployment and use of new-generation BIN of unprecedented scale, heterogeneity, and complexity are not entirely clear. How is BIN involving the emerging technologies best designed? This question pervades all layers of BIN: at the access level (scheduling coding and power level for packet transmissions, etc.); at the network level (providing a personal network space, designing QoS for heterogeneous networks such as wired-cum-wireless, etc.); at the transport level (how to modify TCP for wired-wireless networks, how to multicast in wireless networks, etc.); at the service level (providing intelligent interoperability, automated discovery, etc.); and, at the application level (a homogeneous view of the system in the presence of mobility providing a thin-client view of the environment for mobile units etc). Sounds not that difficult, had there been a standard set of rules to govern the complete ecosystem, given that BIN is not only heterogeneous but also multi-vendor networks (Erl, 2005; Broadband Forum, 2007; TM Forum (2009d). But, unfortunately, there is no standard management guideline for the BIN operations, integration and service provisioning. The lack of global standardization in the OSMS of BIN is a crucial void to fill up as early as possible- more so because many national governments are spending on national BIN projects considering that a country-wide BIN acting as a vital infrastructure can fuel GDP growth. For instance, the Australian Government is committing $4.7 billion of public funds to enable the extension and development of a high-speed National Broadband Network delivering services reaching out to 98% of the population. Korean government initialized the plan for building a nation-wide BIN way back in 2004.

2 TECHNOLOGIES IN BIN

The evolution of BIN began as early as with the advent of ARPANET in 1979, followed by LAN technology leading to distribute computing. Then, we saw the merging of computer networks with communication networks and the emergence of the INTERNET as a seamless internetworking platform. On the other hand, in the arena of infrastructure-based networks, the success of the Internet architecture as the backbone of a global, general-purpose, decentralized, data network is well established. Over the past decade, it has culminated in supporting emerging applications and adapting itself to dramatic changes in access technology.

Apart from developments in mobile telecom space, a silent battle for high bandwidth pipes on the core networks has appeared on the horizon again. High volume content uploading/downloading at a faster pace along with bandwidth intensive services such as Telepresence have fuelled the core broadband demand that can be supplied over Dense Wavelength Division Multiplexing (DWDM) technology. This phenomenon is evident with companies such as Google funding the Unity Consortium, which plans to build multi-terabit backbone linking different parts of the world.

While core network switches will see many fold increases in performance with DWDM on optical networks, major breakthroughs are coming in the near future for access networks too. Shortly, broadband over last mile, may it be copper, fiber, hybrid, or wireless, will be a major thrust in bandwidth growth. Gigabit over copper is already real, and soon we will see prices of 100 Gbps network adapters and switches dropping, further igniting the growth of data from laptops and desktops. Regarding Fiber-To-The-Home (FTTH) access technologies, Passive Optical Network (PON) is considered at present the best candidate for an FTTH solution enabling point-to-multipoint fiber connection. While point-to-point fiber connection for FTTH requires as many optical ports as the number of subscribers, PON has an advantage in reducing this cost as well as cutting down fiber installation cost. Eventually, Wavelength Division Multiplexing PON (WDM-PON) is expected to provide a dedicated 100 Mbps fiber connection to houses.

To accommodate wireless mobility, this wired access networks are now being complemented with a handful of wireless networks, such as cellular nets (GSM/CDMA), mobile ad-hoc nets (Bluetooth, sensor nets), wireless LANs (IEEE 802.11) and overlay nets, which are coming up with differing design goals, operating requirements, and implementation environments. Recent amendments in IEEE 802.11n brings in Multiple Input Multiple Output (MIMO) and OFDM in to home and enterprise LANs with data speeds touching up to 150 Mbps. Wireless networking is also becoming more intrusive using adhoc sensor networking pervasively (for instance RFID technology). Bluetooth can indicate and start proximity interactions with devices within (10-30) meter of distance. Recent advancements in Sensor Nodes Utilizing Powerline Infrastructure (SNUPI) provide ubiquitous radio connectivity within homes. IEEE 802.15.X specification that includes Bluetooth, Wireless Personal Area Networks (WPAN) and other flavors of low-cost, low-power PANs targets to enable 55/100 Mbps multimedia networking using portable devices such as PDAs and smart phones with the required Quality of Service support. Furthermore, if instantaneously reconfigurable wireless mesh networks are in place, and if we have one 802.11-enabled laptop or PDA, we will be able to send, receive and route data even when we are outside hotspot or cellular coverage. WiMax (Worldwide Interoperability for Microwave Access) which provides expanded coverage and high data capacities will probably act as the backbone of these adhoc wireless meshes. This facilitates network access for remote or suburban areas. However, power and spectral efficiency is key to its successful deployment. In 2004, the IEEE 802.16d standard was published for Fixed Wireless Access (FWA) applications. It is being seriously considered for early deployment to promote success in the broadband wireless market. Thanks to the perfect synergy between Wireless Local Area Network (WLAN) and the mobile WiMAX, new scalable wireless distribution system architecture without large investment is envisaged. This all IP based system architecture requires a few WiMAX base-stations for broadband access coverage as well as acting as the back-haul for the WLAN networks in its coverage footprint.

3 ARCHITECTURE OF BIN

At the backbone, there may be multiple constituent networks, such as the Internet, PSTN, satellite networks, packet radio networks and national WANs. At the access shell, there may be Ethernet LANs, wireless LANs, Bluetooth, ad hoc networks, CATV networks, wireless local loops and so on (Saha, et al., 2002).

Emerging wireless technologies along with powerful mobile applications and services are today ushering in an era where users can have access to any information from any device, any time, from any where they want. A whole new crop of Internet-ready portable any-time anywhere devices have become minimal, affordable and at the same time more powerful. This is riding on the falling mobile costs and greater penetration in countries such as Brazil, Russia, China and India. Cell phones, e-mails, instant text messaging and Blackberries are helping mobile, busy citizens link up with relatives, friends and neighbors on their way to work, in the middle of the night and on holiday trips. They provide services within local as well as wider areas on top of the Internet. These developments are making it possible to provide users with a seamless access to services in reality. Users will expect these devices to form dynamic ad-hoc, peer-to-peer networks to automatically around themselves, whether within home or office or in vehicles to connect to the backbone Internet.

Combined together, all these ad-hoc and infrastructure (backbone plus shell) networks, are promising a platform for any-time everywhere computing (Saha & Mukherje, 2003) that seamlessly and ubiquitously aids users in accomplishing their tasks effortlessly by offering unprecedented levels of access to information and assistance from embedded computers. One major component of this platform involves pervasive computation- computers and sensors "everywhere" in devices, appliances, equipment, in

Figure 1. A possible structure of BIN

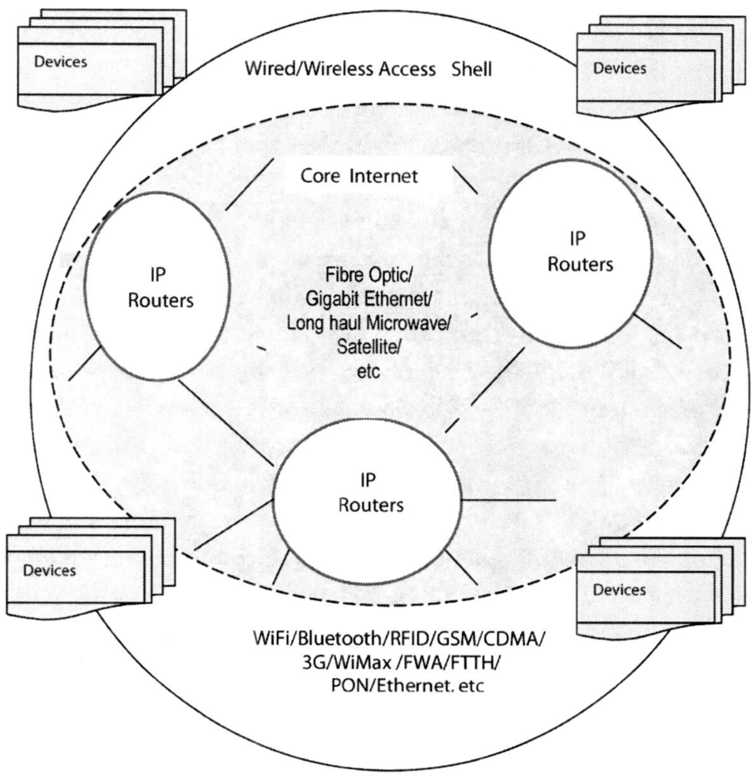

homes, workplaces and factories, and even in clothing. Another equally important component involves pervasive communication- a high degree of communication among devices and sensors through a ubiquitous and secure BIN infrastructure. Thus, BIN is a giant, ad-hoc, mobile, distributed cloud infrastructure with billions of devices hooked together, where services are coming in and going out continuously. Pervasiveness will bring about amazing increases in ability to connect and communicate, and users will be able to exchange information and control their environments from everywhere using this seemingly invisible infrastructure of BIN.

To achieve this goal, BIN has some unique requirements beyond those of traditional networking. Such a network should be cheaper to own and maintain, when each device and service is simple and directly reflects its use. To be accepted by its targeted non-technical consumers, it must be very easy to use too because these users do not have either skill or desire to manage and maintain complex networks. A key aspect of making BIN easy to use is making them self-stabilizing and self-configuring, rendering them virtually transparent to users. For example, networking all appliances and giving them smart software in BRINET environment will enable them to report in when they have problems, so that the agents can show up to replace failing parts, often even before the parts break. Moreover, systems will be able to reconfigure themselves automatically when changes occur. This is certainly not the style that we see in traditional networks today.

Apart from broad bandwidth, two distinctive characteristics, that distinguish BIN from legacy networks, are its *pervasiveness* and *dynamic heterogeneity*. Here, users will work through a wide variety of devices (some with very limited capabilities) and they may attach to an ad-hoc proliferation of wired/

wireless networked appliances. Due to this dynamic and distributed nature, many existing approaches to conventional networking become insufficient, when applied to BIN [8]. Although current network protocols are good for traditional (primarily wireline) networks and offer many technology independent solutions, we must remember that they were not originally designed for even wireless networks, not to speak of BIN. So their limitations are exposed as we move to a BIN like architecture. In fact, researchers strongly believe that the traditional protocols must be enhanced to a great extent in order to support BIN (Saha & Mukherje, 2003).

BIN is also quite different from traditional communication networks that are typically static, or exhibit only host mobility, which is limited to the last hop so far. But BIN will not follow this legacy and may contain ad-hoc proliferation of networks without manual intervention, many of whose nodes will be acting as mobile routers. Clearly, it is not enough to simply configure individual mobile hosts, since the entire network (including intermediate routers) may exhibit an *aggregate mobility*. For instance, consider the example of the train described in Section 1. An imminent OSMS challenge for this scenario is what happens if the train moves to a different territory and the mobile router attaches to a new base station from a different service provider with a different pool of global IP addresses. To ensure seamless connectivity to BIN, the *entire network* that lies below the mobile router, including the LAN access points, the laptops and the Bluetooth devices, may need to obtain *automatically* new (may be care-of) addresses that are part of the new cellular provider's pool. Thus, BIN will be dynamic and self-managed in its true sense. Participants on the network must simply self configure, i.e., network devices and services simply discover each other, negotiate what they need to do and which devices need to collaborate without any manual intervention. Centralized OSMS schemes will not work anymore, as in most traditional architectures and protocols. Moreover, BIN will be extremely heterogeneous in nature, and subnets will not be uniform.

4 BROADBAND ASPECT

It has been studied that with the FTTH that use 600 Mbps/150 Mbps (down/up stream) capacities, the cost of transmission line and cable installation are dominant. To provide the required bandwidth as a universal service, service providers need to upgrade their access network- fiber to the curve (FTTC)-based VDSL and FTTH. Very High-speed Digital Subscriber Loop (VDSL) has already achieved 50 Mbps at 300 m, and since fiber is available for apartment complexes, VDSL service is currently applicable to apartment areas. These trends offer opportunities for researchers to invent methods by which bandwidth intensive applications services are provided effectively over various core and access networks. The mobile broadband wireless industry is in the midst of a significant transition in terms of capabilities and means of delivering multimedia-rich IP services anytime, anywhere. In December 2005 the IEEE ratified the 802.16e amendment, which aimed to support Mobile Wireless Access (MWA) with seamless network coverage. Consequently, as 802.16e is commercialized, WiMAX will become a promising scheme to evolve FWA to MWA. Mobile WiMAX, a next-generation Orthogonal Frequency-Division Multiple Access (OFDMA)-based broadband wireless technology has been developed and published by the IEEE as 802.16e-2005. At present, there is particular interest in mobile WiMAX, since this offers data transfer rates that exceed those of current 3G. The all-IP mobile network specification is being defined by the Network Working Group (NWG) in the WiMAX Forum. The adoption of all-IP functional architecture tenets and IETF standardized protocols enables mobile WiMAX to deliver a wider selection of IP services while lowering capital and operational expenditures with a much faster time to market realization. It is well known that radio relay deployment achieves both coverage enhancement and capacity improvement. Relaying is a promising approach that tackles many of the challenges faced by mobile WiMAX and is currently under development as IEEE 802.16j project. Competing with WiMAX is Long Term Evolution (LTE) developed though the standardization process of Third Generation Partnership Project (3GPP). LTE is claimed to be the first GSM/ 3GPP standard that is fully IP and packet based. It is interoperable with the earlier 2G/3G networks and is touted as the emerging standard for 4G networks with the submission of LTE-Advanced specifications to ITU-T. Along with WiMax, competing technology alternatives such as Long Term Evolution (LTE), along with emerging Software Defined Radios are expected to bring the fourth generation experience to mobile users with full interoperability and anytime anywhere access.

With the above trends in wireless technologies, the successful launch of Apple's iPhone and other similar devices by mobile device manufacturers have enhanced the mobile user experience, resulting in possible increase in adoption of mobile Internet, mobile commerce, and mobile enterprise applications and services. The open movement in mobile software platform development kick started by Nokia's Symbian Foundation, Google's Open Handset Alliance and the LiMO foundation are set to change the mobile software eco system. Social networking sites such as Orkut and Facebook are set to enter in to mobile devices as indicated by the possible acquisition by search companies such as Google and Yahoo!. Microblogs, which enable mobile users to provide location and context sensitive information for sharing amongst the community members, are set to revolutionize the social network paradigm. Augmented Reality applications are becoming increasingly compact and powerful, and requiring nothing more than Smartphones to provide unique user experience. As was evident in the recent Beijing Olympics, video blogs aided by high-speed 3G mobile networks allowed real-time multimedia content to broadcast over the Internet, thus providing viewers in remote parts of the world a visual experience that was not broadcast through the traditional TV networks.

These technology and industry trends provide opportunities for researchers to develop methods by which advanced mobile services can be delivered in a technically efficient and economical way to both individual and enterprise users alike.

5 MANAGEMENT ISSUES OF BIN

As explained in the previous sections, BIN will consist of an extremely large number of diverse devices, numerous networking technologies, and will need to cater for a highly mobile user population.

Until recently, the core or the backbone network had enough bandwidth to accommodate the data traffic originating from the land mobile networks and the Internet. The "best effort" basis of routing on the TCP/IP networks worked well for traditional Internet applications such as emails and basic Internet browsing. Lack of "killer applications" and the narrowband of the access networks did not cause alarming congestions on the core networks. However the access networks have been upgraded through Digital Subscriber Line (DSL), Cable Modems, 3G and soon 4G platforms. Today, multitudes of newer bandwidth hungry applications available especially on the handheld Smartphones and Internet platforms, such as streaming video, telemedicine, on-line video games, and video calling, that are less tolerant to delay and jitter which are characteristics of a capacity constrained TCP/IP based network. The battle between the traditional telcos who controlled the destiny of the end users as to how telecommunication services are provided are being challenged by Internet companies such as Google and Skype who believe in empowering the end users with powerful software and Internet based applications and services.

With the development of such broadband and mobile Internet applications, the "Network Neutrality" principle is coming in to limelight and is being argued (Sridhar &, Venkatesh, Sep 14, 2009). Network neutrality is a principle that says those who operate networks which provide an overall benefit to the public good and rely on public property should not use their ownership to confer discriminatory treatment among their customers (Weiss, 2006). In order for the bandwidth hungry applications not to crowd out the traditional non-data applications, the network proponents suggest the network providers build more bandwidth (Yoo, 2010).

The proponents of "network neutrality", comprising mainly of Internet architects, want prohibition against such blockages of content and applications, especially those that are bandwidth intensive. Even economists argue that content and applications would generate usage of access networks and hence should not be blocked or prioritized. They argue that one key to the Internet revolution is the commitment to an architecture in which the pipes through which the data flows are as simple and general as possible and in which all of the intelligence is concentrated in the computers operating at the edge of the network. This architecture frees content and application providers from the need to obtain permission from network providers before deploying their innovations. In other words, TCP/IP promotes innovation by decoupling content and application providers from the network through which that content and those application travel. Any deviation that creates a tighter integration between the network and the content/applications that the network is carrying would chill innovation by raising the danger that part of the value of any innovations might be captured by the network provider.

However, Yoo (2010) argues that even when adding bandwidth by the network provider is feasible, maintaining sufficient capacity is quite a challenge when the data traffic on the mobile and landline broadband networks is growing exponentially of late. Bandwidth moreover cannot be expanded instantaneously. In these circumstances, it is imperative that the network operators prioritize traffic associated with applications that are less tolerant of delay over traffic with applications that are more tolerant to delay.

Active network management is the key to solve the above problems. Though Differentiated Services (as given in IETF RFC 2474 is implemented to mark and classify packets on the Internet, Quality of Service (Qos) is enforced on a hop-by-hop basis and does not guarantee end-to-end quality assurance. The network providers that are confronted with these constraints have no choice but to look forward to innovative network management practices to conserve bandwidth and better manage capacity. Greater degree of prioritization and integration between networks, applications and devices is required so that the users benefit. Further, researchers have started looking at mechanism design to solve network congestion problems (Anthes, 2010). TCP/IP communication protocols are based on the assumption that when a device on the network sees congestion, it will temporarily delay sending data. However, it does not work when TCP competes for bandwidth with certain non-polite protocols such as UDP. Anthes (2010) cites that all work on communication protocols should consider that the end devices connected to the network are different and selfish entities trying to maximize the use of network values.

The above debate also has prompted National Science Foundation to begin the project on Global Environment for Network Innovations (GENI) dedicated to providing new platforms to avoid the limitations built in to the Internet's current design that inhibit to some extent emerging applications and services. New demands being placed on the network are creating the need for fundamentally different architecture of the TCP/IP network and a very effective approach to network and services management. Provisions must be kept for configuring extremely large numbers of devices, satisfying the different end-to-end transport requirements of new applications, satisfying the special quality of service and security requirements, and also dealing with the high mobility between different network types.

The easy to use services available at any time anywhere along with the high bandwidth have increased the vulnerability of the network infrastructure as well. Enterprise users are mixing personal and business computing, thanks to function rich Smartphones and PDAs, and secure Virtual Private Network connections. These trends in usage increase network vulnerability. Security is a major concern for BIN and will be probably the most interesting research area in the times to come. Many a BIN technology cannot fly simple because of the lack of proper security support. One of the upcoming mobile (with camera and GPS) based solutions in the security space is video surveillance. A new breed of stand-alone camera is also available which offers IP connectivity with many other advanced features. These new video cameras are even powered by the Ethernet network that they are connected to. Some have wireless network interfaces too. Peer-to-peer communication over these devices bypasses the traditional security features of organizations. Exploitation of the Domain Name System vulnerabilities has increased the probability of attacks on Internet root servers and other basic network infrastructure. Experts often say that worm attacks are down, virus up. This appears to be due to more use of WLANs, mobiles, firewalls, and Intrusion Detection Systems. The rate of new virus appearance is also increasing. The IEEE 802.11 wireless network is vulnerable to Denial of Service and Man In the Middle attacks. Increased e-Commerce and m-Commerce activities have spawned more sophisticated social phishing attacks. A few years ago, the term malware, or malicious software, simply referred to viruses. Over the last five years, however, hackers and spammers have developed all sorts of new ways to invade PCs, laptops and 3G phones. Today, malware means any unwanted code or program that embeds itself on a computer without the user's knowledge. Unfortunately, now-a-days malware writers are no longer curious high school or college students pulling a prank. More often, they are professionals using sophisticated techniques, motivated by profit. Hackers have at their disposal vast ammunition to deploy attacks on the telecom and Internet global infrastructure from any location. With the growing black market industry of malware, protections against infection will have to continue to evolve in the coming years. The above trends warrant development of powerful methods and techniques to circumvent and prevent security attacks of global nature. Accordingly, research on cryptographic algorithms, key exchange protocols, antivirus software, spam detectors, firewalls and IDSs need also continue to gain new features to better optimize network-edge performance.

6 CONCLUSION

The long-awaited era of convergence of networks, coupled with overall high capacity and high speed, is fully upon us. As envisioned, it would allow billions of end users to interact seamlessly with billions of interconnected intelligent devices (including conventional desktop, handheld and embedded systems). The past few years have seen PDAs, cell-phones, laptops and handheld computers becoming common commodities. More and more of these portable devices are BIN enabled. The emergence of these user-friendly information appliances and new types of connectivity is spurring new challenges to BIN management: (i) dynamic networks of consumer-oriented devices that spontaneously and unpredictably join and leave BIN which has to maintain optimal network performance and high service levels, (ii) Continual, proactive monitoring and management of BIN must ensure efficient, high-quality broadband-based services and maximum subscriber satisfaction.

Work is being carried out by various organizations such as ITU-T special expert group called Next Generation Network Management Focus Group (NGNMFG), European Telecommunications Standards Institute (ETSI), TeleManagement Forum NGN Management Team on defining and standardizing network operations and service management framework for the next generation BINs (Choi & Hong, 2007). The need for BIN to assimilate traffic from diverse access networks with varying Quality of Service, mobility and security requirements, throws up huge challenges in terms of interoperability, and billing. The operations and services management should focus on areas such as scalable fault management; device configuration management including dynamic devise discovery; a comprehensive accounting management for authentication, accounting charging and billing; robust performance management for negotiating contracts between the customer and the networks; security management to provide appropriate levels of security to the different devices connected; mobility management for handoffs; and finally subscriber management for providing, monitoring and billing various terminal and service configurations.

In the past, the employment of standard platforms was an anathema to the development world because it was believed to limit the targeted technical domain. But times have changed, and the advantages of having a standard architecture for BIN are now well understood.

Traditionally the telcos implemented intelligence in their proprietary networks and managed the technical complexities of the network components, including precise calculations of their capacities. However, addition of features and maintenance of them grew so much that some smart telcos started outsourcing it to the network equipment vendors themselves, who in turn outsourced it to third party service providers. This spawned an entire eco system comprising of firms that researched, implemented and maintained features in the network equipment (Sridhar, 2009). However, in a BIN environment with intelligence moving to Customer Premise Equipments, all the players in this eco-system need to rethink and strategize on their next move about network and services management.

Telecom regulations and policies also need to adapt to this above change to BIN. Age old regulation, that compartmentalizes services with restrictions on ownership and service offerings must give way to a newer form that takes in to account convergence of technologies, networks and services.

We include a set of rich research articles in this collection that deeply probe in to the above issues of network and services management taking in to account technology evolution, economic considerations and regulatory implications.

Varadharajan Sridhar
Sasken Communication Technologies, India

Debashis Saha
Indian Institute of Management Calcutta, India

REFERENCES

Anthes, G. (2010). Mechanism Design Meets Computer Science. *Communications of the ACM, 53*(8), 11–13. doi:10.1145/1787234.1787240

Broadband Forum. (2007). *TR-069 CPE WAN Management Protocol*, Retrieved Sep 18, 2010 from www.broadband-forum.org/technical/.../TR-069_Amendment-2.pdf, Broadband Forum

Erl, T. (2005). *Service-Oriented Architecture (SOA): Concepts, Technology, and Design*. New Jersey: Prentice Hall.

Forum, T. M. (2009a). *New Generation Operations Systems and Software (NGOSS)*, Retrieved Sep 18, 2010 from http://www.tmforum.org/SolutionFrameworks/8428/home.html

Forum, T. M. (2009b). *Business Process Framework- enhanced Telecom Operations Map (GB921): R8.0*, Retrieved Sep 18, 2010 from http://www.tmforum.org/DocumentLibrary/EnhancedTelecomOperations/30660/article.html

Forum, T. M. (2009c). *Multi-Technology Operation Systems Interface (MTOSI) Solution Suite Release 2.0: Service Activation*, Retrieved Sep 18, 2010 http://www.tmforum.org/browse.aspx?linkID=30494&docID=3365

Forum, T. M. (2009d). *Shared Information/Data (SID) Model – Business View Concepts, Principles, and Domains (GB922): R8.0*,Retrieved Sep 18, 2010 from http://www.tmforum.org/DocumentsInformation/1696/home.html.

Saha, D., Mukherje, A.. (2003). Pervasive Computing: A Paradigm for 21st Century. *IEEE Computer*. 37(3)

Saha, D., Mukherje, A., & Bandopadhyay, S. (2002). *Networking Infrastructure for Pervasive Computing: Enabling Technologies & Systems*. Boston: Kluwer Academic Publishers.

Sridhar, V. (2009). Strategic Outsourcing: Opportunities and Challenges for Telecom Operators. In Bose, I. (Ed.). *Breakthrough Perspectives in Network and Data Communications Security, Design and Applications, Volume 1 of the Advances in Business Data Communications and Networking Series*. Hershey, PA, USA: IGI Global, 1-13.

Sridhar, V., & Venkatesh, G. (2009). Let the traffic flow. *Hindu Business Line*. Retrieved on Sep 14, 2009 from http://www.thehindubusinessline.com/

Chapter 1
Fairness Analysis and Improvement of Transport Layer Protocols

D. Tegze
Budapest University of Technology and Economics, Hungary

G. Hosszú
Budapest University of Technology and Economics, Hungary

F. Kovács
Budapest University of Technology and Economics, Hungary

ABSTRACT

The article presents a comparison of fairness properties of different congestion control schemes. It is hard to investigate the various protocol mechanisms implemented in transport protocols; therefore a simulator called SimCast is developed for the analysis of fairness characteristics of transport protocols as well as a network traffic generator and measurement tool called SimTest. This article presents the operation and basic properties of these evaluation systems together with some simulation and measurement results. The article also presents a fairness based bandwidth control mechanism, called the Balancer method, which optimizes resource allocation of busy servers with large amount of outgoing traffic. The efficiency of this control method is presented through simulation results.

INTRODUCTION

Reliability is one of the most important features of all multimedia applications. Due to the increasing deployment of traffic lacking end-to-end congestion control, congestion collapse can arise in the Internet (Floyd, 1999, pp. 458-472). This is resulted by links, sending packets that would only be dropped later in the network. The essential factor behind this form of congestion collapse is the absence of end-to-end feedback. An *unresponsive flow* is failing to reduce its offered load at a router in response to an increased packet drop rate, and a *disproportionate-bandwidth flow* is using considerably more bandwidth than other flows in a time of congestion. In order to achieve correct simulation of streaming media traffic, which is up to this time mostly not TCP-friendly, the effects of TCP protocol's flow control should be determined (Postel, 1981). However, there are many different TCP and other kinds of unicast transport protocol implementations with various flow control mechanisms, which make this investigation rather difficult.

The TCP-friendly congestion control schemes in the Internet were reviewed by Wang (1981) differentiating two groups of the TCP-friendly congestion control algorithms as (a) *end-to-end* and (b) *hop-by-hop* congestion control mechanisms. The end-to-end mechanisms are grouped into *(i)* AIMD-based schemes (AIMD: Additive Increase Multiplicative Decrease) with window- and rate-adaptation schemes, *(ii)* modeling-based schemes, including the equation based congestion control schemes, and the *(iii)* combination of the AIMD-based and modeling-based principle. Mostly this classification is used in our discussion, too.

B. Yu (2001) proposes another important approach about the survey on TCP-friendly congestion control protocols for media streaming applications in which several TCP-friendly congestion control protocols were discussed via a comparison on many important issues that determine the performance and *fairness* of a protocol.

The various mechanisms implemented in different protocols are hard to compare with each other, therefore a modularly structured simulator and measurement system is developed for traffic analysis of transport layer streams (Hosszú, 2001, pp. 369-411 and Tegze 2001, pp. 66-71). To carry out performance analysis of transport layer traffic, a well usable simulation and measurement framework should be applied in order to present statistically acceptable results for transport layer data transfer. The motivation behind the development of a new, custom simulator instead of using a standard framework like *ns* (Breslau, 2000) was the intention to build an integrated and uniform solution for analysis of the transport layer mechanisms and the implementation of some special features. Such a feature is a traffic generator module, which uses the simulator's protocol entities to generate real network traffic. On the one hand this allows the distributed execution of the simulator's congestion control algorithms competing with real transport protocol implementations. In this way fairness properties of real and simulated protocols could be analyzed. On the other hand this mode of operation allows the analysis of the real physical and data link layer effects directly, while using the simulated transport layer protocols. In this way we can avoid the development of complex physical layer models, and use this hybrid method to investigate the effects of lower protocol layers on transport protocols.

Most congestion control algorithms of the Internet are reactive methods. The underlying principle of these algorithms is the closed loop control scheme of control theory. Protocol entities measure some parameters of the network and feed this information back to the sender. This feedback determinates the sending rate of participant flows. Currently most methods make use of per-flow feedback. This kind of feedback cannot give strict guarantees on fair allocation of network resources inherently. One of the most important design and operational objective of transport protocols is the fair and efficient utilization of network resources. This is the main motivation for performance analysis of transport protocols. This article presents the newly developed control method called *Balancer* that is developed to provide the fairer bandwidth allocation for transport layer data flows. The method is implemented in the *SimCast* simulation environment. The mechanism is based on the global knowledge of outgoing sending rate of controlled flows. The article demonstrates the basic properties of endpoint congestion control which is the essence of the *Balancer* method. We prove the efficiency of the developed method through various simulation and measurement based results. Some statistical measures of the simulated and measured protocol entities are also analyzed as well as the most important quantities of protocol fairness and bandwidth distribution of the competing flows. The applied statistical measures include the coefficient of variation and standard deviation of the obtained bandwidth samples. The trace and some statistics of Raj Jain's fairness index (R. Jain, 1999) is also utilized to qualify protocol fairness. To analyze fairness and network efficiency simultaneously bandwidth trajectories of the competing flows are used. This method can represent the above mentioned two important characteristics of protocols very expressively by introducing the notion of fairness line and efficiency lines. To examine the dynamic properties of transport protocols fairness we have analyzed the transient behavior by measuring the time needed to reach a certain threshold value of fairness index. Using some combination of these protocol measures can give us a rather comprehensive estimation of both inter- and intra-protocol fairness and therefore can be used to prove the efficiency of the developed fairness based control method.

Main Features of the Simulator

The developed simulator software follows object oriented design principles, which results in modular structure. It uses a non-preemptive scheduler for the accurate execution of simulated objects. The scheduling is multi layered; these layers correspond to the simulated protocol layers. Most scheduling properties can be set on a layer basis. This scheduling system is capable of the execution of several protocol entities concurrently. The simulator is able to handle complex topologies and detailed network node configurations, especially for the transport layer.

The implemented protocol entities are realized as deterministic finite-state-machines. Some of the simulations carried out by *SimCast* were fully deterministic. However applying non-deterministic loss models, the simulations result in non-deterministic output, of course. *SimCast* has both deterministic and non-deterministic loss models built in. They can be selected arbitrarily at configuration time.

QUANTITATIVE ANALYSIS OF PROTOCOL FAIRNESS

The main goal of congestion control algorithms is to ensure the nearly optimal value of certain protocol parameters for competing flows. These protocol parameters are namely network utilization, delay, packet loss rate and fairness. Fairness is a principle in packet switched networks that determinates the rules of service. In other words it is a regulation that can be used to share available resources among concurrent flows. Quantitative characterization of protocol fairness is very important aspect of development and evaluation of transport protocols (T. A. Trinh, B. Sonkoly, S. Molnár., 2008).

For numerical qualification of fairness *Raj Jain's fairness index is used in a variety of performance analysis procedures for its several favorable properties listed below:*

- It can be used for small number of flows unlike variance for example.
- It is independent of scale and unit unlike standard deviation or other statistical characteristics.
- Its values are limited to the (0-1] interval unlike variance or standard deviation
- It shows positive correlation with fairness. The higher the fairness index is, the higher the degree of fairness is.
- It is continuous. It maps to a non-zero value if one or more flows receive no bandwidth, unlike *min-max ratio.*
- It can be interpreted in an intuitive way. For example bandwidth distribution of (1, 0, 1) gives fairness index of 2/3.
- It can be calculated directly from the coefficient of variation *(Equation 1)*
- It is easy to calculate the per flow fairness index from the global fairness index, which is the fairness perceived by a particular flow.

The calculation method of *Raj Jain's fairness index* is presented below:

$$F = \frac{\left(\sum_{i=1}^{i=N} x_i\right)^2}{N \times \sum_{i=1}^{i=N} (x_i)^2} = \frac{E(X)^2}{E(X^2)} = \frac{1}{1+COV(X)^2}, \quad (1)$$

where X_i is the normalized bandwidth of flow *i*: ($X_i = T_i/O_i$); T_i is the measured bandwidth, O_i is the fair bandwidth; *E(X)* is the expected value of the normalized bandwidth for stochastic analysis.

From the global fairness index presented above the per flow fairness index can be deduced. This shows the degree of perceived fairness for a particular flow. The calculation method defines the fair allocation first then qualifies the actual allocation of current flow proportional to the fair rate.

The global *fairness index presented above can be used to define the per-flow fairness index(Equation 2, 3, and 4). This can be used to determinate how fair the current allocation vector for a particular flow is. This value is greater than 1, if it can allocate more than the fair share, and it is smaller than 1 if the allocated bandwidth is smaller than the fair share. Its calculation method is presented below:*

1. First we define the fair allocation mark (x_f)

$$F = \frac{\left(\sum_{i=1}^{i=N} x_i\right)^2}{N \times \sum_{i=1}^{i=N} (x_i)^2} = \frac{1}{n}\sum \frac{x_i}{x_f} \quad (2)$$

$$X_f = \frac{\sum x_i^2}{\sum x_i}, \quad (3)$$

2. Equation 4 explains the perceived fairness of flow *i*:

$$F_i = \frac{x_i}{x_f}, \quad (4)$$

the ***SimTest*** measurement based procedure for the performance analysis of transport protocols

Measuring network performance is usually a complex procedure. It requires the consideration of several parameters, that determinate the quality perceived by the end users. Various performance measures are originated in some simple parameters that can be obtained by measuring and monitoring. Most performance metrics are related to the following important metrics:

- response time
- utilization
- throughput
- packet loss rate

Evaluation of network performance measures requires the determination of these basic parameters. Obtaining the necessary parameters can be accomplished in several ways:

- querying of intermediate network nodes (eg.: routers and switches)
- monitoring of real network traffic
- generating test traffic

Until now several software solutions were developed for the performance analysis of transport protocols. These software tools can accomplish large range of performance evaluation tasks. These procedures can accomplish comprehensive analysis of transport protocols; however we decided to develop our own solution since none of the studied solutions provided all our requirements together. Some software does not support querying of transport layer state variables or its timing capabilities are not adequate for us. Other procedures cannot handle complex configurations. Integration with our existing simulation procedure was also an important factor; therefore we decided to develop our own network measurement system called *SimTest*. The integration of the two systems made the comparison of simulated and measured results possible. This also allows the validation of protocols entities implemented in the simulator.

The *SimTest* software is able to handle multiple concurrent flows. At the beginning of measurement receiver entities synchronize their timers to the senders' clock. This allows high precision measurement of the generated traffic. The timing module of *SimTest* provides more precise timing properties than other examined measurement solutions.

The *SimTest* procedure can handle transient simulations, since the measured flows can be started and closed at arbitrary times. This allows the analysis of protocol aggressiveness and responsiveness by examining transient behavior of flows after changing network conditions or modifying the number of concurrent flows.

The output formats of the measurement and simulation systems are identical, therefore processing of results can be accomplished by the same procedure. The same software module processes the fairness, bandwidth, protocol state variable and statistical parameters for measured and simulated results.

The *SimTest* procedure supports the following protocols:

- **TCP:** a variety of TCP implementations and their parameters can be selected that is supported by the operating system. Several protocol state variables can be traced like *Cwnd, Twnd, EstRTT* etc. This allows the validation of protocol entities in the developed simulator
- **UDP:** the sender UDP entity can send packets at a configured fixed rate with sequence numbering, it can also easily be extended by more complex traffic models.
- **DCCP:** uses the official Linux kernel's DCCP implementation. This transport protocol is under active development, most performance evaluating measurement software cannot handle querying of its state variables. *SimTest* supports querying protocol variables among other more common features.

We present fairness related results of real TCP and UDP flows below. The measurements were performed with the TCP Reno implementations of Linux kernel version 2.6.21. The simulations were carried out on a host-router-host topology. The topology contained a 100Mbit/sec speed wired *Ethernet* and a 11Mbit/sec speed wireless link. The tests examined the bandwidth share and fairness of concurrent flows. Bandwidth logging frequency is set to 0.5 sec for the measurement.

Measuring Starvation of TCP Flow in the Presence of Fixed Rate UDP

During the measurement presented below a TCP Reno and a fixed rate UDP flow were running concurrently. The sending rate of UDP was set to allocate 90 percent of the available bandwidth. It can be seen in Figure 1 that TCP could only utilize the remaining 10 percent of bandwidth when competing with UDP that lacks congestion control at all. This results in serious deterioration of fairness. The mean fairness index for the 60 seconds of measurement was 0.66262. The minimum value of the fairness index for two competing flows is 0.5 which is experienced in the case of total displacement of a flow. The situation in the presented measurement is hardly better (Figure 2). The starvation of congestion controlled flows is a typical problem in the presence of non congestion controlled media streams. Media streams are usually implemented over UDP in the application layer. The shown results are based on the receiver side data delivery. The smoothness measures are shown in Table 1. In addition to the starvation of TCP, the oscillation of the fixed sending rate UDP flow can also be observed. It is caused by the fact that the two flows must share available bandwidth and buffer capacity at routers and TCP's bandwidth probing algorithm saturates buffers periodically. This causes packet losses in both flows, which can be measured as fluctuations in the delivery. Oscillations of queuing delay caused by the variation of queue length can

Figure 1. Bandwidth distribution of the examined TCP and fixed rate UDP flows

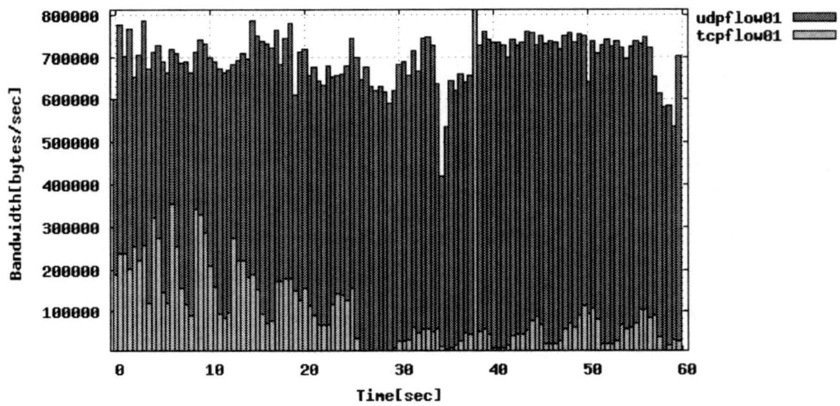

Figure 2. Time function of measured Raj Jain fairness index

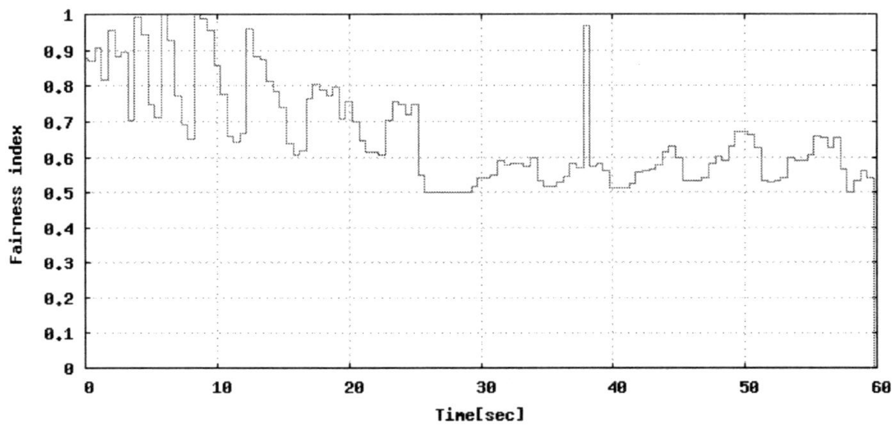

Table 1. Smoothness characteristics of the measured flows

Flow	Standard deviation [byte/sec]	Coefficient of variation
TCP 1	114200	1,078839
UDP 1	103170	0,174562

also cause smaller fluctuations in the measured throughput. The calculated standard deviation and coefficient of variation (Table 1) statistics show that the sending rate of congestion controlled TCP is more variable than that of UDP. This is caused by the AIMD (Additive Increase Multiplicative Decrease) congestion control algorithm of TCP.

It can be seen on Figure 1 that the bandwidth share of TCP decreases gradually due to saturation of queues and consecutive packet losses.

Figure 3 shows the bandwidth trajectory of the measured flows. At the beginning of measurement when the queues weren't saturated the bandwidth distribution was rather fair. After buffers got saturated, the trajectory diverged from the fairness line along the efficiency line. This indicates high utilization, while the fairness deteriorated significantly.

Figure 3. Bandwidth trajectory of the measured flows

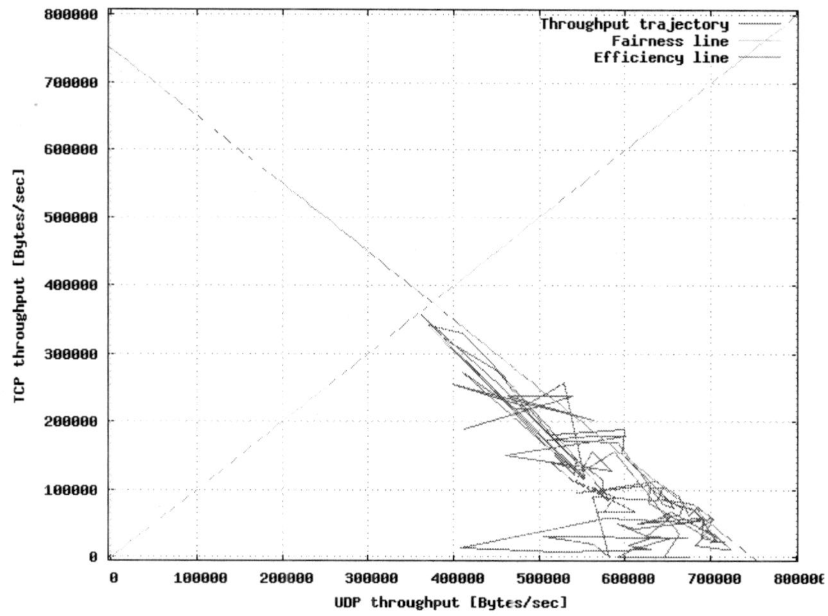

FAIRNESS BASED CONTROL FOR SENDING RATE OF DCCP/TFRC

Nowadays on the Internet a large range of transport layer congestion control mechanisms are operated. The largest part of the traffic is carried by the TCP protocol, which has several variants. In such a heterogeneous environment development of transport protocols that can operate under various conditions is a considerable challenge. Fair sharing of network resources and cooperation among current transport protocols are also of special interest. In many cases the contradictory requirements make it difficult to develop, introduce and disseminate a scalable transport mechanism. It is turned out from the literature, and the results of simulations and measurements that different traffic types will not compete fairly under certain network conditions.

The Essence of Network Congestion

Often the bandwidth requirements of data flows on the Internet exceed the available capacity. This makes the introduction of congestion control mechanisms necessary. On packet switched networks bandwidth requirements arise and disappear randomly, which results in variable traffic parameters. Sometimes traffic bursts reach routers that exceed available capacity of the network. This is often caused by flows that try to use the maximum sending rate available on the network. Because the amount of data that can be delivered is limited by the network, routers can either drop or buffer extra packets until they can be delivered. Since traffic bursts have limited duration, routers apply some kind of buffering. Network nodes often handle buffers according to the FIFO (First In First Out) queuing principle and drop packets when buffers are fully saturated. The basic assumption behind this behavior is that detection of packet losses on the sending node provokes the reduction of sending rate, which causes decreasing queue length. At first glance it makes sense to increase queue length in order to avoid packet losses, but increasing queue length also results in increased delay. For this reason data flows should keep FIFO sizes as low as possible, but large enough to ensure good network utilization (Hirabaru, M. 2006). Traffic of the Internet cannot be modeled using Poisson distribution, which means that on the long term there is no guarantee that the same amount of increment and decrement of bandwidth allocation occurs (Paxson and Floyd 1995). So independently of routers buffer size packet losses can occur, therefore it is worth keeping queue length as low as possible. When the network is congested, growing of queue length can be observed. This can also be inferred by the increasing round trip times or packet loss rate. The goal of congestion control is to achieve high efficiency besides low packet loss rate and delay.

Nowadays networks are usually overprovisioned and the emphasis is on the efficient and fair bandwidth usage rather than eliminating congestion (Crowcroft et al. 2003). Improving efficiency means settling the above mentioned problems, namely to provide nearly optimal throughput, low delay and packet loss.

Characteristics of Internet Congestion Control

Basically the congestion control mechanisms used on the Internet are reactive. The principle behind them is the closed loop control scheme of control theory. These methods are measuring the parameters of the network constantly and provide the sender with the obtained feedback information. The sender controls sending rate based on the incoming feedback. The measurements could be done on the intermediate network nodes of communication path, or at the receiver. The control mechanizm could be realized on routers or at the sender. The main principle of Internet congestion control is the end to end method, which means that the measurement takes place at the receiver side and the control is accomplished on the sender. There are mechanisms that measure the state of buffers on routers (Ramakrishnan et al. 2001), but these are only auxiliary methods, the main principle is the end to end congestion control. End to end control can be used under diverse network conditions, whilst router based methods are generally less scalable. Measuring network characteristics at the receiver can only give implicit feedback, since the receiver cannot obtain information directly about the state of the network along the communication path. In most mechanisms the feedback information is some measure of packet loss. This assumes that the main cause of packet loss is network congestion. This is an indirect inference for the degree of congestion. Under certain circumstances there are special phenomenons that cause packet losses. An example for such a scenario is wireless networking, where bit errors are more frequent because of interference shading and other effects (Cen et al. 2003). Delay information can also be used to estimate buffer saturation and infer the congestion of the network, but other factors like data link layer retransmissions can also be the cause of additional delay, introducing inaccuracy in the inferred degree of congestion.

It can be noticed the above mentioned congestion control mechanisms are based on per flow feedback. Under certain circumstances flows share available network resources rather unfairly. In many cases even protocol entities of the same kind have rather unfavorable fairness properties. One possible reason for this is the phase effect (Floyd, 1992, pp. 115-156) that can result in starvation of certain flows. For example consider a scenario where the buffer size of a router is five packets at the bottleneck link and six flows are competing for bandwidth. In this situation even in the most optimistic case one flow cannot allocate buffer capacity at all. It is shown from simulations and measurements that flows that receive less buffer capacity during the initial phase of communication allocate reasonably smaller amount of bandwidth than those with enough initial buffer capacity even in the long term. This can be explained by the rate reduction of flows that experience packet losses. Later on probability of occupying buffer

Figure 4. The essence of endpoint bandwidth control

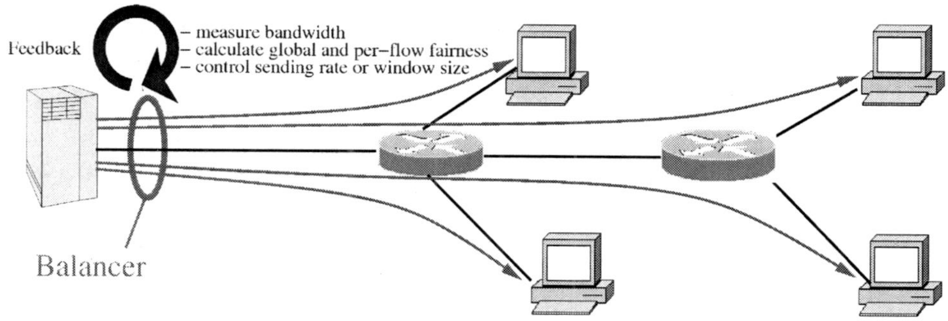

space becoming available at the router is much lower for flows with lower sending rate than for other flows. This results in permanent performance degradation of certain flows. The problem cannot even be solved by increasing buffer size, since the number of concurrent flows is changing constantly, so no optimal queue size can be chosen for all circumstances. Moreover buffer sizes should also be kept as small as possible for the above mentioned reasons.

Fairness Based Tuning of Congestion Control: Endpoint Bandwidth Control

At high traffic links of busy servers the above mentioned problems are of significant interest, especially for the outgoing data flows. Under such circumstances links are often overloaded and the fair allocation of bandwidth is especially important. The per-flow feedback based traditional congestion control mechanisms cannot give strict guarantees on protocol fairness among competing flows because of some problems. Such problem for example is the phase effect. For environments described above the developed *Balancer* method gives effective support to improve fairness of concurrent flows (Figure 4). The method incorporates a terminal bandwidth control mechanism that controls the outgoing traffic of busy servers. The method requires to measure fairness of concurrent flows quantitatively. The basis of the control is the per-flow fairness index (Jain, 1999).

The developed procedure allows numeric qualification of concurrent outgoing flows fairness by means of link monitoring or querying flows state information. Based on the calculated fairness value, the *Balancer* method gives feedback to transport layer protocol entities to improve fairness of their bandwidth distribution by modifying the allowed outgoing sending rate. The basis of the feedback is the per-flow Raj Jain fairness index (Hosszú, 2007 and Jain 1999). This fairness measure shows positive correlation with fairness distribution. The bandwidth control system was implemented in the *SimCast* simulation environment. Currently the TFRC protocol entities support the *Balancer* method. The control method modifies the allowed sending rate of TFRC by a factor, which is defined as the reciprocal of the per flow fairness index. This control modifies the sending rate of controlled flows proportional to the fairness difference from the fair allocation (Figure 5 and Equation 5).

$$B_i = \frac{1}{F_i} = \frac{x_f}{x_i} \qquad (5)$$

The value of the F global fairness index reaches the optimal value of 1 when the per flow fairness indices also converges to 1. From control theory aspect this method implements a closed loop

Figure 5. The Balancer factor as a function of fairness index

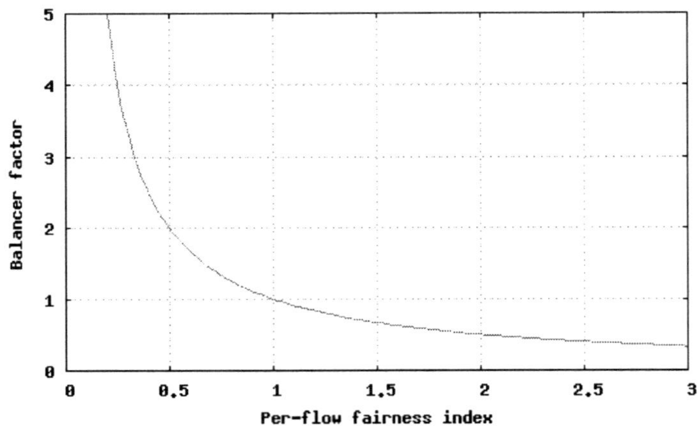

proportional control. More sophisticated control procedures are planned to be implemented in the future. The method is implemented to prove the concept of fairness based terminal control. The advantage of the applied proportional control is its simplicity and the stability. The *fairness manager* module of the *Balancer* method calculates fairness on a per-flow basis.

Gathering bandwidth data can be done by network monitoring or by querying controlled protocol entities. The fairness manager calculates the estimated fairness values periodically. To preserve stability of controlled sending rate and to avoid oscillations we apply an exponentially weighted moving average algorithm on the calculated fairness indices. The calculation method of the moving average is very similar to the algorithm used by most TCP protocols to calculate the estimated round trip time (Equation 6).

$$F_t = \alpha F_{t-1} + (1-\alpha)F_{sample} \quad (6)$$

Here F_t is the current fairness estimation, F_{t-1} is the value of fairness in the previous time step, F_{sample} is the current fairness measurement and α is the parameter of the moving average calculation. The value of α ranges in the interval [0-1]. This means that the fairness estimation of the previous time slice gives the α portion of the current fairness estimate and the rest of the estimation is given by the currently calculated fairness sample. Tipically the value of α is set between 0.8 and 0.9 for most simulations.

The α parameter determinates the basic behavior of bandwidth control. The EWMA control is conceptually similar to the PI control scheme of control theory (Cho, K.-H., 2005), so it corresponds to a control scheme with proportional and integrating terms. The α parameter is proportional to the factor of the integrating term. The α parameter is proportional to the achieved protocol smoothness and inversely proportional to aggressiveness responsiveness. Some kind of oscillations can also be decreased by applying EWMA control. Aggressiveness is not definitely unfavorable, since it can improve network utilization for flows with varying sending rate. An appropriate trade-off should be achieved to ensure proper network utilization and fair bandwidth distribution. The same is true for responsiveness. Generally the greater the value of α, the smoother is the achieved bandwidth. Since certain applications – for example multimedia streaming - require smooth sending rate, for these applications a large value of α is desirable. So when choosing the proper value of α for our control scheme we should take the requirements of our applications into account.

The frequency of bandwidth sampling can be set as a parameter of the procedure. To improve short term fairness the sampling and control period should be set to a value in the magnitude of round trip time. To control long term fairness, longer period should be selected.

The main advantage of the *Balancer* method is that it improves fairness of transport layer protocols without modifying the basic operation. It is also favorable that the method is able to avoid severe congestion in a pro-active way, since it regulates considerably unfair flows before serious congestion evolves. For network monitoring based variants it is sufficient to monitor network links of the sending protocol entities, there is no need to modify the routers or receiver nodes. This corresponds to the endpoint congestion control principle of Internet. The fact that the *Balancer* method does not need the modification of routers facilitates the gradual introduction. Spreading of router based methods is often limited by the requirement that all routers along the path should support the particular mechanism. The fact that only local modification is needed has the advantage that the fairness manager does not need to know the underlying network topology.

Since the basis of the control is the per flow fairness index that can be calculated for arbitrary magnitude of bandwidth and the applied algorithm is the proportional control of control theory which is scalable, the procedure can be used under most network conditions.

Analysis of the *Balancer* Method

The simulation results presented below analyze the efficiency of the *Balancer* procedure. During the simulations we executed a TCP Reno flow concurrently with a TFRC. Both flows start at time $t=0$ sec. End nodes are connected through two Ethernet and a point-point type network link. The simulated network topology contained two routers with FIFO queuing principle.

Table 2. The most important parameters of the simulations

Competing protocol entities:	TCP/RENO, DCCP/TFRC
Network topology	host-router-router-host
Router queuing policy	Drop tail
Bottleneck bandwidth	100kbit/sec
Packet size	1000 bytes
Bandwidth sampling period	1sec
Balancer control period	1sec
Balancer exponential moving average factor(α)	0.8

We have executed two simulations, one without the *Balancer* method and one with it. All other settings were left unchanged. The most important parameters of the simulations are listed below (Table 2).

The figures below present the time functions of bandwidth distribution and the global fairness index for the controlled and uncontrolled cases.

Comparing fairness indices of the two simulations (Figure 7 and 9) it can be seen that with the *Balancer* procedure turned on it takes reasonably less time for the examined flows to reach the fair bandwidth allocation and the mean value of global fairness is considerably better in the controlled case (Table 5). It can also be observed that the procedure eliminates long term oscillations of fairness. The bandwidth figures show that the sending rate of the controlled TFRC flow became less smooth which is affirmed by Table 3 and 4. As it is stated above the period of bandwidth sampling and control determinate if short or long term fairness is controlled. By this we have the control over tradeoff between short time fairness with variable bandwidth and long term fairness with smooth sending rate. The quantitative estimation of smoothness can be performed by calculating statistical characteristics of the time series of per-flow bandwidth data (Table 3 and 4). These tables present the coefficient of variation and standard deviation of examined flows. These

Table 3. Coefficient of variation of examined flows in the controlled and uncontrolled cases

Coefficient of variation (COV)	TCP Reno	TFRC
Without control	0,414159	0,462457
Controlled	0,468698	0,529699

Table 4. Standard deviation of examined flows in the controlled and uncontrolled cases

Standard deviation	TCP Reno	TFRC
Without control	2559,88	1329,27
Controlled	2461,25	2153,40

Figure 6. Bandwidth distribution of TCP and TFRC flows: non-controlled case

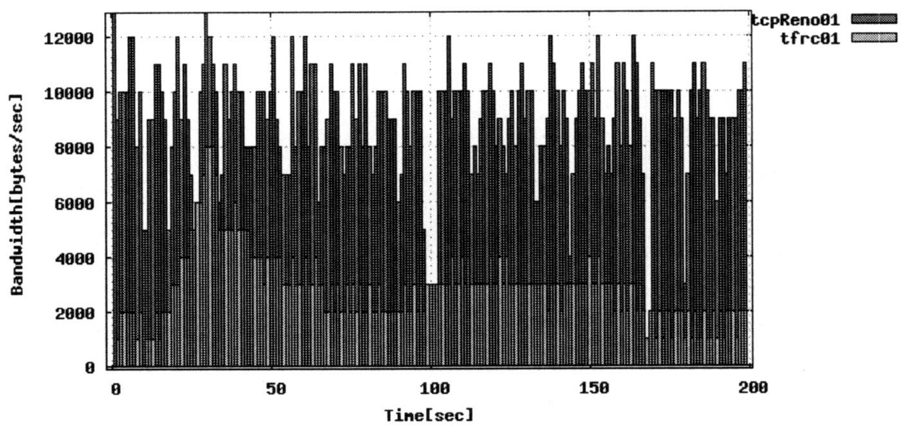

Table 5. Mean value of fairness index samples and the time needed to reach the threshold fairness index of 0.9 in controlled and uncontrolled cases

	Time to reach fairness index of 0,9 [sec]	Mean fairness index
Without control	17,0	0,82428
Controlled	8,5	0,91685

measures are inversely correlated with smoothness (Yang et al. 2001). Since the coefficient of variation means standard deviation normalized to the mean, this is not an absolute measure of flows smoothness; therefore it is not applicable for comparison. For this purpose standard deviation is a better measure, because it shows the typical distance from the mean in terms of the examined quantity. This absolute measure can be used to compare the smoothness of concurrent flows. Coefficient of variation is feasible for the qualification of flow's smoothness of its own, while standard deviation is applicable for the comparison of different participating flows smoothness. Table 3 shows that the coefficient of variation for TFRC is greater than in the case of TCP. This is caused by the above mentioned normalization effect. It can be seen in Figure 6 that the mean bandwidth allocation of TFRC is considerably smaller than that of TCP's in the non controlled case.

Considering the standard deviation data of Table 4, it is obvious that TFRC has smoother sending rate than TCP in both controlled an uncontrolled cases. Examining the influence of control it can be seen that the relative smoothness (coefficient of variation) is deteriorated for both examined protocol entities in the controlled case (Figure 8). The standard deviation increased in the case of TFRC and it decreased in the case of TCP. The decreased standard deviation of TCP is the result of decreased bandwidth interval caused by the *Balancer* method. It can be seen from the results that the applied control algorithm performs

Figure 7. Global fairness: non-controlled case

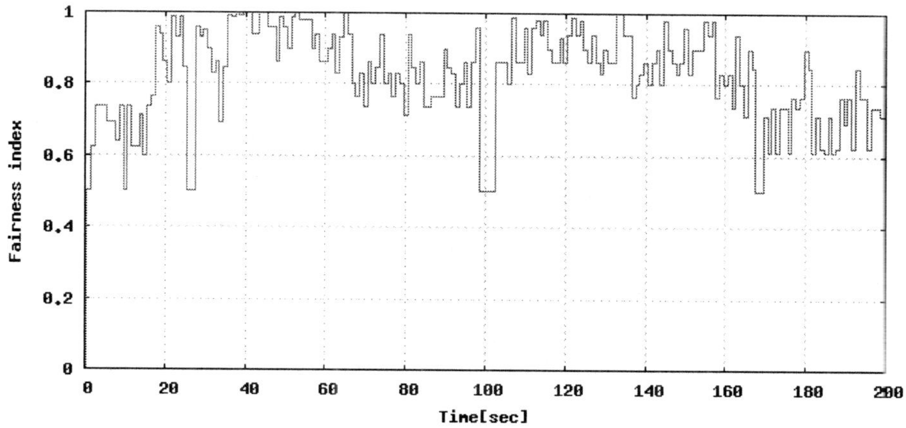

Figure 8. Bandwidth distribution of TCP and TFRC flows: controlled case

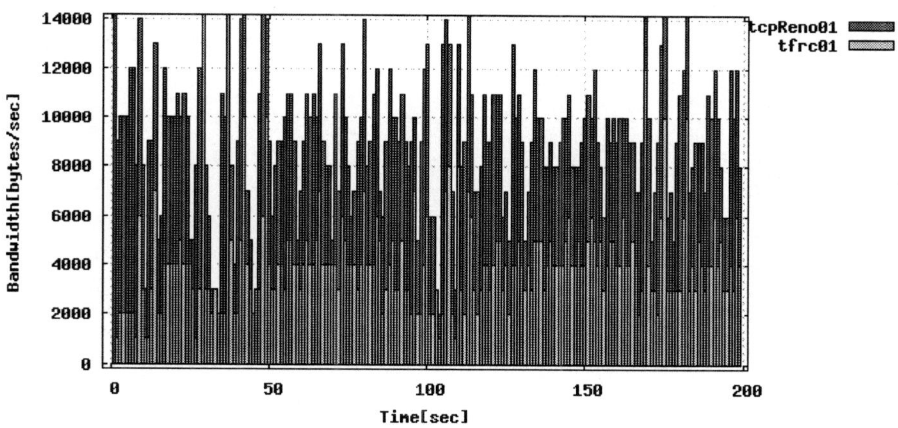

Figure 9. Global fairness: controlled case

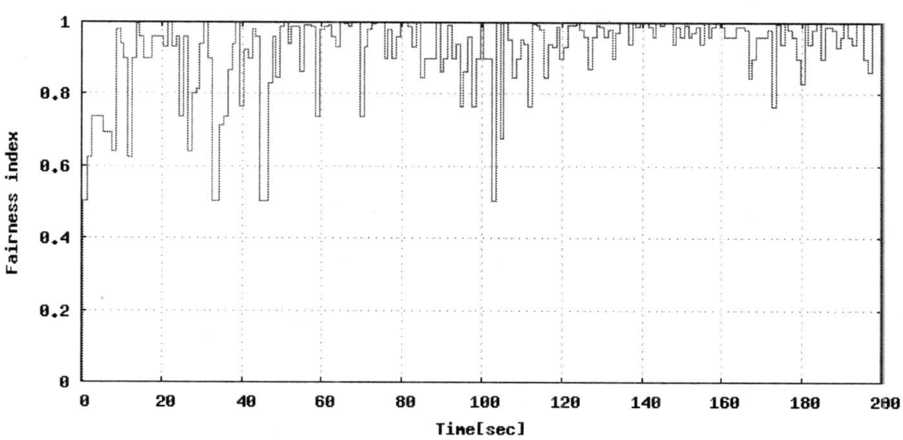

Figure 10. Interpretation of bandwidth trajectories, example of an ideal AIMD system

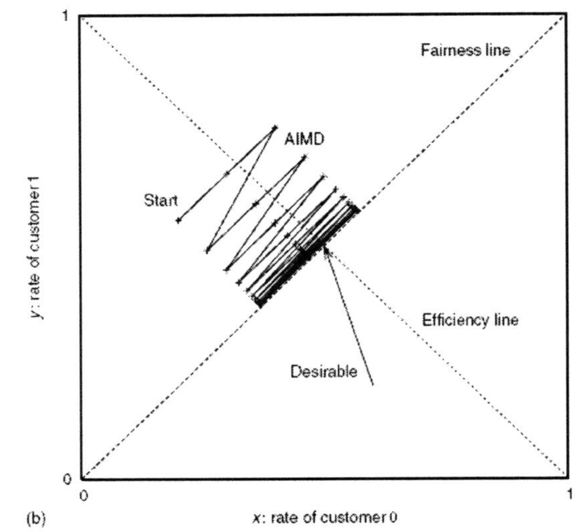

Figure 11. Trajectories of the examined flows in the uncontrolled (left) and the controlled (right) cases

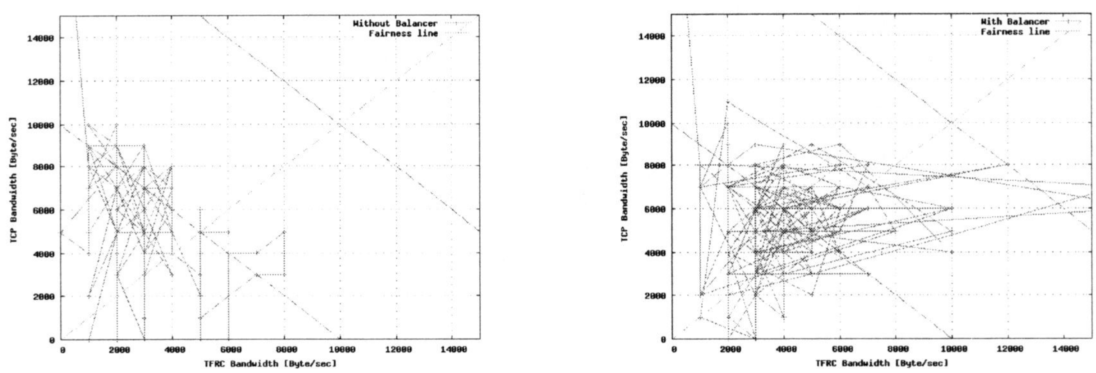

the tuning of tradeoff between smoothness and short term fairness. Deeper understanding of control algorithms and finer tuning of control is an important task for the future. In this simulation we only controlled the sending rate of TFRC, even so the smoothness of TCP changed considerably, since the flows share the available network capacity. Deeper understanding of bandwidth control's additional effects is also an important research task.

Table 5 shows the characteristics of global fairness in the controlled and uncontrolled cases. The mean fairness index values show that the procedure improved global fairness significantly. Dynamic properties of bandwidth control are measured as the time for the global fairness index to reach the threshold value of 0.9. The time period is measured from t=0 sec. The presented results show that it takes half the time to reach threshold value for the controlled case compared to uncontrolled case.

The sending rate trajectories shown below can be used for the analysis of concurrent flows utilization and fairness. Bandwidth trajectory traces are created by plotting the sending rates of one flow as a function of the other flow's sending rate for a particular time interval. Connecting the series of points belonging to the subsequent time intervals shows the evolution of bandwidth distribution over time. This representation can visualize the influence of competing flows to each other expressively. The points representing equal bandwidth distribution are located on the fairness line (Jain, 1989); these points have optimal fairness properties (Figure 10.). The lines crossing the pole form the equi-fairness lines. Points on these lines have equal fairness values, since the ratio of bandwidth values defining the points are constant. *Efficiency lines* connect points with equal aggregate bandwidth. In optimal case flows are operating close to the intersection of fairness line and the efficiency line belonging to the available bandwidth. The goal of congestion control algorithms is to keep flows as close to the intersection as possible. Figure 10 shows the ideal trajectory of two flows using *AIMD (Additive Increase Multiplicative Decrease)* congestion control scheme. It can be seen that in the *additive increase* phase the sending rate moves along a line parallel to the fairness line, while in multiplicative decrease phase it moves along an equi-fairness line.

Below the effect of balancer procedure is shown using bandwidth trajectories. Figure 11 presents the trajectories for the simulations demonstrated above. The figure contains trajectories for both uncontrolled and controlled cases. The figure showed that the Balancer method improved fairness significantly, since the mean distance from the fairness line of points belonging to the controlled case are considerably smaller than in the case of uncontrolled simulation. As a result of the control there are some points with rather unfair bandwidth distribution. This phenomenon deteriorates smoothness of flows significantly. This effect of control on protocol smoothness is justified by the statistical measures of Table 5. These deteriorative effects can be mitigated by finer control like PID control. It can be observed in the figure that the efficiency remained in the same domain in the controlled uncontrolled cases, whilst fairness improved significantly. This is also justified by the results of Table 5.

CONCLUSION

Understanding complex networks is a serious challenge. This is largely assisted by simulation and measurement based procedures. Simulation allows detailed analysis of problematic parts in the models of different protocols. The *SimCast* simulator presented in this article implements several congestion control mechanisms that allow the examination of several aspects transport protocols under development. The presented traffic generator and measurement procedure called *SimTest* can be used to analyze real transport protocols implemented in modern operating systems. Several network characteristics can be determined and the dynamics of transport layer state variables can also be traced for the competing flows via measurements. The results showed that the developed procedures are useful tools of stationary and dynamic investigations of transport layer protocols for both real and simulated environments.

The presented *Balancer* control method allows the fairer allocation of busy servers' outgoing traffic for flows utilizing different transport protocols. The results showed that the method improves flow fairness without the significant modification of transport protocols. As it is shown in the article, the control deteriorates smoothness of controlled flows. The degree of deterioration depends on the parameters of control. By use of different control methods and parameters the tradeoff between fairness and smoothness can be tuned. The implemented control method is a simple application of the unidirectional endpoint control principle presented in the article. By the usage of more complex control schemes like PID control from control theory better control behavior could be achieved. This forms the part of our future investigation. Better understanding of additional effects of endpoint control is also an important area of our research.

REFERENCES

Brakmo, L. S., Peterson, L. L. (1995, October). TCP Vegas: End to end congestion avoidance on a global Internet, *IEEE Journal on Selected Areas in Commun., 13*(8), 1465-1480.

Breslau, L. (2000, May). Advances in network simulation. *IEEE Computer.*

Cho, K.H., Park, S.J., Jung, E.H., Shin, S.W., & Lee, H.H. (2005, October). End-to-end rate-based congestion control using EWMA for multicast services in IP networks Communications, *IEEE Proceedings, 152*(5), 668-672.

Crowcroft, J., Hand, S., Mortier, R., Roscoe, T., & Warfield, A. (2003). Qos's downfall: At the bottom, or not at all! *Proceedings of the ACM SIGCOMM Workshop on Revisiting IP* QoS, 109-114. ACM Press.

Floyd S., & Fall K. (1999, August). Promoting the use of end-to-end congestion control in the Internet, *IEEE/ACM Trans. Networking, 7*(4), 458-472.

Floyd, S., & Jacobson, V. (1992, September). On traffic phase effects in packet-switched gateways, *3*(3), 115-156.

Hirabaru, M. (2006, January). Impact of Bottleneck Queue Size on TCP Protocols and Its Measurement. *IEICE - Trans. Inf. Syst.,* E89-D(1), 132-138.

Hosszú, G. (2001). Introduction to Multicast Technology. In S. M. Rahman (ed.) *Multimedia Networking: Technology, Management & Applications,* , *Idea Group Publishing,* Hershey, USA, pp. 369-411.

Jain, R. (1989) A delay based approach for congestion avoidance in interconnected heterogeneous computer networks. Computer Communications Review, *ACM SIGCOMM,* 56–71.

Jain, R., Durresi, A., & Babic, G. (1999, February). Throughput Fairness Index: An Explanation, *ATM Forum/99-004.*

Miniwatts Marketing Group. (2007, June 30). Internet world stats: Usage population statistics. Retrived from http://www.internetworldstats.com/stats.htm

Orosz, M., & Tegze, D. (2001). The SimCast Multicast Simulator. *Int. Workshop on Control & Information Technology, IWCIT'01,* Ostrava, September 19-20, 66-71.

Paxson, V., & Floyd, S. (1995). Wide area traffic: The failure of poisson modelling. *IEEE/ACM Transactions on Networking* 3(3), 226.

Postel, J. B. (1981, September). Transmission Control Protocol, DARPA Internet Program, Protocol Specification, *RFC 793.*

Ramakrishnan, K., Floyd, S., & Black, D. (2001). The Addition of Explicit Congestion Notification (ECN) to IP *RFC 3168* (Proposed Standard).

Song, C., Cosman, P.C., Voelker, G.M. (2003, October) End-to-end differentiation of congestion and wireless losses, *IEEE/ACM Transactions on Networking* (TON), *11*(5), 703-717.

Tegze, D., Hosszú, G. (2007).Analysis of TCP-Friendly Protocols for Media Streaming. In M. Freire and M. Pereira (eds) *Encyclopedia of Internet Technologies and Applications.* Information Science Reference, Hershey, USA.

Trinh, T.A., Sonkoly, B., & Molnár, S. (2008, May 29-30). Revisiting FAST TCP Fairness, *18th ITC Specialist Seminar on Quality of Experience,* Karlskrona, Sw.

Wang, Q. (2001). TCP-friendly congestion control schemes in the Internet, In *Proceedings of the 2001 Int. Conf. on Inf. Technology and Inf. Networks (ICII'2001)* (vol. B, pp. 205-210), Beijing, China

Yang, Y.R., Kim, N.S., & Lam, S.S. (2001). Transient behaviors of TCP-friendly congestion control protocols. In *Proceedings Twentieth Annual Joint Conference of the IEEE Computer and Communications Societies*. (IEEE Volume 3) (pp. 1716–1725).

Yu, B. (2001, December). Survey on TCP-Friendly Congestion Control Protocols for Media Streaming Applications. Retrieved from http://cairo.cs.uiuc.edu/~binyu/writing/binyu-497-report.pdf

This work was previously published in International Journal of Business Data Communications and Networking (IJBDCN) Volume 5, Issue 1, edited by Varadharajan Sridhar and Debashis Saha, pp. 1-21 , copyright 2009 by IGI Publishing (an imprint of IGI Global).

Chapter 2
Rembassy:
Open Source Tool For Network Monitoring

Vreixo Formoso
University of A Coruña, Spain

Fidel Cacheda
University of A Coruña, Spain

Víctor Carneiro
University of A Coruña, Spain

Juan Valiño
University of A Coruña, Spain

ABSTRACT

Even though monitoring tools are essential to the management of communications networks, Open Source applications still confront their potential users with considerable problems. This work analyses the limitations of the currently existing tools and presents the development of a new tool that solves most of those problems. The tool is based on a new architecture of objects and remote method invocation and allows both centralized and distributed monitoring. Its configuration through web interface, its support to monitoring templates, and its flexibility make it particularly interesting for a large number of users in search of a strong but easily configurable system. The proposed extension system is based on plug-ins and it is highly innovative because of its power and simplicity. Finally, the configuration simplicity and other essential improvements of the proposed system are successfully tested in a real environment.

INTRODUCTION

Monitoring applications are fundamental in network management, because they monitor the state of the various components that make up an information system by notifying the user of distinct problems and incidences. This monitoring can be divided into two main groups, depending on whether the system analysis is carried out to locate future problems (proactive) or if it is limited to locating existent problems (reactive).

However, in the context of Open Source, a complete and flexible application that satisfies the needs of large number of users is still lacking.

DOI: 10.4018/978-1-60960-589-6.ch002

Features that are expected from and provided by modern monitoring applications, such as execution-time configuration through web interface, support for distributed monitoring, monitoring templates, etc, either do not exist in the available open source applications, or present considerable limitations. Also, and in spite of the fact that these are open code applications, it remains very difficult to modify them in order to incorporate some of the above features, partly because many of the applications would require important changes in the application architecture.

This work presents a new monitoring application that solves many of the mentioned problems: the development of Rembassy can be followed on the open web portal http://rembassy.sourceforge.net.

Firstly, Rembassy simplifies the interaction of the user by means of an intuitive web interface that allows us to both configure the system and consult the state of the monitored elements. It does not rely on a complex configuration based on text archives that is used in most of the existing applications. Instead, it provides great flexibility, supports both centralized and distributed monitoring schemes, and offers the remote access and configuration of agents that allow us to monitor elements that are not available through the network.

Rembassy's distributed monitoring features convert it into a system that can be scaled up to large networks. Also, thanks to its hierarchical object structure and the monitoring templates at various levels, configuring Rembassy is easy, even in environments with a large number of machines and services to monitor.

Finally, we would like to mention its plug-ins system, which is remarkably superior to that of other Open Source tools and simplifies the extension of the tool and its adaptation to the needs of each individual user.

The present article is structured as follows: we start by analysing the most important characteristics of the existent Open Source monitoring systems and identifying their major shortcomings; we continue by describing the architecture proposed to resolve these problems and presenting a real case study; finally, we present the conclusions and the future lines of research.

STATE OF THE ART

In spite of numerous efforts in recent years, during which the Open Source monitoring applications have improved considerably, a number of limitations persist and essential features are still lacking or are too complex for many users.

This work analyses various open source applications, in particular a subset of the most complete and/or popular ones, whose characteristics can be seen in Table 1.

We have been able to detect a set of limitations that are common to most of the currently existing Open Source applications:

- **Monitoring limitations:** distributed polling remains one of the weak points of Open Source monitoring systems. Many tools do not even support it, and those that do are hindered by configuration difficulties or characteristics limitations.
- **Complex configuration:** In most applications, the configuration is carried out by means of text archives. Creating or modifying these archives requires a more or less profound knowledge of their syntax, and even an expert user may need to consult a handbook to configure the application. The more complete an application is, and the more elements it allows us to monitor, the more complex are its archives. Also, applications that support distributed polling require the configuration of each node separately, which complicates the configuration even more. Finally, in these applications configuration is a static process that requires the re-initiation of the system.

Table 1. Open Source monitoring tools

	Monitoring interface	Polling	config	SNMP	extensible?	historic?
Angel Network Monitor[1]	Web C	C	Text C		√	
Big Sister[2]	Web C	D	Text D	√ [trap]	√	√
Ganglia[3]	Web C	D	Text D		√	√
Mars[4]	Windows C	C	Windows C		√	
Nagios[5]	Web/Wap C	C/D	Text C/D	√ [traps]	√	√
OpenNMS[6]	Web C	C	Text C	√		√
Pandora FMS[7]	Web C	C	Web C	√ [lim]		√
Sysmon[8]	Web /Commands/ Text C	C	Text C			
Zabbix[9]	Web C	C/D	Text D / Web C	√ [trap/poll]		√
Zenoss[10]	Web C	C	Web C	√ [traps]	√	√

Legend: C: Centralized; D: Distributed

- **Limited extensibility:** even though many of the studied applications support a system of modules or plug-ins, extending an application beyond small features, such as monitoring a new element, is really quite complicated. Also, the extensibility provided by the existing tools is either rather complex or very limited. In general, many tools are extensible in that they allow the monitoring of a new element, but not one single tool supports the extension of features or capacities. With regard to the elements that can be monitored, the tools range from those that only monitor TCP/IP services to those that monitor a large number of system applications or parameters. Tools based on agents even allow us to monitor elements that are not accessible through the network. Logically, since they are free applications, those that are widely used and dispose of a simple extension mechanism will benefit from the contributions of their users and provide support for many more elements. This is the case of applications such as Nagios.
- **Limited SNMP support:** SNMP (Simple Network Management Protocol) is a standard for the monitoring of network devices [William Stallings (1998)]. It is a widely extended protocol that interacts with many monitoring tools. However, due to certain limitations in representing management information and a series of security problems, especially in versions 1 and 2c, most applications are not based on this standard but rather define their own protocol to obtain, transmit, and store monitoring data. In most cases, the support of SNMP is limited to the reception of SNMP traps. Among the analysed applications, only OpenNMS is actually based on SNMP.
- **Few analysis tools:** It would be advisable for a monitoring application to provide not only the notification of incidents or the state of monitored services, but also certain analysis tools, such as incidences registry, graphics with the evolution of the services workload and output, etc. Such tools are extremely useful for the anticipation of potential problems and the planning of systems growth. However, support for this type of utilities is rather limited and in many cases reduced to the mere recollection of historic data.

- **Orientation towards Unix platforms:** Given the narrow relationship between GNU/Linux systems and Open Source software, most tools are logically designed for Unix platforms. However, since they tend to dispose of agents for various platforms and generally have devices dedicated to monitoring, this is a minor problem. Even so, a multiplatform architecture is highly recommendable.

There is no doubt about the fact that, at the present moment, Nagios is the most popular open source tool because of its flexibility and support to monitoring large amounts of elements [E. Galstad (2007)]. On the other hand, the complex configuration of Nagios, which is based on text archives, may constitute an important problem for many users. Furthermore, its design based on structured programming and the shell scripts complicate its maintenance and extension, even with such an ample user base. Other projects that have appeared recently, such as Zenoss or Pandora FMS, simplify the configuration of the system by basing the interaction with the user on a web interface, but they continue to present considerable deficiencies.

ARCHITECTURE

General Vision

Rembassy is executed by means of a background process or daemon. The Rembassy daemon groups two tasks that were traditionally kept in separate execution units: obtaining the monitoring information, which is a task for the agent, and managing that information, which is done by the manager. In Rembassy, one single process is in charge of both tasks. The main advantage of this system is its simplicity, both in developing the system and making it available on the network (all the devices are installed and configured in the same way).

Generally, when monitoring a network or a set of servers, various daemons are deployed in various computers that interact. The fact that all the elements are homogeneous simplifies the interaction. In Rembassy, the communication between the manager and the agent does not differ from the communication between managers, because manager and agent are one and the same process and as such share the interaction with the outside. This simplifies the deployment of the system, both centralized (one manager) and distributed (several managers).

Object-Based Management

One of the most important aspects of Rembassy is how it represents the management information. The Management Information Base (MIB) of most applications (such as SNMP) is restricted to maintaining the value or state of each monitored element, a value updated periodically by an agent. In Rembassy, the management information acquires a new dimension thanks to its representation through objects, which not only store the state of the element, but also define, through a set of configurable methods and parameters, their own behaviour: how and when the management information is recovered or updated, how it is stored and transmitted, etc.

Apart from that, the traditional more or less static tables are replaced by a dynamic structure that is completely modifiable in execution time and in which the objects are stored hierarchically. Each daemon maintains its own objects structure while managing the creation and elimination of objects and the invocation of methods for them. The structure is based on a set of containers and subcontainers that, naturally, are also objects and define the necessary methods to add and eliminate objects in execution time.

Some objects of this hierarchy are persistent, i.e. they maintain information that must be conserved even though the execution of the daemon is interrupted. Rembassy maintains such information in a relational database. Each object is responsible for defining which information must be conserved in a persistent medium. On the other hand, certain objects simply define behaviour, they do not store information, and therefore do not need persistent storage. Rembassy loads these objects during the start-up and maintains them in a virtual container (which functions similarly to the files system/ proc of the Linux systems) during the execution of the system.

In Rembassy, objects are important to the degree that the system is entirely based on the invocation of methods in the various objects. The next section will show that one of the most remarkable characteristics of Rembassy is its capacity to invoke methods in remote objects. This objects system presents an additional advantage in that it is able to build a new object from an existing one. This means that we can specialize a given object so as to make it carry out a determined function, extend its capacities, or modify its behaviour. This extension capacity, which is unique in Rembassy, will prove itself to be especially interesting for the development of new modules or plug-ins. As a matter of fact, most objects in Rembassy are created from a basic object that is extended.

Rembassy defines three basic monitoring objects: sensors, probes, and services.

Sensors are monitoring objects of the lowest level, whose only function is to consult the state of a determined element. Normally, they receive various parameters that will determine the element that is to be consulted and the consultation options. For instance, the "ping" sensor, in charge of checking whether a computer responds to an ICMP ECHO_REQUEST message, could take the IP address of the computer as a parameter and return the "OK" or "Error" state depending on the response. But since it is also important that the message is produced in a reasonable lapse of time, we could add a second parameter that allows the user to select that time-lapse. In that case, the sensor would return the "Alert" state if the computer responds, but only after the passing of that time. It is important to note that the sensor does not store the state internally, it simply returns it. The sensors are, therefore, non-persistent objects that do not store information.

Probes, on the other hand, do store information, and that is precisely the feature that distinguishes them from sensors, since in all other respects they are identical. In fact, probes can be seen as a special type of sensors in which certain parameters have been determined beforehand and are stored persistently as part of their internal state. From any other viewpoint probes have the same function as sensors, i.e. consult the state of a given element, and this function will be delegated towards a previously configured sensor. In other words, probes are a type of sensors that encapsulate a sensor for which certain parameters have been established. Following up on the above example, if we had several computers to monitor, we could wish to maintain the same maximum response time for each computer. By using a probe, we could determine that parameter, and if subsequently we wished to change the response time, we would not have to change it for each element but could simply modify the probe. We can therefore conclude that the main asset of the probe is that it helps us create monitoring templates.

Finally, the services are in charge of monitoring a specific element. The information request concerning the state of that element is delegated towards a sensor (or a probe, which, as explained above, are functionally equivalent). The services plan the periodic consultation of that state, and communicate any possible problems to the user. In order for such a consultation to take place automatically, a service must have established all the parameters that are obligatory. Rembassy allows the use of monitored entities to help establish those parameters: objects that are useful for the configuration of various services for one single element. Let us imagine the case of a host of which we wish to monitor various elements. The previous example shows that the probes will probably need the IP address of that host as a parameter. Since we are going to monitor various services, we will need various probes, each of which will monitor different elements. However, the IP address is common to all, so we can define it in a monitored entity. The service passes the IP address on to the probe; if the IP address changes, we only need to modify the monitored entity and the parameter will be upgraded in all the services, which is particularly helpful in networks of a certain size.

Communications System

Most of the power and flexibility of Rembassy is due to its communications system, i.e. the system that allows the exchange of information between the daemons that form part of the deployment of Rembassy to monitor a given network. This system is based on the remote invocation of methods. Similarly to how a Rembassy daemon functions through the invocation of methods in its objects, a set of daemons interacts through the remote invocation of methods in other daemons (see Figure 1).

This remote invocation is carried out by means of the XML-RPC protocol [Dave Winer (1999)], which was chosen for its simplicity over other alternatives such as SOAP [W3C (2003)], whose complexity entails an excessive, and in our case unnecessary, communications cost.

Since the communication is based on methods invocation, it can be used for all kinds of purposes, including the transmission of management information, configuration parameters, alerts, etc.

Also, the Rembassy core guarantees that such a system is transparent for the monitoring objects, so that, for instance, there is no difference for a service in using a local sensor or a remote one. It is precisely this abstraction in method invocation that makes Rembassy an extraordinarily flexible system, adequate for various monitoring schemes including distributed monitoring.

Finally, it is important to note that this system is not restricted to communication between the daemons. Other types of XML-RPC clients can be developed that interact with the Rembassy daemon, such as, for instance, the Web interface, which was developed independently from the daemon and interacts with it through the remote invocation of methods in its objects.

Components

As can be observed in the next diagram (Figure 2), three fundamental components constitute the Rembassy architecture: the core and plug-ins, which are part of the Rembassy daemon; the

Figure 1. Communication through remote methods invocation

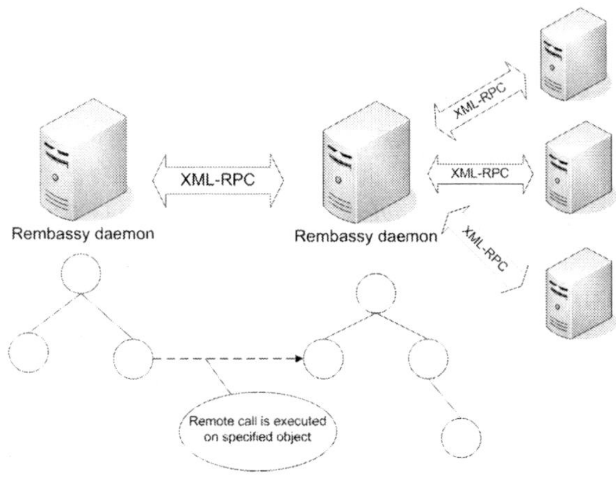

Figure 2. Components of the Rembassy architecture

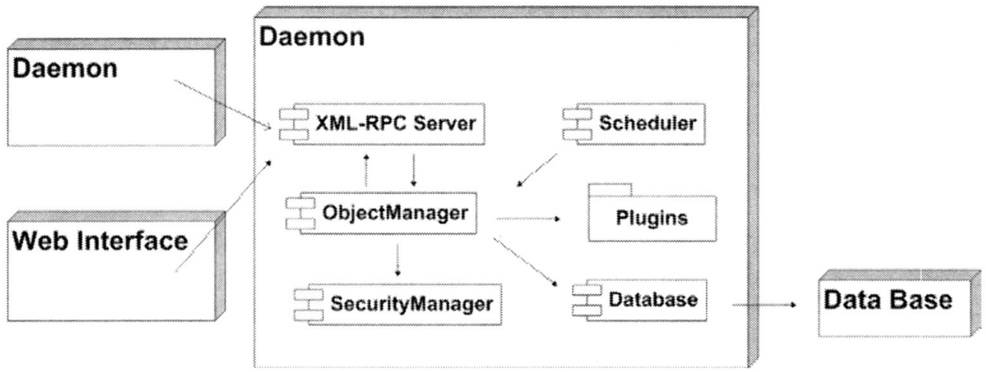

user interface, a Web application that is entirely independent from the daemon which, as mentioned, communicates through XML-RPC; and an external database, which stores the persistent information.

Core

The core is the heart of Rembassy. It implements the basic functions and tools that are used by the application. The purpose of the core is to equip the various objects, which are mainly implemented by means of plug-ins, with a simple interface that enables them to carry out recurring tasks that range from simple operations (e.g. invoke a method in a local object) to complex tasks (e.g. persistence management, remote methods invocation, etc.).

The core is divided into a series of components:

- **Object Manager:** Central component of the core, manages the hierarchy of daemon objects: creation and elimination of objects, methods invocation, persistence management, etc.
- **Database:** Administrates the system database at a low level, by managing, among other things, the persistent information. The database is accessed through the object-relational mapper SQLAlchemy [Michael Bayer (2007)], whose use simplifies the access to data and therefore improves the extensibility of the application. New components do not need to implement the data access layer, which means that their developer can focus on other aspects. In fact, a new persistent object only has to indicate the relationship between its attributes and the fields in the database. The Object Manager is in charge of recuperating the values of the Database and storing them again each time they change their value.
- **XML-RPC Server:** Implements the communications system based on the XML-RPC protocol. As mentioned above, the interaction consists in invoking methods in remote objects. The XML-RPC server encapsulates the details of the communication. When a method is invoked in a remote object, the Object Manager delegates in the XML-RPC server, which establishes the communication with the remote daemon and sends the necessary messages. The XML-RPC server receives the query in this remote daemon and passes it on to the Object Manager, so that it can be executed in the indicated object. For both the invoking and the invoked objects, the query is executed as if it were local. Finally, to guarantee the security of the communications, the XML-RPC server supports the use of SSL (Secure Socket Layer).

- **Security Manager:** Controls the access and guarantees the security of the remote calls. Rembassy's security system is based on users, roles, and permissions. The user is authenticated by a password that is stored in a database encrypted with the SHA-1 hashing algorithm. A user is associated to a series of roles. Each role provides the user with a predetermined set of permissions, which establish the tasks that he/she is allowed to execute. On the other hand, each object associates one or several permissions to its methods. In order to execute a given method, a user must possess one of the necessary permissions. This scheme is tremendously flexible, because we can create global permissions that affect various objects, or permissions that are specific for an object or a method. Rembassy offers two predefined roles: admin (which allows system configuration) and users (which allows only the viewing of the created objects); but it also permits the creation and configuration of new roles.
- **Scheduler:** Plans the monitoring of the different services by periodically checking their state. The user selects the checking interval for each service. The planner distributes the check-ups in the available time, minimizing the load both in the monitoring and the monitored computers, but at the same time guaranteeing that the check-ups take place in the selected intervals.

Plug-Ins

The purpose of the plug-ins system is to extend the basic functionality of the core in order to satisfy the user's needs. As mentioned above, the core carries out tasks of a very low level. These are the plug-ins that implement the monitoring activities for the user on the basis of the low level functions he/she implements.

This design makes it extremely easy to elaborate new plug-ins. We know that an important aspect of a monitoring application is its ability to be extended by the user. Since it is based on plug-ins, Rembassy's architecture can easily be extended through its simple and very powerful extension mechanism. The plug-ins add object classes to the system. If we wish to create a new monitoring object, we can extend one of the objects that are already present and reduce the development time. Also, complex questions such as persistence, remote methods invocation, objects management, or planning, are automatically managed by the core.

As an example, let us suppose that we wish to implement the sensor of the previous example, which executes a "ping" in a computer, and verify the response time. To do this, we extend the base sensor defined in the core. This base sensor implements the logic that manages the relation between the sensor and the rest of the core, which simplifies the creation of the new sensor. We only need to implement two methods: one method to return the parameters that support the sensor (IP address and the maximum response time), and one method to receive these parameters and return the corresponding state, after executing the "ping" in the indicated computer.

Web Interface

The third and last element of Rembassy is the Web interface, which allows the user to interact with the system. This interface is implemented as an application that is entirely independent from the daemon, and communicates with it through the query system XML-RPC. This design guarantees the independence between the application and the system daemon and facilitates the future elaboration of other types of user interfaces.

One of the main features of the Rembassy user interface is that it allows us to both monitor and configure the system. As mentioned above, this constitutes a significant improvement with

Figure 3. Network scheme

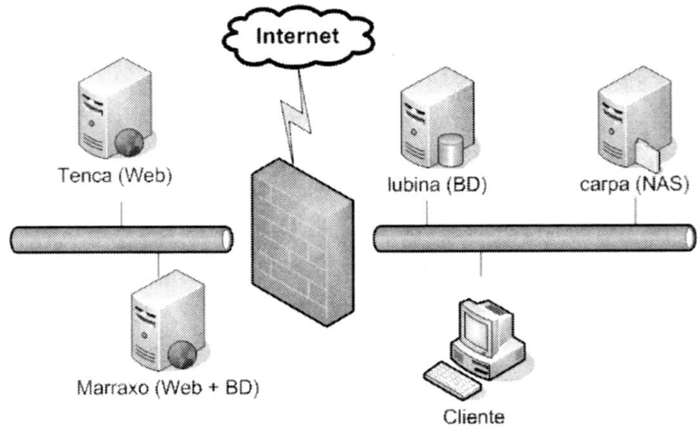

respect to the usual mechanism in Open Source monitoring applications, whose configuration is based on text archives. The Web interface not only allows configuration from virtually any device, it also provides a much more intuitive and accessible user interface.

Finally, we equipped the interface with a system of adaptors for the integration of the plug-ins. In a certain way, the adaptors are the user interface of the plug-ins, and we can interact with them through the web interface.

The developers who want to define a specific user interface for their plug-ins have to implement the corresponding adaptor and register it in the system. Just like the plug-ins, the implementation of adaptors is extremely simple. It is also entirely optional, because Rembassy possesses a default interface that allows the user to execute the methods of any plug-in.

IMPLEMENTATION OF THE SYSTEM

The main motive for the development of Rembassy was the need for a powerful and flexible tool that could monitor the network of the Department for Information and Communications Technologies of the Faculty of Informatics at the University of A Coruña, Spain. This section briefly presents that particular environment and the steps that were taken to configure Rembassy.

Characteristics of the Environment that is to be Monitored

Figure 3 shows a small scheme of the network that will be monitored with Rembassy. A Linux client will monitor several servers:

- A web applications server (Tenca), which gives support to the network of the School of Computer Sciences at the University of A Coruña and to various management applications, all of them implemented with J2EE technology. Tomcat is used as a J2EE applications server, and Apache as an HTTP server. The persistence in the web applications is implemented by means of a remote DB, which is accessed through JDBC (Java Database Connectivity).
- A web server (Marraxo) which hosts a small web page implemented with LAMP technology. It operates an Apache server and a MySOL data base, used as persistent support of the web page's dynamic information. It allows public access to the web via HTTP and private access via

SSH (Secure Shell) in order to manage the contents.

- An Oracle Data Base server (Lubina), which gives support to the persistent information of the Tenca web applications.
- A NAS —Network Attached Storage— (Carpa) which is used to store the backup files of the other three servers. A script in the Lubina server is responsible for the backup process, which gains access to the Carpa hard disk via SAMBA.

Rembassy Configuration

We start by selecting the sensors that are necessary to monitor the various elements. In this case, we use three sensor types, according to the element that is to be monitored:

- **Local parameters (disc, memory, etc).** Since these system parameters cannot be consulted through the network, we must install a Rembassy agent in each computer, but the configuration can nevertheless be centralized. The Rembassy daemon uses the adequate XML-RPC calls to establish the configuration parameters that are selected by the user. The user can manage the remote objects thanks to the ``Proxy'' plug-in, which integrates objects from other daemons into the hierarchy of the local daemon so that the configuration is exactly the same. He or she only needs to indicate the IP address and the remote daemon port, together with the access password. Rembassy allows us to configure an unlimited number of remote daemons in this way.
- **TCP services.** In this case, the sensors can in principle access the monitored services through the network. It is therefore not necessary to install a Rembassy agent, for the sensors of the monitoring device can directly monitor the required services. However, if this access is not possible (e.g. due to the existence of a firewall), Rembassy allows the installation of a second daemon, located behind the firewall, that does provide access. Again, the communication between the daemons is transparent for the user.
- **Applications that are accessible through the network:** similarly to the above, these applications consult the state of the service to be monitored by means of the network. The main difference is that while the former monitor standard services, which are well known and supported by the operating system, the latter will require the installation of drivers or specific clients in the monitoring system. This is for instance the case in database monitoring.

After selecting the sensors, we can configure the services we prefer. However, since we are going to monitor similar services in various computers (we have two HTTP servers), we make use of Rembassy's facilities for creating monitoring templates, more concretely a probe that groups the parameters that are common to both servers (e.g. the sensor that is to be used, the return code for certain queries, etc.).

Finally, we define the services that Rembassy has to monitor on the basis of the probes and sensors. Since the state of the services is checked automatically, we must previously configure correctly all the parameters (see Figure 4), including the desired checking interval.

Monitoring with Rembassy

After configuring the services that are to be monitored, Rembassy periodically plans and checks their state. The user can consult the system's situation at all times by means of the tactical view (see Figure 5), which shows the state of the different services in real time.

Figure 4. Parameters that are determined for a service and used to monitor an SMB server

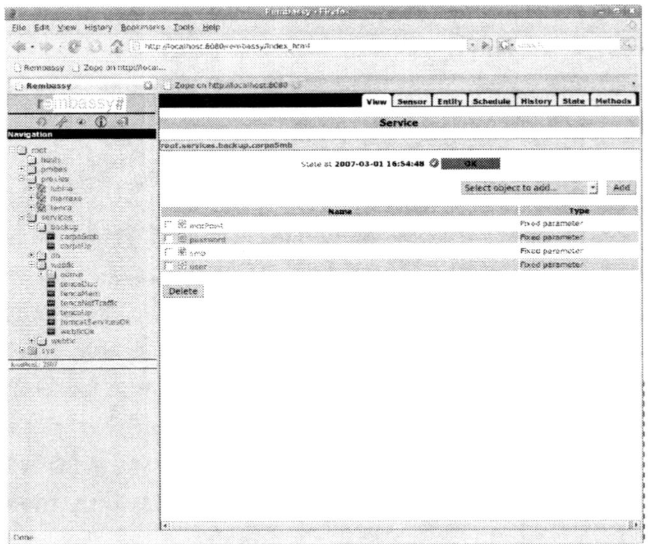

If we want more information on the evolution of a service's state, Rembassy provides a view that shows the history of that state (Figure 6).

If we wish to analyse the history of a service with more detail, Rembassy can provide us with a graphic view of that history (see Figure 7). This view is particularly interesting for the study of the service's evolution in the course of time, and more so if we wish to know not only its state but also its output. For instance, in many occasions we are not only interested in knowing that a server is accessible, we also want to know its load and output, and its evolution in the course of time. This analysis can help us to anticipate future problems and detect possible bottlenecks in the network.

CONCLUSION AND FUTURE WORKS

This work has presented a new monitoring application that solves the main problems and short-

Figure 5. The tactical view allows us to consult the current state of the different services

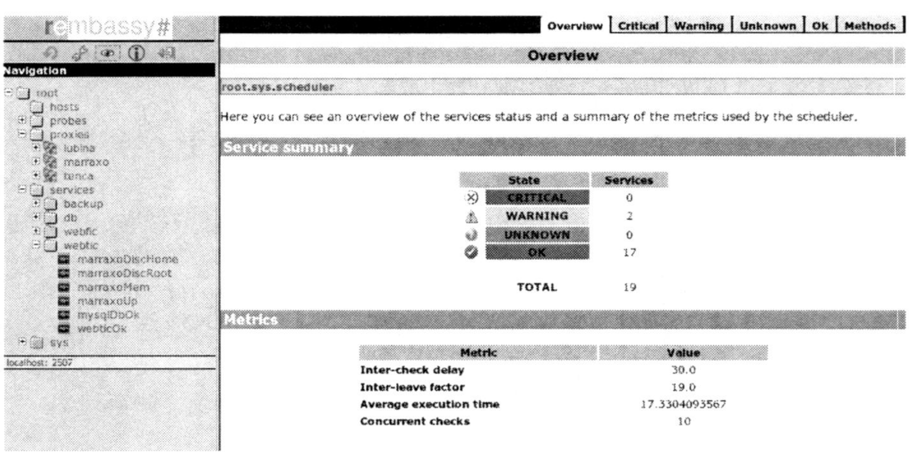

Figure 6. Rembassy keeps the history of a service state

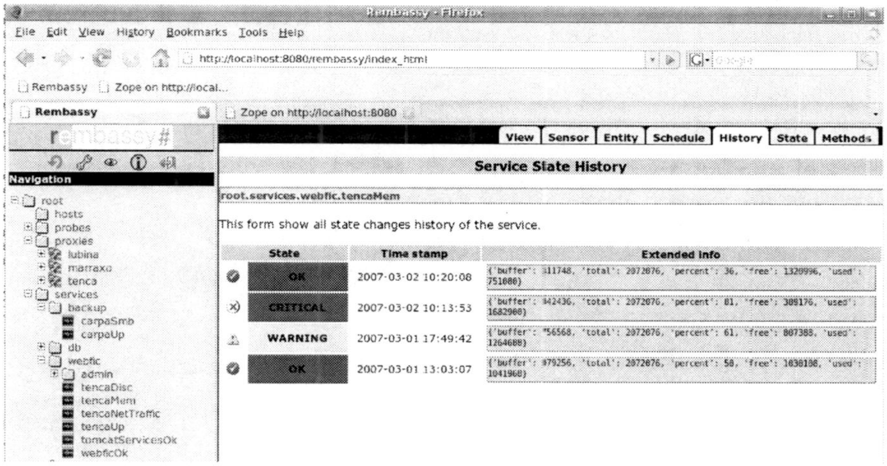

Figure 7. Rembassy allows the graphic representation of a service's evolution

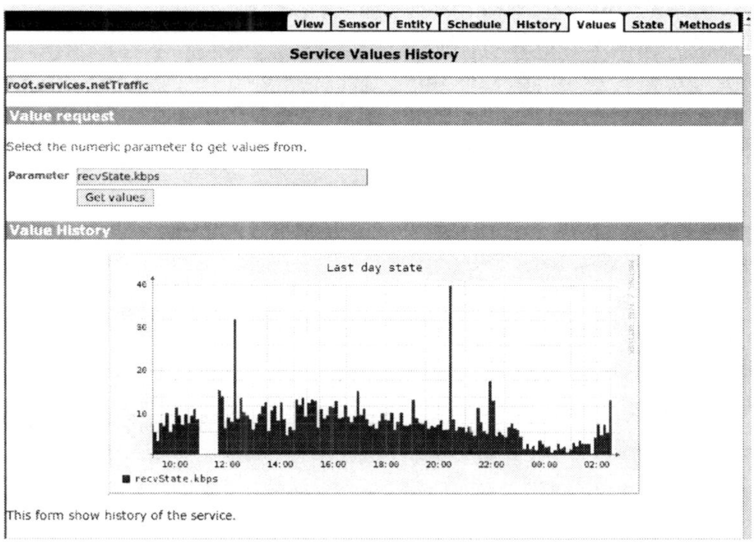

comings of the existing Open Source monitoring applications. Its innovative architecture provides functional and design functions that are unknown in this field and efficiently solve the existing problems.

Firstly, the design is oriented towards the extensibility and scalability of the architecture. Its plug-ins system is particularly innovative, with considerable more power and simplicity than other Open Source applications.

Rembassy introduces two important novelties to the system architecture: the monitoring objects system, which is based on objects orientation techniques and totally manageable in execution time, and the communications system, which is based on remote methods invocation and has proven to be especially efficient.

With regard to user interaction, Rembassy incorporates various important novelties that simplify the interaction and improve the user experience: a complete web interface, which allows centralized system configuration, a monitoring templates system at various levels, etc.

With respect to future developments, Rembassy will be improved in the following aspects:

- **Notification of incidences.** A system will be implemented for the real-time notification of possible incidences by means such as electronic mail or SMS. Some Open Source tools already support these types of notifications. In Rembassy, this functionality is currently in development, and it will be implemented with a new family of objects (notifier) which they will be able to associate with services according to its alert state.
- **Improvement of the analysis capabilities.** At the present moment, the Rembassy representation system is limited. The development of a more advanced system that permits the comparison of various parameters or a statistical analysis of the data would be very useful.
- **Dependencies among services.** The creation of dependencies among services facilitates the monitoring, because if one of them fails, there is no need to check the state of others that depend on it. The majority of Open Source tools do not provide this feature.
- **Development of plug-ins.** Until now, the main work objective has been the development of a suitable architecture that could solve the most urgent problems. Once this task is completed, the focus will be increasingly put on the elaboration of plug-ins that increase the capacity of the system and its usefulness in different environments.

REFERENCES

W3C (2003). *SOAP Version 1.2 Part 0: Primer.* http://www.w3.org/TR/soap12-part0/

Bayer, M. (2007). *SQLAlchemy 0.3 Documentation.* http://www.sqlalchemy.org/docs/

Galstad, E. (2007). *Nagios.* http://www.nagios.org/

Stallings, W. (1998). *SNMP, SNMPv2, SNMPv3, and RMON 1 and 2 (3 ed.).* Boston, MA, USA: Addison-Wesley Longman Publishing Co., Inc.

Winer, D. (1999). *XML-RPC Specification.* http://www.xmlrpc.com/spec

ENDNOTES

1 http://www.paganini.net/index.cgi/angel/angel.html
2 http://www.bigsister.ch/
3 http://ganglia.sourceforge.net/
4 http://leapfrog-mars.sourceforge.net/
5 http://www.nagios.org/
6 http://www.opennms.org/
7 http://pandora.sourceforge.net/en/index.php
8 http://www.sysmon.org/
9 http://www.zabbix.com/
10 http://www.zenoss.com/

This work was previously published in International Journal of Business Data Communications and Networking (IJBDCN) Volume 5, Issue 1, edited by Varadharajan Sridhar and Debashis Saha, pp. 22-39, copyright 2009 by IGI Publishing (an imprint of IGI Global).

Chapter 3
Fourier-Based Assessment Strategies for Simulated Ad Hoc Networks

M. Fazio
Università di Messina, Italy

M. Villari
Università di Messina, Italy

A. Puliafito
Università di Messina, Italy

ABSTRACT

An ad hoc network comprises mobile devices with limited computing and energy resources together with wireless communication, which have to cooperate to provide networking services. This communication scenario presents many specific challenges that make ad hoc networks very different from traditional wired and wireless data networks. It makes classical approaches for network analysis insufficient. To deal with the design, implementation and test of this innovative communication paradigm, simulation techniques are of primary importance, since they allow to specify the level of detail of the simulated model. At the same time, the complex interaction among different entities make the performance evaluation of real ad hoc systems through simulation very hard. This chapter discusses traditional simulation strategies for ad hoc networks, highlighting their limits, drawbacks and possible overcoming. It presents efforts of the research community in improving the quality of simulation analysis according to different aspects, such as metrics definition, model design and simulation tools extensions. Then, the chapter focuses its attention on the benefits that the Discrete Fourier Transform analysis can produce if it is applied on simulation data. It describes a detailed methodology to gather and elaborate simulation measurements in order to avoid loss of information on rare events that occur in simulations. The presented methodology gets advantages (such as simplicity and flexibility) from simulative investigation approaches and, at the same time, offers a new analysis tool suitable for both protocol debugging and system performances evaluation. In fact, it transfers time-dependent measurements into the frequency domain, allowing to point out the occurrence of events which take place only under particular conditions and to detect occasional misbehaviors of the system.

DOI: 10.4018/978-1-60960-589-6.ch003

INTRODUCTION

In the next generation of wireless systems, a main target will be the ability of rapid deployment of the communication infrastructure, independent from fixed access points and specific hardware devices. In a such dynamic scenario, ***ad hoc networks*** represent the emerging technology which allows establishing a communication system without relying on pre-existing agreements.

An ad hoc network is a set of mobile nodes which cooperate in a distributed fashion, in order to provide network functionalities, such as delivering messages, service discovering and so. Since nodes are mobile, the network topology may change rapidly and unpredictably over time. So, ad hoc networks needs to automatically reconfigure the communication system to offer an undisrupted networking service.

The application of ad hoc networks successfully cover a wide range of scenarios: collaborative and distributed mobile computing (sensors, conferences, conventions), disaster recovery (such as fire, flood, earthquake), law enforcement (crowd control, search and rescue) and tactical communications (digital battlefields) (Haas 2002). However, they presents unique challenges which differentiate them from traditional wireless and wired systems (Toh 2002; Giordano 2002; Sesay et al. 2004). Ad hoc networks are self-organizing and nodes have to be highly cooperative: management tasks are distributed over nodes and any service is the result of collaboration among them. To increase the network capacity, nodes relay traffic on behalf of one another to reach distant stations along multi-hop paths. In many cases, nodes are battery-driven and it makes the power budget tight for all the power-consuming components in devices. Also, wireless links have limited bandwidth and are not reliable. All these features affect CPU processing, memory size/usage, signal processing and transceiver output/input power. Additional issues result from node mobility. In fact, users can connect or abruptly disconnect from the network or move in the surrounding space, causing continual changes in the network topology. Nonetheless, end-to-end communication services have to operate seamlessly to provide good experience to users.

Such a complex communication paradigm requires new networking approaches and specific functionalities, that have to be accurately investigated in order to guarantee efficiency and robustness. In the past years many research groups have developed a lot of protocols and architectures to support ad hoc scenarios and most of them have been tested through simulative methodologies. In fact, simulations allow to study easily ad hoc networks through a model of the system. Working conditions (e.g., node mobility, transmission range, etc...) can be modified by simply tuning network parameters and a large spectrum of network scenarios can be considered. However, the simulation analysis suffers from approximations in measurements due to simplified network models and limited observation periods. From this consideration, the necessity to improve the analysis based on simulation tools arise. This necessity is particularly felt in the ad hoc network context, where constrains in available resources and complexity in the communication system make the efficiency of networking services a main target. The behavior of the network depends not only on active protocols, but also on the placement of the nodes in the network, interference arising from neighboring communications and mobility patterns.

The necessity of accuracy in the simulation of ad hoc networks is proved by the great interest of the scientific community in improving the quality of simulation results. Proposed solutions cover different issues, since several aspects of a simulative analysis can be investigated and improved. They range from the definition of meaningful metrics, able to capture the actual behavior of ad hoc system, to the design of realistic simulation models, from the implementation of additional features in simulation tools, to have more robust and exhaustive measures of performance, to the parallelization of runs, in order to increase the amount of numerical results on which to perform the analysis. Following, we provide a survey of the different solutions presented in literature.

An interesting approach to cover deficits of the simulative analysis in ad hoc networks makes use of the Fourier Analysis. The Fourier transforms have many scientific applications — in physics, number theory, combinatory, signal processing, probability theory, statistics, cryptography, acoustics, oceanography, optics, geometry, and other areas (Bracewell 2000). In this chapter, we describe how the Discrete Fourier Transform (DFT) can be applied for performance evaluation in ad hoc networks. It allows to gather information on the evolution of the system through the classic simulative approach and then to assess simulation results in the frequency domain along with the Fourier Analysis. It can be used to estimate goodness of ad hoc protocols and, in case, to test errors in their implementation.

The purposes of this chapter are:

1. to compare simulation approaches against other analysis approaches and present the most common simulation tools for ad hoc networks;
2. to show limits and drawbacks in using simulation tools to test, analyze and evaluate ad hoc networking systems;
3. to discuss current trends in improving the quality of results in simulation studies;
4. to explain how the Fourier Analysis can improve the knowledge of a system behavior, along with an opportunistic assessment of metrics;
5. to present a new methodology based on the Discrete Fourier Transform (DFT) that is able to overcome the main limit of Monte Carlo simulations, that is loss of information on rare events occurred in simulations.

SIMULATION TOOLS

There are three main traditional techniques for analyzing the performance of wired and wireless networks; analytical methods, computer simulation, and physical measurement or a testbed measurement. Traditionally, formal modeling of systems has been via a mathematical model, which attempts to find analytical solutions to problems and thereby enable the prediction of the behavior of the system from a set of parameters and initial conditions. However, it is widely known that comprehensive models for wireless ad hoc networks are mathematically intractable (Mehta et al. 2009). On its own, each individual layer of the protocol stack may be complex enough to discourage attempts at mathematical analysis. Interactions between layers in the protocol stack magnify this complexity. The construction of real testbeds for any predefined scenario is usually an expensive or even impossible task, if factors like mobility, testing area, etc. come into account. Additionally, most measurements are not repeatable and require a high effort. Simulation is, therefore, the most common approach to developing and testing new protocol for a wireless network. Simulation analysis is a standardized, mature, and flexible modeling tool to study ad hoc protocols with different network scenarios. A survey on ad hoc network simulation studies (Kurkowski et al. 2005) shows that simulation is the most used tool to analyze ad hoc networks, illustrated by the fact that 84 out of the 111 MobiHoc papers published in 2000-2004 (75.7%) used simulation to describe esperimental results. The wide usage of simulative models is also proved by the large number of simulation models and protocols that have been developed to study ad hoc networks under different scenarios (number of nodes, mobility patterns, transmission rates, etc.) (Cavin et al. 2002). The most popular simulator tools are OPNET (OPNET Modeler n.d.), OMNeT++ (OMNeT++ Discrete Event Simulation System n.d.), NS2 (Fall and Varadhan 1999), Glomosim

(GloMoSim n.d.) and its commercial version Qualnet (Technologies n.d.).

However, the simulation analysis suffers from approximations in measurements due to simplified network models and limited time intervals to observe the system evolution. To reduce such approximations and to achieve more general results, several deterministic runs of a simulation are necessary (Lewis and Orav 1988). Therefore, to characterize a system (i.e., network protocol, Internet service, etc.) along with a simulation tool it is needed:

1. to choose one or more metrics that can be representative of the system (a metric is a sensitive parameter of the simplified network models);
2. to gather results achieved by several runs characterized by different initial network conditions;
3. to perform a statistical analysis of the results (e.g., normal distribution - Gaussian: mean, standard deviation, etc.). It provides a quantitative evaluation of the metrics and, consequently, a description of the behavior of the system during the time.

This analysis approach, often defined Monte Carlo method, is useful for an early stage of study of the system, but it is not able to assess rare events, which occasionally affect the network. Under particular conditions, the system could show an unstable behavior, causing problems in the communication infrastructure, such high overhead, increasing delay and so on. If the occurrence of such conditions is occasional, information on them are lost in the simulative analysis because of the "Law of Large Numbers". To illustrate some practical problems related to the simulative analysis, we describe our experience in the code developing phase of our protocol AIPAC (Automatic IP Address Configuration) (Fazio et al. 2006) into the ns2 simulator (Fall and Varadhan 1999). AIPAC is a protocol for automatic IP address configuration in ad hoc networks, which supports merging and partitioning in the system. The goal of the protocol is to maintain a stable configuration of the IP address and the Network Identifier (NetId) and to minimize the reorganization time whenever a change in the address configuration occurs. The chief metrics we have used to evaluate the protocol are: configuration time and overhead. Also, we have made use of the graphical interface NAM (Network AniMator) supported by ns2 to view the initial configuration of the network and its evolution with time. After a first stage of test, we have obtained interesting results from simulations, which prove low configuration time and low overhead. In NAM we have identified nodes belonging to the same network (that is with the same NetId parameter) through a specific color (pair Color-NetID). We notice that, in consequence of a merging among different networks, some nodes change their colors in order to conform their NetId to the prevalent one in the system. However, the most of nodes maintain the same color till to the end of simulations. It is a proof of an uniform and stable configuration of nodes. So the protocol seems to satisfy the requirements imposed. Fortunately, we have found a particular run where NAM shows frequent changes of node colors, which implies high instability in address configuration. Really there has been an heavy misbehavior by the protocol and so we have fixed a bug in the code. However, picking up a fluke run is difficult and the protocol implementation should not depend on luck and randomness.

In ad hoc networks, node resources are constrained. So, burst of traffic to manage networking services should be avoided. Also, ad hoc systems are distributed and autonomous, so ad hoc protocols should seldom run recovery tasks for supporting network requirements. As a consequence, reactions to the instability of the system are critical issues for ad hoc protocols. Another important aspect arises from of autonomy of each node and their interaction. In distributed networking scenarios, the probability that the action of an entity rouses unexpected reactions among nodes is higher in comparison with a centralized architecture, where coordinator points can identify all possible failures.

RELATED WORKS

Huge effort in ad hoc networking is devoted on how to improve results of performance analysis, in order to suit special challenges that are not present in more traditional data networks. So, specific simulators for wireless networks have been implemented, such as OPNET, Omnet++, openWNS (Open Source Wireless Network Simulator (openWNS) n.d.) and Prowler (Prowler: Probabilistic Wireless Network Simulator n.d.). They have been designed to meet specific requirements of wireless communication modeling and to offer a simulation platform for the developing of wireless systems.

Even though simulation is a powerful tool, it is still occupied with potential pitfalls, mainly due to three issues:

1. difficulties in the definition of meaningful metrics, able to capture properties of the system;
2. limits of models implemented into the simulator (e.g. communication protocols, MAC layer models, etc.), which do not match realistic systems;
3. features of simulation tools, which do not match requirements necessary to evaluate performance of wireless networks.

In this Section we present research contributions in improving the quality of simulation analysis, classifying them according to the problems they try to overcome. So we talk about three types of solutions: metric-oriented, model-oriented and simulator-oriented.

Metric-Oriented Solutions

Solutions oriented to the metric look for more meaningful metrics, able to capture properties of the system. Most of related works define additional metrics for performance evaluation that are strictly related with the new ad hoc paradigm. In (Boleng et al. 2002) the authors introduce the "mobility metric" problem. A mobility metric permits to quantify the effect of node movement on the system and to measure the communication potential in the mobile network. The paper defines a set of requirements that a mobility metric must meet in order to provide a meaningful description of node movement. Then it introduces a new metric, called link duration metric, that combines the link change rate with the weight of such changes by measuring their stability. The potentiality of link duration as a mobility metric is demonstrated through the NS-2 simulator (Fall and Varadhan 1999).

The metric proposed in (Ghassemian et al. 2005) for ad hoc protocols is called "link stability" metric and it captures both longevity and link changes rate features in order to better represent the impact of mobility on protocol performances. The paper compares it with two classic performance metrics: 1) the "average link duration" metric, that is the average link duration between two nodes and indicates the longevity of the links, and 2) the "average number of link changes" metric, that is the number of counts that link state changes. The random waypoint mobility model has been selected to evaluate the precision of the proposed mobility metric and results show that the link stability metric overcomes limitations in evaluating separately the average link duration and the average number of link changes metrics, since it depends on both transmission range and speed of mobile nodes.

In (Tsumochi et al. 2003) the authors discuss several metrics for routing protocols based on mobility information such as node speed, direction, position, transmission range and pause time.

The main metrics presented in the paper are: relative speed, link expiration time, connectivity, frequency of link state changes. All such metrics depend on geometric considerations in relation to the relative speed. Then the impact of these mobility metrics has been evaluated on three ad hoc routing protocol through simulation: AODV, DSR, DSDV.

Model-Oriented Solutions

This class of solutions try to overcomes limits of models implemented into the simulator (e.g. communication protocols, MAC layer models, etc.), which do not match realistic systems (Heidemann et al. 2001). Due to the broadcast nature of communications, ad hoc networks suffer a lot of issues at the physical and mac layers, such as interference, hidden terminal problem, etc. So, a careful modelling of radio channels and communication protocols is strategic. Also, the de-facto standard used for communications in ad hoc networks is IEEE 802.11, with all its specifications (a, b, g, e, n, ...). For these reasons, many works in literature try to improve models for IEEE 802.11 in different simulation environments. In this direction, (Ma et al. 2006) implement a new model for the Clear Channel Assessment of the IEEE 802.11 Wireless Mesh networks model in the OPNET simulator, to conform better to the IEEE 802.11 specifications. In a similar way, a channel model for 802.11 links to simulate indoor wireless propagation environments has been proposed in (Agüero et al. 2010) and integrated within the NS2 platform. (Ryu et al. 2008) analyzes different features for several popular IEEE 802.11a chipset vendors and compares them with simulation models implemented in NS-2 and QualNet. Then, it modifies the model of the RX process in presence of interference in the QualNet simulator by augmenting two components: the SINR-based preamble detection and the capture algorithm for RCV frames.

Another important issue in ad hoc and, in general, in wireless multi-hop networks is the limited energy supply of nodes. So, also models of energy consumption in the system should be well designed. Aiming this purpose, (Chen 2009) presents a model implemented in Omnet++, which allows to evaluate the energy performance (in terms of energy consumption or network lifetime) of sensor networks (or in principle any wireless network) taking into account the energy consumption of both the radio transceiver and the CPU.

Simulator-Oriented Solutions

Traditional network simulators are not designed to support specific requirements of wireless ad hoc networks. This leads to unrealistic results of performance analyses, which can vary widely between different simulators. So, it is important to implement new features in simulation tools, to have more robust and exhaustive measures of performance of wireless networks.

In (Fehnker et. al. 2009), the authors propose to enhance performance analysis of wireless networks based on simulation by using a graphical specification style. They try to couple a graphical interface to easily set up a simulation environment together with a model checking supply. The work provides a more robust and exhaustive measures of performance which eases the study of the effect of topologies in performance analysis by visualizing both the spatial characteristics of the network as well as critical measures of performance that they imply.

The contribution of (Yamaguchi et. al. 2007) aims to increase the speed up of simulation runs by improve the parallelization degree in parallel discrete-event simulations. A discrete event simulator has an event queue in which events are sorted in the incremental order of timestamps. The simulator dequeues the event with the earliest timestamp, proceeds the simulation clock to the timestamp, and executes the event. The basic idea of Yamaguchi's solution is to predict the lower bounds of timestamps of potential upcoming events by estimating minimum multi-hop propagation delay. This prediction is done in run-time by seeking all the currently scheduled events and geographic location of these events. By this prediction method, it is possible to enlarge the lookahead of parallel processes, and their concurrency can be enhanced. This solution indirectly improve performance evaluation if Montecarlo simulations are performed, since it gathers results from a large number of runs in a short period, so reducing the confidence interval of the measurement.

In this chapter we present an innovative solution to improve performance evaluation of simulation results, which does not belong to any of the three above classes. It deals with limits of simulations in detecting rare events by applying an additional elaboration of results. With this methodology, we can benefit from new ad hoc metrics presented in literature and overcome drawbacks of simulative study at the same time.

PERFORMANCE ANALYSIS THROUGH SIMULATION TECHNIQUES

To perform a simulation-based study (Lewis and Orav 1988), the first step is to choose one or more metrics meaningful for the system in exam; these metrics have to be observed during all the simulations. For simplicity we consider one metric (e.g. throughput, protocol overhead, delay, jitter, power consumption). The performance function for such metric is estimated during a simulative runs. It should be noted that the performance function is an abstraction of the measured parameters. The probabilistic nature of function is given by several realizations of the simulation scenario that are used to estimate its mathematical expectation. Each simulation experiment is characterized by random values for the simulation parameters (network topology, node mobility, traffic generators, communication wireless model and so on). That reflects the randomness in the evolution of the modeled system. The reliability of the estimated measure is generally stated in the form of ***confidence interval***. The width of the confidence interval gives an idea about the uncertain of the simulative results. The higher is the number of runs, the smaller is the confidence interval and hence the uncertainty on the measurement.

Such statistical methodology allows to evaluate the behavior of a protocol, or of the system, on average, that is the average result of the metrics measurements with the system in different conditions. However, further considerations should be done. Let us consider the protocol X_{prot}. Under particular conditions, X_{prot} could be unstable. Wireless links and mobile nodes are characterized by limited and unreliable resources. Then, it is very important to save these resources avoiding bursting of load, otherwise networking services will be degraded. If X_{prot} has been well designed, it will manage events in the network with low signaling traffic on links, loop-free services and a steady-state behavior. On the contrary, it can overload resources in the system causing troubles in communications for a meantime. In these intervals of time users will experience a degradation of the networking service. Unfortunately such misbehaviors can not be detected with the previously described simulative-study. Let us consider the overhead that X_{prot} generates as a performance metric. Suppose that to the $event_x$ event, X_{prot} reacts by sending a lot of control packets, overloading wireless links and then degrading the communication service in the network (particularly if the control packets flood the network). If $event_x$ seldom occurs during the whole simulation time, the overall estimated overhead of the protocol is low and the final evaluation of the protocol will be positive.

To capture protocol behaviors in time and to detect rare events that characterize the system, an analysis in the frequency domain can be helpful. For this reason, in this chapter we present a solution for the performance evaluation of ad hoc networks based on Fourier Analysis. At the beginning of the study, a network metric is selected and information on its evolution under many possible simulation scenarios are gathered into a simulation output function, that is the "metric measurement". Rather than to estimate the performance function directly from the metric measurements, it will be investigated in the frequency domain in order to point out occasional misbehaviors and the occurrence of events which take place only under particular conditions. As *event* we adopt the same definition used in the common event-based simulators. An event is defined as an incident which causes the system to change its state. Examples of events are arrival of packets, alarms, timeouts, etc.

FOURIER ANALYSIS

The Fourier Transform (FT) (Howell 2001) defines a relationship between a signal in the time domain and its representation in the frequency domain. Being a transformation, no information is created or lost in the process, the original signal can be recovered from its Fourier transform, and vice versa. The Fourier transform of a signal is a continuous complex valued signal capable of representing real valued or complex valued continuous time signals.

Most popular simulators are discrete event-based simulators. In a discrete event-based simulation, time is not viewed as a constant flow, but as separate points in time in which events occur. Through discrete event-based simulators we can not get a metric measurement as a continuous function $x(t)$, but we can evaluate it at regularly-spaced intervals, with arbitrarily small spacing, as a discrete function $x[n]$. To perform a Fourier Analysis on the output of simulations, we need to operate in the discrete time domain. For this reason, the Continuous Fourier Transform (CFT) is not suitable for our purpose.

The discrete-time Fourier transform (DTFT) is one of the specific forms of Fourier analysis that works on discrete input functions. However, the DTFT output is defined on a continuous domain. So, it can not be numerically evaluated for every unique frequency. A very useful approximation can be made by evaluating it at regularly-spaced intervals, leading to this simplification, called Discrete Fourier Transform (DFT). DFT (Grafakos 2003; Sundararajan 2001) defines a relationship between a discrete, periodic (with period T) finite duration signal in the time domain $x[n]$ and its representation in the discrete frequency domain $X[\omega]$.

DFT is extremely useful because it reveals periodicities in input data, as well as the relative strength of any periodic component. It has been successfully applied in many research fields for several types of studies and we are going to show that it can be successfully applied also for performance evaluation in ad hoc networks.

The output of a simulation study is a discrete function $x[n]$ that represents the measurement of a performance metric $x(t)$. Then $x[n]$ has to be processed through the DFT. To have a meaningful response in the frequency domain, it is useful to generate $x[n]$ in a particular form.

The DFT of a square wave consists of the fundamental frequency of the square wave itself plus odd harmonics, that are odd multiples of the fundamental frequency. The amplitude of the Hth harmonic is $1/H$ of the amplitude of the fundamental component. In (Kreysig 1992) a mathematical treatment of DFT for a square wave can be found. Let us consider two square waves with unity amplitude: $s_1(t)$ with period T_1=250s and $s_2(t)$ with period T_2=40s. The first signal is characterized by the fundamental frequency $\omega_1=1/T_1=0{,}004Hz$ plus its odd harmonics and the second signal has $\omega_2=1/T_2=0{,}025Hz$. Now let us consider a signal $s(t)$ in a time interval T which is constituted of $s_1(t)$ and $s_2(t)$ so that $s_1(t)$ characterizes the first $T/2$ interval and $s_2(t)$ characterizes the last $T/2$ interval. $s(t)$ and its DFT ($S(\omega)$) are plotted in Figure 1. The location of peaks along the horizontal axis in Figure 1(b) identifies the frequencies that characterize the system. The height of each peak represents the magnitude of

Figure 1. DFT applied to a not periodic signal s(t)

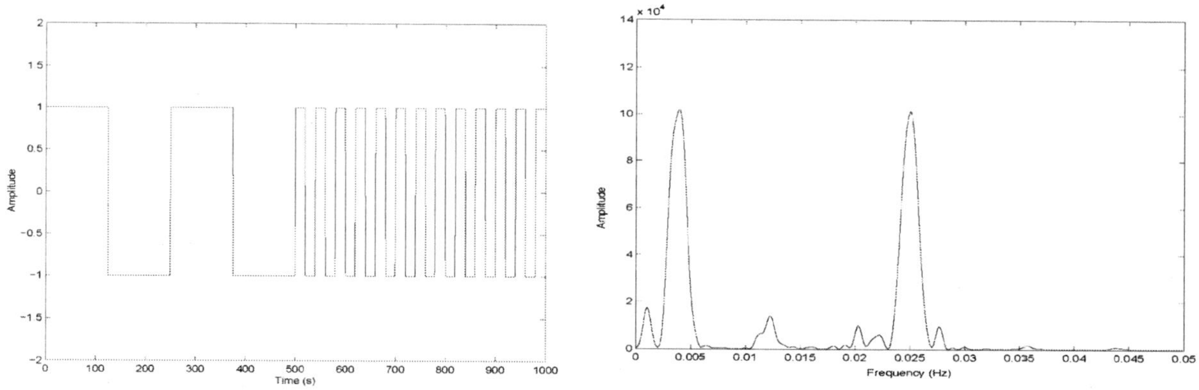

(a) Square waves *s(t)* (b) $S(\omega)$ = DF T of *s(t)*

the contribution from that particular frequency. The most important contributions for the system in Figure 1(a) are respectively at frequencies ω_1 and ω_2, corresponding to the fundamental frequencies of s_1 and s_2.

Other considerations can be done if s_1 and s_2 does not have the same duration. For example, Let us consider that s_1 duration is (¾)T and s_2 duration is (¼)T, as in Figure 2(a). In this case, the contribution at frequency ω_2 considerably decreases because the information of the input signal is for the most part stored in the component at the ω_1 frequency. Such result is shown in Figure 2(b) and it indicates that the contribution in amplitude of a frequency component is proportional to the duration of the component action in the time domain. On these considerations we base our evaluation procedure.

PROPOSED METHODOLOGY

In some applications, such as digital signal processing, there are functions, *x[n]*, that are defined

Figure 2. DFT applied to a not periodic signal s(t) with asymmetric components

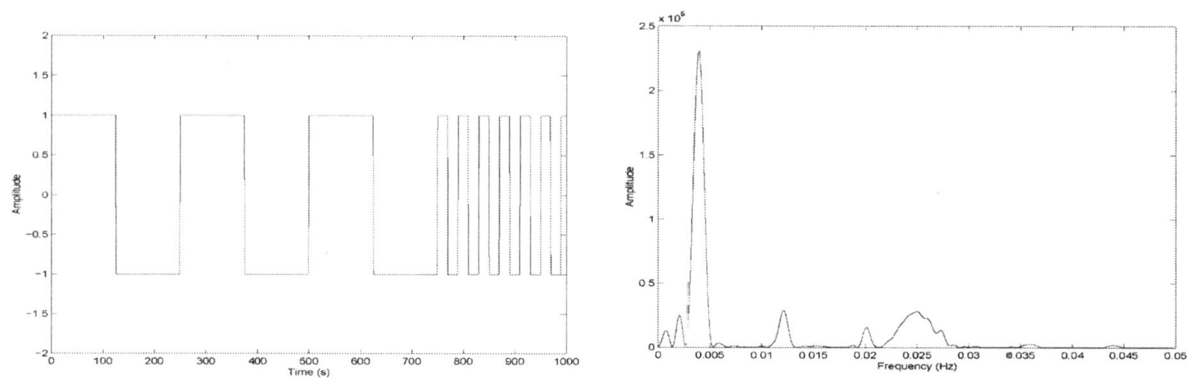

(a) Square waves *s(t)* (b) $S(\omega)$ = DFT of *s(t)*

Table 1. Summary of the notations used in the chapter

Notation	Meaning
$x(t)$	metric measurement in continuous time
$x[n]$	metric measurement in discrete time
n	index of samples in the metric measurements
N	number of samples in the metric measurements
ω	frequency
d	index for fundamental frequencies
D	maximum number of fundamental frequencies for each simulation iteration
i	index for iterations in the simulative study
I	number of iterations in the simulative study
M	matrix to gather metric measurements of several iterations
Th	threshold to reduce effects of leakage
α	constant used to calculate Th
A_{max}	amplitude of the DFT signal at the frequency that provides the highest contribution
$J(\omega)$	curve drawn on the DFT output to catch fundamental frequencies
F	matrix to gather fundamental frequencies detected in the system
$\Delta\omega$	frequency interval to construct $H(\omega)$
k	index for frequency intervals to construct $H(\omega)$
K	number of frequency intervals $\Delta\omega$
$H(\Delta\omega)$	histogram derived from the DFT analysis for the evaluation of systems in ad hoc networks
A_c	the continuous component in the DFT spectrum

for discrete rather than for continuous domains, again finite or periodic. A useful discrete-time function can be obtained by sampling a continuous-time function with a frequency that satisfies the Shannon Sampling theorem (Higgins 1996). In our context, a discrete signal $x[n]$ is the output of the simulator (the metric measurement). If $x[n]$ takes up a square shape, we are able to discuss the DFT output of $x[n]$ like in the examples in Figure 1 and Figure 2. Any network metric (overhead, throughput, delay,...) can be investigated if it produces a suitable $x[n]$ output. Square waves can be drawn by measuring transient parameters or tracing the occurrence of particular events in the system (indicating if and when they affect the network). To this aim, thresholds in the expected values of metric measurement can be helpful. For example, to investigate QoS constrains in video data transmission, it is possible to evaluate how the jitter impacts the packets transfer. If packets arrive to the client with very different delays, the client will experience degradation in the multimedia flow. Let us define the jitter J at time n as:

$$J[n] = d[n] - d[n-1] \tag{1}$$

where d is the delay in the packet transmission. To have good performance, slight variations in the delay are desirable, that is:

$$|J[n]| \leq d[n-1]/2 \tag{2}$$

To provide jitter measurement, we can use the following $x[n]$ function:

$$x[n] = \begin{cases} 1 & if\ x[n-1] = -1 \\ & and |J[n]| > d[n-1] \ /2 \\ -1 & if\ x[n-1] = 1 \\ & and |J[n]| > d[n-1] \ /2x \\ [n-1] & Otherwise \end{cases} \quad (3)$$

x[n] can only have values -1 and 1, thus drawing a square wave. The analysis of *x[n]* through the DFT will provide information on the occurrences of high values of jitter experienced during simulations and, hence, the level of quality in data transmission.

Other possible metrics directly describe the occurrence of an event and they do not need to be managed through thresholds. For example, if a logical hierarchy is assumed among nodes, a switching of the status of each node, between leader or normal conditions, is a metric of interest. Another example is the link breakage among nodes as well as switching in the active-sleep status of transceivers in the study of protocols for energy saving. If the system is perfectly stable, changes in *x[n]* occur periodically with period T. So *x[n]* is characterized by a single frequency. On the contrary, if the system is not stable, as expected in a mobile ad hoc environment, *x[n]* is composed of several frequency components, due to the time-varying behavior of the system. In particular, if flickering in the system occurs, *x[n]* will be characterized by components with high frequencies. Flickering is a frequent switching between two different states of the system and it is an undesirable effect, because it causes waste of network resources. By examining the DFT function of a signal *x[n]*, it is possible to catch the frequencies that characterize it and, hence, to draw the system behavior.

The use of DFT to manipulate simulation results is limited by two issues: 1) the reliability of simulation results depends on the way the network components are modeled and their models should be as close as possible to the reality; 2) results from simulation are not true by themselves (this is the main difference with an analytic approach). They depend on the particular scenario that has been analyzed. For this reason, to obtain meaningful results, a lot of iterations need to be performed with several scenarios. Furthermore, it is not easy to determine the number of simulations needed to assert the general validity of the analysis. These issues make useless an in depth quantitative study of the DFT response on the simulation output, because the simulation output itself is affected by inaccuracy. For this reason, we provide a methodology to improve the simulation analysis through DFT based on qualitative considerations.

The methodology presented in this chapter goes through two stages. The first is the classical simulative analysis of a model of the system that has to be investigated. It is performed through several iterations with different initial conditions in order to get general results. Each iteration *i* provides a metric measurement $x[n]_i$, with $0 \leq n \leq N-1$. So, at the end of the simulation stage with I iterations, the output of the simulator is the matrix *M*:

$$M = \begin{bmatrix} x[0]_0 & x[1]_0 & \dots\ x[N-1]_0 \\ x[0]_1 & x[1]_1 & \dots\ x[N-1]_1 \\ \dots & & \\ x[0]_{I-1} & x[1]_{I-1} & \dots\ x[n]_{I-1} \end{bmatrix} \quad (4)$$

The second stage of the methodology is the DFT analysis. It post-processes the simulation results stored in *M* once the simulative stage is finished. It draws the fundamental frequencies that characterize the system and their impact on the system performance. In our work, we apply the DFT analysis by using the Matlab environment and the functions it supplies. In the following, we describe in detail the steps necessary to perform the DFT analysis on the results collected from a simulator.

Figure 3. Construction of the J(ω)

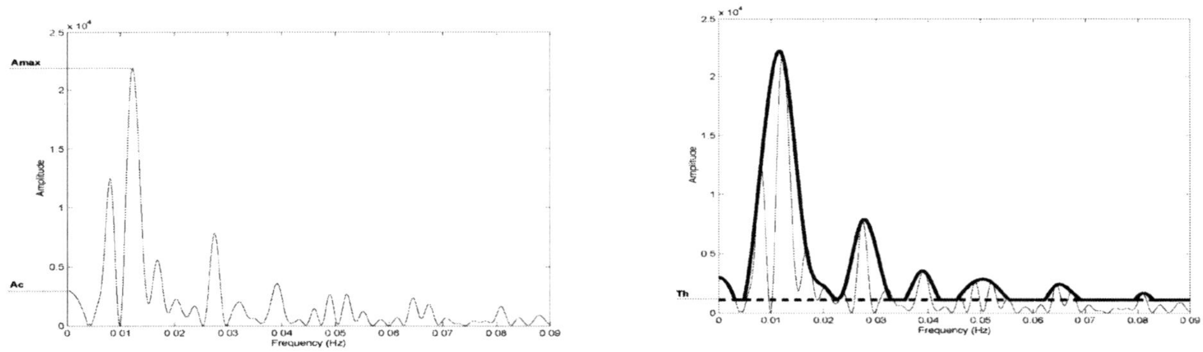

(a) DFT of a generic input signal

(b) Interpolation of *J(ω)*

Step 1: Recording of Fundamental Frequencies

To get information on frequencies that characterize the system, each metric measurement $x[n]_i$ is processed through the DFT. DFT describes how the fluctuating components of a finite duration signal are distributed over frequencies. However the DFT computation assumes that a signal is periodic, that is it repeats over and over again and it is identical every time. When the DFT of a non periodic signal is computed, the resulting frequency spectrum suffers from the *Leakage effect*. Leakage results in the signal spreading over a wide frequency range in the DFT, when it should be in a narrow frequency range. The dispersed shape of the DFT makes it very hard to identify the frequency content of the input signal.

In a simulation-based study we usually manage non-deterministic signals, because they come from chaotic evolutions of the system. Also, since there are not periodic components in the network, the resulting signals usually will be non-stationary. The leakage effect is amplified in presence of non-stationary waveforms. Unfortunately, the level of indeterminacy due to the leakage effect can not be quashed. Nevertheless, we believe that frequency analysis can help us in studying ad hoc networks. Since the spreading of frequency components affect information on the input signal, we can not perform a quantitative assessment, but a qualitative evaluation of results.

Due to the Leakage effect, when DFT is applied on the $x[n]_i$ measurement, spurious peaks exists which do not express a real behavior of the system. Since the amount of smeared spectrum from the true frequencies of the signal is a small fraction of the whole signal, we overcome such noise source by discarding peaks under a minimum amplitude threshold *Th*. For example, let us consider a function $x[n]_i$, with the DFT output in Figure 3(a). The cut at threshold *Th* is shown in Figure 3(b).

The value of *Th* is not set a priori, but it depends on the magnitude of the frequency contributions. In particular:

$$Th = \alpha A_{max} \quad 0 \leq \alpha < 1 \tag{5}$$

where A_{max} is the amplitude of the DFT signal at the frequency that provides the highest contribution into the DFT graph (see Figure 3(a)) and α is a constant.

In our analysis we set $\alpha = 0.05$, but this parameter can be tuned according to the accuracy level of the analysis. Higher values of α reduce the error due to the leakage effect, but it could cut off information on the metric. On the contrary, low values of α could affect measures with ambiguous information. Unfortunately it is impossible to know the effects of leakage on a signal and we have to manage results with a level of indeterminacy that can not be set aside.

Since we are interested in a qualitative analysis, we have to catch frequencies in the spectrum that mostly characterize the system. To remove information on frequency contributions with low impact or noise, such leakage, it is necessary to pick up frequencies with higher amplitude in the frequency spectrum. It means to detect the highest relative maximums in the DFT. To better understand the importance of this procedure, let us consider Figure 1(b). There are a lot of peaks due to the fundamental frequencies, their harmonics and distortion. In the example we easily detect the fundamental frequencies at $4mHz$ and $25mHz$ by considering frequencies with the greatest amplitudes in the spectrum. To generalize such approach, we have to modify the DFT output by ignoring information of contributions around the fundamental frequencies. Practically, we come to such result if we draw a curve, $J_i(\omega)$, related with the metric measurement $x[n]_i$, that connects all the relative maximums in the DFT graph of $x[n]_i$. The corresponding $J_i(\omega)$ curve for DFT in Figure 3(a) is shown in Figure 3(b).

Connecting a fundamental frequency and its harmonics leads to a bell-shape curve in $J_i(\omega)$, because the higher is the degree of harmonics, the lower is the contribution in amplitudes. The maximum of such bell-shape correspond to the fundamental frequency. So, we can detect the fundamental frequencies of a system $[\omega_0, ..., \omega_D]_i$ for the iteration i by picking up the relative maximums of $J_i(\omega)$.

Step 2: Elaboration of Information

At the end of Step 1, the process calculates the fundamental frequencies ω_{di}, with $0 \leq d \leq D-1$ and $0 \leq i \leq I-1$, that characterize the system for each iteration run at the simulation stage. The result is a matrix F:

$$F = \begin{bmatrix} w_{0\,0} & w_{1\,0} & \dots\ w_{D-1\,0} \\ w_{0\,1} & w_{1\,1} & \dots\ w_{D-1\,1} \\ \dots & & \\ w_{0\,I-1} & w_{0\,I-1} & \dots\ w_{D-1\,\,I-1} \end{bmatrix} \tag{6}$$

All these information have to be merged in order to take out a general evaluation of the system. At this purpose, we gather information from all the iterations in an histogram $H(\Delta\omega)$ consisting of: in the abscissa the frequency spectrum divided in a set of K intervals $[\Delta\omega_0, ..., \Delta\omega_{K-1}]$ and in the ordinate the weight of each set of frequencies on the system behavior from all the simulation iterations. To construct $H(\Delta\omega)$, we pick up all the frequencies ω_{di} stored in F. The amplitude of $\Delta\omega_k$, with $k=1,2,..K$, is increased by $A(\omega_{di})/A_{ci}$ if $(k-1)\Delta\omega < \omega_{di} < k\,\Delta\omega$. $A(\omega_{di})$ is the amplitude of DFT at frequency w_{di} and A_{ci} is the continuous component in the DFT spectrum of signal $x[n]_i$ (see Figure 3(a)). The normalization of each contribution $A(\omega_{di})$ to the continuous component A_{ci} in the correspondent DFT spectrum allows us to quantify the impact of occasional events on the steady state behavior of the system. To show results of the procedure to build the $H(\Delta\omega)$ histogram, we draw in Figure 4 the histogram that corresponds to the spectrum in Figure 3. It shows how the component in the interval $\Delta\omega=[0,0115;0,128]$ has the more important role in comparison with the other ones. With this approach, we do not loose information on occasional events and we can detect if an instance never happens, or if there is a low probability that the instance occurs, or if the instance strongly characterizes the system.

The $H(\Delta\omega)$ histogram is divided in three areas: low, medium and high frequencies areas. The contribution of frequency components for each area can be extracted by applying a FIR filter. In

Figure 4. Histogram of frequency contribution with reference to Figure 3

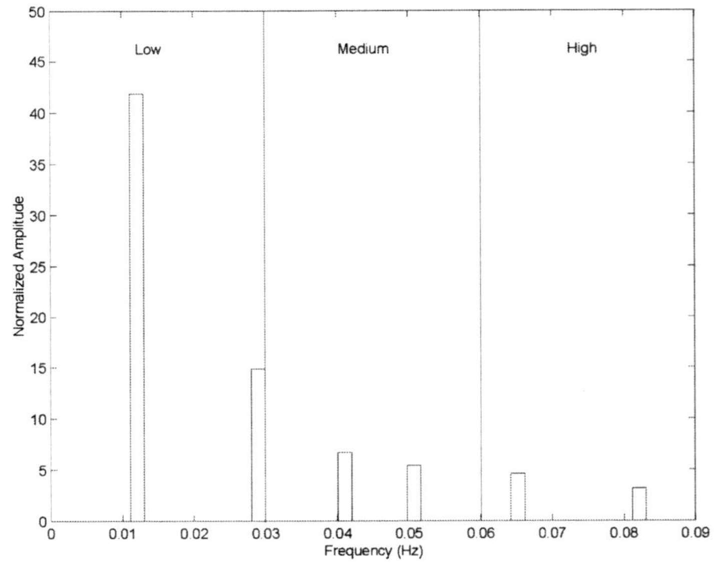

our work, we use FIR filter implemented in Matlab specifying the ends of the band.

To perform this Step of the methodology, some parameters have been introduced, such the $\Delta\omega$ intervals and the bands of FIR filters. These parameters are set from the user before starting the DFT analysis. In fact, their values depend from the metric under evaluation and the necessary accuracy of the analysis. For example, let us consider an user that carries out a study on a protocol for the IP address configuration in ad hoc networks (such our work named AIPAC) and, in particular, he analyzes the time of reconfiguration of the node address. To have a stable system, a node can not change configuration every few minutes. So, the frequency spectrum in the DFT analysis is in the range of *μHz-mHz*. The study of a routing protocol for ad hoc networks is very different. If the user analyzes the expiration period of entries in the routing tables, the frequency spectrum in the DFT analysis is in the range of *Hz*. $\Delta\omega$ intervals and bands of FIR filters have to be set in accordance with the expected frequency spectrum.

Step 3: Evaluation of Results

The $H(\Delta\omega)$ histogram presents three frequency bands, but to wisely explain the results, we need to know the sense of such areas. The design of efficient protocols for ad hoc networks should follow two main guidelines: 1) limited signaling traffic and 2) stability in the system behavior. We can assert that the above guidelines are satisfied if events in the system occur with low frequency. Whenever nodes deal with network issues, they generate traffic on communication links for synchronization with other nodes or they try to adapt their state to the new conditions. Such activities cause an increment in the number of events carried out into the system, which are monitored through the $x[n]$ metric. In stable conditions, events occur occasionally. It means that switching in the output signal $x[n]$ recurs with low frequency and hence there are many contributions into $H(\Delta\omega)$ in the low frequency area. On the contrary, if the system is characterized by an unstable behavior, the amount of events registered in a small time interval is high and hence there are important contributions into $H(\Delta\omega)$ in the high frequency area.

Another consideration comes from the shape of the histogram $H(\Delta\omega)$. Whether the system is stable, measurements from several runs are quite similar, even if initial conditions of the ad hoc scenario and, hence, the evolution of the system with time are very different. It means that frequency components that characterize the system are very close each other and they produce the histogram $H(\Delta\omega)$ with a smoothed shape. On the contrary, whether the behavior of the system is chaotic, many different frequency components in the three areas of $H(\Delta\omega)$ are present. So a picked shape of $H(\Delta\omega)$ indicates instable reactions of the network to environmental solicitations.

A very interesting instance comes whether the protocol is affected by a bug and the system evolution presents a persistent behavior independent from the initial conditions of the simulation. It is detected in $H(\Delta\omega)$ since it shows a peak for a particular frequency component $\Delta\omega_k$ that describes such misbehavior. It is a feature of the protocol that is independent from the whole system, but that characterizes the protocol itself. Unless special conditions can be identified, this misbehavior is a consequence of mistakes in the protocol implementation. This analysis procedure allows not only to evaluate the performance of a protocol, but also to verify the correctness of its implementation.

Methodology Application Example

In this subsection we test the proposed methodology with a practical example. We want to show the limitations of the results from a simulation analysis and to prove the impact of our methodology in the performance evaluation phase.

To this aim, we consider to carry out a simulation analysis to evaluate a generic $X_{prot}(t)$ protocol, characterized by a $x(t)$ metric. Each simulation takes 10000s and 50 runs are performed. The output of the simulations we get are the discrete metric measurements $x[n]_i$, with $i=0,2,..,49$. Let us assume that the system under analysis is quite stable, that is in most of the iterations the metric measurement during the time presents slight variations around a mean value. To describe such behavior, we assume that 47 runs over 50 in the simulation procedure describe the steady state behavior of the system. They provide as output a square wave with frequency in the range *[0.0005, 0.002]Hz*, with a random phase and amplitude equal to 1. Then we assume that the 3 remaining experiments are characterized by rare events that make the system unstable for short periods of time. During these time intervals, the metric measurements assume values in a large range. Rare events cause jamming in the output function for limited time periods and they can be represented through a square wave with 0,006 Hz as frequency and 4 as amplitude. In Figure 5(a) we show the output functions for the iterations with a jamming.

In the traditional simulation study, in order to evaluate the X_{prot} metric with time varying we calculate the mean value $\overline{x}[n]$ over the 50 runs. The result is depicted in Figure 5(b) and it shows that information on rare events are lost. If we apply our methodology, we obtain the histogram in Figure 5(c). It presents components at low frequencies derived from the cosine functions that describe the steady state behavior of the system. Then, there is a component at high frequency. It derives from the rare events that have effected the system 3 times over 50 simulations for *800s* each time, with a total amount of *3x800s=2400s* over *10000x50=50000s*.

CONCLUSION

In this chapter, we have presented a new methodology to deal with performance analysis of ad hoc networks. It couples a traditional simulation study with the Fourier Analysis, in order to overcome the limitations of a simulation-based study.

At first we have discussed problems in simulation, i.e., lost of information if a protocol

Figure 5. Comparison between simulative analysis and simulative-Fourier analysis

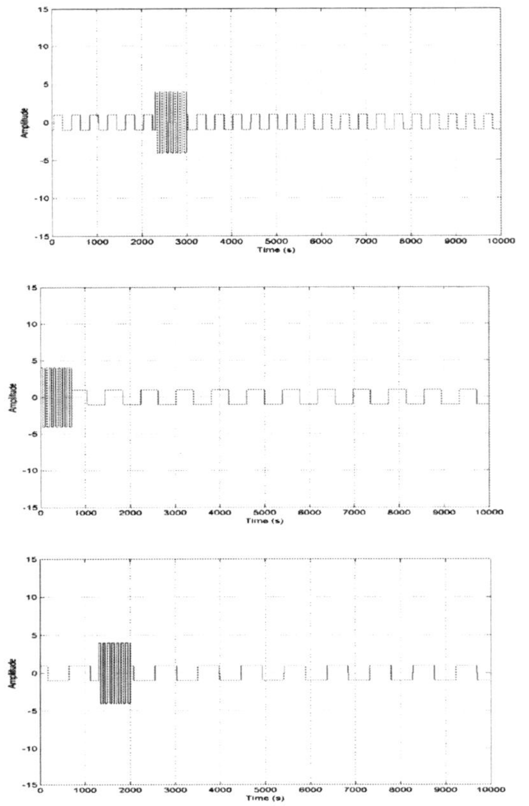

(a) Simulation outputs for runs with rare events

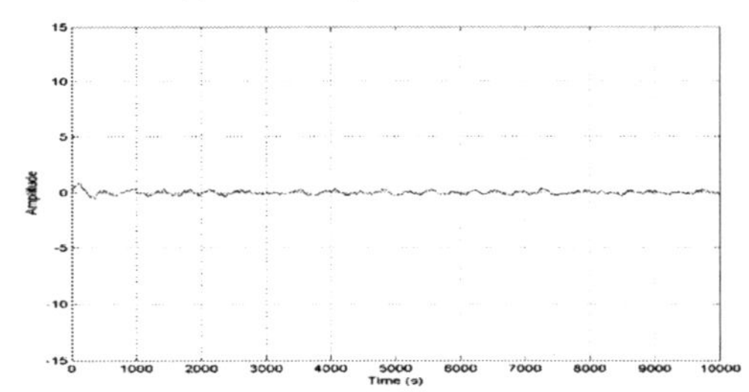

(b) measurement of metric $X_{prot}(t)$ on average over 50 runs

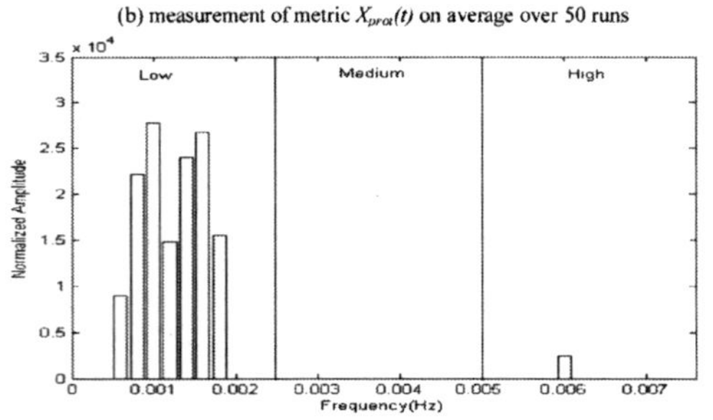

(C) $H(\Delta\Omega)$ HISTOGRAM

presents occasional misbehaviors. In fact, we can be interested in their occurrences if we are testing the protocol or if we are evaluating its stability in time. To solve such drawbacks, we have introduced the Fourier Analysis that allows us to characterize events in the system through the frequency of their occurrences.

The final result of our work provides a detailed methodology to carry out a meaningful analysis of the performance of ad hoc networks through both simulation and Fourier Analysis. Due to limits of simulative models and the leakage effect in the Fourier analysis, we can not provide a quantitative assessment, but a qualitative evaluation methodology. We demonstrated the validity of the proposed solution with a practical example.

Our future goal will be to provide an in-depth analysis of a well-known ad hoc protocol through the presented methodology, thus proving its quality and efficiency.

REFERENCES

Agüero, R., García-Arranz, M., & Muñoz, L. (2010). Accurate Simulation of 802.11 Indoor Links: A "Bursty" Channel Model Based on Real Measurements. *EURASIP Journal on Wireless Communications and Networking*, 2010.

Boleng, J., Navidi, W., & Camp, T. (2002). *Metrics to enable adaptive protocols for mobile ad hoc networks*. In Proc. of the International Conference on Wireless Networks (ICWN'02), pp. 293–298.

Bracewell, R. N. (2000). *I* (3rd ed.). New York: McGraw-Hill.

Cavin, D., Sasson, Y., & Schiper, A. (2002). *On the accuracy of MANET simulators.* In Proc. of the second ACM international workshop on Principles of mobile computing table of contents'. Toulouse, France.

Chen, F., Dietrich, I., German, R., & Dressler, F. (2009). An Energy Model for Simulation Studies of Wireless Sensor Networks using OMNeT++. [PIK]. *Praxis der Informationsverarbeitung und Kommunikation*, *32*(2), 133–138. doi:10.1515/piko.2009.0023

Fall, K., & Varadhan, K. (1999). *ns Notes and Documentation*. The VINT Project, UC Berkeley, LBL, USC/ISI, and Xerox PARC. Available http://www-mash.cs.erkeley.EDU/ns/.

Fazio, M., Villari, M., & Puliafito, A. (2006). IP Address Autoconfiguration in Ad Hoc Networks: Design, implementation and Measurements. *Computer Networks Journal Elsevier Science*, *50*(7), 898–920. doi:10.1016/j.comnet.2005.04.017

Fehnker, A. Fruth, M. McIver, A. K. (2007). *Graphical modelling for simulation and formal analysis of wireless network protocols*. In Proc. Workshop on Methods, Models and Tools for Fault-Tolerance (MeMoT'07).

Ghassemian, M., Friderikos, V., & Aghvami, A. H. (2005). *On mobility metrics applied for ad hoc network protocol evaluation.* In Proc. of the 7th IFIP International Conference on Mobile and Wireless Communications Networks (MWCN 2005), Marrakech, Morocco.

Giordano, S. (2002). Mobile ad-hoc networks. I. Stojmenovic (Ed.), *Handbook of Wireless Networks and Mobile Computing* (pp. 325–343, 371–391.) New York: Wiley.

GloMoSim. (n.d.). *Global Mobile Information Systems Simulation Library*. Retrieved from http://pcl.cs.ucla.edu/projects/glomosim/.

Grafakos, L. (2003). *Classical and Modern Fourier Analysis*. Upper Saddle River, NJ: Prentice Hall.

Haas, Z. J., Deng, J., Liang, B., Padadimitratos, P., & Sajama, S. (2002). Wireless Ad Hoc Networks. In Proakis (Ed.) *Encyclopedia of Telecommunications*.New York: John Wiley.

Heidemann, J., Bulusu, N., Elson, J., & Intanagonwiwat, C. Lan, Ya Xu, K.-C-, Ye, W., Estrin, D., & Govindan, R.(2001). *Effects of Detail in Wireless Network Simulation.* In Proceedings of the SCS Multiconference on Distributed Simulation, pp. 3-11. Phoenix, Arizona, USA.

Higgins, J. R. (1996). *Sampling Theory in Fourier and Signal Analysis*. New York: Oxford Science Publications.

Howell, K. B. (2001). *Principles of Fourier analysis*. Boca Raton, FL: Chapman and Hall/CRC.

Infrastrutture Software per Reti Ad-hoc Orientate ad Ambienti Difficili (2002-2007). *Consiglio Nazionale delle Ricerche (CNR) with the Strategic IS-MANET Project.*http://zeus.elet.polimi.it/is-manet/.

Kurkowski, S., Camp, T., & Colagrosso, M. (2005). Manet simulation studies: the incredibles. *ACM SIGMOBILE Mobile Computing and Communications Review*, *9*(4), 50–61. doi:10.1145/1096166.1096174

Lewis, P. A. W., & Orav, E. J. (1988). *Simulation Methodology for Statisticians. Operations Analysts and Engineers* (*Vol. I*). New York: Wadsworth & Brooks/Cole Advanced Books & Software.

Ma, H., Alotaibi, E., & Roy, S. (2006). *IEEE 802.11 Mesh Networks, OPNETWORK 2006.* Washington, D.C.: Analysis and Simulation Model of Physical Carrier Sensing.

Mehta, S., & Ullah, N. Kabir Md. H., Sultana Mst. N., Kwak K. S. (2009). *A case study of networks simulation tools for wireless networks.* Third Asia International Conference on Modelling & Simulation (AMS'09). OMNeT++ Discrete Event Simulation System (n.d.). Retrieved from http://www.omnetpp.org/.

Modeler, O. P. N. E. T. (n.d.). Retrieved from http://www.opnet.com/products/modeler/home.html.

Open Source Wireless Network Simulator. (openWNS) (n.d.). Retrieved from http://www.openwns.org/

Prowler: Probabilistic Wireless Network Simulator (n.d.). Retrieved from http://www.isis.vanderbilt.edu/Projects/nest/prowler/

Ryu, J., Lee, J., Lee, S.-J., & Kwon, T. (2008). *Revamping the IEEE 802.11a PHY Simulation Models*. In Proc. of the 11th ACM/IEEE Intl. Symposium on Modeling, Analysis and Simulation of Wireless and Mobile Systems (MSWIM'08), Vancouver, British Columbia, Canada, October.

Sesay, S., Yang, Z., & He, J. (2004). A Survey on Mobile Ad Hoc Wireless Network. *Information Technology Journal, 3*(2), 168–175. doi:10.3923/itj.2004.168.175

Sundararajan, D. (2001). *The Discrete Fourier Transform-Theory, Algorithms and Applications*. New York: World Scientific Publishing Company. doi:10.1142/9789812810298

Technologies, S. N. (n.d.). *Qualnet family of products*. Retrieved from http://www.scalable-networks.com/products/qualnet.php.

Toh, C. (2002). *Ad hoc Mobile Wireless Networks Protocols and Systems*. Upper Saddle River, NJ: Prentice Hall, Inc.

Tsumochi, J., Masayama, K., Uehara, H., & Yokoyama, M. (2003). *Impact of mobility metric on routing protocols for mobile ad hoc networks.* In Proc. IEEE Pacific Rim Conference on Communications, Computers and Signal Processing(PACRIM03)'.

Yamaguchi, H. Konishi, K.& Higashino, T. (2007). *Efficient Parallel Simulation of Mobile Wireless Networks by Run-time Prediction of Multi-hop Propagation Delay*.In Proc. of TridentCom 2007.

Chapter 4
Comparison of Policies for Epidemic Broadcast in DTNs under Different Mobility Models

Francesco Giudici
Università degli Studi di Milano, Italy

Elena Pagani
Università degli Studi di Milano, Italy

Gian Paolo Rossi
Università degli Studi di Milano, Italy

ABSTRACT

The broadcast diffusion of messages in Delay Tolerant Networks (DTNs) is heavily dependent on nodes mobility, since protocols must rely on contact opportunities among devices to diffuse data. This work is a first effort to study how the dynamics of nodes affect both the effectiveness of the broadcast protocols in diffusing data, and their efficiency in using the network resources. The paper describes three control mechanisms. The mechanisms characterize a family of protocols able to achieve some awareness about the surrounding environment, and to use this knowledge in order to keep the broadcast overhead low, while ensuring high node coverage. Simulation results allow to identify the winning mechanisms to diffuse messages in DTNs under different conditions.

INTRODUCTION

The mobile nodes of a delay tolerant network, or DTN (DTN Research Group, 2008), experiment intermittent connectivity and network partitions. So far, in such a critical scenario, the research mainly focused on the problem of providing unicast communications (e.g. (Burgess, Gallagher, Jensen, & Levine, 2006; Davis, Fagg, & Levine, 2001; Jones, Li, & Ward, 2005; Juang, Oki, Wang, Martonosi, Peh, & Rubenstein, 2002; Spyropoulos, Psounis, & Raghavendra, 2004)). By contrast, the one-to-all communication scheme has not received the same attention despite the fact that its service is strategic to support protocols at both application and routing levels. For instance,

DOI: 10.4018/978-1-60960-589-6.ch004

a broadcast service is required to diffuse scoped advertisements—e.g. about available services or events—and summaries (Lee, Magistretti, Zhou, Gerla, Bellavista, & Corradi, 2006), to support podcasting (Lenders, Karlsson, & May, 2007), or to diffuse acknowledgments, or cure, packets (Harras, Almeroth, & Belding-Royer, 2005). In a DTN, broadcast can be designed by adopting one of the gossip-based mechanisms that have been proposed in the literature in a few slightly different alternatives by starting from the following *PUSH-based* scheme: when a node has a message *m*, it forwards *m* to one or more (possibly all) neighbors it happens to encounter while moving. The forwarding, elsewhere called infection or epidemic, can be either performed periodically (Montresor, Jelasity, & Babaoglu, 2005) or whenever a contact occurs (Vahdat, & Becker, 2000). Infection can continue up to either the message lifetime or a given number of transmissions. The PUSH-based algorithm is effective in achieving a node coverage arbitrarily close to 1 with low latency, but fails in doing this efficiently. Indeed, nodes perform epidemic forwarding with no knowledge about the state of the encountered nodes and, as a consequence, they often happen to forward a message to already infected nodes.

The primary focus in the design of a—both efficient and effective—broadcast protocol is to increase the node likelihood of delivering a message only to uninfected nodes. There are several, growing levels of knowledge a node could achieve about the neighbors state and to approximate the global system state. Whatever is the followed approach to achieve efficiency, the performance of the algorithms is greatly influenced by the mobility patterns that nodes follow (Camp, Boleng, & Davies, 2002). The comparative analysis of the effects of mobility on protocols deserves more attention and, as far as we know, has never been applied to broadcast. This paper moves into this research track and provides some interesting contributions to understand broadcast delivery over DTNs under different mobility conditions. Firstly, the paper defines a family of broadcast protocols, obtained from the PUSH-based approach through the introduction of adaptive mechanisms whose purpose is the improvement of the node awareness about the level of infection in the neighborhood. Secondly, the performances of the different mechanisms are analyzed in three basic mobility model: the classical random waypoint, the swarm mobility, and the aggregation model where nodes move throughout aggregation points. The main contribution of the paper is twofold: *(i)* it identifies and characterizes the winning mechanisms to diffuse messages in DTNs under different conditions, and *(ii)* it is a first attempt to move toward the design of an autonomic and situational algorithm able of autonomously adapting its parameters according to the mobility context the node is moving through, in order to optimize performances.

SYSTEM AND MOBILITY MODELS

System Model

The scenario we consider in this paper includes people walking in a limited urban area, such as a *campus area*, and equipped with wireless portable devices. No base stations are assumed and the communication between a source *s* and a destination *d* may eventually occur through either direct contact, when, for instance, node *d* moves into the range of *s*, or indirect contact, when one or more relaying nodes help to create the multi-hop path toward the destination and the last of them finally enters the range of *d*. The devices have a unique identifier, are not required to have positioning capabilities on board and, to meet resource saving requirements, are supposed to adopt a short radio range to communicate. This latter point, together with the fact that devices can be sparsely distributed over a large area, makes high the probability of network partitions and link disruption. Throughout the paper we only assume that each mobile device, or node, periodically broadcasts a *beacon* message in its radio cell. Beacons are used to discover other devices in the neighborhood and their content is limited to the device identifier. In such a scenario, people mobility might follow either a Random Waypoint (RWP) model (Camp et al., 2002) or a more structured motion.

Mobility Models

A great deal of research is currently ongoing in order to characterize mobility models suitable for DTNs. Mobility could be extracted from traces of nodes contacts in real settings; several traces are for instance provided by the CRAWDAD community (CRAWDAD, 2008). Yet, the use of traces with simulators creates some problems. Their timescale is hardly scaled to the simulation time and they generally model the specific behavior of a given mobility scenario, thus loosing the generality required during the protocol design phase. Traces are more likely useful during the validation process than during the design and performance analysis phase.

In this work, we analyze how three basic mobility models affect the performance of the epidemic protocols described in the next section. The classical *Random Waypoint* (RWP) model is not realistic, but it is simple, often provided within network simulators, and the most commonly used in the literature. In the *aggregation* (AGG) model, a node moves toward an aggregation point (*ap*), chosen from a set according to a certain probability distribution P_{AGG}, and once there it pauses for a time t_{AGG} before selecting the next *ap*. This model reproduces the mobility of users who may group in interest points according to some spatial or functional rule. Parameters of this model are also the number and position of *ap*s over the area, the speed range $[v_{min}, v_{max}]$ and the radius of the *ap*s. The distance of a node from the center of an *ap* is determined according to a Rayleigh distribution with standard deviation σ_{AGG}. In the *swarm* (SWR) model, nodes move in a coordinated way. Each swarm has a logical center, which moves toward a destination chosen randomly. Once there, the nodes in the swarm stop for a pause time t_{SWR} before moving to a new destination. Each swarm has a number of nodes determined by a probability distribution with mean μ_{SWR} and standard deviation σ_{SWR}. Nodes in a swarm move randomly around its logical center, within maximum distance d_{SWR}. If the regions of two swarms overlap, nodes in the intersection may choose to migrate to the other swarm according to a probability M_{SWR}.

The RWP model reproduces sporadic encounters of two (or a few) nodes. The AGG model reproduces the encounters of many nodes, with nodes experiencing relevant neighborhood changes every time they enter an *ap*. The SWR model reproduces nodes maintaining the same neighborhood for a long time, with sporadic encounters with other (groups of) nodes. Analyzing these three basic models allows to bring into evidence the effects each of them separately has on message diffusion. However, mobility of people in a DTN is more likely modeled by the combination of the above three patterns. A current research topic is producing a synthetic model that addresses these issues or adopts statistical distributions inferred from real traces (CRAWDAD, 2008). A first attempt in this direction is represented by (Pedersini, Grossi, Gaito, & Rossi, 2008); as a future work we intend to conduct performance measures with this model.

Whatever is the adopted mobility model, it should be assumed that the following *mobility assumption* applies: when a contact occurs, the reciprocal speed is such that the two nodes can set up a communication channel and a significant amount of data is exchanged before they become disconnected. This assumption is reasonable according to results achieved by observations reported in (Hui, Chaintreau, Scott, Gass, Crowcroft, & Diot, 2005; Su, Chin, Popinova, Goely, & de Lara, 2004), but can be occasionally violated in our simulations.

FAMILY OF BROADCAST PROTOCOLS

Given a general DTN scenario as described in the previous section, purpose of this paper is the comparative analysis of a family of topology-independent broadcast protocols under different mobility conditions. Protocols are obtained by the incremental introduction of adaptive and situational mechanisms that progressively augment the protocol capability to adapt to changing conditions. Purpose of the paper is the evaluation of the role played by the different mechanisms and parameter settings to ensure broadcast *effectiveness*, i.e. the capability of the protocol of eventually achieving node coverage arbitrarily close to 1, and *efficiency*, i.e. the capability of the protocol of keeping the generated broadcasts-per-message as close as possible to $O(n \ln n)$ (Cooper, Ezhilchelvan, & Mitrani, 2004).

Each broadcast message *m* is supposed to have a *"scope"* that is defined by the source and specified through a *lifetime*; when this time expires, a node deletes its copy of *m* and stops its diffusion. In the following, we assume long lived messages to better understand the broadcast behavior independently of other constraints.

The basic broadcast protocol to start with is represented by the PUSH-based algorithm:

P-BCAST: a node p, holding a message m, starts a forwarding of m with probability $Prob_p=1$ *whenever a node enters its radio range.*

This algorithm guarantees node coverage approximately close to 1 with low latency, but generates a redundant load of duplicates. This is motivated by the fact that nodes perform epidemic forwarding with a very limited knowledge about the state of the encountered nodes and, as a consequence, they often happen to forward the message to already infected nodes. It is possible to improve P-BCAST behavior by adding some autonomic capabilities that extract an approximation of global state from locally observed data. We characterize a taxonomy of approaches, dividing them in two classes according to the amount of knowledge about the system state the nodes maintain.

Zero-Knowledge Approaches

In order to reduce duplicate diffusions, the protocol design must respond to the following questions: *(i)* how to provide a node with information about the state of its neighbors to enable the forwarding control? *(ii)* how to ensure the termination of the forwarding algorithm (*stop condition*) when the message has been delivered to the entire population of nodes? A viable approach to simply answer to point *(ii)* is to stop the forwarding when a message copy count has been reached:

CC-BCAST: a node p, holding a message m, starts a forwarding of m with probability $Prob_p=1$ *whenever a node enters its radio range. Every time m is forwarded, a local copy count is incremented; when it reaches a threshold* τ*, forwarding is stopped.*

The choice of τ influences the capability of achieving high coverage; it can be obtained either from the analysis of mobility traces in the target environment (Chaintreau, Mtibaa, Massoulie, & Diot, 2007), or—under restrictive assumptions—derived analytically. Cooper et al. (2004) showed that, under the condition of *a well known node cardinality n and RWP*, each node can stop forwarding when a copy count of $\tau = 2\lceil \ln n + \gamma \rceil$ broadcasts is reached, with γ the Eulero-Mascheroni's number.

The drawback of the previous approach is that nodes do not keep trace of the infected nodes: if *p* contacts *q* multiple times, it forwards the message every time, thus generating duplicates and wasting its budget of transmissions. As an alternative, a *self-adaptive mechanism* could be adopted such that a node *p* could send *m* only when the percentage of neighborhood changes with respect to the previous diffusion exceeds a threshold *Nth*, and it should be able to tune the value of $Prob_p$ according to the delivery status of *m* in its current area: when most nodes in the neighborhood have delivered *m*, *p* should reduce $Prob_p$ and vice versa. In a zero-knowledge paradigm, *p* is unable to obtain this information; however, *p* can derive the symptoms of the delivery status from sensing the events generated by its encounters, i.e. counting the received duplicates. Each node maintains a local view of its neighborhood at the time it performs a forwarding, by exploiting the underlying beaconing. The mechanism works as follows:

SA-BCAST: a node p, holding a message m, starts a forwarding of m with probability $Prob_p$ *whenever the number of neighborhood changes w.r.t. the previous forwarding is greater than a threshold Nth.* $Prob_p$ *is dynamically adapted within a range [MINP, MAXP] according to a function F that decreases* $Prob_p$ *every time a duplicate is received, and increases* $Prob_p$ *when m is received from a process that has been infected less than t time units before.*

In (Giudici, Pagani, & Rossi, 2009), it has been shown that this simple mechanism works properly to limit the number of duplicates under high coverage conditions. However, SA-BCAST might to not reach a *stop condition* (in case *MINP*>0), when the approach has to ensure a drip feed of message transmissions to manage late node joins and temporary partitions. A non-terminating protocol has severe effects on the efficiency, which can be only mitigated by adding more knowledge to the nodes about their neighborhood.

Local-Knowledge Approaches

The introduction of an encounters' history can provide the extra information required. A history mechanism has been recently adopted either to compute utility functions to control unicast forwarding (Boldrini, Conti, Iacopini, & Passarella, 2007; Burns, Brock, & Levine, 2005; Dubois-Ferriere, Grossglauser, & Vetterli, 2003), or as the log of infection events to control broadcast forwarding. In this work, we assume that the node cardinality *n* is known (we discuss this assumption in the Performance Evaluation section). The basic history-based mechanism is obtained from P-BCAST by adding the following statement:

HP-BCAST: whenever a node p forwards or receives the message m, it adds to a local data structure $history_p$ *the list of its current neighbors or m's sender respectively. Initially,* $history_p$ *only contains p itself. p broadcasts a message m if its current neighborhood has nodes other than those held in* $history_p$*. It skips forwarding otherwise. p terminates the epidemic algorithm when* $history_p$ *contains n entries.*

In the basic HP-BCAST (indicated as HP-$BCAST_0$), each node maintains a local history. The simple evolution of this algorithm is obtained by enabling a node to exchange its local history with its neighbors. This policy accelerates the node's awareness about the epidemic diffusion and should provide a more effective forwarding control. We obtain this new algorithm from HP-$BCAST_0$ by adding the following statement:

HP-$BCAST_{100}$*: when p forwards m then it piggybacks (the 100% of) its* $history_p$ *on m. When p receives m from q, it merges* $history_p$ *with* $history_q$*.*

It is worth to notice that the described HP-$BCAST_{100}$ algorithm is not as aggressive as it could in its attempts of propagating the infection knowledge. In fact, nodes could achieve a much quicker awareness about the evolution of the infection by piggybacking the history on the beacons.

Several aspects have yet to be explored. There is the need of defining the amount of history a node should maintain and/or exchange, of evaluating the real performance advantages it guarantees, of verifying its validity under mobility models other than RWP, and of exploring the scalability issues the history involves. Part of these arguments are considered in the Performance Evaluation section.

Local Suppression

The described mechanisms have been mainly designed to properly work under both RWP and AGG mobility models. By contrast, the SWR model generates some new issues that deserve specific attention. In fact, with this setting, we will show in the next section that 100% coverage is hard to reach, while a high number of duplicates might be generated. In order to achieve high effectiveness while maintaining an acceptable efficiency, a mechanism of *local duplicate suppression* (LS) can been added to HP-BCAST$_{100}$. When a beacon from a new neighbor is received, a node *p* sets a timer slightly larger than the beacon period. When the timer expires, the message is sent, with the history *including all the nodes that are new neighbors*. If another infected node with the timer set receives this message, it schedules its own transmission only if it has new neighbors not included in the history. This way, diffusion is not prevented for nodes that connect two swarms or two different groups in a swarm, but multiple infections of the same nodes are avoided.

Merging the Approaches

All the described basic mechanisms can be combined, thus leading to the complete family of broadcast protocols shown in Figure 1, where unlabeled vertices correspond to combinations not analyzed in the next section. In particular, we did not test the combination of copy count with the self-adaptive mechanism, because the latter starts working when a high number of duplicates is generated, i.e. when many nodes are already infected. At this time, the copy count stops diffusions; hence, the combined effects of the two mechanisms cannot be seen. The impact of the LS mechanism is shown only with HSA-BCAST; its effects are comparable for the other history-based algorithms. The characteristics of the basic mechanisms are summarized in Table 1. Both the self-adaptive mechanism and the history aim at controlling the number of duplicates, while history and copy-count aim at achieving a *stop condition*; SA-BCAST can stop when *MINP*=0, but in this case infection of late joining nodes is not guaranteed. The history requires additional bandwidth only in case of sharing. All mechanisms require some additional memory. Only the copy-count mechanism—when used according to (Cooper et al., 2004)—assumes that nodes move according to a known model and their number is known. In the

Figure 1. Family of broadcast protocols

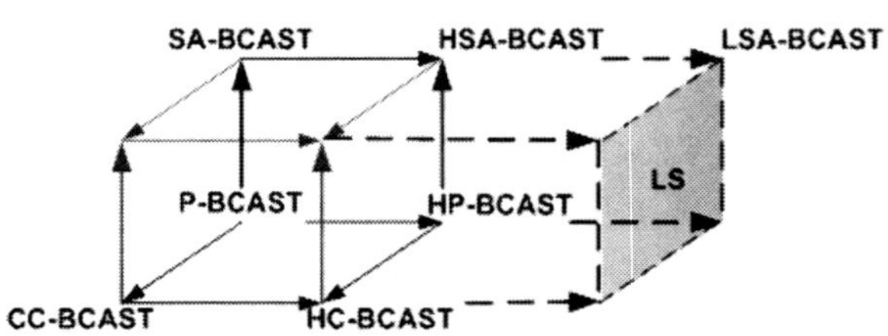

Table 1. Characteristics of the broadcast protocols

	memory	*bwth*	*stop condition*	*duplicates*	*mobility model*	*node cardinality*
P-BCAST	–	no	no	Yes	no	no
CC-BCAST	counter	no	Yes	Yes	RWP	known
SA-BCAST	neighbors	no	if MINP=0	decreased	no	no
HP-BCAST	encounters	if sharing	Yes	decreased	no	no

next section, we compare the above algorithms to identify their contribution to achieve the control on forwarding.

PERFORMANCE EVALUATION

Simulation Environment

We implemented the described protocols in the framework of the GloMoSim 2.03 (UCLA, 2008) simulation environment. The simulation setting considers a system of 50 nodes sparsely distributed over a 1000 × 1000 m. area. Nodes move at a speed in [1, 2] m/s, thus reproducing a pedestrian environment. They are equipped with a low power 802.11 radio device with 10 m. communication range and DCF at the MAC layer. Beaconing is performed every 1 sec.; after 3 missing beacons, the corresponding neighbor is removed from the neighbor list.

For SA-BCAST, the simulations run different values of *Nth* and two different functions F: a linearly decreasing function (or *Lin10*) and an inverse exponential function (or *InvExp*). When an infected node *p* receives a duplicate from a node that is infected from less than 3 min., then *p* sets $Prob_p$ to *MAXP*=1. Otherwise, $Prob_p$ is either decremented of 0.1 or halved in line with functions *Lin10* or *InvExp*, respectively. $Prob_p$ has a lower bound defined by *MINP* = 0.01.

For CC-BCAST, according to (Cooper et al., 2004), τ=10 is adopted.

We consider long lived broadcasts, with simulations lasting up to 6 hours. All simulation results are averaged over 50 simulations performed with variable random seed. The mobility models are provided by BonnMotion (de Waal, & Gerharz, 2008); in order to allow movements to reach a steady state, the first 1000 sec. of the traces are not considered for the measures. The parameters of the different mobility models are presented in the next subsections.

Main performance indexes are the *coverage*, i.e. the percentage of nodes infected, and the *duplicate messages* (a message is a duplicate if it is received by an already infected node).

Simulation Results

Random Waypoint Model

In the measurements presented in this work, the pause time for RWP is 0. In Figure 2, the basic P-BCAST is compared with the zero-knowledge approaches. SA-BCAST effectively reduces the number of duplicates with respect to P-BCAST (Figure 2(b)); yet, when coverage is high and several duplicates are generated, the self-adaptive mechanism decreases the infection aggressiveness and, as a side effect, a higher latency is observed in reaching 100% coverage (Figure 2(a)). CC-BCAST has the same latency as SA-BCAST with F=*Lin10*, because nodes that exhaust their budget stop diffusions, thus reducing the number of relays. CC-BCAST stops diffusions 1.5 hours after full coverage is achieved, and from this point on no more packets are generated.

HP-BCAST (Figure 3) provides a twofold advantage over P-BCAST: it does not affect coverage (the white circle (Rc 100%) evidences the point where full coverage is reached in the worst case), and it provides an effective mechanism to control the forwarding. The higher is the global state awareness (HP-BCAST_{100} vs. HP-BCAST_0), the lower the number of duplicates. However, HP-BCAST is very slow in reaching the stop condition: our simulations show that full coverage is achieved in less than 1 hour of simulated time. At this point, each node on average knows the 15% of infected nodes with HP-BCAST_0 and 65% with HP-BCAST_{100} (Figure 8(b)). With HP-BCAST_0 all nodes stop within 47 hours (although only 2 packets/hour overall are sent after 24 hours); with HP-BCAST_{100}, only 1 packet/hour is sent after 24 hours and all nodes stop within 41 hours. In (Cooper, Ezhilchelvan, Mitrani, & Vollset,

Figure 2. (a) Coverage and (b) cumulative number of duplicates vs. time for P-BCAST, CC-BCAST and SA-BCAST in the RWP model

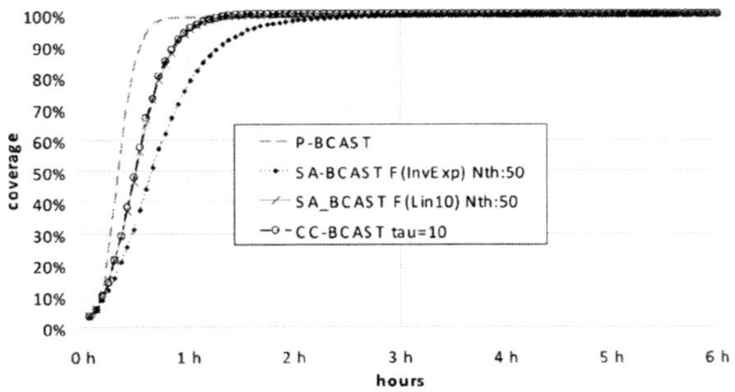

2(a)

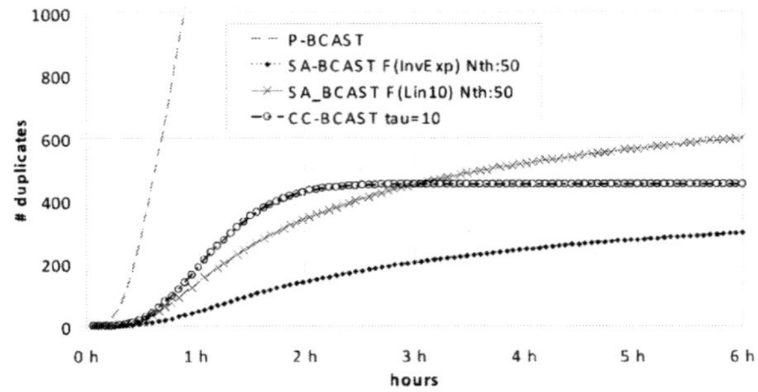

2(b)

2005), the CC-BCAST approach has been enhanced by adding a history either locally maintained by nodes or completely shared; let us indicate it with HC-BCAST$^{\tau}_{\alpha}$ with τ the copy count threshold and α the amount of history exchanged. When a node *p* encounters a node *q* already in its history, *p* suppresses the transmission but increments its copy count. HC-BCAST (Figure 3) achieves full coverage, although with a slightly higher latency than HP-BCAST, due to nodes that exhaust their available copy count before all nodes have been infected, thus decreasing the number of relays and slowing down the diffusion. Yet, thanks to the bound on the number of diffusions, HC-BCAST is optimal in RWP and is thus much more effective than HP-BCAST in limiting useless traffic: full coverage is reached after 3672 sec., and message diffusion stops after 6696 sec. with HC-BCAST$^{10}_{100}$, and after 7560 sec. with HC-BCAST$^{10}_{0}$.

We studied how the algorithm behavior is affected by the broadcast nature of the radio channel where, to correctly deliver *m* to an uninfected node *q*, *m* may be duplicated in a node *p* that happens to be in range. A good forwarding control should identify the presence of uninfected nodes and refrain from forwarding otherwise. The *broadcast success rate bsr* properly captures this ability; it is defined as the ratio *hitting broadcasts/total broadcasts*, where a "hitting broadcast" is a

Figure 3. (a) Coverage and (b) cumulative number of duplicates vs. time for HP-BCAST and HC-BCAST in the RWP model

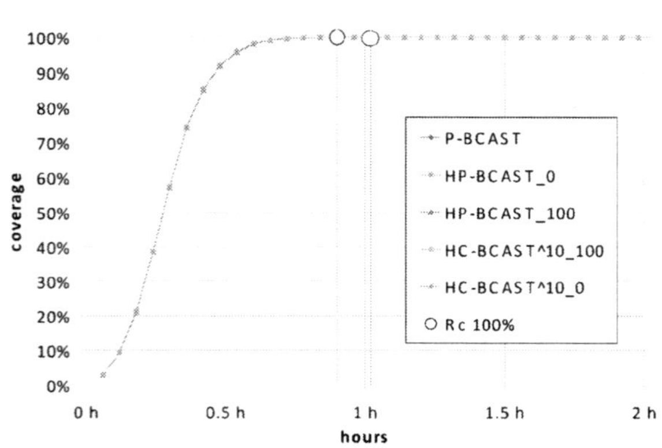

3(a)

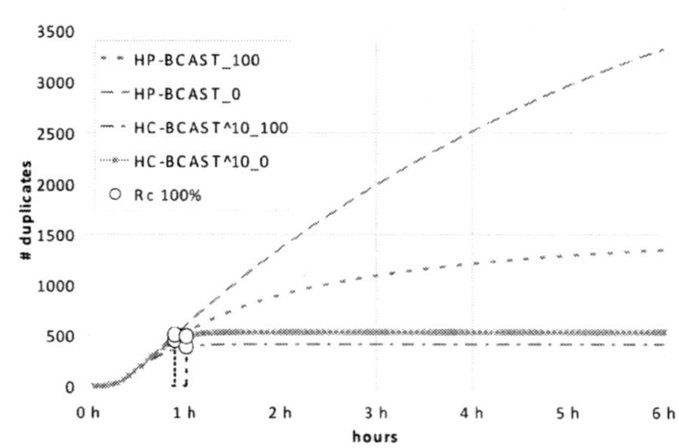

3(b)

Figure 4. (a) Broadcast success rate for HP-BCAST and HC-BCAST in the RWP model. (b) Coverage and number of generated packets vs. time for SA-BCAST in the RWP and AGG model with 10 aps and σ_{AGG}*=10*

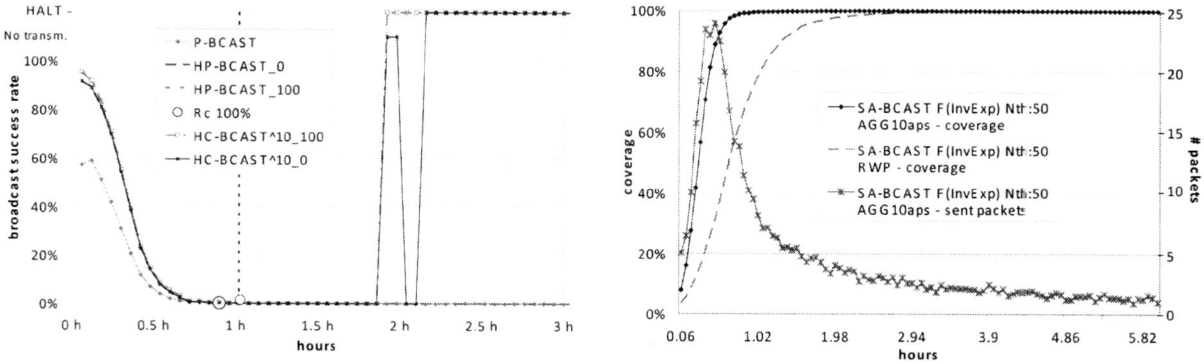

4(a) 4(b)

broadcast that delivered *m* to at least one uninfected node. Of course, *bsr=1* indicates that all the broadcasts hit the mark. The index drops to 0 when full coverage is reached. The behavior of *bsr* (Figure 4(a)) confirms the remark above. The *bsr* index before full coverage is the same for HP-BCAST and HC-BCAST, confirming the effectiveness of the history in suppressing useless transmissions. However, the node's knowledge does not grow as quickly as the nodes infection and this influences the efficiency of the forwarding control. With HP-BCAST nodes continue performing transmissions till their histories are full; with HC-BCAST no useless transmission is anymore generated after all nodes halt. With HC-BCAST$^{10}_{0}$, though, the last infected nodes could issue sporadic transmissions before halting.

Lesson learnt: SA-BCAST is more effective than HP-BCAST in reducing duplicates. HC-BCAST is more effective than HP-BCAST in both implementing a stop condition and decreasing duplicates.

Aggregation Model

With AGG, we performed experiments with a number of *aps* variable from 3 to 10, uniformly distributed in the area. A node stops in an *ap* for 10 minutes. The next *ap* is chosen according to a uniform probability distribution. The distance of a node from the *ap* center is determined by σ_{AGG} in [0,15]; for σ_{AGG} =10 it follows the probability distribution shown in Figure 5. When switching to the AGG model, SA-BCAST fruitfully uses the contact opportunities in *ap*s to speed up the infection (Figure 4(b)). The collateral effect is that SA-BCAST, although able of smoothing down the generated traffic as soon as a high coverage is reached, is too aggressive in diffusing when nodes are in an *ap*, thus generating a high number

Figure 5. Rayleigh distribution for σ_{AGG}*=10*

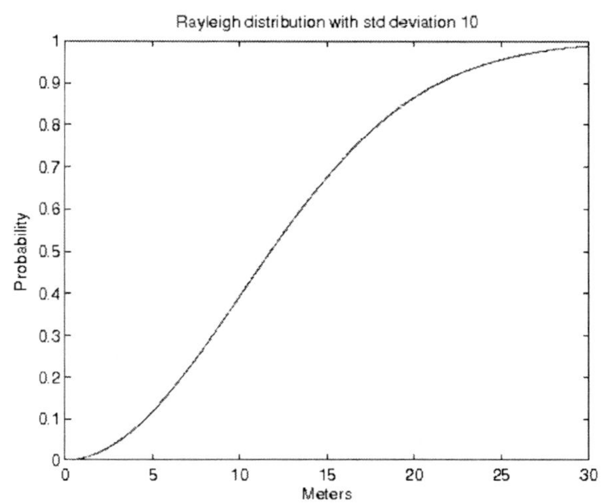

Figure 6. (a) Cumulative number of duplicates after 6 hours of simulated time for SA-BCAST in the AGG model. (b) Progress of coverage vs. time in a single simulation, with $\sigma_{AGG}=10$

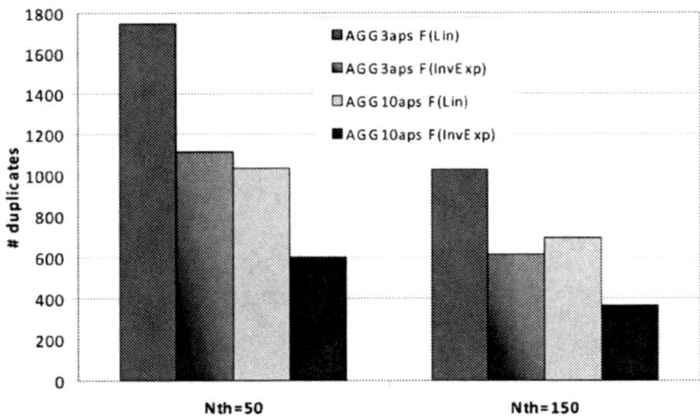

6(a)

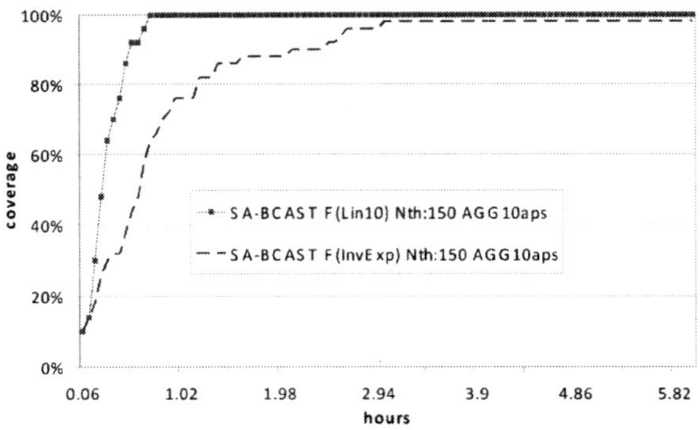

6(b)

of duplicates (Figure 6(a)). Some improvements can be achieved with a more stringent F, as shown. With *Nth*>100 some contact opportunities may be missed, thus further reducing traffic; coverage is anyway achieved, although with a higher latency, thanks to the existence of multiple relays. The number of *ap*s also has impact: when it tends to ∞, the RWP and AGG models converge. By contrast, with 3 *ap*s we observed more duplicates independently of F and *Nth*, because *(i)* nodes are more dense in *ap*s, and *(ii)* there is a higher probability of encounters during movements between two *ap*s, which are used for dissemination. For larger *ap*s (higher σ_{AGG}), nodes in an *ap* are not all in mutual communication range. When a node enters an *ap*, the messages exchanged do not affect all nodes and, at the same time, reduce $Prob_p$ thus preventing excessive diffusions; as a result, we observed a lower number of duplicates. An interesting aspect is shown in Figure 6(b), where the network coverage is shown for a single simulation; in order to emphasize the behavior, with F =*InvExp* the $Prob_p$ is decreased by dividing the current value by 20. With 10 *ap*s, where encounters during movements are more sporadic, coverage increases in steps, which correspond to relevant membership changes in the *ap*s. Steps become less high with the progress of the simulation, because the probability of entering an *ap* with already infected nodes increases. This behavior is much more evident for $Prob_p$ decreasing more quickly. The broadcast nature of the channel, although nodes are unaware of being either in an *ap* or on the road, allows to effectively exploit node density to increase the coverage. However, this is achieved at the expenses of efficiency.

The weakness of the copy-count approach as used by (Cooper et al., 2004) is its dependence on the uniform distribution of contacts as in RWP. We simulated HC-BCAST in AGG with 10 *aps* and σ_{AGG}=10. In this mobility scenario, HC-BCAST suffers multiple contacts between the same pair of nodes (Figure 7(a)); this does not lead to duplicate generation but forces the nodes to waste their broadcast budget every time they re-encounter a node seen in the past. As a consequence, the algorithm is too conservative and full coverage is not achieved: in the conditions shown in the figure, all nodes terminate the algorithm when, on average, the 2% of them is still uninfected; the minimum coverage observed is 94%. By exasperating the non-uniformity of contacts till considering a swarm mobility, the coverage drops to 17% in the worst case. By contrast, under the same conditions the HP-BCAST algorithm guarantees full coverage independently of the mobility model, although this result is paid with a higher number of duplicates. The above argu-

Figure 7. (a) Coverage for HP-BCAST and HC-BCAST, and (b) cumulative number of duplicates for HP-BCAST and HSA-BCAST, in the AGG model with 10 aps and σ_{AGG}=10

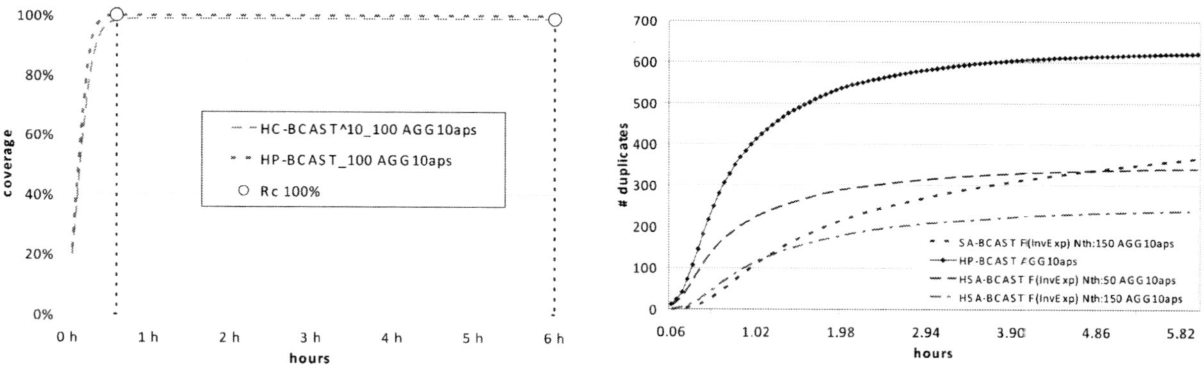

7(a) 7(b)

ments lead to say that—when nodes are unaware of the node cardinality and the mobility model—the copy count mechanism as used according to (Cooper et al., 2004) fails.

We then measured the performance achieved with HSA-BCAST: a node can schedule a diffusion only when has one or more neighbors not in its history; the diffusion is performed or suppressed according to $Prob_p$. Merging history and self-adaptive mechanism favors the control of the number of generated duplicates (Figure 7(b)), because useless transmissions are suppressed both towards already known nodes and when duplicates indicate that coverage in the region is already high. The latency obtained by HSA-BCAST is intermediate between HP-BCAST and SA-BCAST, and higher for higher *Nth*. Yet, the number of duplicates diverges for all algorithms, indicating that HSA-BCAST too is still far from reaching a stop condition within the 6 hours of simulated time.

In order to compare the efficiency of the approaches, we consider as performance index the *target ratio T = (msgrecv – dups)/msgrecv*, with *msgrecv* the total number of messages received, and *dups* the total number of duplicates among them. *T* is a measure of efficiency in using the network resources. Of course, *T* is optimized by

Figure 8. (a) Target ratio vs. coverage with and without history in RWP and AGG model with 10 aps and σ_{AGG}=10. (b) Cumulative knowledge about the coverage status

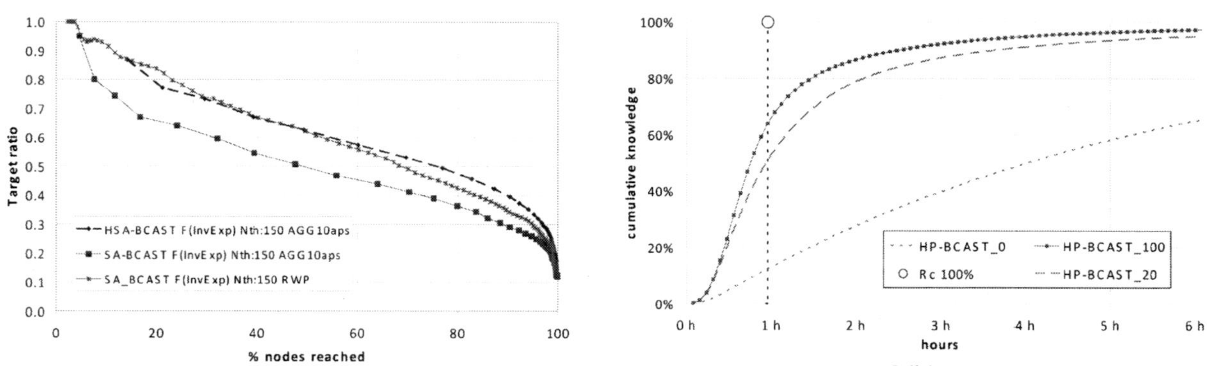

8(a) 8(b)

dups=0 and is affected by the number of the encounter nodes and by the progress of the infection in the neighborhood. In fact, packets are broadcast to the nodes in range, let us say *k*; so that, for any message sent, we count *msgrecv=k* and the *dups* value depends on the level of infection among the *k* neighbors. The measure in Figure 8(a) confirms that HSA-BCAST in the AGG model behaves even better than SA-BCAST in the RWP model.

Since history proves to be an efficient mechanism, a trade-off could be characterized between the amount of history exchanged and the bandwidth saved by suppressing duplicates. The comparison among different percentages of history exchanged and different policies to extract the history entries to be exchanged (Gamberini, Giudici, Pagani, & Rossi, 2008) yields that—by indicating with 100% the full history condition and exchanging the most recent entries—the knowledge growth by sharing 20% of the history (HP-BCAST_{20}) well approximates the behavior achievable with full sharing (Figure 8(b)); this effect is obviously reproduced by the number of generated duplicates. However, in the following measures the nodes exchange the whole history.

Lesson learnt: the copy count mechanism fails in case of non-uniform mobility model. History and self-adaptive mechanisms are efficient in suppressing duplicates. However, the stop condition is reached too late.

Swarm Model

The SWR model used in the experiments has pause time of 10 minutes, σ_{SWR} =1.73, μ_{SWR} of 4 or 15, d_{SWR} =15 m., and M_{SWR} =0.2. With this model, coverage can be incremented when two swarms partially overlap, one of which has already been infected. On the other hand, once a swarm has been infected, the nodes belonging to it should refrain from transmitting again till the swarm membership does not change. Better performance is achieved with *Nth* low, which promptly detects swarm overlapping. In Figure 9(a), the performance of SA-BCAST is reported for F =*Lin10*; with F =*InvExp* the coverage achieved is worse. Yet, in the latter case a lower number of duplicates is generated (Figure 9(b)). Hence, the *InvExp* function has been adopted for experiments with HSA-BCAST and LSA-BCAST. The local suppression mechanism does not produce benefits (nor drawbacks) in the aggregation model. However, in the SWR model it is able to improve both coverage and—above all—efficiency. This derives from the small delay before diffusing: if two swarms A and B are overlapping, such that nodes in A own *m* while nodes in B do not, an

Figure 9. (a) Coverage and (b) cumulative number of duplicates vs. time for different algorithms in the SWR model

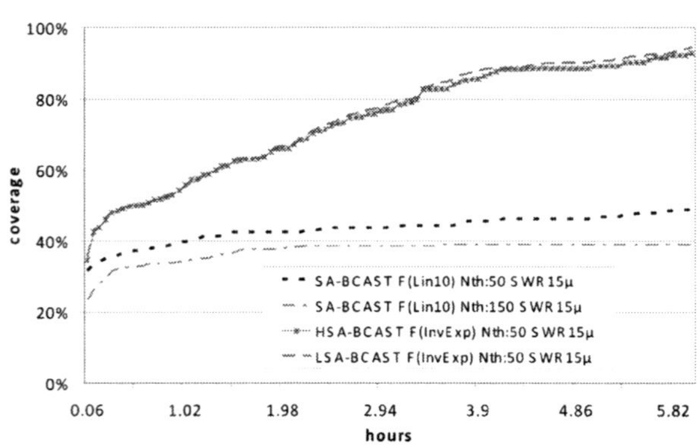

9(a)

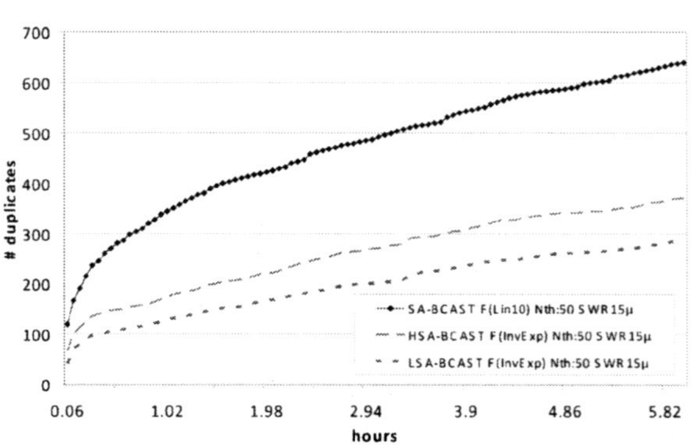

9(b)

infected node in swarm A is likely to observe a sequence of new neighbors appearing at a short interval one after another. One "late" transmission allows to infect more new neighbors at one time. At the same time, the history mechanism allows infection propagation in swarm B: the newly infected nodes in the intersection have empty histories. They see all their neighbors in swarm B as not being in their histories, thus starting message diffusion. This repeats recursively till the whole swarm B is infected.

Lesson learnt: with the LS mechanism, HSA-BCAST seems to achieve the best performance in all basic mobility models. However, it is unable to stop promptly.

CONCLUSION

In this work, three mechanisms for message broadcasting in DTNs have been analyzed, together with their combinations, in different mobility models. The most promising approach seems to be the adoption of both a self-adaptive policy and a history of past encounters, optimized with a local duplicate suppression mechanism. This combination is independent of the mobility model, but is slow in reaching a stop condition thus being possibly inefficient. On the other hand, with a copy count approach nodes risk of being too prompt in halting diffusions, thus not reaching a satisfying coverage nor infecting late joining nodes. Moreover, estimating an appropriate copy count threshold could be a hard task.

Future research work can proceed along different directions. We intend to analyze real mobility traces in order to develop a synthetic model that captures real human behaviors. This model will be useful for both more accurate performance evaluation of communication protocols for DTNs via simulations, and analysis aiming at optimizing the protocols. The LSA-BCAST protocol must be improved by including an adaptive stop condition. A promising approach could be exploiting the *bsr* index: if nodes could (quite accurately) estimate *bsr* using local observations, they could stop when *bsr* approximates 0, which should be the time at which full coverage is achieved. Finally, the analytical model proposed in (Cooper et al., 2004) could be extended in order to compute a value for τ more appropriate for realistic mobility, to then re-evaluate the performances obtained.

ACKNOWLEDGMENT

This work has been partially funded by the Italian Ministry of University and Research in the framework of the "Context-Aware RouTing Over Opportunistic Networks (CARTOON)" PRIN Project.

REFERENCES

Boldrini, C., Conti, M., Iacopini, I., & Passarella, A. (2007). HiBOp: a history based routing protocol for opportunistic networks. *Proc. IEEE WoWMoM.*

Burgess, J., Gallagher, B., Jensen, D., & Levine, B. N. (2006). MaxProp: routing for vehicle-based disruption tolerant networks. *Proc. IEEE INFOCOM.*

Burns, B., Brock, O., & Levine, B.N. (2005). MV routing and capacity building in disruption tolerant networks. *Proc. IEEE Infocom.*

Camp, T., Boleng, J., & Davies, V. (2002). A survey of mobility models for ad hoc network research. *Wireless Communication and Mobile Computing, 2*(5), 483–502. doi:10.1002/wcm.72

Chaintreau, A., Mtibaa, A., Massoulie, L., & Diot, C. (2007). The diameter of opportunistic mobile networks. *Proc. ACM CoNEXT Conference*, (pp. 1-12).

Cooper, D. E., Ezhilchelvan, P., & Mitrani, I. (2004). High coverage broadcasting for mobile ad-hoc networks. *Lecture Notes in Computer Science*, *3042*, 100–111. doi:10.1007/978-3-540-24693-0_9

Cooper, D. E., Ezhilchelvan, P., Mitrani, I., & Vollset, E. (2005). Optimization of encounter gossip propagation in mobile ad-hoc networks. *Proc. 13th MASCOTS*, (pp. 529-532).

CRAWDAD. (2008). *CRAWDAD – A community resource for archiving wireless data at Dartmouth.* Retrieved from http://www.crawdad.org

Davis, J. A., Fagg, A. H., & Levine, B. N. (2001). Wearable computers as packet transport mechanisms in highly partitioned ad-hoc networks. *Proc. 5th IEEE Intl. Symp. on Wearable Computers.*

de Waal, C., & Gerharz, M. (2008). *BonnMotion – A mobility scenario generation and analysis tool.* Retrieved from http://web.informatik.uni-bonn.de/IV/Mitarbeiter/dewaal/BonnMotion/.

DTN Research Group. (2008). *Delay Tolerant Networking Research Group.* Retrieved from http://www.dtnrg.org/wiki

Dubois-Ferriere, H., Grossglauser, M., & Vetterli, M. (2003). Age matters: efficient route discovery in mobile ad hoc networks using encounter ages. *Proc. 4th ACM Intl. Symp. MobiHoc*, (pp. 257-266).

Gamberini, G., Giudici, F., Pagani, E., & Rossi, G. P. (2008). Impact of history on epidemic broadcast in DTNs. *Proc. 1st IFIP Wireless Days Conference.*

Giudici, F., Pagani, E., & Rossi, G. P. (2009). Self-adaptive and stateless broadcast in delay and disruption-tolerant networks. *Proc. Italian Networking Workshop.*

Harras, K. A., Almeroth, K. C., & Belding-Royer, E. M. (2005). Delay tolerant mobile networks (DMTNs): controlled flooding in sparse mobile networks. *Proc. IFIP Networking.*

Hui, P., Chaintreau, A., Scott, J., Gass, R., Crowcroft, J., & Diot, C. (2005). Pocket switched networks and human mobility in conference environments. *Proc. ACM SIGCOMM Workshop on Delay Tolerant Networking.*

Jones, E. P. C., Li, L., & Ward, P. A. S. (2005). Practical routing in delay tolerant networks. *Proc. ACM SIGCOMM Workshop on Delay Tolerant Networking*, (pp. 237-243).

Juang, P., Oki, H., Wang, Y., Martonosi, M., Peh, L., & Rubenstein, D. (2002). Energy-efficient computing for wildlife tracking: design tradeoffs and early experiences with Zebranet. *Proc. ASPLOS.*

Lee, U., Magistretti, E., Zhou, B., Gerla, M., Bellavista, P., & Corradi, A. (2006). MobEyes: smart mobs for urban monitoring with a vehicular sensor network. *IEEE Wireless Communications, 13*(5).

Lenders, V., Karlsson, G., & May, M. (2007). Wireless ad hoc podcasting. *Proc. 4th IEEE Conf. SECON*, (pp. 273-283).

Montresor, A., Jelasity, M., & Babaoglu, O. (2005). Gossip-based aggregation in large dynamic networks. *ACM Transactions on Computer Systems*, *23*(3), 219–252. doi:10.1145/1082469.1082470

Pedersini, F., Grossi, G., Gaito, S., & Rossi, G. P. (2008). Experimental validation of a 2-level social mobility model in opportunistic networks. *Proc. 1st IFIP Wireless Days Conference.*

Spyropoulos, T., Psounis, K., & Raghavendra, C. S. (2004). Single-copy routing in intermittently connected mobile networks. *Proc. 1st IEEE SECON*, (pp. 235-244).

Su, J., Chin, A., Popinova, A., Goely, A., & de Lara, E. (2004). User mobility for opportunistic ad-hoc networking. *Proc. 6th IEEE Workshop on Mobile Computing Systems and Applications.*

UCLA Parallel Computing Laboratory. (2008). *GloMoSim – Global Mobile Information Systems Simulation Library*. Retrieved from http://pcl.cs.ucla.edu/projects/glomosim

Vahdat, A., & Becker, D. (2000). Epidemic routing for partially connected ad-hoc networks. Technical Report CS-200006, Duke University.

This work was previously published in International Journal of Business Data Communications and Networking (IJBDCN) Volume 5, Issue 2, edited by Varadharajan Sridhar and Debashis Saha, pp. 1-15, copyright 2009 by IGI Publishing (an imprint of IGI Global).

Chapter 5
On the Impact of Network Dynamics on a Discovery Protocol for Ad-Hoc Networks

Fei Liu
University of Twente, the Netherlands

Geert Heijenk
University of Twente, the Netherlands

ABSTRACT

Deliberate exploitation of natural resources and excessive use of environmentally abhorrent materials have resulted in environmental disruptions threatening the life support systems. A human centric approach of development has already damaged nature to a large extent. This has attracted the attention of environmental specialists and policy makers. It has also led to discussions at various national and international conventions. The objective of protecting natural resources cannot be achieved without the involvement of professionals from multidisciplinary areas. This chapter recommends a model for the creation of knowledge-based systems for natural resources management. Further, it describes making use of unique capabilities of remote sensing satellites for conserving natural resources and managing natural disasters. It is exclusively for the people who are not familiar with the technology and who are given the task of framing policies.

INTRODUCTION

Context-aware ad-hoc networks adapt their behavior based on the context in which they operate. For this purpose, nodes use information from context sources. To discover these sources, a context discovery protocol is needed. Such a protocol disseminates information on context information that can be provided by nodes to nodes that might want to use the information. Ad-hoc networks are severely limited in resources, such as communication bandwidth, energy usage, and processing power. To save communication resources, we have proposed to perform context discovery using attenuated Bloom filters (ABFs) (Liu & Heijenk, 2007). We have proven that using ABFs our discovery protocol can provide discovery, while exchanging far less information than conventional approaches.

Another important feature of ad-hoc networks is dynamics in connectivity. In this paper, we present an investigation of the impact of network dynamics on our ABF-based discovery protocol through an analytical approach. In general, three categories of causes of network dynamics can be identified: nodes may be mobile; battery-supplied devices might exhaust their batteries; the quality of the wireless transmissions might be varying due to varying propagation conditions.

Because of the random position and movement of the nodes, it is not feasible to quantify the network traffic of our discovery protocol in a mobile environment mathematically. Thus, simulation is a good approach to study this problem. (Goering, Heijenk, Haverkort, & Haarman, 2007) has examined the network traffic generated by updating the ABFs while nodes are moving in a low density network, and the reachability of the required services through simulations. In this paper, we present an analytical modeling of the dynamics due to the limited battery-supply and unstable transmission quality in very high-density networks. First, we will consider node disappearance and appearance. When a node is powered off, it disappears from the network. After it switches on again, it joins the network again. We quantify the network load through analytical study and verify obtained results with simulations. Further, we observe a special case where the packets transmitted by a node get lost for certain time due to the poor propagation conditions. In this scenario, the node is considered as disappearing and reappearing in the network. We obtain simulation results for various packet loss periods. Finally, we study the effect of node movement on network traffic for various network densities using simulations.

This paper is structured as follows. Section 2 gives a brief introduction of the ABF-based context discovery protocol for ad-hoc networks. Section 3 discusses network structures, the assumptions we use in our analysis, and an approximation for the basic notion of i-hop node degree. Section 4 presents the analysis of network traffic when nodes appear and disappear in the network, when a series of consecutive advertisement packets are lost, or when nodes are moving. Section 5 concludes the study and discusses the future work.

A DISCOVERY PROTOCOL FOR AD-HOC NETWORKS

Attenuated Bloom Filters (ABF)

Bloom filters (Bloom, 1970) have been proposed in the 1970s to represent a set of information in a simple and efficient way. They use b independent hash functions to code the information. The hash results are over a range $\{1..w\}$, where w denotes the width of the filter. In the filter, which has a length of w bits, every bit is set to 0 by default. Only the bit positions associated with the hash results will be set to 1. The resulting Bloom filter can be used to query the existence of certain information. If all the bit positions related to the hash results of the queried information are 1 in the filter, the information exists with small chance of false positive.

Attenuated Bloom filters (ABFs) are layers of basic Bloom filters. We use ABFs to represent information regarding the presence of context sources on a hop-distance basis (Liu & Heijenk, 2007). The ith layer of an ABF ($0 \leq i < d-1$) aggregates all information about context sources i hops away. The depth of the ABF, d, also stands for the total propagation range of the information. Note that context sources reachable in i hops may also be reachable via longer paths. As a result, hash results at layer i will often be repeated in lower layer j ($j>i$).

Figure 1 exemplifies the context aggregation operation for a node with two neighbors. In this example, each node has an ABF with 8 bits width (w=8) and a depth of 3 (d=3). The node uses two

Figure 1. An example of ABF aggregation

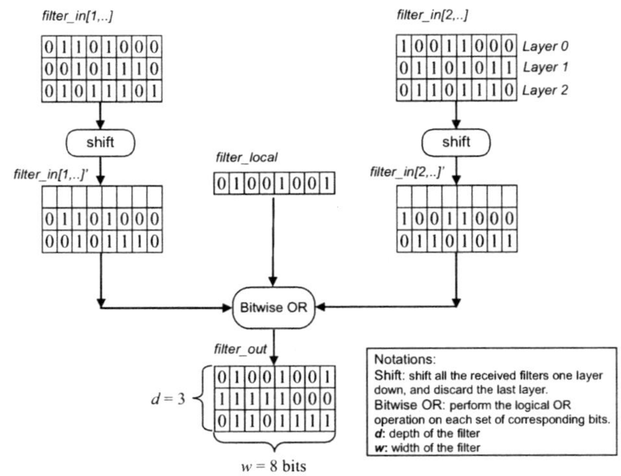

hash functions (*b*=2) to encode its local context sources "temperature" and "humidity" into {2,8} and {2,5} respectively. If we set the corresponding bit positions, we can obtain *filter_local* as shown in Figure 1. When the node receives the incoming filters *filter_in[1,..]* and *filter_in[2,..]* from its neighbors, it shifts the received filters one layer down and discards the last layer. Thus, *filter_in[1,..]'* and *filter_in[2,..]'* are obtained. We perform a logical OR operation on each set of corresponding bits of *filter_local*, *filter_in[1,..]'*, and *filter_in[2,..]'*. *filter_out* can be obtained as the ABF that the node broadcasts to its neighbors. This filter contains the local information of the node on layer 0; one hop neighbors' information on layer 1; and two hop neighbors' information on layer 2.

Protocol Specification

Our ABF-based context discovery protocol distinguishes 3 phases (Liu, Goering, & Heijenk, 2007): context exchange, context query, and context update and maintenance.

- **Context exchange:** every node stores two kinds of ABFs: incoming ABFs for each neighbor and an aggregated outgoing one with all local and neighboring information. When a new node joins the network, it will broadcast an ABF with only the local information first. This broadcast will be received by the nodes within communication range. Any neighboring node receiving this ABF will update its outgoing ABF with the new information and broadcast it. Once the newly joined node receives the neighboring information, it updates its outgoing ABF and broadcasts it. Every neighboring node will aggregate this update into its outgoing ABF as well. If there is any change in the outgoing ABF, the updated ABF will be broadcast to the network. After the exchange of ABFs, every node will have a clear view of the context information present within *d* hops.
- **Context query:** whenever a node is looking for a specific context source, and a query is generated in that node, the node first looks for presence of the information locally. If the required information is not available locally, it will hash the query string and check it against the stored neighboring ABFs. If there is no match, the query will be discarded. If there is any match, a query message will be unicasted to that neighbor

Figure 2. Network structure

(a) grid structure *(b) circle structure*

with a *hop-counter* set to *d*. The neighbor will perform the same action. It checks the query against the locally available context sources. If there is any match, a response message will be sent back to the querying node. If nothing matches, it will check the stored neighboring ABFs. Whenever there is a match, the query will be propagated to that node with the *hop-counter* decreased by 1. When the *hop-counter* equals 0, the query will stop spreading. If a node receives the same query multiple times, as identified by a unique query ID, the query will simply be dropped. State information will be temporarily stored to be able to route responses back to the query initiator.

- **Context update and maintenance:** if there is no change in the context sources offered by a node, a keep-alive message will be sent out periodically. A keep-alive message is a short message with a *generation-id* of the last broadcasted ABF from this node. A node can identify the freshness of the stored ABFs by comparing *generation-ids*. Once it notices the *generation-id* is different from that of the stored ABF for this neighbor, an update request is sent out. The neighbor replies back with its latest ABF. If a node does not receive keep-alive messages from a certain neighbor for two consecutive *keep-alive periods*, it considers the node has left and removes its neighbor's information.

MODELING PRELIMINARIES

In this section, we introduce background knowledge regarding modeling of connectivity in multi-hop ad-hoc networks. We start with the introduction of the network structures we are assuming in our modeling, i.e., a grid structure and a circular structure. Further, we discuss connectivity in ad-hoc network models and continue with the assumptions for circular structured networks. Finally, we conclude with an approximation of the mean multi-hop node degree in circular structured networks, which is an essential component of our further analysis.

Network Structures

To be able to analyze the performance of the proposed system, we model two typical network structures, which we will refer to as grid structure and circle structure. The grid structure represents a simple and regular network in which each node has four neighbors. In the circle structure placement of nodes, and connectivity of nodes is based on the notion of a random geometric graph.

Grid Structure

In grid structures, each node has 4 direct neighbors within communication range. The distance of each pair of connected nodes equals to the communication range of the nodes r. The connections between nodes can be considered as series of intersecting vertical and horizontal axes that form a two-dimensional grid structure with $r \times r$ size square-shaped grids. The structure is shown in Figure 2a. The regularity of the grid structure and the fixed number of neighbors for each node enables a rather simple analysis of models using this structure.

Circle Structure

To model the network structure of an ad-hoc network, based on random location of nodes, we can model it as a random geometric graph. Given a graph $G=G(V,E)$, where nodes are vertices (V), while links between nodes are edges (E), a unit disk graph (Clark, Colbourn, & Johnson, 1990) models an ideal network where the two-dimensional radio coverage of a mobile node is a circle (with range r). This can be extended to a random geometric graph, where vertices (nodes) are located at random, uniformly and independently in a region, and an edge between two vertices exists if and only if the distance between them is at most r. Random geometric graphs are often used to model ad-hoc networks (Penrose, 2003), and we will use them as the basis for our analysis.

In Section 3.4, we will argue that, from the point of view of a selected node, the set of areas in which nodes at i hops distance can be located can be approximated by a set of concentric circles, one for each value of i. For this reason, we will refer to this model as circle structure, shown in Figure 2b.

Connectivity in Ad-Hoc Network Models

Using graph theory can help us to analyze some specific network characteristics. Most studies have been done in the area of node degree and connectivity. The degree of a vertex can be defined as the number of edges incident to it. Further, a graph is called k-connected if the graph remains connected when fewer than k vertices are removed from the graph. (Clark, Colbourn, & Johnson, 1990) has discussed the connectivity problems of unit disk graph. (Bettstetter, 2002) has investigated the relationship between required range r, node density, and almost certainly k-connected networks, assuming random geometric graphs. The results provide the principles to choose practical values of those parameters for simulations and design. (Hekmat & Mieghem, 2003) has shown that the degree distribution in wireless ad-hoc networks, modeled as a random geometric graph, is binomial for low values of the mean degree.

Besides using graph theory, some other approaches have also been used to investigate the connectivity problems in ad-hoc networks. (Albero, Sempere, & Mataix, 2006) has examined the connectivity of a certain number of mobile nodes within a certain area by using a stochastic activity model. However, the study is limited to low-density networks due to the limitation of the stochastic model. Besides this analytical analysis, some studies have been done by means of simulation (Trajanov, Filiposka, Efnuseva, & Grnarov, 2004) and test bed (Lenders, Wagner, & May, 2006).

We can generalize that the current research of connectivity mostly focuses on the following two major questions: (1) how to achieve a k-connected network; (2) what is the degree distribution of a node. However, we have not found any research describing the degree distribution multiple hops away.

In here, we define the i-hop node degree to be the number of nodes a selected node can reach in exactly i hops, but not fewer than that. Let us denote the i-hop node degree as D_i.

For grid-structured network, it is straightforward to determine the i-hop node degree, D_i^g. By definition, we define the node degree at 0 hops, i.e. the number of nodes 0 hops away as

$$D_0^g = 1. \quad (1)$$

Here the superscript g denotes the grid structure. For larger i, it can easily be seen from Figure 2a. that $4i$ nodes become reachable when increasing the maximum number of hops from the central node from i-1 to i. The node degree of ith hop, D_i^g, can be derived as

$$D_i^g = 4 \cdot i. \quad (2)$$

This value increases linearly with the distance (the number of hops). The total number of nodes reachable in at most i hops, N_i^g, including the central node can be easily found by summation:

$$N_i^g = \sum_{j=0}^{i} D_j^g = 1 + 2i(i+1). \quad (3)$$

For a random geometric graph, the distribution of D_i^c can be derived from D_{i-1}^c (the superscript c denotes the circular structure), conditioned on the position and the number of nodes reachable in i-1 hops. Theoretically, we can derive D_i^c in this way. However, the expression is going to be computationally infeasible. Therefore, we will take a step back and observe the upper bound of this problem.

Modeling Assumptions of Circular Structured Networks

As we mentioned above, we model our network based on random geometric graph, where nodes are uniformly and independently distributed in a certain area. To observe the entire network as one graph, we assume the graph is connected, which implies that no node is isolated. For a given node density n, total number of nodes N, and the probability that the network has no isolated node p_c, we can obtain the minimum communication range r_0 for which there is no node isolated in the network (Bettstetter, 2002):

$$r_0 \geq \sqrt{\frac{-\ln\left(1 - p_c^{1/N}\right)}{n\pi}} \quad (4)$$

In our model, we abstract from the fact that communication between two nodes is subject to various kinds of time- and place-dependent propagation effects, which would imply that the communication range is also varying with time and place. Therefore, we assume for each node a fixed communication range r. To simplify our analysis, and to achieve with high probability a network without isolated nodes, we assume a very high-density network.

i-hop Communication Range in Circular Structured Networks

In line with our assumptions, a node can reach all the nodes located within the circle with the radius of r whose center is the position of the node A as shown in Figure 3a. Node B, which is located within the annulus R_2 with outer circle radius $2r$ and the inner circle radius r, will reach the center node A, if and only if there is a node C located within the intersection area S_{AB} of the communication range of A and B, as shown in Figure 3a. Because the distance between A and B is between r and $2r$, the intersection of circles A and B, S_{AB}, is between 0 and

$$\frac{2}{3}\pi r^2 - \frac{\sqrt{3}}{2} r^2.$$

Since we have assumed that nodes are uniformly and independently distributed in the network, the number of nodes located in the intersection area S_{AB} fits the Poisson distribution with $\lambda_{AB} = S_{AB} \cdot n$. We set the probability that there is at least one

Figure 3. (a) A and B are connected through C; (b) A and F are connected through E and D.

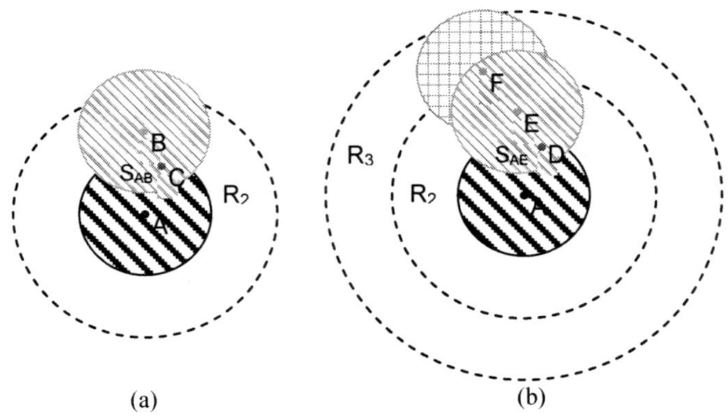

node located in the area S_{AB} as $P(N_{AB}>0)$, which is also the probability of having a path between A and B. This probability equals 1 minus the probability that no node is located in the area S_{AB}:

P(B is a 2-hop neighbor of A|B is in R2)
= P(NAB > 0) = 1 - P(NAB = 0). (5)

Since the number of nodes is Poisson distributed, Eq. 5 can be rewritten as:

$$P(N_{AB}>0)=1-e^{-\lambda_{AB}}=1-e^{-S_{AB}\cdot n}$$
$$\left(0<S_{AB}<\frac{2}{3}\pi r^2-\frac{\sqrt{3}}{2}r^2\right). \quad (6)$$

We can observe that if n is sufficiently large, $P(N_{AB}>0)$ goes to 1. This implies that with almost 100% probability there is a path between node A and B if the node density is sufficiently high:

$$\lim_{n\to\infty} P(B \text{ is a 2-hop neighbor of } A|B \text{ is in } R_2)\to 1 \quad (7)$$

Let us now have a look at node F in Figure 3b, which is located within the annulus R_3 with outer circle radius $3r$ and the inner circle radius $2r$. Node F can reach the center node A, if and only if there is at least one node E located within the communication range of node F, and that node has a connection to node A. Therefore, the probability that node F is a 3-hop neighbor of A can be derived as:

$$P(F \text{ is a 3-hop neighbor of } A|F \text{ is in } R_3)$$
$$=P(\exists E\text{: d}(E, F)\le r \wedge E \text{ is a 3-hop neighbor of } A|F \text{ is in } R_3) \quad (8)$$

From Figure 3b, we can observe that if node E is located outside ring R_2, the probability that E is a 2-hop neighbor of A is 0. Moreover, when the network density goes to infinite, from Eq. 7, we can obtain:

$$\lim_{n\to\infty} P(F \text{ is a 3-hop neighbor of } A|F \text{ is in } R_3)$$
$$\to \lim_{n\to\infty} P(\exists E\text{: d}(E,F)\le r\wedge E \text{ is a 3-hop neighbor of } A|F \text{ is in } R_3). \quad (9)$$

Since n goes to infinity, we have:

$$\lim_{n\to\infty} P(\exists E\text{: d}(E, F)\le r\wedge E \text{ is in } R_2|\ Fs \text{ in } R_3)\to 1. \quad (10)$$

In a similar way, we can deduce the formula for the probability of node X located in R_i is a i-hop neighbor node A as:

$$\lim_{n\to\infty} P(X \text{ is a i-hop neighbor of } A|X \text{ is in } R_i)$$
$$\to \lim_{n\to\infty} P(\exists Y\text{: } d(Y,X)\le r\wedge Y \text{ is (i-1)-hop neighbor of } A|X \text{ is in } R_i)$$
$$\to \lim_{n\to\infty} P(\exists Y\text{: } d(Y,X)\le r\wedge Y \text{ is in } R_i|X \text{is in } R_i)\to 1 \quad (11)$$

This shows that in a high-density network, when node X is located in R_i, X is an i-hop neighbor of A with almost 100% probability. Therefore, we assume a high-density network in this paper. This implies that with very high probability, the network is connected. Further, adding or removing a node in the network will not influence the length of the shortest path between any two other nodes in the network.

Eq. 11 shows that in a high-density network, the probability of any node located in R_i being node A's i-hop neighbor goes to 1. That implies that the i-hop communication range of node A goes to ir in a high-density network, which can be represented as:

$$\lim_{n\to\infty} (\text{i-hop communication range of } A)\ ir. \tag{12}$$

Therefore, in this paper we can use ir as our approximate i-hop communication range of node A with the assumption of a high-density network. The accuracy of this approximation depends on the actual network density, being highest at very high density.

Mean Multi-Hop Node Degree of Circular Structured Networks

For a circular structured network, we denote the number of nodes that can be reached within i hops, but not fewer than i hops, as random variable D_i^c. Given Eq. 11, those nodes are located in the annulus with outer circle radius as ir and the inner circle radius as $(i\text{-}1)r$ in high density networks. The expected value of D_i^c can be written as:

$$\lim_{n\to\infty} E\left[D_i^c\right] = \begin{cases} 1 & i = 0 \\ n\pi r^2\left(i^2 - (i-1)^2\right) = (2i-1)n\pi r^2 & (i > 0) \end{cases}.$$

The total number of reachable nodes in i hops can be derived as

$$N_i^c = \sum_{j=0}^{i} D_j^c.$$

The expected total reachable nodes in i hops can be derived as:

$$\lim_{n\to\infty} E\left[N_i^c\right] =$$

$$\lim_{n\to\infty} E\left[\sum_{j=0}^{i} D_j^c\right] = \lim_{n\to\infty} \sum_{j=0}^{i} E\left[D_j^c\right] = 1 + n\pi r^2 i^2.$$

ANALYSIS OF DYNAMIC CONNECTIVITY

In this section, we analyze the effect of dynamic connectivity, e.g., due to the limited battery supply and unstable transmission on the performance of our ABF-based discovery protocol. For grid structures, we study the effect of a single node disappearing, appearing, and moving across the network, analytically. We study the same phenomena for circular structures, and add also the case where packets from a specific node are lost for a period of time. We quantify the network load generated due to the dynamic connectivity through analytical study and verify obtained results with simulations. Nodes disappearing and (re-)appearing can be caused by a (temporary) lack of energy supply, e.g., in case the needed energy has to be harvested from the environment. When for some reason the propagation conditions of the wireless medium are bad, some packets of a node might get lost. If these packets are a number of consecutive keep-alive messages, other nodes in the network will consider this node disappeared. After some time, the propagation conditions may improve so that the node's packet will be received again. As a result, it reappears in the network. When a node is moving across the networks, the set of neighbors it can reach with its broadcasts is changing continuously.

In this section, we will first present the analysis for the grid structure. Next, in Section 4.2, we will present the analysis, and both simulation and analytical results for the grid and circular structure, respectively. We conclude the section with a discussion of our findings in Section 4.3.

Grid Structure

Node Disappearance

Besides insufficient battery supply, a node might disappear from the current network, due to various other reasons, such as un-functional antenna, system crash, personal opinions like switching off the mobile or leaving the network.

In the current protocol, the disappearance of a neighbor is noticed if no keep-alive message has been received from this node for two consecutive keep-alive periods. The neighbors of this absent node will generate a new ABF, removing the hash results relating only to the absent node from layer 1, and broadcast it. Due to the change, all nodes that receive the updated filter will also regenerate a filter and broadcast it. This process will continue till *(d-1)* hops away from the absent node. Note that there is always more than one path from any node within range to the disappearing node due to the special structure of grid network. Especially, any node within range can reach the disappearing node in every two additional hops starting from the shortest path length. Therefore, nodes need to clean up every second layer after the first time of cleaning up one by one.

When removing the context sources of some node A from the ABF at a certain layer, there is a slight chance that no changes to the ABF are required, because the bits that would have to be set to zero have to remain one, as they also represent other context sources in other nodes. This is the same property that causes a false positive when querying context sources. Liu and Heijenk (2007) have already defined and derived this probability. We use $P_{fp,i}$ to represent the false positive probability of layer *i*. Liu and Heijenk (2007) has proven that

$$P_{fp,i} \approx \left(1 - e^{\frac{-bx_i}{w}}\right)^b, \tag{15}$$

where x_i denotes the total number of represented context sources in layer *i*. We use x_i^g for the grid structures and x_i^c for the circular structures. We can obtain that $x_i^g = s \cdot N_i^g$ and $x_i^c = s \cdot (1 + n\pi r^2 i^2)$. $P_{fp,i}$ is the probability that no changes have to be made to layer *i* of an ABF, upon the disappearance of node A, provided that node A has only one context source advertised. If node A advertises *s* context sources, the probability that no changes have to be made to layer *i* is raised to the power *s*, i.e. $P_{fp,i}^s$.

Therefore, the number of updates for different maximum hop count *d* can be represented as follows:

$$N^g_{update\_ndisa}(d) = \sum_{i=1}^{d-1}(1 - P^s_{fp,i}) \sum_{j=0}^{\left\lfloor \frac{i-1}{2} \right\rfloor} D^g_{i-2j} \tag{16}$$

In this equation, for a certain value of *i*, we count all the transmissions, done *i* nodes away from the node that disappeared, and also the transmissions done at *i-2*, *i-4*, etc., nodes away, because of alternative longer paths to the node that disappeared. The number of transmissions depends on the number of nodes present at the relevant distance, and weighed by the probability that there is no false positive. This is done for every possible value of *i*.

Node Appearance

In this subsection, we continue with the analysis of a node appearing in the network. In here, we assume that every node, including the new one which just appears, knows about the format and hash functions of attenuated Bloom filters used in the network. Further, the new node does not have any information about the neighbors. In here, we will refer to the appearing node as the new node.

Figure 4. Node appears in a grid network

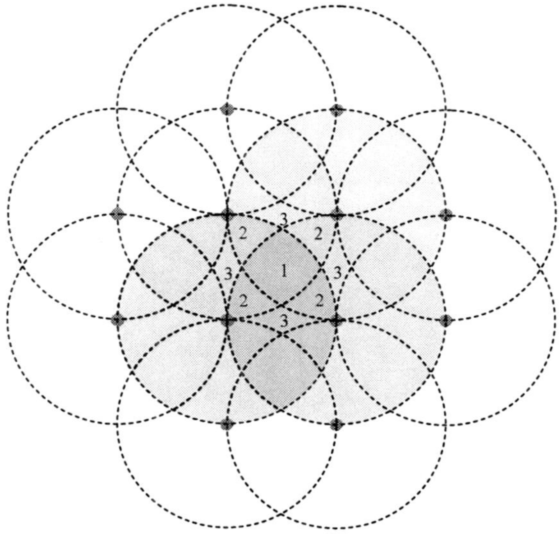

The new node broadcasts keep-alive message periodically. The direct neighbors who receive a keep-alive message from an unknown neighbor initialize a link with this new neighbor and send an update-request back to the new node. As the answers to the update-request, the new node will broadcast its filter (of size $w \times d$ bits) with its local services. The direct neighbors who receive this filter will update theirs. Those new filters will be broadcasted around. The new node waits for a short moment, till it receives all the neighbors' replies. It aggregates all incoming filters, updates its own filter and broadcast it. The direct neighbors update their filters after receiving the filter, and broadcast them. The other nodes within range will also update their filters and broadcast them.

Please be aware that because of the duplication (see Section 2.2), every node in range only needs to update once. The network traffic depends on how many neighbors the appearing node has, which is directly related to the location of the new node appearing. In a grid structure, we can divide the space of a grid into 3 different areas as shown in Figure 4, based on the number of surrounding neighbors that have the area within their transmission range. Area 1, can be reached by all 4 surrounding nodes of the grid, area 2 can be reached by 3 nodes, and area 3 can only be reached by 2 nodes.

Therefore, the network traffic can be obtained as the sum of the number of updates $N^g_{updates\_na,i}$ when the new node is appearing in area i ($i \in \{1, 2, 3\}$) with the probability p_i that the node appears there. We have:

$$N^g_{update\_na} = N^g_{update\_na,1} \times p_1 + N^g_{update\_na,2} \times p_2 + N^g_{update\_na,3} \times p_3 \tag{17}$$

When the node appears in the area 1, it has 4 direct neighbors. All the nodes within range need to update their filters. We can obtain the number of updates with the probability that no changes are need to certain layers in ABF in this situation as:

$$N^g_{udpates\_na,1} = \sum_{i=0}^{d-1} D^g_i \left(1 - P^s_{fp,i}\right) \tag{18}$$

Similarly, we can obtain the number of updates when the node appears in the area 2 as:

Figure 5. Horizontal move

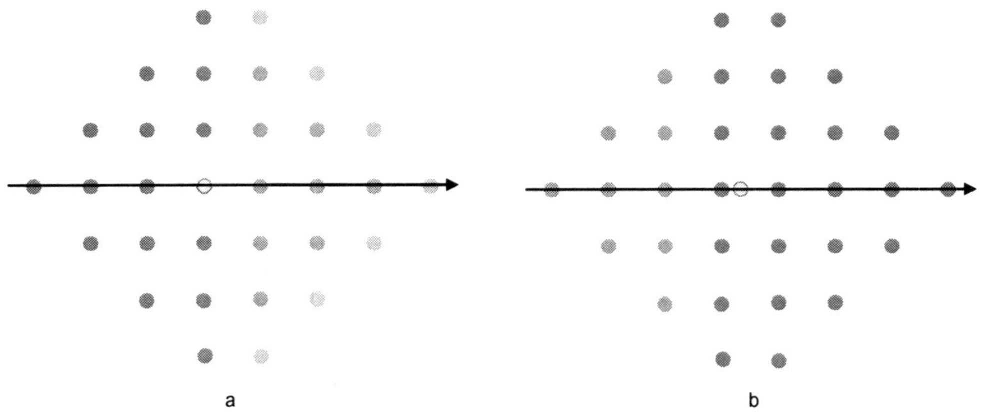

$$N^{g}_{updates\_na,2} = 1 + \sum_{i=1}^{d-1} \left(D_i^g - 1\right)\left(1 - P^s_{fp,i}\right) \tag{19}$$

The number of updates when the node appears in the area 3 can be obtained as:

$$N^{g}_{updates\_na,3} = 1 + \sum_{i=1}^{d-1} \left(D_i^g - 2\right)\left(1 - P^s_{fp,i}\right) \tag{20}$$

The probabilities of a node appearing in area 1, 2, and 3 equal the ratio of these three specific areas over the entire grid. They can be denoted as:

$$\begin{aligned} p_1 &= \left(\frac{\pi}{12} - \frac{1}{4}\right) \times 4 + \left(2 \times \sin\frac{\pi}{12}\right)^2 \approx 0.3151 \\ p_2 &= \left(\left(\frac{\pi}{4} - \frac{1}{2}\right) \times 2 - p_1\right) \times 2 = \frac{\pi}{3} - 8\sin^2\frac{\pi}{12} \approx 0.5113 \\ p_3 &= 1 - p_1 - p_2 = 2 - \frac{2\pi}{3} + 4\sin^2\frac{\pi}{12} \approx 0.1736 \end{aligned} \tag{21}$$

One Node Moving

Now we study the scenario that a node is moving horizontally through the network, crossing one of the rows of nodes. Theoretically, when the mobile node is in the same location as one of the nodes in the row, it will establish three new direct connections to the nodes located right on top of, under, and next to it along the direction of movement. However, we assume the side of the grid is exactly 300 meters, which is also the maximum transmission range of a node. Therefore, when the mobile node is in the same location as one of the nodes along its trace, it is the only position where the mobile node is in reach of the nodes above and under it vertically. Theoretically, the mobile node only spends 0 second staying in that exact position. It is not possible for the mobile node to establish a direct connection to the nodes right above and below it in the grid within 0 second. Therefore, we assume only one new direct connection will be established. That is the link between the mobile node and the next node right along the moving direction.

This new connection results in the nodes in orange in Figure 5a to update (for $d = 3$). Similarly, when the mobile node passes the point mentioned above, it loses the direct connection with the node behind in the direction of moving. This also results in the updates of those orange nodes in Figure 5b. Each orange node needs to update once, because the alternative paths of longer length are already/still available, and represented in the Bloom filters. In cases, the number of the updates is the same, and is given by:

$$N^{g}_{udpates\_hori} = \frac{(1+(2d-1))\cdot d}{2} = d^2. \tag{22}$$

Circular Structure

Simulation Setup

Besides the approximate analysis, based on the assumptions stated in Section 3, the circular structured model has also been implemented in the discrete event simulator OPNET Modeler version 11.5 ("OPNET Modeler", n.d.). We observe the node density influence on the traffic load in first three experiments below. We place 25, 61, 100, 125, 150 nodes randomly into a $1700 \times 1700 m^2$ area with 300 meters communication range, r, for every node. Note that the 61 nodes scenario generates networks that are 1-connected graphs with 90% probability (see Eq. 4). We consider the 25 nodes scenario as a low density network, and the 150 nodes scenario as a high-density network. We expect that simulation results obtained for this high-density network are close to our analytical analysis, as this analysis was based on the assumption of a very high node density. The node that disappears, or appears, or is temporarily unreachable, is located in the center of the area to avoid border effects. For each parameter setting of the simulations introduced below, 30 independent runs will be done to calculate a 90% confidence interval.

Some basic ABF parameters are set as follows: number of hash functions per service, $b = 10$; ABF width, $w = 1024$ bits; ABF depth, $d = 3$; number of context sources advertised per node, $s = 1$.

Node Disappearance

We start the analysis again with the case of one node disappearing from the network. We assume that a node disappears at the moment the network has reached the stable state, i.e. all ABFs are up-to-date at the moment the node disappears. Possible reasons for node disappearance are insufficient battery supply, entering deep power saving mode, system shut down, un-functional antenna, system crash, etc. The absence of node A will be discovered by its direct neighbors when no keep-alive messages have been received for two consecutive keep-alive periods. Since the keep-alive period is unsynchronized, one of the direct neighbors B will notice this first. This node will remove the incoming ABF from node A. As a result, the representation of node A's context sources will be removed from layer 1 of B's outgoing ABF. For ease of explanation we will write that a node is advertised (or removed), where we actually mean that the context sources of a node are advertised (or removed). Since the other direct neighbors of A have advertised A in layer 1 of their ABF, node B will continue to advertise A in layer 2 of its ABF. As a matter of fact, we duplicate the local context sources of each node to every lower layer from layer 2 in the advertisement and maintenance phase of our protocol. This is because in a very high density network, if a path exists to a node, there are always longer paths to the same node. By duplicating a node's context sources to all lower layers in the ABF, we avoid that extra advertisements are exchanged to announce these longer path. So in the situation above, node B will still think that it can reach the services of node A via other neighbors. Therefore, those absent services are only removed from Layer 1 of the outgoing ABF from B.

There are two kinds of nodes that are the direct neighbor of B: direct neighbors of A and two-hop neighbors of A. The other direct nodes which receive this information will not take any action since they still think there are other routes to A. As the last direct neighbor of A notices the disappearance of node A, it will realize that it cannot reach the absent services within one or two hops. It will start sending out an updated ABF with layer 1 and layer 2 cleaned up. Nodes which are two-hop neighbors of A that receive the updated ABF from direct neighbors of A will also take no action, until all their direct neighbors which are also the direct neighbors of A notice the absence of A. Only then these nodes will realize there is no path to A with 2 hops. They will send out an ABF with layer 2 cleaned up (Note that these two-hop neighbors do not have node A's information on layer 1). This clean up will be spread till *(d-1)*-hop neighbors in a similar way.

It is a very complex procedure to clean up the services. Every node *i* hops away has to clean up *(d-i)* layers in total. It will clean up layer *j* $(i \leq j \leq d-1)$ once it realizes there is no route to the service within *j* hops. Every node cleans up the service layer by layer. Only the last node in the *i*th hop that realizes the absent service will clean up two layers at once. Similar to the analysis of the grid structure, we have to take into account the probability that no changes are needed to a certain layer in the ABF, because the relevant bits have to remain set, because of context sources of other nodes. We have defined the false positive probability in Eq. 15. Further, we have defined for the circular structure $x_i^c = s \cdot \left(1 + n\pi r^2 i^2\right)$.

Therefore, we can derive the expected number of clean-up updates while one node disappears, $E[N^c_{updates-disa}]$, as:

$$E[N^c_{updates-disa}] \approx$$
$$\sum_{i=1}^{d-2}\left(\left(E[D_i^c]-1\right)\cdot\sum_{j=i}^{d-1}\left(1-P_{fp,j}^s\right)+\sum_{j=i+1}^{d-1}\left(1-P_{fp,j}^s\right)\right)+$$
$$E[D_{d-1}^c]\cdot\left(1-P_{fp,d-1}^s\right) \qquad (23)$$

We verify the results of this approximation with simulations. In here, we study the traffic load generated by one node disappearance under different network density see Figure 6. We found that the analytical results are slightly higher than the simulation results. This is because our analytical analysis is based on the assumption of a very high-density network, so that the *i*-hop node degree is slightly overestimated. In that respect, our analysis provides an upper bound to the expected number of updates. In our scenarios, there are fewer nodes involved in updating and more nodes clean up more than one layer of filter at once. This results in fewer update packets sent out than we estimate in Eq. 23. We also observe that the higher the network density, the smaller difference there is between analytical and simulation analysis as we expected from our assumption.

Figure 6. Node disappears

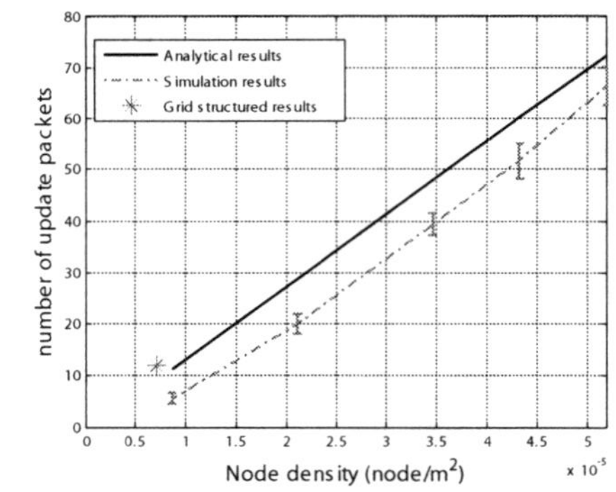

We obtained the number of updated packets for the grid structure from Eq. 16, with the related network density estimated as in (Liu & Heijenk, 2007). The result is 12.0 updates transmitted, at a node density of 0.7×10^{-5}. This has also been plotted in Figure 6, and matches the analytical circular structured results.

Node Appearance

In this section, we again consider the scenario that one new node appears in the network. The reason could be that the node just switches on. We assume that the new node is familiar with the standardized format and hash functions of the attenuated Bloom filters used in the current network. However, the node does not have knowledge about nodes and context sources in the network. First of all, this node will broadcast its filter (size of $w \times d$ bits) with its local services. The direct neighbors, who receive this filter, will update theirs. Those new filters will be broadcasted around. The new node waits for a short moment, till it receives all the neighbors' replies. It aggregates all incoming filters, updates its own filter and broadcasts it. Note that, the network is assumed to have a high node density, which means that the appearance of the new node will not generate any shorter path between any pair of existing nodes. Therefore, for any node up to *d*-1 hops away from the new node, only the appearance of the new node will be added into the existing filters. Further, since the local information is duplicated to every layer of the ABF before it is sent out, there will not be a loop between neighbors to add the information layer by layer. After the initial broadcast, every node, including the new one, will only update once. Similarly, with the probability that no changes are needed to certain layers in the ABF, the expected number of updates can be quantified as the total number of nodes within range plus one:

$$E\left[N^{c}_{update-a}\right] = \sum_{i=0}^{d-1} E\left[D^{c}_{i}\right]\cdot\left(1 - P^{s}_{fp,i}\right) + 1 \qquad (24)$$

We observe the traffic load generated by one node appearing in networks with various densities. It turns out that the load is quite accurately predicted by the analytical model. This is because in a very high density network as we assumed, one nodes' appearance will not generate new shortest paths between any pair of nodes. In the lower density networks we simulated, quite some nodes update more than once, because extra new indirect neighbors are discovered due to the appearance of one node. Interestingly, as shown in Figure 7, we found that the number of extra updates compensates the number of extra nodes we estimated in the analytical study. Of course, the higher the network density is, the more accurately the *i*-hop node degree is approximated, and fewer new neighbors are discovered due to one node appearance.

The corresponding results of the cost for the grid structure, as obtained from Eq. 17, with again a node density of 0.7×10^{-5} has been plotted in Figure 7. This result is also very close to the analytical results for circular structured network.

Figure 7. Node appears

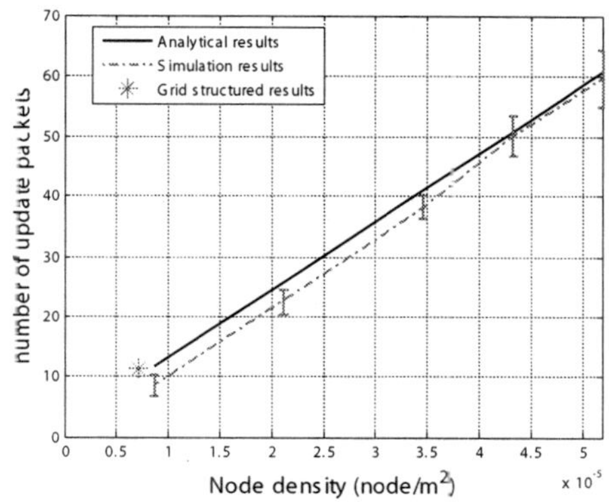

Packet Loss

In this section, we study the situation where some packets of a node get lost, e.g., due to unfavorable propagation conditions. If at least two keep-alive messages from a node are lost consecutively, the other neighbors of the node consider the node disappeared. They start cleaning up their ABFs as we described in Section 4.2.1. After some time, the transmission quality of the node gets better, and keep-alive messages will be received again. The neighbors think there is a new node appearing in the network. The actions as addressed in Section 4.2.2 will be taken. Here, we assume that the packet loss only occurs in one direction, i.e., we assume the node can still receive packets from its neighbors. Therefore, it keeps updated information of the neighbors. This is a slight difference from the scenario in Section 4.2.2. In Section 4.2.2, the appearing node does not have any knowledge about the network. Therefore, in here, one update less is generated than in section 4.2.2. The number of updates generated in this scenario, $N^{c}_{packet\_loss}$, can be obtained as:

$$N^{c}_{packet\_loss} = N^{c}_{update-disa} + N^{c}_{updates-a} - 1 \qquad (25)$$

We compared this with simulation results in networks of various densities, as shown in Figure 8. The keep-alive period, i.e. the time between two consecutive keep-alive messages is distrib-

Figure 8. Effect of Packet loss

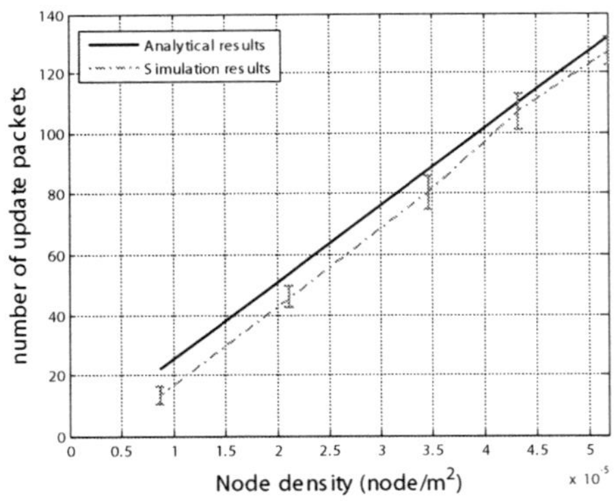

Figure 9. Effect of Packet loss periods

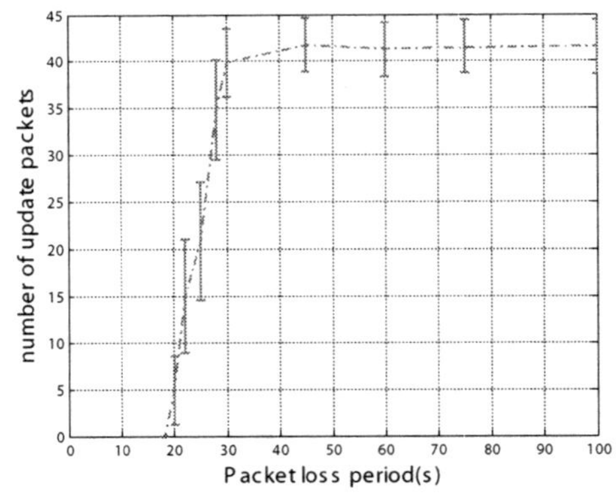

uted uniformly in the interval [15, 17] seconds. Packet loss period is 45 sec, which guarantees at least two continuous keep-alive messages are lost. Simulation results are again slightly lower than analytical results. All the reasons we mentioned above in Section 4.2.1 and 4.2.2 apply to the results of this experiment as well.

We also did an experiment to study the effect of different packet loss periods, which is shown in Figure 9. We use the 61-node scenario. We vary the packet loss period from 18 sec to 100 sec. When there is only one packet lost, there is no update needed in the network, since nodes are only considered disappeared if no keep-alive messages have been received for two keep-alive periods. When there are at least two packets lost consecutively, updates are generated. When reappearance period is between 20 and 45 sec, some of the nodes notice the node disappearance, but the node reappears before its information has been removed from the network totally. The longer the period is, the more updates can be done before the reappearance of the node. Therefore, the number of updates is growing in this period. If the packet loss period is larger than 45 sec, the number of packets generated is almost constant. This is because the nodes have enough time to complete the updates for disappearance within this time period.

A Moving Node

A moving node is a more complex action, which extends beyond the actions of nodes appearing and disappearing. In this section, we continue our study of the extra updated traffic generated by a moving node via simulations. The node is moving from one spot to another in a straight line with steady speed. The start and end point of the journey are far enough that two nodes located in both positions do not share any neighbour within *d* hops. This guarantees that the nodes in range (*d* hops) of the moving node need to update their filters at last. This offers the chance for us to compare the simulation results with analytical results in the extreme case of a node disappearing in one location, and reappearing in a different location. The simulation has been done in a two-dimensional area of 4200×1800m^2. We assume *d* is equal to 3. The mobile node with 300 meters communication range is set to move 2400 meters from point (900m, 900m) to (3300m, 900m). That guarantees the moving node to avoid border effects. The additional update traffic is highly related to the speed of the node and the network density with fixed keep-alive period. Therefore, we first fix the network density and observe the traffic load under different speed. Afterwards, we observe the extreme case in which the mobile node is moving with extremely fast speed, so that the other nodes along the path will not have time

Figure 10. One node moving at different speeds

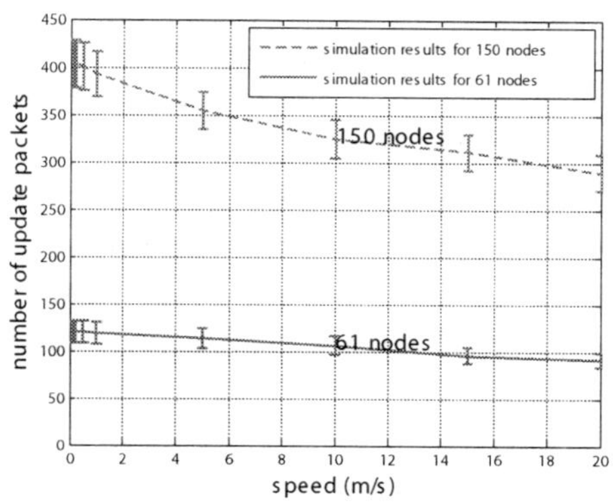

Figure 11. One node moving for various densities

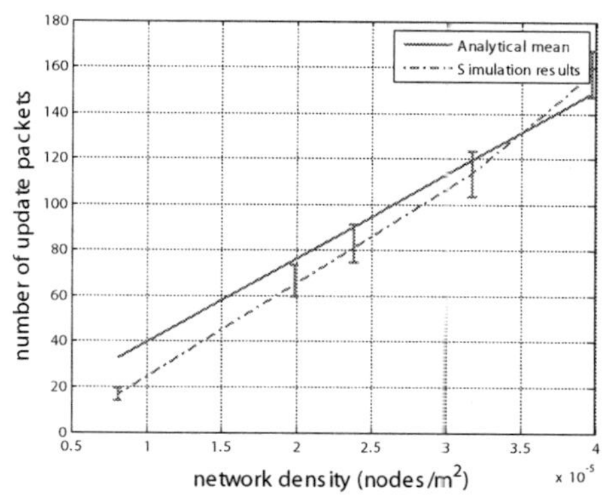

to notice the movement. The network behaviour resembles the case where the mobile node disappears from one point and reappears at the other point far away from the previous position. We compare the simulation results with our analytical results for different network density.

First of all, the simulations have been done with two different densities 8.07×10^{-6} (61 nodes in the experiment area) and 1.98×10^{-5} node/m^2 (150 nodes in the experiment area). The mobile node is moving with different speeds from 0.1m/s to 20m/s. 20m/s can be considered the average speed of a car; 5m/s as the average speed of a bicycle; 1m/s as the average speed of a walking adult. The results are shown in the Figure 10. The traffic load decreases as the speed increases. This is because the faster the node is moving, the more nodes it misses when updating information. Therefore, fewer updates will be generated. Further, the higher the network density, the more nodes are involved in the update.

Finally, we simulate the extreme case with a very high speed of 24000 m/s. Ideally, at this speed, the nodes along the path will not have time to notice the move. This scenario can be considered as the node disappears from one position and reappears at another. The effect of the disappearance of the node is exactly the same as discussed in Section 4.2.1. However, the reappearance is slightly different from Section 4.2.2. In Section 4.2.2, we assumed that the node appears in a new environment without any previous information stored in ABF. In here, the node appears in a new environment with an ABF filled with the information of the previous position. Therefore, the updates will take place twice. When any node notices the existence of the mobile node, the first round of updates will happen to add information still present in the ABF of the mobile node. After two continuous keep-alive periods, the mobile node notices the loss of the connection to the neighbors at the previous position; it will clean up those neighbors from the filter. The second round of updates will take place. Therefore, we can generalize the total number of updates due to the moving node, $N^c_{update-move}$, as:

$$N^c_{update-move} = N^c_{update-disa} + 2 \cdot N^c_{update-a} \tag{26}$$

We compare the simulation results with the analytical results for different network densities: 8.07×10^{-6}, 1.98×10^{-5}, 2.38×10^{-5}, 3.17×10^{-5}, and 3.97×10^{-5} nodes/m^2. The results are shown in Figure 11. We can see that the higher the network density, the more updates are generated. When the network density is 8.07×10^{-6}, the analytical results are higher than the confidence interval of simulation results. This is because this is a low density scenario, there are much less nodes connected as we expected in the model. Therefore, the simulation generates less updates as we expected. As we can see, higher the network density, more accurate is our analytical analysis. In high density scenarios of 2.38×10^{-5}, 3.17×10^{-5}, and 3.97×10^{-5} node/m^2, the analytical results are always within the confidence interval of the simulation results.

SUMMARY AND DISCUSSION

We have derived analytical expressions for the number of additional broadcasts in an ad-hoc network using ABF-based context discovery. We have done this for two different network models, a grid structure, and a circular structure, based on the assumption of a random geometric graph. For the grid structure, we found exact expressions for the expected number of update messages caused by the disappearance and appearance of a node, and by the movement of a node along a straight, horizontal line. For the circular structure, the expressions found approximate the expected number of additional broadcasts, in case the network density is sufficiently high. From the comparisons above, we observed that the analysis is indeed more quite accurate for high-density networks. This fits our hypothesis. The higher the network density is, the more accurate our approximation of the i-hop node degree. Further, since the proposed protocol automatically duplicates the Bloom filter representing its own context sources to all lower layers of the ABF, the appearance of a new node can be handled in a single pass of advertisements. No advertisements have to go up and down to propagate the availability of indirect paths to the new node into the lower layers of the ABF. However, in the case of removal of a node, multiple passes are needed to remove its representation completely from all ABFs. Removal of context sources has to be done layer by layer, as the equivalent of duplication cannot be performed. Therefore, adding a node generates less traffic than removing a node.

An important issue to improve the performance of ABF-based ad-hoc networks in dynamic environments is to reduce the traffic while removing context sources. One of the possible solutions to improve the protocol is to be more conservative when adding information regarding new context sources to the ABFs. We could add certain policies to restrict adding new context sources, based on the quality level of the source, such as stability, bandwidth, and distance, etc. Only "good quality" and "valuable" information will be added into ABFs. By restricting incoming information, we can reduce the traffic for removing context sources that are most probably not used during their presence.

In Section 4.2.2, we have studied the update traffic caused by node appearance without outdated information that needs to be cleaned up. However, in reality this is not always the case. Nodes might appear in an environment where its advertised services may still be present somewhere in some of the ABFs, especially in the case when a node is moving. Based on our study, we found it is mathematically infeasible to quantify the update traffic load caused by this type of clean up.

We studied a simple mobile case of one node moving in a straight line in the network. For increasing speed of the moving node, less traffic is generated. However, network density has more influence on the number of updates generated than the speed of the mobile node. In the extreme case of node movement at very high speed, the higher the network densities, the better our analytical results fit the simulation results.

CONCLUSION AND FUTURE WORK

In this paper, we have extended the performance analysis of ABF-based ad-hoc networks to a dynamic environment where nodes appear and leave the network, are temporarily unreachable due to poor propagation conditions, or move through the network. The analysis has been performed for two different network models. We have found exact expressions for a grid-structured model where nodes are located at the line crossings of a regular grid, whereas the transmission range is such that two neighboring nodes on a line of the grid can just reach each other. The second network model assumes that nodes are located and connected, based on a random geometric graph. In order to be able to approximate the *i*-hop node degree accurately for this network model, we assume that the network density is very high, which results in a circular structure, consisting of concentric circles to denote the *i*-hop node degree. The study has been done analytically to quantify the update traffic, measured in the number of broadcast packets with ABFs. For this circular structure, we verified the approximate results with simulations. The analytical expressions give more accurate results when the network density is higher. We discovered that it is easier to add context information than to remove it. Especially, in the case when context information moves out of the range of some nodes but still can be reached by other nodes, there are many dynamic parameters, such as node positions and network topology, needed to compute the exact network traffic. This part of work cannot be done analytically.

General conclusions with respect to the performance of the ABF-based discovery protocol in a dynamic environment are that network load increases linearly with the node density. Furthermore, the network load decreases slowly with increasing node speed.

In the experiments, we observed that there is much less traffic generated by adding context information than removing. Therefore, reducing broadcast traffic for removing context sources is an important topic of further study.

REFERENCES

Albero, T., Sempere, V., & Mataix, J. (2006). A Study of Mobility and Research in Ad Hoc Networks Using Stochastic Activity Networks. *Proc. 2nd Conference on Next Generation Internet Design and Engineering.*

Bettstetter, C. (2002). On the Minimum node Degree and Connectivity of a Wireless Multihop Networks. *Proc. 3rd ACM International Symposium on Mobile Ad Hoc Networking and Computing.*

Bloom, B. H. (1970). Space/Time Trade-offs in Hash Coding with Allowable Errors., *Communications of the ACM, 13*(7), 422-426.

Clark, B., Colbourn, C., & Johnson, D. (1990). Unit Disk Graphs. *Discrete Mathematics, 86*(1-3), 165-177.

Goering, P. T. H., Heijenk, G. J., Haverkort, B., & R. Haarman (2007). *Effect of Mobility on Local Service Discovery in Ad-Hoc. Proc. Performance Engineering Workshop.*

Hekmat, R., & Van Mieghem, P. (2003). Degree Distribution and Hopcount in Wireless Ad-hoc Networks. *Proc. 11th IEEE International Conference on Networks.*

Lenders, V., Wagner, J., & May, M. (2006). Analyzing the impact of mobility in Ad Hoc Networks. *Proc. 2nd international workshop on Multi-hop ad hoc networks: from theory to reality.*

Liu, F., & Heijenk, G. (2007). Context Discovery Using Attenuated Bloom filters in Ad-hoc Networks. *Journal of Internet Engineering, 1*(1), 49-58.

Liu, F., Goering, P., & Heijenk, G. (2007). Modeling Service Discovery in Ad-hoc Networks. *Proc. 4th ACM International Workshop on Performance Evaluation of Wireless Ad Hoc, Sensor, and Ubiquitous Networks.*

OPNET technologies Inc., *OPNET Modeler Accelerating Networks R&D.* (n.d.). Retrieved 2008, from http://www.opnet.com/products/modeler

Penrose, M. (2003). *Random Geometric Graphs.* New York, USA: Oxford University Press Inc..

Trajanov, D., Filiposka, S., Efnuseva, M., & Grnarov, A. (2004). Ad Hoc Networks Connection Availability Modeling. *Proc. 1st ACM international workshop on Performance evaluation of wireless ad hoc, sensor, and ubiquitous networks.*

This work was previously published in International Journal of Business Data Communications and Networking (IJBDCN) Volume 5, Issue 2, edited by Varadharajan Sridhar and Debashis Saha, pp. 16-34, copyright 2009 by IGI Publishing (an imprint of IGI Global).

Chapter 6
Models and Architecture for Autonomic Network Management

N. Van Wambeke
Centre National de la Recherche Scientifique and Université de Toulouse, France

C. Chassot
Centre National de la Recherche Scientifique and Université de Toulouse, France

F. Armando
Centre National de la Recherche Scientifique and Université de Toulouse, France

K. Guennoun
Centre National de la Recherche Scientifique and Université de Toulouse, France

A. Abdelkefi
Centre National de la Recherche Scientifique and Université de Toulouse, France

K. Drira
Centre National de la Recherche Scientifique and Université de Toulouse, France

ABSTRACT

This paper presents a model-based framework to support the automated and adaptive deployment of communication services for QoS. The application domain targets cooperative group activities applied to military emergency operation management systems. Various models are introduced to represent the different levels of cooperation (applicative / middleware / transport). The adaptation decision process relies on structural model transformations while its enforcement is based on the dynamic composition of micro-protocols and software components. Automated deployment is performed both at the transport (i.e. UDP-TCP level) and middleware level. The architecture to support automated network management based on these models is introduced and its performance is evaluated through the use of a Java prototype.

INTRODUCTION

Cooperative group activities using wireless mobile communicating systems constitute an increasingly evolving application domain. It is likely to be one of the most important directions that may enable reliable and efficient human and machine-to-machine cooperation under the current networking systems and software, and may deeply shape their future deployment. Such activity-support systems have to deal with dynamically evolving activity-level requirements under constantly changing network-level unpredictable constraints. Maintaining reliable connectivity and QoS in such a communication context is difficult. Adaptive service provisioning should help the different provisioning actors to achieve this goal and constitutes a challenge for different research communities.

Ad hoc solutions are not likely to be applicable to solve such a complex problem. Providing a basic framework for automated services and QoS deployment may constitute an important contribution towards solving such a problem.

Aiming to answering this problem, we propose a model-based framework for adaptability management. Our framework has been elaborated in the context of network management systems with service provisioning at the transport and network layers of the TCP/IP stack as the final objectives.

Our approach provides, refines and exploits different models, each one representing a different point of view on the context. The models that represent other aspects of communication are automatically generated from higher level models representing the cooperation requirements and the communication constraints. Our research efforts have been developed to cover communication at the transport layer as well as the network layer.

Our paper is organized as follows. Section 2 describes related work. Section 3 describes the different models of the framework. Section 4 presents an architecture to support the use of these models for automated network adaptation management as well as an example of their use in response to a change of collaboration. This architecture is currently under study and development within the European NETQoS Project. Section 5 provides an evaluation of the system's performances under different loads. Finally, section 6 presents conclusions and future works.

RELATED WORK

Classification of Context Adaptation Solutions

This section studies and classifies the main facets of adaptation: its objectives, techniques and properties.

Adaptation Objectives

Adaptation targets several objectives depending on the context in which it takes place. QoS aspects such as access bandwidth issues in roaming scenarios are considered by Kaloxylos (2006). End to end QoS optimization for the Best Effort Internet makes heavy use of adaptation techniques such as considered by Akan (2004). Security in wireless networks, such as firewalls activation and deactivation, can also benefit from adaptability as studied in Perez (2004). Resources optimization related to device power, computation or storage capability are presented by Marshall (2001).

Adaptation Techniques

Application layer – Wu (2001) addresses adaptation of video streaming applications for the Best-Effort Internet. The proposed techniques are based on two mechanisms: an applicative congestion control (rate control, rate-adaptive video encoding) and time aware error control with FEC.

Middleware layer – Reflexive architectures such as OpenORB or Xmiddle as presented by Capra (2003) are good supports for adaptation as they allow run-time modification of the architecture.

Transport layer - TCP's congestion control is a well-known adaptation example. Akan (2004) presents various types of mobile applications in wireless Internet are studied. Adaptation consists in parameterization of congestion control mechanisms using context information. Exposito (2003) and Hutchinson (1991) define the architectural adaptation of transport protocols by dynamic composition of protocol modules are presented, these approaches are detailed in the following sections.

Network layer – DaSilva (2004) addresses QoS-aware routing problems within mobile networks. Perez (2004) studies dynamic provision of IP services for military wired/wireless networks is considered. In a policy-based networking management context, the need for self-adaptation is considered by Samaan (2005), using a learning-based approach.

MAC layer - The solutions handle connection and access QoS problems for mobile users using different terminals and roaming. Kaloxylos (2006) provides a solution for optimizing the handover latency but the other QoS requirements are not considered.

Adaptation Properties

The adaptation is *behavioral* when a service can be modified without modifying its structure. TCP and protocols such as the one proposed by Akan (2004) provide behavioral adaptation. This easy to implement approach limits adaptability because the components have to be recompiled to be extended. Adaptation cannot be performed during run-time.

The adaptation is *architectural* when the services' structure can be modified. The replacement components can be implemented following a plug and play approach where the new component has the same interfaces as the replaced one.

Finally, adapting components can be distributed or centralized. In the first case, adaptation is *vertical* as changes are local. In the second case, it is horizontal and synchronization between adapting peers has to be managed.

Dynamically Configurable Protocol Architectures and model based adaptation

Dynamically configurable protocol architectures are based on the *protocol module* concept introduced by Hutchinson (1991). A protocol is then viewed as the composition of various protocol modules in order to provide a given service. These architectures can be classified depending on their internal structure: the event based model and the hierarchical model. The Enhanced Transport Protocol (ETP) detailed in Exposito (2003) follows a hybrid approach combining both models.

These protocol architectures are a good choice for *self-adaptation* as they provide *run-time* architectural reconfiguration. The modules composing them can change during communication. This *run-time* architectural adaptation raises many problems such as: (1) synchronization of peers; and (2) the choice of the best composition.

Adaptation management still remains a complex problem, particularly when it is required at several layers (Transport, Middleware ...) simultaneously as stated by Landry (2004). In such cases, the need to ensure coherency of the adaptation choices, both within and between layers clearly appears.

Informal methods lead to suboptimal solutions, often specific to a problem. This is due, in part to the complexity of the problem. To overcome these limitations, graph based formal approaches are appropriate to coordinate architectural adaptation at different layers of the stack. Chassot (2006) illustrates this approach by using graph based models and graph transformation rules. In the present paper, we complement this initial work by an architecture to manage these models.

Figure 1. Relationship between context elements and framework models

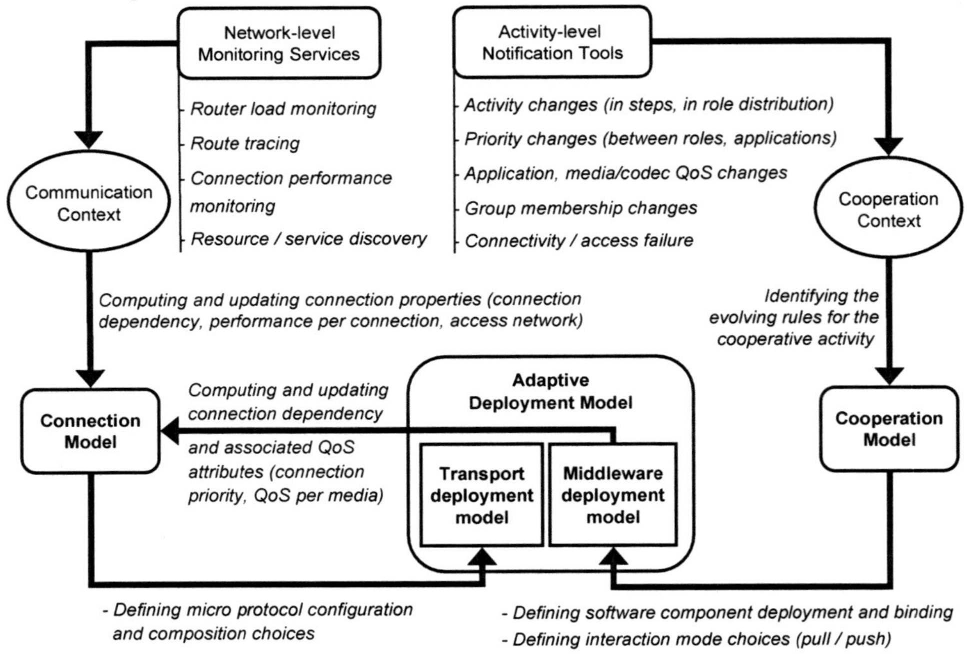

THE PROPOSED MODEL-BASED FRAMEWORK

The framework is composed of three main models: the Connection Model, the Cooperation Model and the Adaptive Deployment Model (ADM). The communication context is captured by the Connection Model while the Cooperation Model captures the activity cooperation context. The relationships between these models presented hereafter are summarized on Figure 1.

General Overview of the Framework Models

The activity requirements are derived from the cooperation context and captured by the Cooperation Model which captures the changes occurring at the activity level. These include modifying activity phases, role distribution, modifying priorities between roles and applications, modifying QoS parameters of applications, media and codecs, dynamic group membership, and access and connectivity failures.

The communication context includes connection dependencies, connection performances and the characteristics of the access network which are captured by the Connection Model. This model expresses the connection dependencies and the associated Quality of Service (QoS) attributes including connection priority and per media QoS parameters. The communication context changes are monitored by a set of network-level monitoring services. Changes occurring at this level include router load, routing choices, connection performances, and resource and service discovering.

The Adaptive Deployment Model (ADM) is generated from the above two context models. It is composed of two sub-models, the Middleware Deployment Model and the Transport Deployment Model (TDM).

The Middleware Deployment Model (MDM) represents the different software components supporting the information exchange between the different actors of the cooperative activity. Such components are event producers, event consumers, and channel managers interacting following

Figure 2. Example of cooperations in MEO

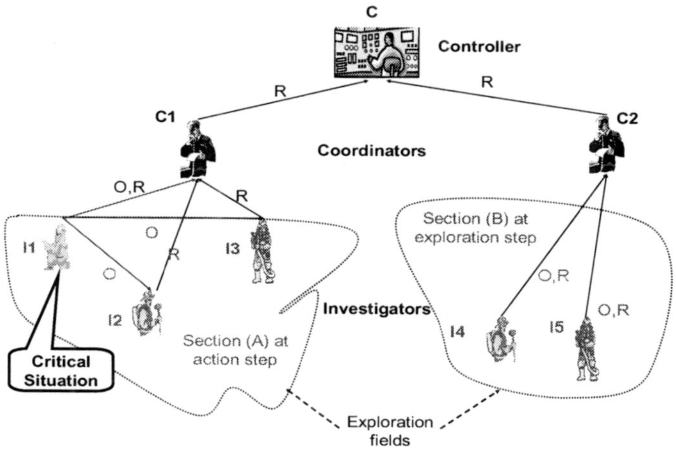

the publish/subscribe paradigm or simple clients and servers interacting through direct message exchange. The different bindings of information requesters to information providers and the different interaction modes are also elements of the MDM. These elements can change for adaptability purposes at runtime.

The Transport Deployment Model (TDM) is deduced from the Connection Model and the Middleware Deployment Model. It represents the transport level decisions. In the case of dynamically configurable protocol architectures, the different protocol modules as well as their configurations are represented.

Framework Instantiation Example

In order to illustrate the use of the previously presented models, their application to Military Emergency Operation (MEO) management is presented in the next paragraphs.

Application Context: Military Emergency Operation Management

We consider the context of Military Emergency Operation (MEO) management systems (see Figure 2) which involve structured groups of communicating actors that cooperate to manage a given crisis. The cooperating actors have roles and use communication devices with unequal communication, processing, energy and storage resources. Devices are fixed or mobile, and communicate through wired and/or wireless networks.

Cooperation is based on data exchange between members: Observation data (O) and Report data (R) are produced periodically or immediately after a particular event. An activity controller supervises the teams, receives the coordinator reports summarizing the current situation and mission progress. According to actions and objectives assigned by the controller, a coordinator manages a team of investigators by giving orders and assigning tasks to be performed.

Investigators explore the field; they observe, analyze and submit reports of the situation to coordinators. For each team, two phases are considered. During the exploration phase (section B), investigators communications have the same priority. The action phase (section A) corresponds to the discovery of a critical situation. The investigator who discovers it is given high priority for communications.

Figure 3. Techniques for automated model adaptation in the framework

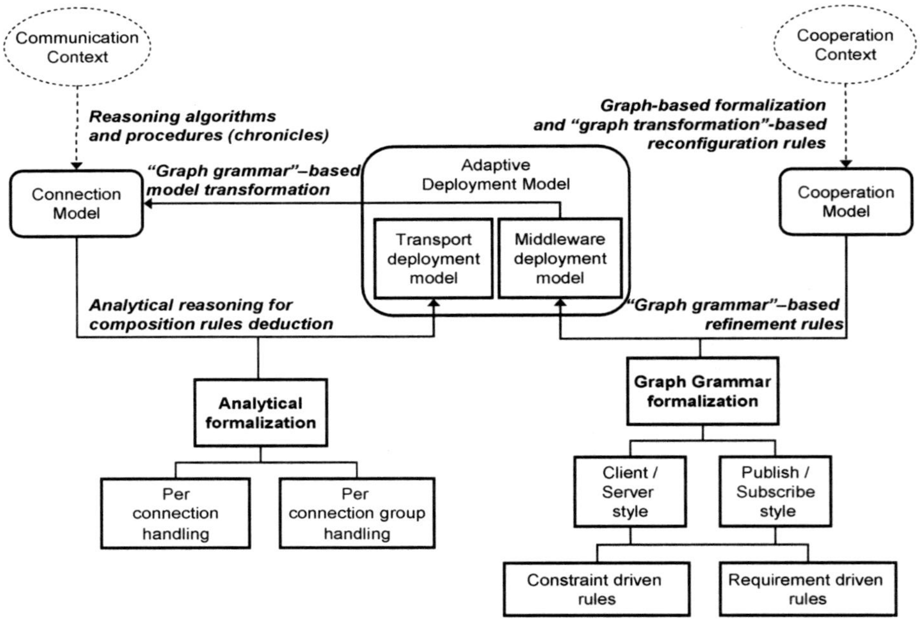

Description of the Elaborated Models

In the context presented above, the various models introduced in our approach are detailed in the following paragraphs. As a summary, Figure 3 provides a global view of the elaborated models, their relationship as well as the different techniques used for automating their implementation.

The Cooperation Model (CoopM) is deduced from the cooperation context which is subject to changes, e.g. from one phase of the activity to another. When such changes happen, *reconfiguration rules* are used to automate the model's adaptation. These rules are written as *graph transformation* rules introduced by Chassot (2006).

The CoopM involves (1) characterizing all valid configurations as a graph grammar, (2) describing valid configurations, (3) defining all possible structural changes at the cooperation level as graph transformation rules. This formally provides the valid reconfiguration rules and actions to be performed in reaction to changes in the cooperation context.

For cooperation level graphs, node labels represent actor identifiers, cooperation roles, hosting devices, and mission phases. Edge labels represent exchanged data, priority, and the required level of QoS.

The Middleware Deployment Model (MDM) is deduced from the CoopM using model *refinement rules* expressed as *graph-grammar* productions. It supports two architectural paradigms: the Client/Server and the Publish/Subscribe.

For graphs handled at the middleware deployment level, nodes represent the deployment elements. They are labeled by parameters such as type (e.g. P/C/G on Figure 4), and hosting devices. Edges are labeled by related communication characteristics such as QoS and priorities.

Refinement from the CoopM into the MDM is implemented using extended *edNCE graph grammars* presented by Rozenberg (1997). The system allows the generation of all the deployment configurations with respect to a given cooperation model. On the other hand, it also allows automatic conformance verification of dynamically evolving cooperation and deployment instance models.

The Connection Model (ConnM) is deduced from the MDM following a set of model transformation rules expressed as *graph-grammar productions*.

The connections are represented by graph nodes as first level elements of the ConnM. Dependant connections are immediate neighbors in the graph.

In practice, a dependency relationship means that the two connections share at least one common resource, such as access networks or routers. The higher the dependency degree is, the more it will be suitable to coordinate the connections to improve their performance. Dependency also results from topology properties such as sharing of the sending and/or receiving hosts, or *n* common routers. Such information may be useful to *estimate* the probability of a common bottleneck when its presence cannot be determined by the monitoring services.

Node labels, such as (c1, R, QoS_R, high, AN_{MC1}, AN_{MC}, $Perf_{c1}$) on Figure 4 represent, respectively, the the connection id, the transported data type, the QoS required, the priority of the connection, the access network of the sender and the receiver, and the observed performances (i.e. delay, loss rate).

Edge labels refine the dependency degree between each pair of dependant connections, using values deduced from the MDM and monitoring information.

The Transport Deployment Model (TDM) is built in two stages. First, a *per-connection* decision is taken. This decision is based on reasoning procedures that take the ConnM as input and output the protocol modules composition to be used in a dynamically configurable transport protocol in order to optimize the QoS. This approach of our work detailed by Van Wambeke (2008). Then, a *per-group of connections* decision is taken in order to consider dependency and priority properties. This decision refines the compositions and adds modules for managing priorities.

The reasoning process is based on two models, the *composition model* and the *decision model*. The composition model is used to define the *conditions* of the validity of the assembled protocol modules. By such, it reduces the size of the set of potential composition candidates for the reasoning process. The decision model is used to guide the process of choosing a composition among all the valid ones in order to maximize the overall user perceived efficiency (i.e. the required QoS). In previous works by Van Wambeke (2008), we have shown that this problem is equivalent to a multi-criteria optimization problem given a proper formal description of the candidate protocol modules composition.

IMPLEMENTING AN ARCHITECTURE TO SUPPORT MODEL BASED ADAPTATION

In this section, the details and benefits of using the previously introduced models in the provisioning process is described in the context of the NETQoS IST project which addresses the problem of QoS management using a policy-based approach.

The general architecture of the NETQoS system distinguishes four main entities (see Figure 5).

The *Policy Description* is used to specify the actor-level policies, the operational policies, etc. The *Automated Policy Adaptor* (APA) does not provide QoS by itself, but decides upon and dispatches operational policies. It is responsible for the provisioning process in which the models are used. The *Actor Preference Manager* (APM) provides NETQoS GUI/API allowing users to define policies. These policies (e.g. requirements, preferences, profile, quality reporting...) may be expressed before or during the communications. This information is used by the APA as input to the CoopM. The *Monitoring and Measurement* (MoMe) captures context evolution, (e.g. evolving actor's policies, end systems/network resource change). This information is used by the APA as input to the ConnM.

Figure 4. Example of refinement / transformation from Cooperation model to Connection model

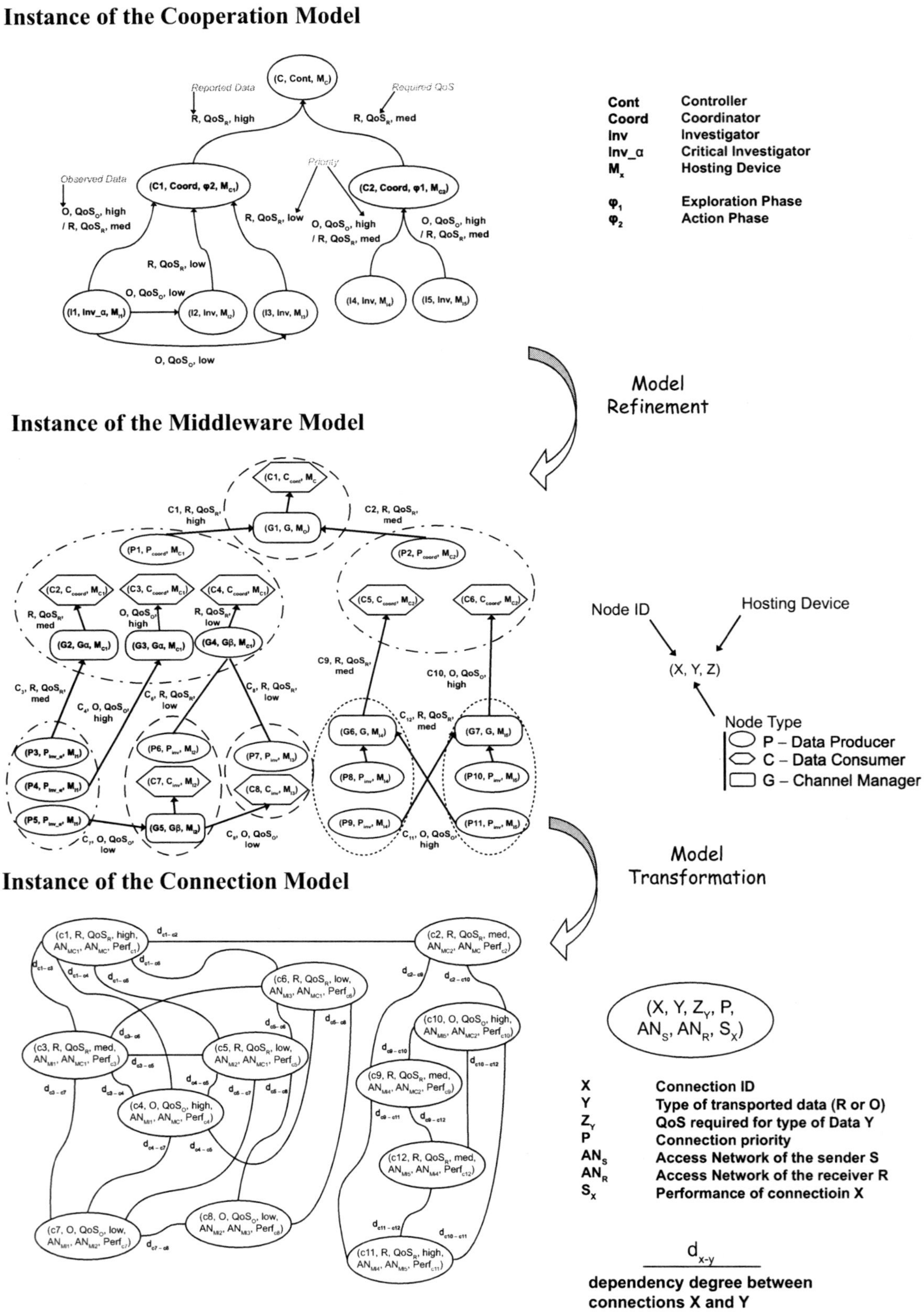

Figure 5. General NETQoS architecture

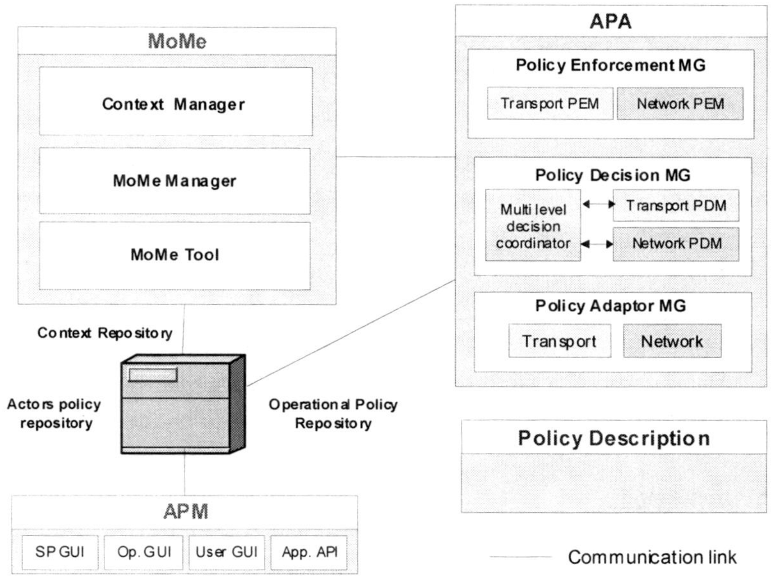

Models for the Automated Policy Adaptor (APA)

The APA decides, dispatches, and adapts the operational policies that take into account the actors dynamic requirements as well as the evolving context. It is composed of three main components. The *Policy Decision Manager* (PDM) is in charge of the decision process. This process is based on the use of the models presented in section 3. The *Policy Enforcement Manager* (PEM) is in charge of the deployment of the policies decided by the PDM on the policy enforcement points. The *Policy Adaptation Manager* (PAM) is in charge of the adaptation, individually or by groups, when the communication or the cooperation context changes. The following paragraphs detail the three APA components implementation.

Policy Decision Manager (PDM)

The PDM decides an optimal set of policies to be settled at the Network and/or at the Transport level to satisfy the set of actor-level policies. This *provisioning* is performed using information contained on the MDM for network provisioning as well as the TDM for transport provisioning. The models are derived from the CoopM and ConnM which are constructed using context monitoring information. Each time the PDM takes a decision, it transmits it to the PEM presented below.

Policy Enforcement Manager (PEM)

The PEM is in charge of dispatching the PDM decisions to the actual policy enforcement point (PEP). For instance, for a transport level adaptation, the PEM dispatches the transport protocol configuration rules to be applied on the end nodes.

The PEM is independent of the network and transport technologies that are used to enforce the policies, (i.e. the PEM provides decisions in a generic language). Consequently, adaptors have to be provided on the PEP themselves to translate the generic PEM rules into specific technology-dependant rules.

Policy Adaptation Manager (PAM)

The PAM is in charge of the adaptation when the context changes. It may decide to adapt the policy by re-creating a completely new one or simply amending the one in place.

The PAM mainly acts when it receives alarms from the MoMe component informing it that the communication or cooperation contexts have changed. It performs adaptation by applying the *graph-grammar* transformations to the actual Connection and Cooperation Models and re-generating derived (MDM, TDM ...) models.

Example: Model Based Adaptation

In this section, the Military Emergency Operations context presented in section *3.2.1* is used to illustrate the adaptation steps that take place on the models constructed by the APA when the cooperation goes from the exploration step to the action step. This adaptation corresponds to the discovery of a critical situation by one of the investigators taking place in the activity.

For instance on the MDM presented on Figure 6a, the investigators M_1 and M_2 are in the exploration step. Two channels are implemented: each one is in charge of a specific (data, priority) couple. Assuming a mobile participant is allowed to host only one event service component, the two channel managers (CMs) are deployed on participants M_1 and M_2.

When an investigator discovers a critical situation, the policies/preferences change and MoMe component informs the APA of this event. The APA then modifies the Cooperation Model according to the user's preferences. In this case, the user has specified that in exploration step, each investigator reports directly to its coordinator while in the action step, the investigator who discovered a critical situation reports to both his direct coordinator as well as his fellow investigators.

To support the changes of the Cooperation Model when the mission goes from the exploration step (step 1) to the action step (step 2), several adaptation actions are performed on the MDM. These actions lead to different possible architectures, one of them is illustrated on Figure 6b.

Architecture transformation is guided by rules that consider not only changes of the Cooperation Model, but also changes of resource-oriented parameters such as machines' energy and storage/computation capacity. For instance, in the action step, four CM are implemented: one per (data, priority) couple. In order to save energy, two CM are deployed on the controller's machine M_3, and only one CM is deployed on each of the investigators' machine M_1 and M_2. Such transformation may also be caused by resources parameters only. In such cases, a "good" transformation should have no impact on the upper Cooperation Model.

This new MDM is used to automatically update the Connection Model, both models will then be used to adapt the policy in place and deploy the updated decision on the different network and transport PEP resulting in optimized QoS to all users in this new step of their mission.

EVALUATING AN ARCHITECTURE TO SUPPORT MODEL BASED ADAPTATION

In this section, the evaluation of the previously presented framework implementation is presented. Particularly, the focus is made on the APA component. The evaluation deals with several metrics related to the service time and the resources consumption.

Measured Metrics

The APA behavior is triggered by the reception of events from the Context Manager such as *ApplicationLaunchEvent* in the case of a provi-

Figure 6. Graph transformations from exploration step to action step – MDM

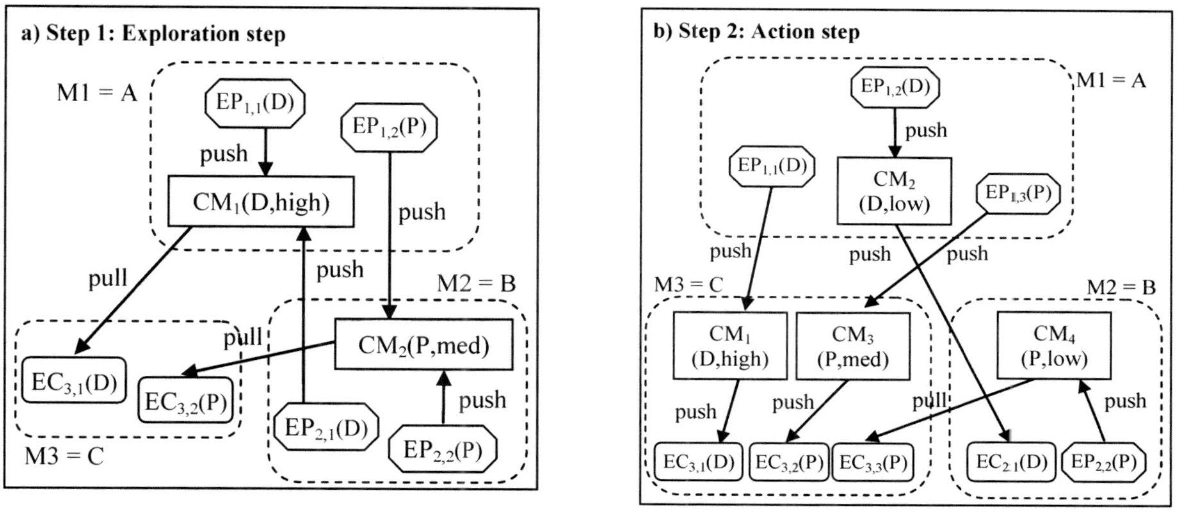

sioning request, and *PolicyViolationEvent* when adaptation is required.

In order to evaluate the APA performances, the following metrics have been observed for several values of the arrival rate of both *ApplicationLaunchEvent* and *PolicyViolationEvent*: mean service time per request (event), internal components event queues length. These metrics have been taken for different values of the event's arrival rate (λ) ranging from 20 events/s to more than 50 events/s.

The following paragraphs present and discuss the results obtained for the above metrics for both the provisioning and the adaptation respectively. Finally, conclusions on the evaluation are presented.

Processing Time of Provisioning Requests

The results obtained for the evaluation of the provisioning phases are presented on Figure 7 as well as 8 and 9. Two different behaviors can be observed depending on the arrival rate of *ApplicationLaunch* events:

From 20 to 39 event/s, the APA is unloaded: the best performances are obtained and empty queues are noticed for all the involved components. For instance, at an arrival rate of 20 event/s, almost 60% of the requests are served in less than 50 ms which is the value of the mean inter arrival time of the requests.

When the arrival rate exceeds 40 request/s (see Figure 8 and 9), the APA starts to be overloaded: this is indicated by the progressive increase of the internal queues. Additionally, one can notice that the APA is able to resist to a high number of requests per second only during a short period (less than 1 second). This statement can be explained by the massive use of multi-threading in the decision process (one PDM worker thread is created per request): after the creation of a given number of threads, the Java Virtual Machine (JVM) is not able to create new threads and the system is blocked until old threads have died.

Let's finally note that the histograms and the mean adaptation request transit times do not take into account values measured after blocking.

Figure 7. Histogram of the ratio of served provisioning requests per time interval

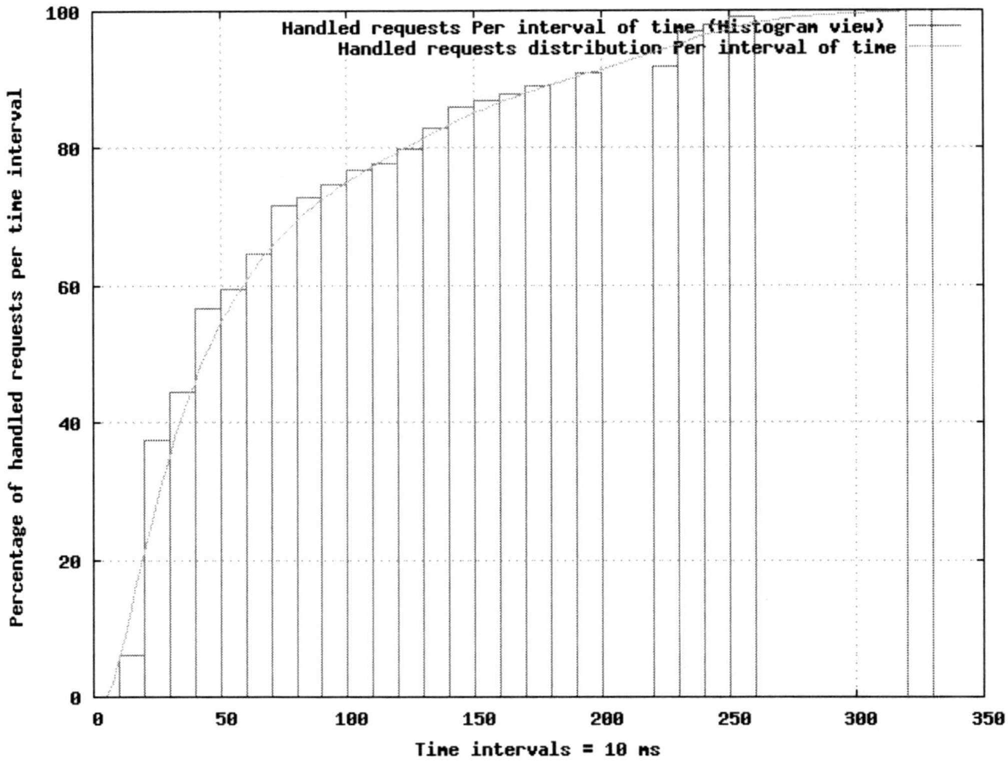

λ = 20 events /s

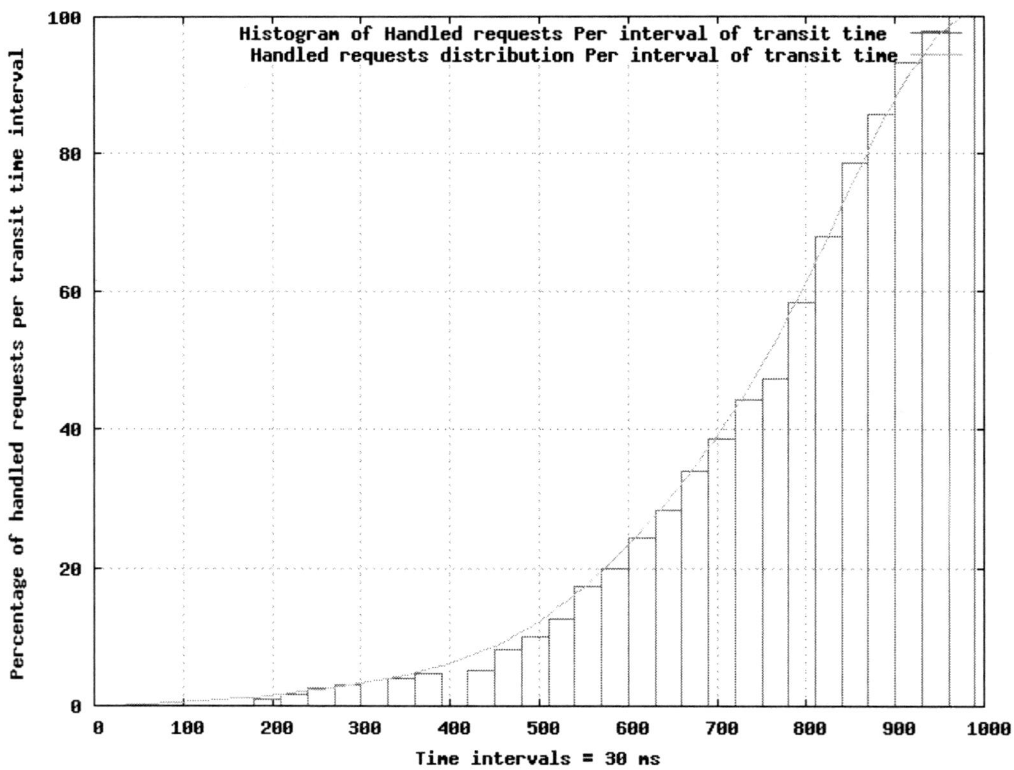

λ = 40 events /s

continued on following page

Figure 7. Continued

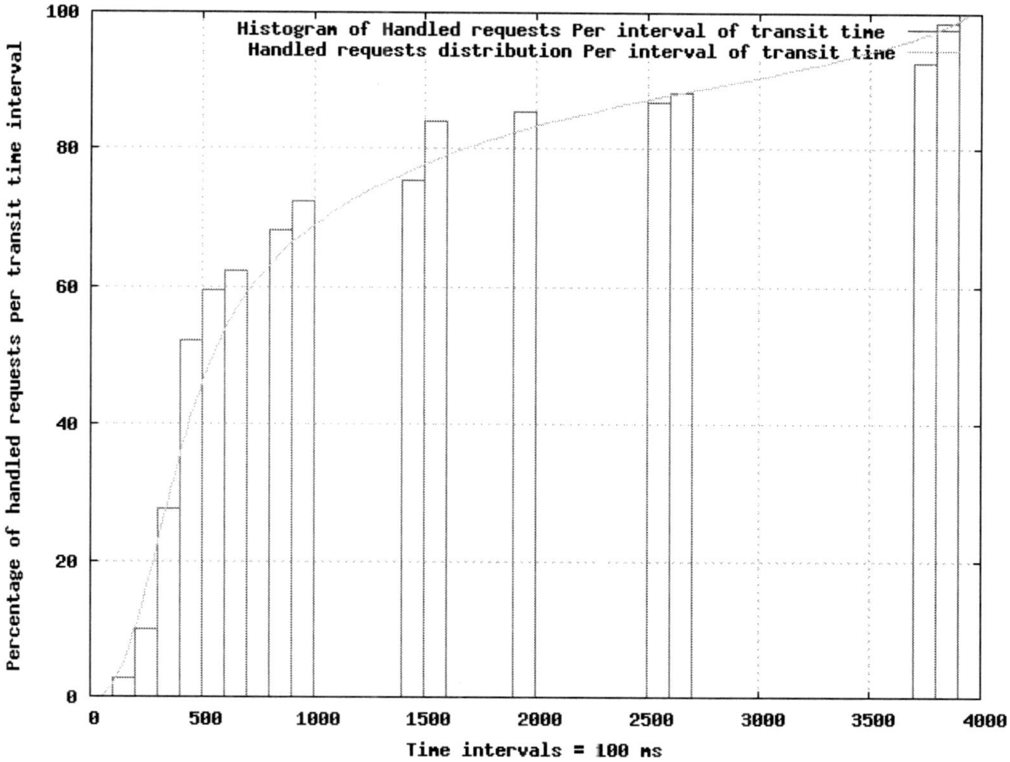

λ = 50 events /s

Figure 8. PAM: Transit time per PolicyViolation request

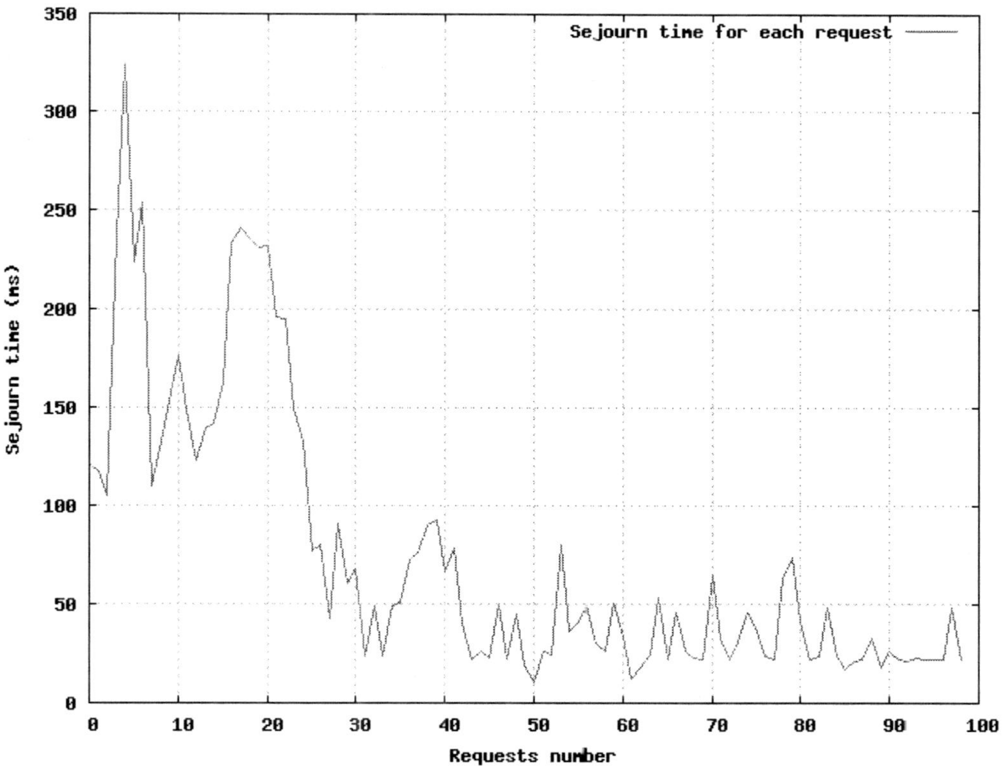

λ = 20 events /s

continued on following page

Figure 8. Continued

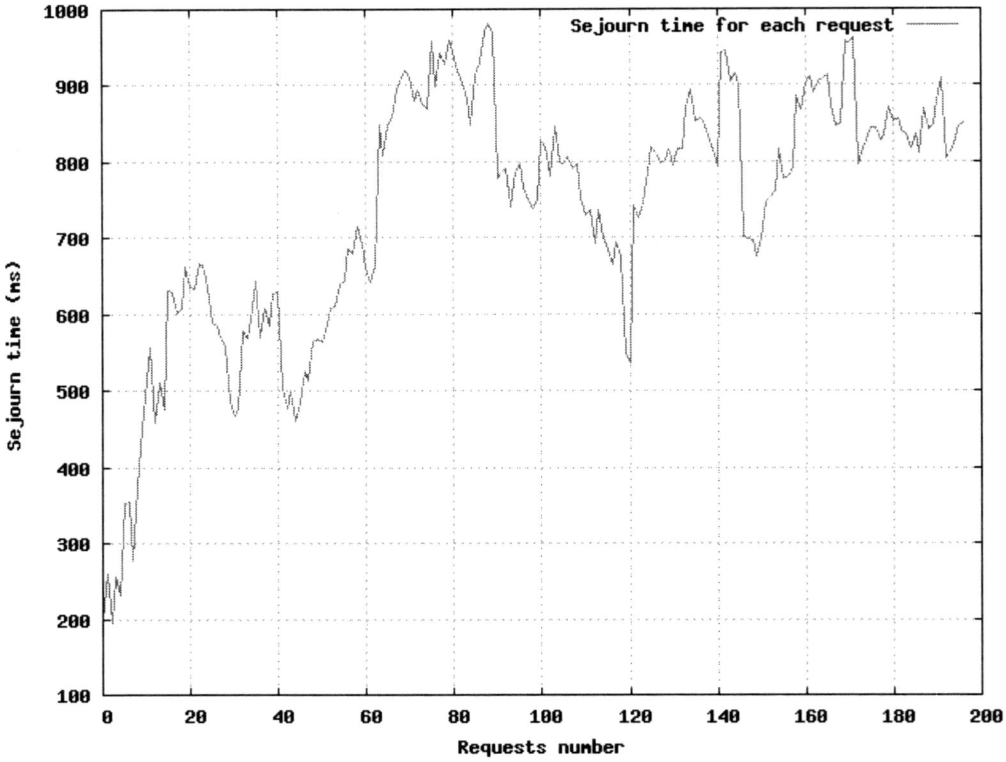

λ = 40 events /s

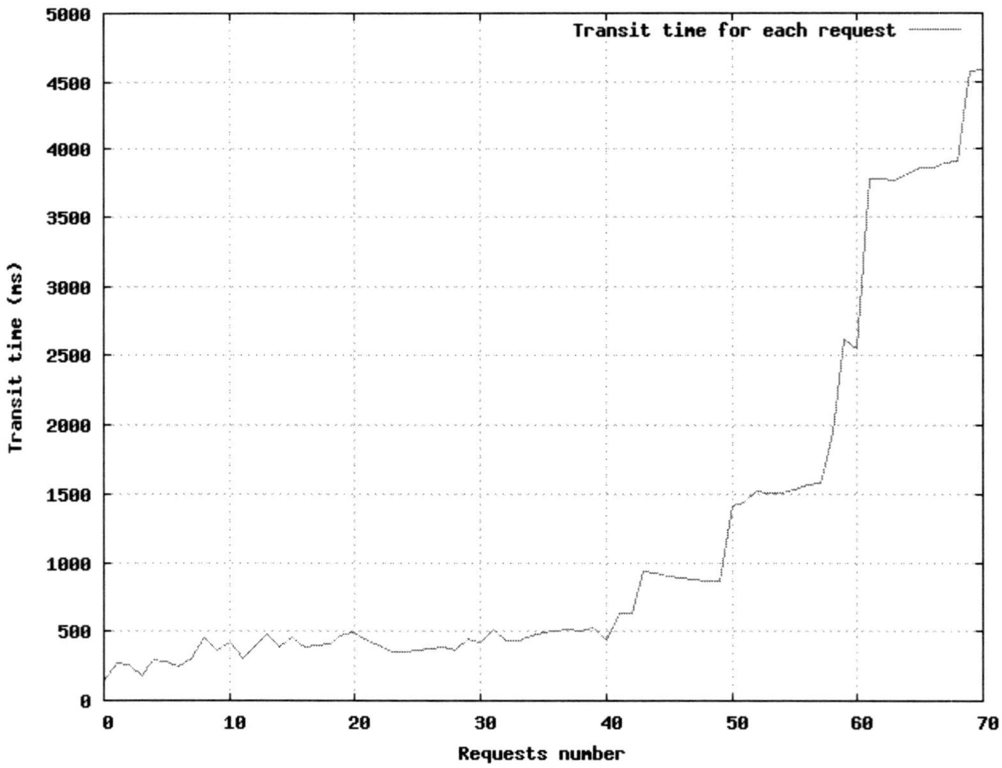

λ = 50 events /s

Figure 9. Mean provisioning request transit time vs. request arrival rate

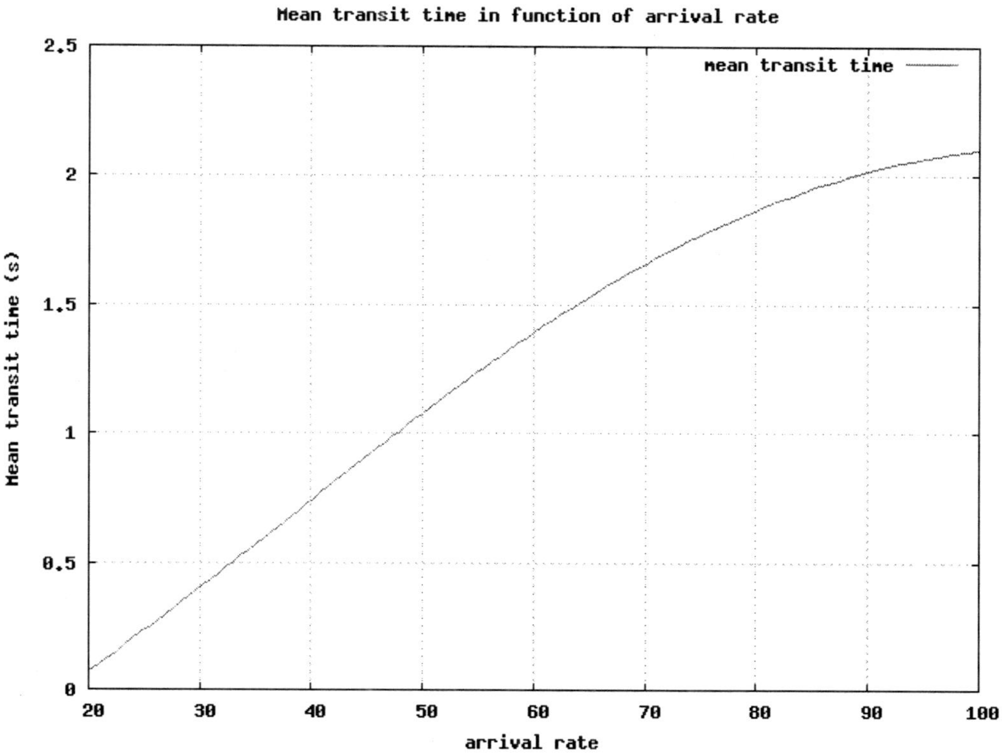

Processing Time of Adaptation Requests

The results for the performance evaluation of the APA's adaptation process are presented on Figure 10, 11 and 12. As in the provisioning-oriented measurements presented previously, two different behaviors have been observed depending on the arrival rate of *PolicyViolation* events:

From 20 to 39 event/s, the APA is unloaded: the best performances have been obtained and empty queues have been noticed for all the involved components. For instance, for an arrival rate of 20 event/s, more than 75% of the requests are served in less than 50 ms.

When the arrival rate exceeds 40 events/s (see Figure 11 and 12), the APA starts to be overloaded: this is indicated by the progressive increase of the internal queues length. Similarly to what has been observed for the provisioning-oriented measurements, one can notice on Figure 11 that the APA is able resist to a high number of requests per second during a given period, then the transit delay grows very rapidly. This statement is once again explained by the massive use of multi-threading in the decision process (one PAM worker thread is created per request): after the creation of a given number of threads, the JVM is not able to create new threads and the system is blocked until old threads have died.

Let's finally note that the histograms and the mean adaptation request transit times do not take into account values measured after blocking.

Conclusion on the Evaluations

The performances of the T-APA have been evaluated and discussed in this section. Several metrics related to the service time and the resources consumption of the APA have been analyzed. One specific goal was to validate the choice of the multi-threading in the implementation of the APA's computation intensive operations.

Figure 10. APA: Histogram of the % of served adaptation requests per time interval

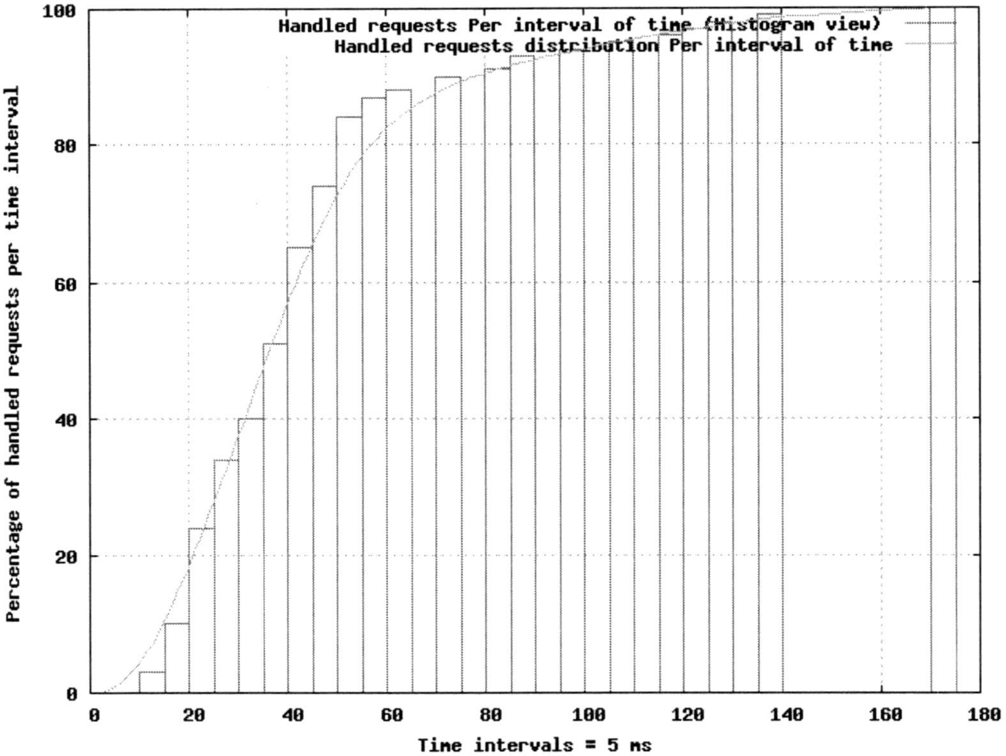

λ = 20 events /s

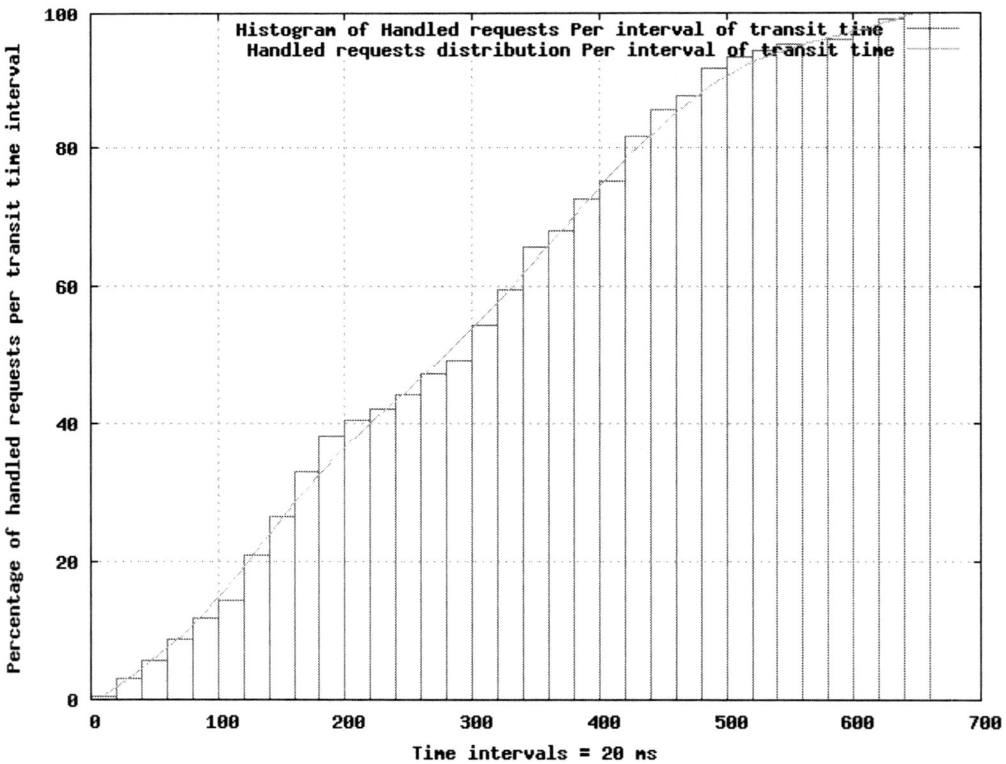

λ = 40 events /s

continued on following page

Figure 10. Continued

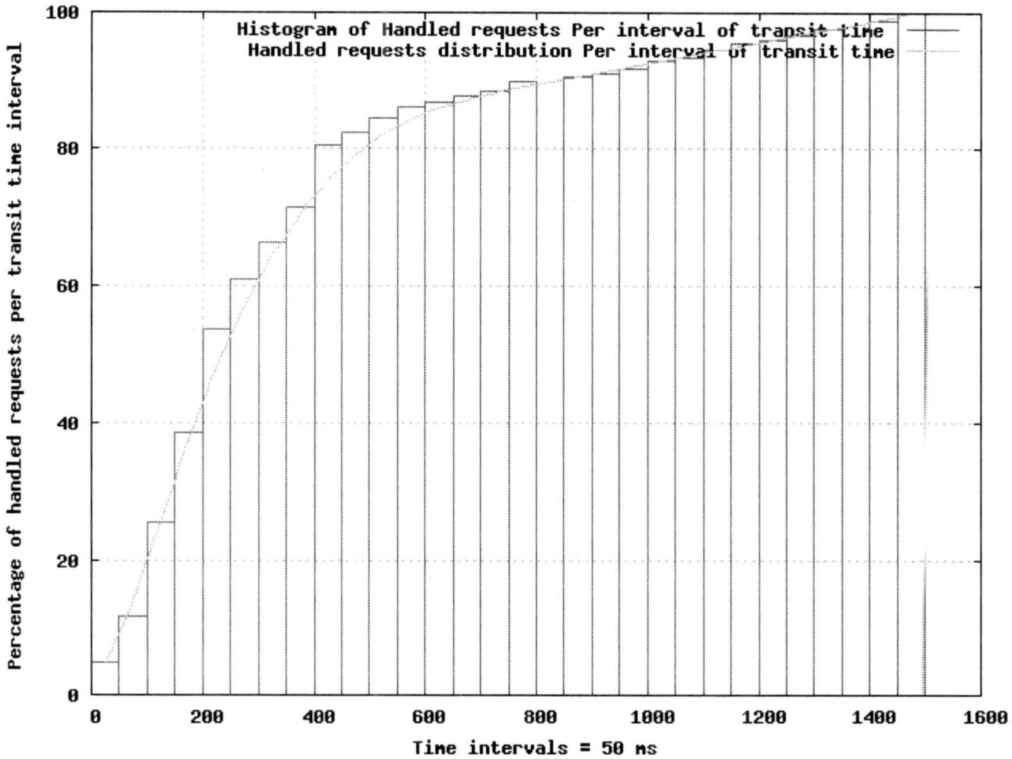

λ = 50 events /s

Figure 11. PAM: Transit time per PolicyViolation request

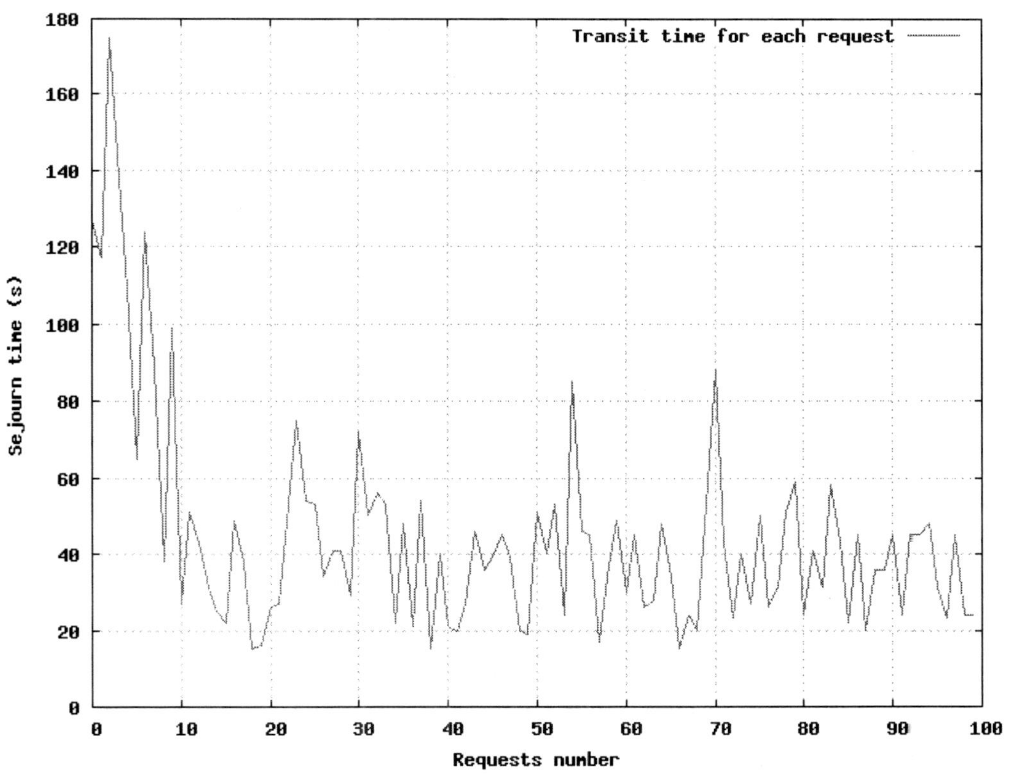

λ = 20 events /s

continued on following page

Figure 11. Continued

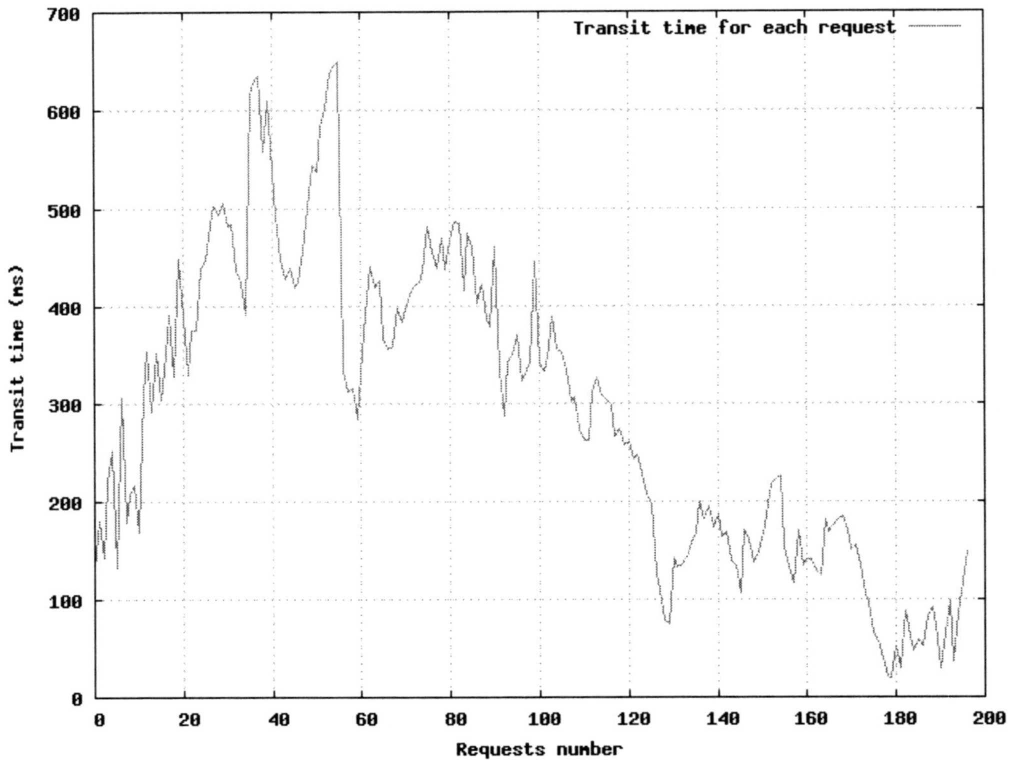

λ = 40 events /s

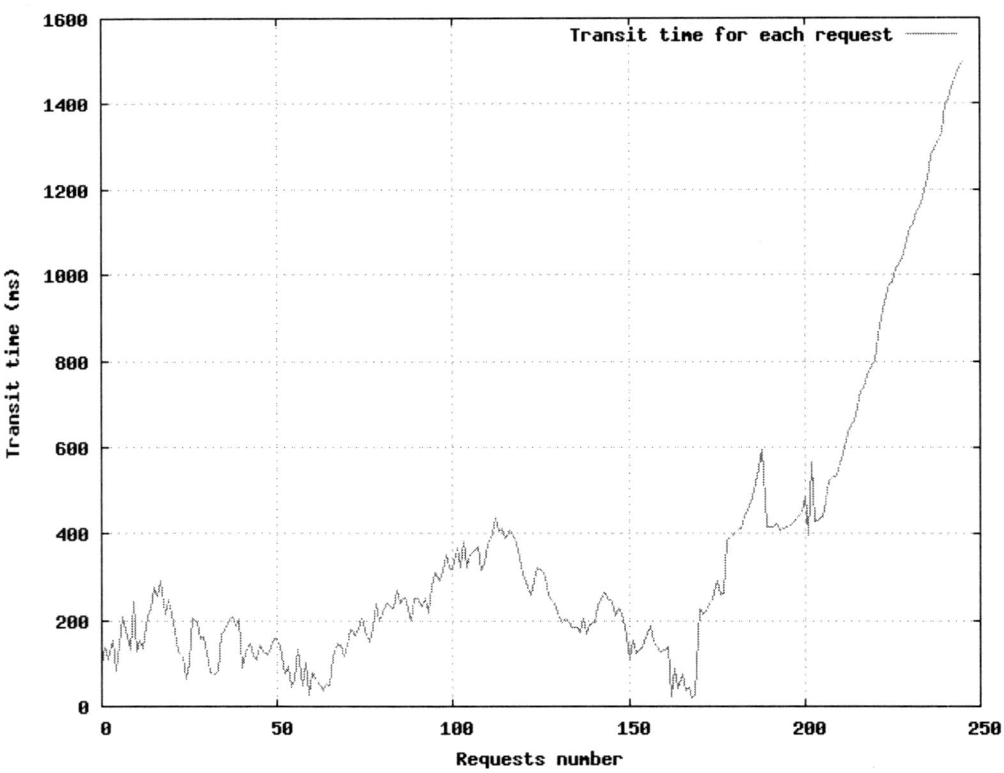

λ = 50 events /s

Figure 12. APA: Mean adaptation request transit time vs. request arrival rate

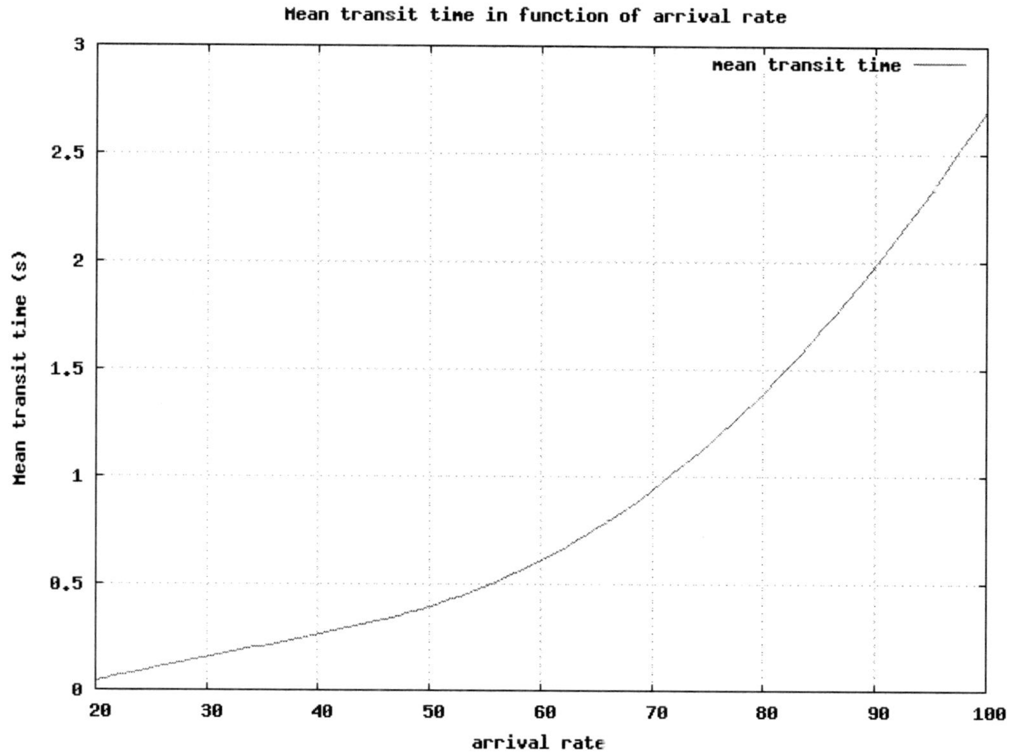

As a general conclusion, it clearly appears that the APA is able to process in an acceptable delay a correct number of provisioning / adaptation requests per second. When this number exceeds a given value, the APA is still able to process the requests but only during a very short period. This phenomenon results from a massive use of the multi threading, which conducts to the blocking of the system due to JVM limitation. A solution to be studied could be to manage a hybrid solution involving both mono and multi threading, in order to maintain the number of created PDM/PAM Worker threads under a given value.

CONCLUSION AND PERSPECTIVES

In this paper, different models have been elaborated and implemented to help automating adaptive deployment for QoS management. Different points of view have been considered to capture the influence of the cooperation as well as the communication contexts. The interest of our approach resides in its capacity to support the full automation of the network management tasks in evolving contexts. An architecture that uses these models in the context of automated policy based network management has been presented. This architecture allows activity-level requirement to be expressed by the user and have the system behave accordingly. The use of the models in the decision process has been illustrated in the context of simple Military Emergency Operations. The evaluation of the architecture's prototype has been presented and showed the existence of limits in terms of scalability. However, the convergence and speed of the decision algorithms have shown to be good with response times under normal conditions that do not exceed 50ms.

Extending the work presented here, future work includes the adaptation of the architecture and algorithms to other environments such as ad hoc networks in which no infrastructure is available.

ACKNOWLEDGMENT

Part of this work is done in the context of the EU IST FP6 NETQoS (STREP) project and under a French DGA (Direction Générale de l'Armement) grant.

REFERENCES

Akan, O. B., & Akyildiz, I. F. (2004). ATL: An adaptive transport layer suite for next-generation wireless Internet. *IEEE Journal on Selected Areas in Communications, 5,* 802–817.

Capra, L., Emmerich, W., & Mascolo, C. (2003). Carisma: Context-aware reflective middleware system for mobile applications. *IEEE Transactions on Software Engineering, 10*, 929–945.

Chassot, C., Guennoun, K., Drira, K., Armando, F., Exposito, E., & Lozes, A. (2006). Towards autonomous management of QoS through model driven adaptability in communication-centric systems. *Int Transactions on Systems Science and Applications, 3*, 255–264.

DaSilva, L. A., Midkiff, S. F., Park, J. S., Hadjichristofi, G. C., Davis, N. J., Phanse, K. S., & Lin, T. (2004). Network mobility and protocol interoperability in ad hoc networks. *IEEE Communications Magazine, 11*, 88–96.

Exposito, E. J. (2003). *Design and implementation of quality of service oriented transport protocol for multimedia applications.* PhD thesis, National Polytechnic Institute of Toulouse.

Hutchinson, N. C., & Peterson, L. L. (1991). The x-kernel: An architecture for implementing network protocols. *IEEE Transactions on Software Engineering, 1*, 64–76.

Kaloxylos, A., Lampropoulos, G., Passas, N., & Merakos, L. (2006). A flexible handover mechanism for seamless service continuity in heterogeneous environments. *Computer Communications, 6*, 717–729.

Landry, R., Grace, K., & Saidi, A. (2004). On the design and management of heterogeneous networks: A predictability-based perspective. *IEEE Communications Magazine, 11,* 80–87.

Marshall, I., & Roadknight, C. (2001). Provision of quality of service for active services. *Computer Networks, 1*, 75–85.

Perez, G., & Skarmeta, A. G. (2004). Policy-based dynamic provision of IP services in a secure vpn coalition scenario. *IEEE Communications Magazine, 11,* 118–124.

Rozenberg, G. (1997). *Handbook of Graph Grammars and Computing by Graph Transformation.* World Scientific Publishing, ISBN 981-02-2884-8.

Samaan, N., & Karmouch, A. (2005). An automated policy-based management framework for differentiated communication systems. *IEEE Journal on Selected Areas in Communications, 12*, 2236-2248.

Van Wambeke, N., Armando, F., Chassot, C., & Exposito, E. (2008). *A model-based approach for self-adaptive Transport protocols.* Comput Commun. doi: 10.1016/j.comcom.2008.02.026.

Wu, D., Hou, Y. T., Zhu, W. (2001). Streaming video over the internet: Approaches and directions. *IEEE Transactions on Circuits and Systems for Video Technology, 11*, 282-301.

This work was previously published in International Journal of Business Data Communications and Networking, Volume 5, Issue 2, edited by Varadharajan Sridhar and Debashis Saha, pp. 35-51, copyright 2009 by IGI Publishing (an imprint of IGI Global).

Chapter 7
Reservation MAC Protocols for Ad-Hoc Networks:
Analysis of the Approaches

Ghalem Boudour
IRIT - Paul Sabatier University, France

Cédric Teyssié
IRIT - Paul Sabatier University, France

Mammeri Zoubir
IRIT - Paul Sabatier University, France

ABSTRACT

Multimedia and real-time applications require bandwidth guarantees, which may be achieved by resource reservation. Several researches were done to propose efficient reservation MAC protocols for ad-hoc networks. In these schemes, channel is segmented into super-frames composed of fixed number of slots. They allocate slots to each traffic source, and make sure that neighbor nodes record the reservation in order to ensure consistency of reservations between neighbor nodes. However, resource reservation in ad-hoc networks remain a very challenging task due to the instability of radio channels, node mobility and lack of coordination between mobile nodes. Proposed reservation MAC protocols like CATA, FPRP, R-CSMA and SRMA/PA have limitations and are suitable only for particular situations. In this paper, we propose a comparative analysis of the most representative reservation MAC protocols. We identify the major issues unresolved by reservation MAC protocols. A performance evaluation and comparative analysis with the IEEE 802.11e are achieved through the NS-2 simulator.

INTRODUCTION

Mobile ad-hoc networks (MANETs) are collections of mobile nodes forming temporary networks without any infrastructure support. They can be set up anywhere anytime owing to their easy deployment and self-organization ability. As a result, MANETs become the primary mean of communication in several domains where the deployment of wired infrastructure is difficult.

DOI: 10.4018/978-1-60960-589-6.ch007

Such domains include battle fields, forestry fire, and disaster recovery.

The characteristics of MANETs like the lack of centralized coordination, node mobility and resource availability make the Quality of Service (QoS) support in MANETs a very challenging task. MAC protocols for MANETs define the manner channels are shared between mobile nodes. They have significant impacts on the overall system performances and their design is a very challenging issue.

Many solutions have been proposed to support QoS at the MAC sub-layer. Those solutions attempted to improve the channel access mechanism to provide QoS guarantees to multimedia and real-time applications. Proposed solutions may be classified into two categories: contention-based and reservation-based schemes.

Contention-based protocols are non deterministic and nodes compete to get access to the channel for the transmission of each data packet. The aim of these protocols is to avoid packet collisions, and resolve the hidden and exposed terminal problems. This is achieved through carrier sensing, handshaking and backoff mechanisms. Carrier sense ensures that nodes compete to access the channel only when the channel is detected idle. The handshake mechanism uses short control frames (RTS/CTS) exchange between the sender and receiver prior to data transmission in order to avoid the hidden and exposed terminals issues. The IEEE 802.11 MAC protocol is the most known example of contention-based protocols.

Reservation protocols seem to be attractive solutions for QoS provisioning in ad-hoc networks. Their characteristics such as the contention free medium access and the reduced collision rate are very interesting for MANETs. In this paper we provide a comparative analysis of these protocols and the major issues encountered in designing such protocols. Particularly, we analyze the effects of mobility on the performance of reservation MAC protocols. We also compare these protocols with the IEEE802.11e standard.

In order to face the limited bandwidth issue in MANETs, MAC protocols should ensure high bandwidth utilization, and reduce bandwidth wasting.

The rest of this paper is organized as follows. In section 2, we give an overview of the IEEE 802.11e standard and reservation MAC protocols. In section 3 we highlight the major challenges and limitations of reservation MAC protocols. In section 4 we give a performance evaluation of reservation MAC protocols. Section 5 gives our conclusions.

BACKGROUND AND RELATED WORK

Channel access protocols in MANETs can be classified into two categories: contention-based and reservation-based protocols. Contention-based protocols are non-deterministic and nodes compete to get access to the wireless channel. The IEEE 802.11 is the most known example of contention-based protocols.

The IEEE 802.11 (IEEE Std. 802.11, 1999) standard is considered as the de-facto MAC protocol for wireless networks. The DCF mode is based on the Carrier Sense Multiple Access with Collision Avoidance (CSMA/CA). It uses two mechanisms to avoid collision: the physical carrier sensing and the virtual carrier sensing. The physical carrier sensing is used to detect the presence of signal on the common physical channel. The virtual carrier sensing uses the duration field of the MAC frame header to indicate the duration during which a node will reserve the channel.

DATA transmission in DCF is accomplished following the RTS / CTS / DATA / ACK handshake. A station which has a DATA packet to send waits the channel to be idle for the duration of DIFS (DCF Inter Frame Space). If the channel lasts idle for DIFS, the station transmits an RTS packet. Otherwise, the station enters in a backoff period, by choosing a backoff timer uniformly distributed in [0, CW], where CW is the Contention Window (CW) size. The backoff timer is decremented for each idle time-slot. The station transmits its RTS packet when the backoff timer expires. When the receiver receives successfully the RTS packet, it waits for SIFS (Short InterFrame Space) before replaying with a CTS packet. Both the RTS and CTS packets contain the Duration field which is used in order to prevent neighbours from accessing the channel during the RTS / CTS / DATA / ACK handshake.

Unfortunately, the contention-based access of the IEEE 802.11 makes it unable to fit the requirements of multimedia applications over multi-hop networks. In (Xu & Saadawi, 2001), authors discovered that the IEEE 802.11 did not function well in a wireless multi-hop environment. The results revealed that the standard suffers from serious throughput degradation and unfairness. Performance degradations are mainly due to the hidden and exposed terminals problems, and the binary exponential backoff scheme. The same conclusion was drawn in (Hsieh & Sivakumar, 2002). The new IEEE 802.11e standard (IEEE 802.11WG, 2005) enables deterministic QoS guarantees through MAC level service differentiation. However, the throughput of IEEE 802.11e is expected to degrade at high traffic load. Authors in (Romdhami & Turletti, 2003; Zhu & al, 2004) showed that the performance of MANETs running EDCF are not optimal, and the collision rate increases quickly when the number of contentions to access the medium is high.

On the other hand, reservation MAC protocols seem to be very suitable for multimedia and real-time applications since they reserve the required bandwidth to each source. The basis of these protocols is to give to each node a guaranteed periodic access to the wireless channel. In these protocols, channel is segmented into super-frames, and a global synchronization between nodes is assumed. The MAC protocol reserves a slot to each real-time node. Once the reservation is done, the node uses the same slot in subsequent super-frames without contention. Examples of protocols in this category are FPRP, D-PRMA (Shengming & al, 2002), CATA, and R-CSMA. These protocols mainly differ in the super-frame structure and the medium access control mechanism adopted to reserve time-slots.

In FPRP (Zhu & Corson, 1998), the super-frame is composed of a reservation frame (RF) followed by several information frames (IF). Each RF is composed of N reservation slots (RS), and each IF is composed of N information slots (IS). In order to reserve an IS, the nodes must make reservations during the corresponding RS. Each RS is composed of M reservation cycles (RC), and, in each RC, a five-step reservation process is followed to make a reservation in the current RS. These five steps are: Reservation Request, Collision Report, Reservation Confirmation, Reservation acknowledgement, and Packing and Elimination. These five phases are undertaken by each node to compete to reserve a time-slot, and to inform neighbors about the result of the competition (reservation success of failure). A node which fails in reserving the slot in a RC, enters in competition to reserve the slot in another RC. However, FPRP incurs a significant amount of overhead for slots reservation.

CATA (Tang & Garcia-Luna-Aceves, 1999) protocol divides time into equal size super-frames, and each super-frame is composed of S slots. Each slot is composed of four control mini-slots and one Data mini-slot (DMS). Control mini-slots are used to establish reservations, and prevent neighbors from using already reserved slots. The advantage of CATA over other reservation protocols is it permits to establish unicast / multicast / broadcast reservations. Its major drawback is the waste of bandwidth due to control mini-slots. Reserving four mini-slots in each slot reduces the available bandwidth dedicated for the transmission of data packets.

SRMA/PA (Ahn & al, 2003) adopts the same concepts as CATA. The added feature is that it distinguishes higher-priority nodes from lower-priority nodes. It permits to a higher-priority node to grab reservation from lower-priority nodes.

In R-CSMA (Inwhee, 2004), time is segmented into super-frames. Each super-frame is composed of a contention period (CP) and a set of TDMA slots. A node which wants to establish a reservation follows a three way handshake during the CP in order to negotiate reservations with the receiver. Neighbor nodes record the reservation thus preventing any collision during reserved slots. The major advantage of R-CSMA against FPRP, CATA and SRMA/PA is that it doesn't reserve any bandwidth for control packets. R-CSMA doesn't allocate any control slot since control packets are transmitted only once at the reservation request step.

In (boudour & al, 2008), we have proposed the ER-CSMA protocol, which is an extension of R-CSMA to resolve the reservations clash due to mobility. The reservation clash happens when two nodes which are far away from each other and which have reserved the same slot move. If one of them enters in the transmission range of the other, collisions happen in reserved slots and one or both of them loses its reservation.

RTMAC (Manoj & Siva, 2002) is a reservation MAC protocol that doesn't need global synchronization between mobile nodes. Each super-frame consists of a number of reservation-slots (resv-slots). The duration of each resv-slot is twice the maximum propagation delay. A node that has real-time packets for transmission, reserves a block of consecutive resv-slots, which is called connection-slot on a super-frame and uses the same connection-slot to transmit in successive super-frames. The reserved connection-slot is repaired using relative times of starting and ending times of the connection-slot. With relative time of connection-slots, RTMAC eliminates the need of time synchronization. Each node maintains a reservation table that records for each reservation the pair of sender and receiver identifiers, and the starting and ending time of the reserved connection-slot. When the real-time node finishes its transmission, or if a path break is detected by an upstream node (i.e., the sender associated with reservation), the node release the reserved connection-slot by sending Reservation Release RTS (ResvRelRTS) packet. When the receiver receives the ResvRelRTS packet, it sends a ResvRelCTS packet. The purpose of the ResvRelRTS and ResvRelCTS packets is to request neighbours of the sender and receiver to release the reserved connection-slot.

Despite they alleviate efficiently the effects of packets collision during the reservation phase thanks to the use of control mini-slots and collision resolution schemes, reservation MAC protocols have many limitations that make their deployment difficult. The most challenging issue with these protocols is mobility of nodes. These protocols are suitable only for static ad-hoc networks since no mobility considerations are taken into account. When nodes are mobile, collisions may occur during reserved slots. This phenomenon is called reservation clash and must be handled at the MAC sub-layer. The other issues with reservation MAC protocols are the important control traffic overhead, the support of multimedia applications with different QoS requirements, and the lack of fairness between traffic flows. These issues and possible solutions are discussed in the following section.

DISCUSSION OF RESERVATION MAC PROTOCOLS

Reservation protocols provide some bandwidth guarantees for real-traffic sources. However, they suffer some drawbacks: the waste of bandwidth due to control traffic, reservation clash in case of mobility, lack of support of heterogonous classes of traffic, inefficiency of the reservation release scheme, and lack of fairness. These five issues will be discussed in detail in this section.

Table 1. TSpec parameters for the considered traffic classes

Traffic Models				
Parameters	G.711	G.723	MPEG-4	H.263
ρmean (kbps)	64	6.4	150	270
ρmax (kbps)	64	6.4	1600	2300
average frame size (bytes)	160	24	770	1278
Mean inter-frame arrival time (ms)	20	30	40	40

Heterogeneous Classes of Traffic Support

The first drawback with almost all the reservation MAC protocols is that they consider that real-time traffic sources have the same QoS requirements, and the varying requirements of heterogeneous sources of traffic are not considered. They reserve a slot to each real-time traffic source, with the assumption that the traffic source will use the reserved slot in each frame to transmit its data packets.

Reserving a slot to each real-time traffic source is not efficient, especially when heterogeneous traffic streams are characterized by different QoS requirements. According to the encoding and compression techniques used to represent multimedia sources, traffic streams will have widely varying traffic characteristics (bit-rate, delay). Reserving one slot to each traffic stream (TS) results in a waste of bandwidth mainly when the inter-packets arrival time is greater than the super-frame length. Table 1 shows QoS requirements of some multimedia applications.

A well designed MAC protocol should provide an efficient mechanism to share the limited bandwidth resource and satisfy the heterogeneous and usually contradictory QoS requirements of each traffic class (voice, video, data …). The reservation MAC protocol should ensure that each reserving node will be allocated exactly its required share of bandwidth. To achieve such adaptive scheme, we need QoS mapping scheme which determines the quantity of bandwidth to reserve to each class of traffic in function of the considered channel structure (i.e. super-frame length and the number of slots per super-frame).

Another issue with multimedia traffics in reservation MAC protocols is the support of VBR traffics. Due to its high burstiness and nonstationarity, VBR traffics like video introduce extra difficulty in efficient channel access as well as effective bandwidth allocation. Figure 1 shows the length

Figure 1. Frame length of MPEG-4 high quality Jurassic Park I film

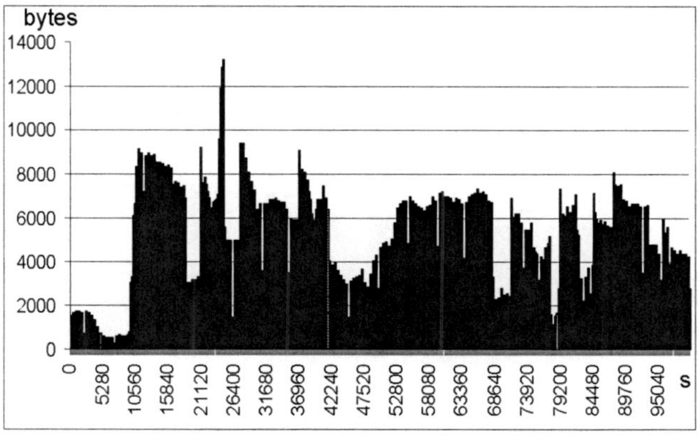

of video frames of a high quality MPEG-4 video film available at (Seeling & al, 2007).

As the figure shows, there is high variation of the bit rate. A simple solution to respond to the bandwidth requirements of these VBR delay-sensitive applications is to reserve a bandwidth corresponding to the peak transmission rate. While this scheme is simple and offers high QoS for individual traffic sources, it doesn't ensure efficient channel utilization as it is bandwidth consuming. Bandwidth is wasted when VBR sources generate packets at low rate, especially when the ratio of peak to the mean rate is high. In addition, the number of slots available for reservation by other neighbour nodes will be small. Another allocation scheme consists of allocating a bandwidth corresponding to the average bit rate to each traffic stream. This scheme has the advantage that it is not bandwidth consuming, and more reservations can be accepted since more slots will be available for reservation compared to the peak allocation scheme. However, this scheme can't satisfy bandwidth requirements of bursty VBR sources, especially when the peak to the average bit rate is high.

Impacts of the Super-Frame Structure

Unlike the IEEE 802.11e where nodes are enabled to transmit each time they win contention to the wireless channel, nodes in reservation MAC protocols can transmit only on their reserved slots. A node which has the opportunity of transmission at time t, can transmit the next packet only after $t+T_{super\text{-}frame}$, where $T_{super\text{-}frame}$ is the super-frame length. The super-frame length (in term of number of slots per frame) affects the bandwidth and delay offered to real-time and multimedia traffic sources.

There is a trade-off between the super-frame length, delay, and bandwidth requirements of real-time traffic sources. On one hand, choosing a small number of slots per super-frame guarantees a small delay. This scheme is suitable for multimedia traffic sources with a short inter-packets arrival time and stringent delay requirements. However, the call acceptance ratio (the ratio of accepted reservations) is low since each real-time traffic source reserves exclusively a slot on the super-frame. On the other hand, choosing a too large super-frame length results in more established real-time connections, but does not meet the delay requirements of multimedia applications having stringent delay requirements.

Another factor which should be taken into account is the constraint of fixed slot length. There is a compromise between slots length and packets lengths of multimedia traffic sources. On one hand, considering short slots is suitable for real-time traffic sources which generate short periodic frames like voice sources. However, the use of short slots is not suitable for traffic sources which generate long frames like video MPEG. These traffic sources are required to fragment their data packets into several small packets, consequently there is an increase in the control traffic (due to encapsulation of packets). On the other hand, considering long time-slots is suitable for high data rate traffic sources, while it represents a waste of bandwidth when traffic sources with small-size packets are considered. A way to reduce the waste of bandwidth and the fragmentation overhead consists in considering small slots, and making sources with long frames like video reserve a set of contiguous slots. These contiguous slots will be considered as one long slot, and the video source transmits its video frames on it without fragmentation.

Performances of reservation MAC protocols are strongly affected by the super-frame length and the slot length. The impact of the super-frame structure should be carefully taken into account at network configuration step.

Figure 2. Reservation clash due to mobility

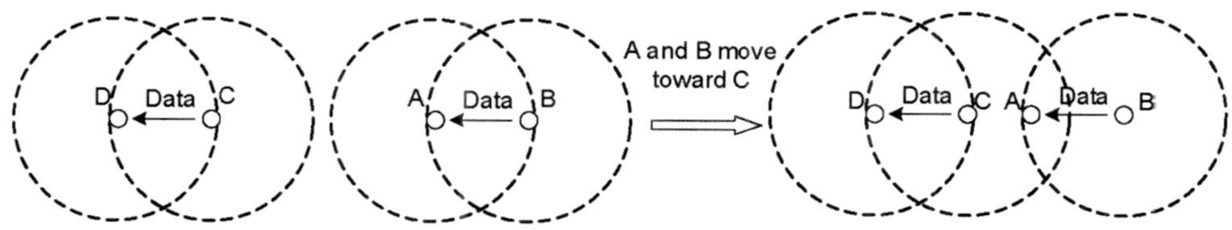

Mobility Handling and Reservation Break Detection

Unlike contention based protocols where mobility of nodes has not a strong impact on the MAC protocol performance, the mobility factor is a challenging issue in the design of reservation MAC protocols. When nodes are mobile, conflicts between reservations and collisions may occur during reserved slots. This phenomenon is called reservation clash and must be handled at the MAC layer. The reservation clash phenomenon due to mobility is illustrated in the following scenario.

In Figure 2, nodes B and C establish reservation with A and D respectively on the same slot s. As long as A and C are far away from each other, no collision occur in reserved slots. If nodes C and D move toward A, reservation clash will occur at A. Both of B and C transmit on the reserved slot s and collision occur during slot s. Reservation clash has drastic consequences on the QoS, especially in highly mobile nodes. Reserving nodes affected by reservation clash will suffer excessive packets collisions and dropping.

Almost all proposed reservation MAC protocols are suitable only for static ad-hoc networks. They consider that nodes are static and no mobility considerations are taken into account. Reservation MAC protocols must provide efficient mechanisms to face mobility of nodes, and reduce the degradation of performance in dynamic ad-hoc networks. Particularly, reservation MAC protocols should provide reservation clash detection mechanisms. In addition, efficient reservations recovery mechanisms must be defined in order to permit to nodes that lost their reservations due to mobility to release their reservations and establish new reservations rapidly. In (boudour & al, 2008), authors proposed a solution to handle this reservation clash phenomenon through defining a basic reservation loss detection and reservations recovery schemes. When the receiver of a reservation detects collision on reserved slots, it informs the sender about the reservation loss. The sender cancels its reservation and stops sending data packets during reserved slots. Afterward, the sender restarts the reservation negotiation process with the receiver in order to reserve new slots. However, the efficiency of this solution is conditioned by the velocity of nodes. At high mobility, the topology changes very frequently, and reservations disruptions occur more frequently, and the dropping rate increases.

Control Traffic Overhead

One important parameter in the performance of reservation MAC protocols is the control traffic overhead. The control traffic overhead determines the amount of control packets transmitted by mobile nodes in order to maintain coherent reservations. The transmission of control traffic results in an increase of energy consumption. In addition, it decreases the effective bandwidth offered to real-time traffic sources to transmit their data packets.

CATA allocates four control mini-slots (CMS1, CMS2, CMS3, and CMS4) on each slot. After reservation is successfully established, CMS1 is used by the receiver to provide a "busy tone" to senders attempting to reserve the slot for transmission. CMS2 is used by the sender to jam any possible

RTS addressed to its neighbors. CMS3 and CMS4 are used only at the reservation setup. Once the reservation is established, these two slots are not used. However, the use of four control mini-slots in each slot incurs a significant overhead.

In FPRP, each RS is composed of M reservation cycle (RC), and a five control mini-slots are associated with each RC to establish reservation. If a node successfully reserves a slot during one of the RCs, the remaining RCs are not used any more for contention. Hence, depending on the number of RC associated with each reservation slot (RS), the control traffic overhead of FPRP may be high, and the waste of bandwidth may be significant.

Like CATA, SRMA/PA allocates four control mini-slots (SR, RR, RC, and ACK) in each slot. The SR is used by the sender to indicate the reservation to its neighbors once the reservation is established. Hence, only the SR slot is used to indicate the slot reservation in subsequent frames, the other control slots (RR, RC) are used only during the reservation handshake. However, allocating three control mini-slots in each slot to coordinate reservations results in a significant overhead.

The major advantage of R-CSMA and RTMAC against FPRP, CATA and SRMA/PA is that they don't reserve any bandwidth for control packets. Control packets are transmitted only at the reservation request step. Instead of allocating control mini-slots to prevent neighbor nodes from reserving already reserved slots, R-CSMA and RTMAC use reservation tables that include for each slot its state "reserved" or "available".

As bandwidth is limited in MANETs, the effective bandwidth offered to real-time traffic sources must be increased, and the wasted bandwidth must be reduced as much as possible. Protocols like CATA, FPRP, and SRMA/PA are characterized by high control traffic overhead, and a significant amount of bandwidth is wasted because of control traffic. An efficient MAC reservation protocol should permit to maintain coherence of reservations with less control traffic overhead.

Efficiency of Reservation Release Scheme

The reservation release scheme is a key component in reservation MAC protocols. Reservation release is required when the source of real-time traffic has finished its data transmission. At the end of a real-time session, the sender should inform the receiver and its neighbors about the end of transmission. The receiver also is required to inform its neighbors that the slots are no more reserved for reception.

The role of the reservation release scheme is to permit neighbors of the sender and receiver to reserve the slots that have been released. However, the efficiency of the reservation release scheme impacts performance of the reservation protocol. A flaw in the reservation release scheme may result in a saturation of the network where slots can not be reserved while they where released.

The reservation release schemes proposed by reservation protocols are inefficient because there exist situations in which some nodes (receiver, sender neighbors, or receiver neighbors) are not informed about the reservation release. Authors of R-CSMA consider that reserved slot is released automatically when it is left empty. However, this scheme is not efficient because reserving nodes may not use all their reserved slots periodically to send data packets, especially when the inter-packets arrival time is greater than the super-frame length. If a reserving node has no data packet to transmit in the current super-frame while it has not finished the transmission, the node loses its reservation, and the slot is available for reservation by other nodes. The node is required to re-establish reservation each time the reserved slot is not used for transmission. Another issue with the reservation release of R-CSMA is that only the sender is able to signal reservation release to its neighbors by leaving the reserved slot empty. The receiver has no way to indicate the reservation release to its neighbors. The slot will remain reserved from the viewpoint of the receiver's neighbors.

Authors of RTMAC use explicit reservation release packets to inform neighbors about the reservation release. At the end of transmission, the sender informs the receiver and its neighbors by sending a ResvRelease packet. When the receiver receives the ResvRelease, it informs its neighbors by sending a ResvRelease packet. However, since the ResvRelease packet is transmitted using contention, it may collide with other transmitted packets. Consequently, there may be situations in which either the receiver or neighbors of the sender/receiver don't receive the ResvRelease packet. In FPRP, CATA, and SRMA/PA no reservation release scheme is defined.

An efficient reservation MAC protocol should ensure that at the end of real-time session both the receiver and all nodes around the sender and the receiver receive correctly the reservation release. In addition the MAC protocol should ensure reuse of slots once these slots are released.

Fairness

Fairness is another parameter in the performances of MAC protocols. Proposed reservation MAC protocols lack the definition of mechanisms to ensure fairness between traffic flows, and between different service classes. In FPRP, CATA, SRMA/PA, and RTMAC there is no limit on the maximum bandwidth that can be reserved by a real-time traffic source. In addition, there is no limit on the amount of bandwidth that can be reserved to the real-time traffic class. Real-time traffic sources are allowed to reserve time slots as long as there are free slots in the super-frame. However, this scheme is not efficient since real-time traffic sources can monopolize all the available bandwidth leading to starvation of other classes of traffic like best effort traffic sources. Unlike, FPRP, CATA, and SRMA/PA, R-CSMA allocates a fraction of the super-frame for the transmission of best effort packets. Best effort traffic sources have always the chance to transmit their data packets during the contention period regardless the offered traffic load since no traffic class is authorized to monopolize the contention period.

Much attention should be paid on the amount of bandwidth that can be allocated to real-time applications. Reservation MAC protocols should define a limit on this bandwidth, and the available bandwidth should be well partitioned between the different classes of service to avoid starvation of low priority traffic classes.

This can be achieved through an admission control scheme. Each Class of traffic C is characterized by an acceptable required bandwidth *acceptableBW(C)*. The bandwidth allocated to a connection of class C at any time must be at least equal to *acceptableBW(C)* for acceptable reception quality. In addition to the status (reserved or available) of each slot, each node records the acceptable bandwidth and the actual reservations made by each one of its neighboring connections. A new reservation request for a connection of class C is accepted if there are available slots to satisfy the acceptable bandwidth of class C. If the set of available slots is not sufficient to satisfy the acceptable bandwidth of the connection, the node checks whether there is some neighbors which have exceeded their acceptable required bandwidth. If it is the case, the node grabs the slots which are reserved by neighboring sources which have exceeded their acceptable required bandwidth. The bandwidth allocated to these neighboring sources will decrease. Through such mechanism each admitted connection receives at least its acceptable bandwidth, and bandwidth allocated to each connection is adapted to the traffic load in the neighborhood.

SIMULATIONS

We compare the performance of previously presented reservation protocols (R-CSMA, CATA, FPRP, and IEEE 802.11e). Particularly, we are interested in analyzing how these protocols provide QoS guarantees to voice and video traffic flows. Particularly, we analyze their control traffic overhead, their efficiency in regard to mobility and their control overhead. The comparative analysis is performed through a set of tests using the network simulator NS-2. We use the IEEE 802.11e simulation model of Wiethölter and Hoene available at (Wiethölter & Hoene, 2003; Wiethölter & al, 2004).

Simulation Model

We consider an ad-hoc network composed of 100 nodes randomly distributed on a 1 km^2 area. The wireless channel is 11Mb/s. We assume that the wireless channel is noise and distortion free. Nodes are considered equipped with omni-directional antenna with a 250 meters transmission range.

In our simulations we consider two voice traffic models (G.711 and G.723 models), and two video models (MPEG-4 and H263 video models). Table 1 summarizes the TSpec parameters for the various classes of considered traffic for simulation. The TSpec parameters of G.711 and G.723 are taken from (Sharafeddine & al, 2003; Stewart, 2005). For MPEG-4 and H263 TSs, the TSpec parameters are extracted from video traces available at (Seeling & al, 2007). Each station can generate a G.711 audio, a G.723 audio, an MPEG-4, or an H263 video flow.

The maximum payload of a TDMA slot is set to 160 bytes in our simulations. Each slot consists of the transmission time of a real-time packet (including different layer overheads), and the round trip propagation time. With 11Mbps channel bit-rate, the slot length is 0.18ms. Simulation parameters are shown in Table 2.

Because video frames are larger than the payload of a TDMA slot, video frames are fragmented into several packets. After fragmentation, MPEG source generates one packet every 10ms, H263 generates a packet every 5ms, G.711 generates one packet every 20ms, and G.723 generates one packet every 30ms. In order to satisfy bandwidth requirements of the considered traffic classes, we set the super-frame length to the smallest inter-packets arrival time (i.e. the inter-packets arrival time of H263 source) which is 5ms.

Table 2. Simulation parameters

Parameters	Value
Channel bit rate (Mbps)	11
Slot payload size (bytes)	160
UDP/IP header (bytes)	8+20
MAC header (bytes)	38
PHY layer overhead (PLCP header+preamble) (bits)	8+48
Slot length (ms)	0.18
Guard time between slots (μs)	20
Super-frame length (ms)	5
Number of slots per super-frame	25
Simulation time (s)	1000

Simulation Results

Analysis of the Impact of the Super-Frame Length and Traffic Load

We analyze the impact of the traffic load on the performances of the considered protocols in a static ad-hoc network. In this analysis we increase the traffic load by increasing the number of BE and RT sessions (MPEG, H263, G711, and G723) in equal numbers. The maximum number of sessions is 100, and sessions are uniformly distributed among the 100 nodes.

Figure 3 shows the reservation acceptance ratio of CATA, FPRP, and R-CSMA versus the increase of traffic load. The figure shows that the reservation acceptance ratio remains above 90% as long as the number of sessions is less than 40 sessions. When the number of sessions exceeds 40 sessions, the reservations acceptance ratio decreases linearly because the number of sessions become much higher than the number of slots per super-frame. Some sessions will be rejected because of the unavailability of resources. R-CSMA

Figure 3. Reservations acceptance ratio with the increase of traffic load

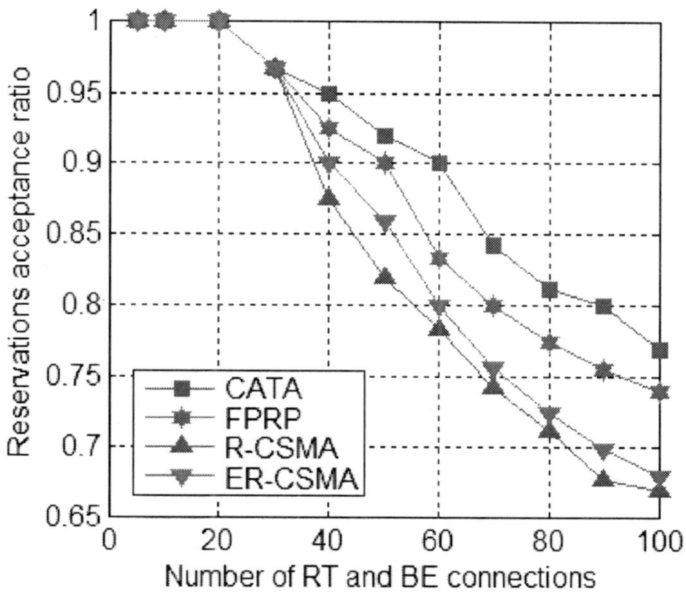

has a lower reservations acceptance ratio than the other protocols at high traffic load because of the portion of the super-frame allocated to the contention period. We don't give the reservation acceptance ratio of the IEEE 802.11e because this protocol doesn't make explicit reservations.

Figure 4 shows the throughput achieved by FPRP, CATA, R-CSMA, and the IEEE 802.11e versus the increase of traffic load. The figure shows that at low traffic load the considered protocols have approximately the same throughput. At high traffic load, reservation protocols achieve higher throughput than the IEEE 802.11e.

Figure 4. Throughput of RT traffic with the increase of traffic load

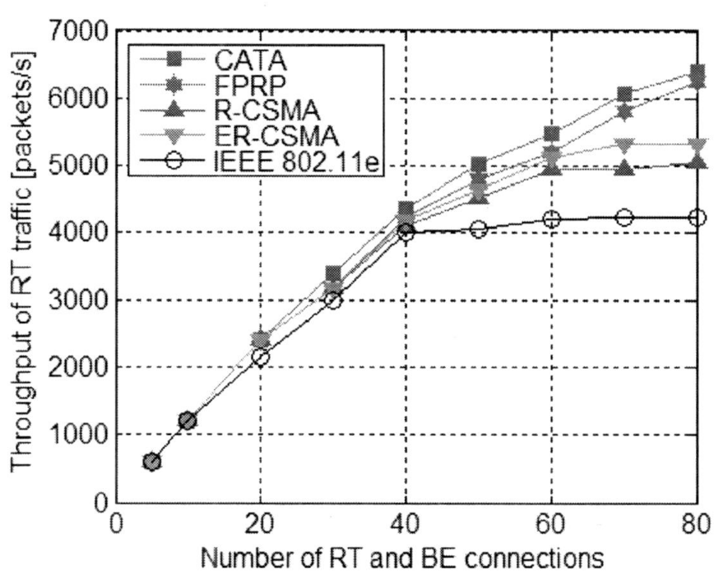

Figure 5. RT packet delivery ratio with the increase of traffic load

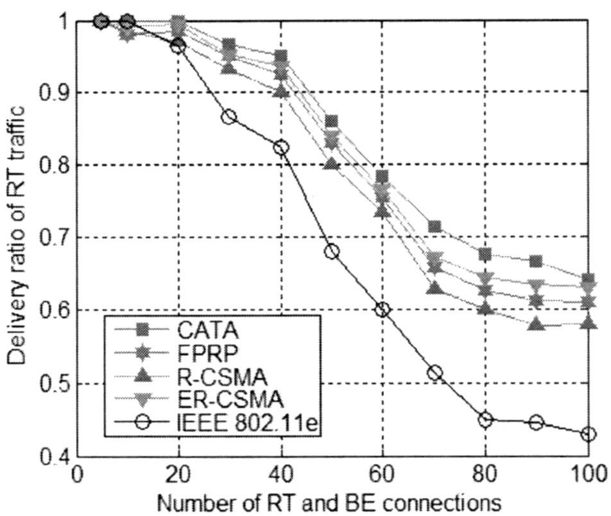

Figure 5 shows the packets delivery ratio (i.e. the percentage of packets received by their destinations) of RT packets offered by FPRP, CATA, R-CSMA, and IEEE 802.11e versus the increase of traffic load. The figure shows that reservation protocols offer better packets delivery ratio than the IEEE 802.11e at high traffic load. The low throughput and packets delivery ratio of the IEEE 802.11e is due to the increase of contention and collision rate at high traffic load. The high delivery ratio of FPRP, CATA, and R-CSMA results from that packets in these protocols are transmitted periodically on reserved slots in collision-free way. Consequently, the probability of collision and packet dropping is very low.

Figure 6 shows the average RT packets delay with FPRP, CATA, R-CSMA, and IEEE 802.11e versus the increase of traffic load. The figure

Figure 6. Average delay of RT traffic with the increase of traffic load

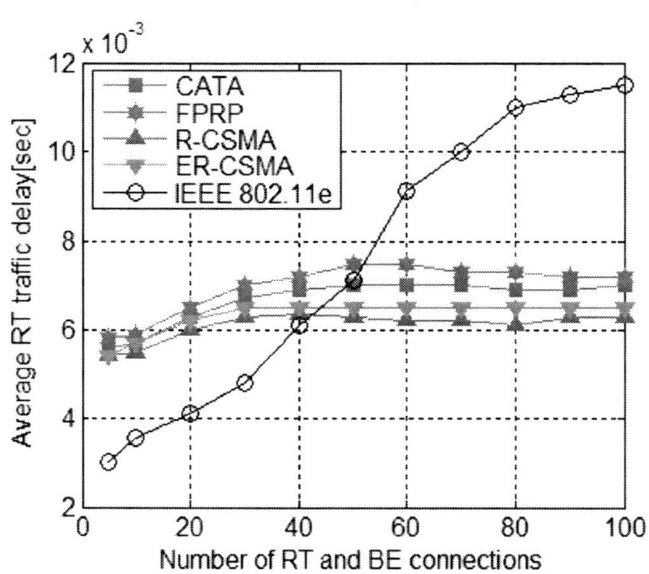

Figure 7. Packet dropping rate with the increase of number of mobile nodes

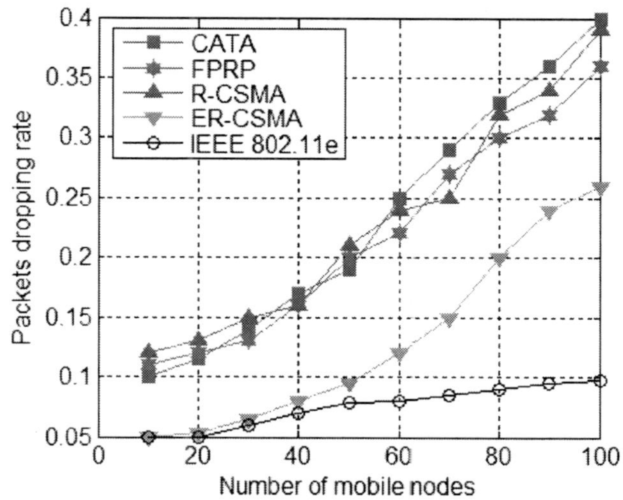

shows that reservation protocols give deterministic delay regardless the traffic load, while IEEE802.11e diverges with the increase of traffic load. At low traffic load, the IEEE802.11e outperforms the other protocols because low level of contention results in a small number of collisions and short backoffs. At high input load, IEEE 802.11e nodes experience more contention, and thus more collisions and wider backoff windows, and consequently the access delay increases drastically. Reservation protocols provide quasi-constant delay because real-time packets are transmitted at regular intervals once the reservations established.

Analysis of Mobility Impact

For mobility of nodes, we use the RWP (Random Walk Point) model. Each mobile node chooses randomly its next position and moves toward that position with a velocity uniformly distributed between *Vmin* and *Vmax*. In our simulations, we choose *Vmin=1 m/s and Vmax=10 m/s*. The node stays in its new position for a time *dt* (set to 30 seconds in our simulations) after witch it chooses another position.

Figure 7 shows that the packet dropping rate with CATA, FPRP and R-CSMA increases drastically with the increase of the number of mobile nodes. The packet dropping rate of IEEE 802.11e remains very low compared to other protocols. The drastic packets dropping ratio of FPRP, CATA, and R-CSMA is due to the reservation clash, and the high number of collisions. As mobility increases, reservations clashes increase and nodes start losing their reservations. Since no reservation recovery mechanism is defined, reserving nodes have no way to establish new reservations, and reserving nodes continue sending their data packets on their reserved lost slots. The IEEE 802.11e is less affected by mobility of nodes because nodes are required to compete and acquire the channel for the transmission of each data packet no matter of their positions.

Figure 8 shows the throughput with the increase of mobility. The throughput with R-CSMA, CATA and FPRP decreases drastically with the increase of mobile nodes. Like the packets dropping rate, the reduced throughput of these protocols is linked to the increase of the number of collision slots.

This section has shown that the IEEE 802.11e is more efficient than reservation protocols in the case of high mobility of nodes.

Figure 8. Throughput with the increase of number of mobile nodes

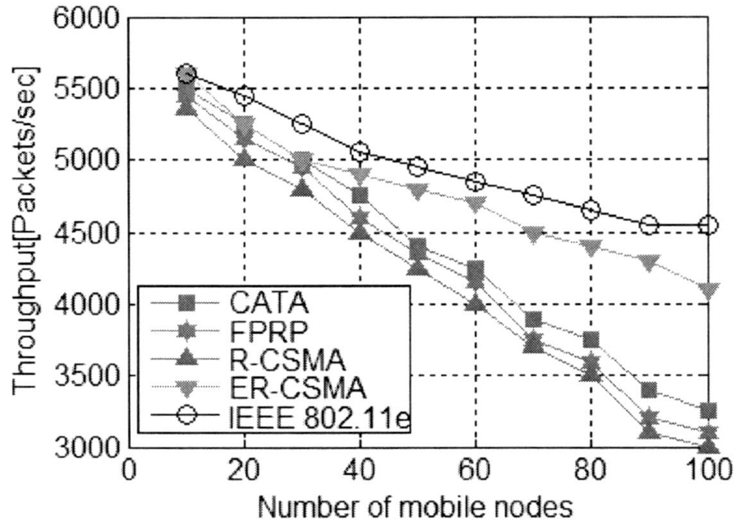

Bandwidth Wasting Analysis

We analyze the waste of bandwidth incurred by reservation schemes, especially when heterogeneous classes with different QoS requirements are considered. Figure 9 shows the ratio of unused reserved slots with the increase of the number RT sessions. With FPRP, CATA, and R-CSMA the ratio of unused slots increases linearly with the increase of the number of RT sessions. This waste of bandwidth is due to the low rate of voice sources. G711 and G723 sources consume only 3/12 (2/12 respectively) of their reserved slots. The IEEE 802.11e doesn't suffer waste of bandwidth

Figure 9. Ratio of unused slots with the increase of number of RT sessions

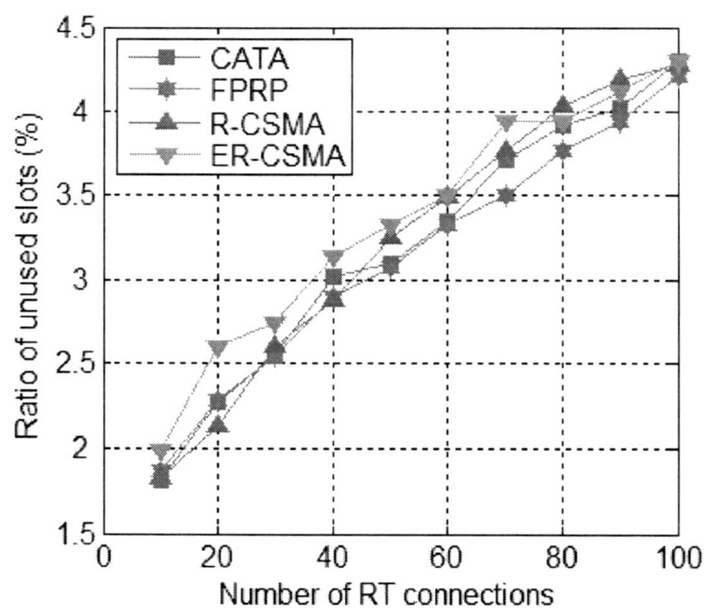

because bandwidth is shared between all nodes. The bandwidth unused by some node is available for utilization by other nodes.

This section points out the need to define a more efficient and flexible reservation MAC protocol. The MAC protocol should distribute the available bandwidth to reserving nodes based on their QoS requirements so that bandwidth wasting is reduced. Low data rate sources (like G.711 and G.723 voice) should be allocated less bandwidth than the high data rate sources such as video.

Control Traffic Overhead

In this section we analyze the control traffic of CATA, FPRP, and R-CSMA. Figure 10 shows the control traffic generated by nodes as a function of the number of RT sessions. On one hand, we observe that the amount of control traffic generated by R-CSMA remains very low. This is because R-CSMA does not use any control slots to coordinate reservations. Control packets are transmitted only at the reservations setup. On the other hand, we observe that CATA and FPRP and IEEE 802.11e generate high quantity of control traffic. CATA requires each reserving node to transmit RS and RTS packets on each reserved slot. FPRP requires the repetition of the five-phase reservation steps on each Reservation Frame. The IEEE 802.11e requires the transmission of the RTS and CTS packets before the transmission of each data packet.

This section reveals that CATA, FPRP, and IEEE 802.11e suffer from significant control traffic overhead when the number of traffic streams in the network is high. R-CSMA has the advantage that it uses less control traffic.

CONCLUSION

In this paper we have analyzed the most advantages of reservation protocols against their counter-part contention-based protocols, especially the IEEE 802.11e standard. Also, we prcvide a detailed analysis of the main drawbacks, and challenging issues unresolved by these protocols. In addition, we have analyzed the situations in which reserva-

Figure 10. Control traffic overhead with the increase of number of RT sessions

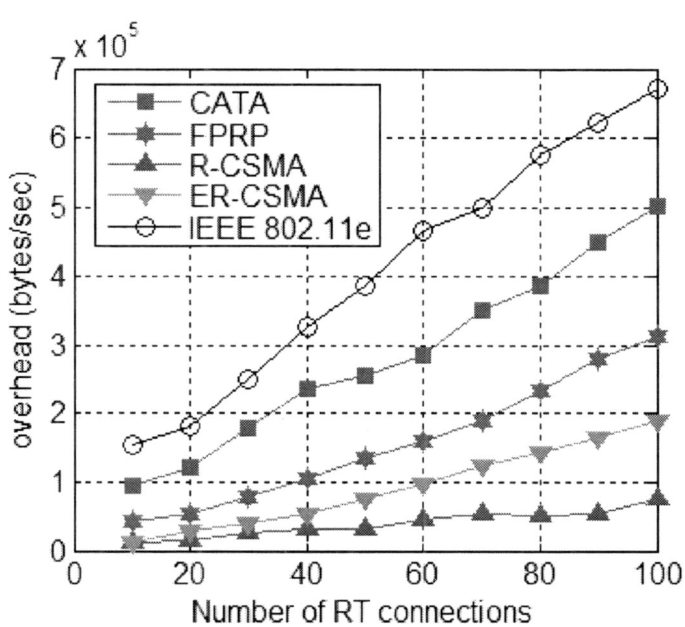

tion protocols are more efficient than contention based protocols.

First, we have found that reservation MAC protocols perform well in static ad hoc networks. Simulation results show that these protocols outperform the IEEE 802.11e standard in low mobility scenarios. However, the performances of these protocols are expected to degrade as mobility of nodes increases. All reservations which are being built since the initialization of the network overlap one with each other and collisions during reserved slots appear. In these situations the IEEE 802.11e is more efficient since no permanent transmission scheduling is established.

Second, we found that some protocols like FPRP, CATA, and IEEE 802.11e suffer from significant control traffic overhead. R-CSMA have the advantage that it generates less control traffic overhead since control packets are transmitted only at the reservations establishment step.

Finally, we conclude that reservation MAC protocols can be a promising solution to provide QoS in ad-hoc networks provided that degradation of performance due to the mobility of nodes is reduced. However, the other issues related to the waste of bandwidth, fairness, and the control traffic overhead must be also resolved. The waste of bandwidth can be reduced by allowing neighbors of reserving node to use slots when these slots are not used for transmission. Fairness can be ensured through defining a limit on the amount of bandwidth that can be allocated to each class of traffic.

REFERENCES

Ahn, C. W., Kang, C. G., & Cho, Y. Z. (2003). Soft Reservation Multiple Access with Priority Assignment (SRMA/PA): A Distributed MAC Protocol QoS-guaranteed Integrated Services in Mobile Ad-hoc Networks. *IEICE Transactions on Communications. E (Norwalk, Conn.), 86-B*(1), 50–59.

Boudour, G., Teyssié, C., & Mammeri, Z. (2008). Reservation Clash Handling to Optimize Bandwidth Utilization in MANETs. *In proceedings of the International Conference on Communication Theory, Reliability, and Quality of Service (CTRQ'2008),* Bucharest, Romania, (pp. 77-82).

Hsieh, H. Y., & Sivakumar, R. (2002). IEEE 802.11 over Multi-hop Wireless Networks: Problems and New Perspectives. *In proceedings of the IEEE Vehicular technology Conference 2002 Fall: Vol(3).* (748-752).

IEEE 802.11WG. (2005). Drat supplement to standard for telecommunications and information exchange between systems-LAN/MAN specific requirements-part11: MAC enhancements for Quality of service (QoS). *IEEE 802.11e standard Draft/D13.0,* January 2005.

Inwhee, J. (2004). Qos-Aware MAC With Reservation For Mobile Ad-Hoc Networks. *In proceedings of the IEEE Vehicular Technology Conference 2004 Fall: Vol(2).* (1108-1112).

Lin, C. R., & Gerla, M. (1997). Asynchronous Multimedia Multi-hop Wireless Networks. *In proceedings of the IEEE Conference on Computer Communications (INFOCOM'97),* Kobe, Japan: *Vol(1).* (118-125).

Manoj, B. S., & Siva, R. M. C. (2002). *Real*-time traffic support for ad hoc wireless networks. *In proceedings of the* IEEE *International Conference on Networks (ICON 2002),* Singapore. (pp. 335-340).

Romdhami, L., Ni, Q., & Turletti, T. (2003). Adaptive EDCF: Enhanced Service differentiation for IEEE 802.11 Wireless Ad-Hoc Networks. *In proceedings of the IEEE Wireless Communications and Networking Conference (WCNC 2003),* New Orleans, Louisiana, USA: *Vol (2).* (1373-1378).

Seeling, P., Fitzek, F. H. P., & Reisslein, M. (2007). *Video Traces for network Performance Evaluation, A Comprehensive Overview and Guide on Video Traces and Their Utilization in Networking Research*. Netherlands: Springer.

Sharafeddine, S., Riedl, A., Glasmann, J., & Totzke, J. (2003). On Traffic Characteristics and Bandwidth Requirements of Voice over IP Applications. *In proceedings of the IEEE International Symposium on Computers and Communication (ISCC '03),* Antalya, Turkey: *Vol (2)*. (1324-1330).

Shengming, J., Jianqiang, R., Dajiang, H., Xinhua, L., & Chi, C. K. (2002). A Simple Distributed PRMA for MANETs. *IEEE Transactions on Vehicular Technology*, *51*(2), 293–305. doi:10.1109/25.994807

IEEE Std. 802.11. (1999), Part11: Wireless LAN Medium Access Control (MAC) and Physical Layer (PHY) Specifications, Reference number ISO/IEC 8802-11:1999(E), *IEEE Std 802.11,* 1999 edition.

Stewart, B. (2005). *How Many Voice Callers Fit on the Head of an Access Point?* from http://www.oreillynet.com/etel/blog/2005/12/how_many_voice_callers_fit_on.html

Tang, Z., & Garcia-Luna-Aceves, J. J. (1999). A Protocol for Topology-Dependent Transmission Scheduling in Wireless Networks. *In proceedings of the IEEE Wireless Communications and Networking Conference (WCNC 1999)*: *Vol(3)*. (1333-1337).

Wiethölter, S., & Hoene, C. (2003). *Design and Verification of an IEEE 802.11e EDCF Simulation Model in ns-2.26.* (Tech. Rep. No. TKN-03-19). Berlin, Germany: Technische Universität Berlin.

Wiethölter, S., Hoene, C., & Wolisz, A. (2004). *Perceptual Quality of Internet Telephony over IEEE 802.11e Supporting Enhanced DCF and Contention Free Bursting*. (Tech. Rep. No. TKN–04-11). Berlin, Germany: Technische Universität Berlin.

Xu, S., & Saadawi, T. (2001). Does the IEEE 802.11 MAC Protocol Work Well in Multihop Wireless Ad Hoc Networks? *IEEE Communications Magazine*, *39*(6), 130–137. doi:10.1109/35.925681

Zhu, C., Corson, & M. S. *A* five-phase reservation protocol (FPRP) for mobile ad hoc networks. *In proceedings of the IEEE INFOCOM'98*, San Francisco: *Vol(1)*. 322-331.

Zhu, H., Cao, G., Yener, A., & Mathias, A. D. EDCF-DM: A Novel Enhanced Distributed coordination Function for Wireless Ad Hoc Networks. *In proceedings of the IEEE International Conference on Communications (ICC2004),* Paris: *Vol(7)*. 3886- 3890.

This work was previously published in International Journal of Business Data Communications and Networking, Volume 5, Issue 2, edited by Varadharajan Sridhar and Debashis Saha, pp. 52-67, copyright 2009 by IGI Publishing (an imprint of IGI Global).

Chapter 8
Slot Allocation Algorithms for Minimizing Delay in Alarm-Driven WSNs Applications

Mário Macedo
INESC-ID, Portugal

António Grilo
INESC-ID, Portugal

Mário Nunes
INESC-ID, Portugal

ABSTRACT

Energy-efficiency and latency requirements in alarm-driven Wireless Sensor Networks often demand the use of TDMA protocols with special features such as cascading of timeslots, in a way that the sensor-to-sink delay bound can stay below the duration of a single frame. However, this single TDMA frame should be as small as possible. The results presented in this paper, point to the conclusion that a largest-distances-first strategy can achieve the smallest single frame sizes, and also the lowest frame size variations. A quite simple distributed version of this algorithm is presented, which obtains the same results of its centralized version. Simulations also show that this discipline presents the best results in terms of sensor-to-sink slot distance, even if they require a few more slots than breadth-first in multi-frame scenarios.

INTRODUCTION

Wireless sensor Networks (WSNs) are geographically distributed, self-organized and robustly networked micro-sensing systems that can be readily deployed and operated in environments in which more conventional infrastructure-based systems and networks are impractical, or cost-ineffective. WSN nodes are interconnected by means of a wireless communications technology, collaborating to forward the sensorial data hop-by-hop from the source node to the sink nodes.

In this paper, we mainly address critical alarm-driven WSN applications, such as surveillance

DOI: 10.4018/978-1-60960-589-6.ch008

of sensitive areas (e.g., intrusion detection and tracking). In this kind of WSN applications, traffic generation can be characterized as very sporadic, but the generation of an alarm report demands an immediate response to the event, which makes this kind of traffic very delay-sensitive. However, as WSNs devices have limited energy resources, low duty-cycles are also required. These two goals are usually contradictory, but TDMA protocols can provide low latency in the convergecast of data from the nodes to the sink, while still providing low duty-cycles. The quick convergecast is usually achieved by building a routing tree routed on the sink node, and by ordering the timeslots in the path from a node to the sink, in such a way that the receiving slot(s) number(s) of a given node is lower than its transmitting slot number, while the slot distance is kept as low as possible (a procedure that is called "cascading of timeslots"). On the other hand, low duty-cycle can be achieved by TDMA protocols, since each node only needs to be active during its reception and transmission slots, while staying asleep for the rest of the TDMA frame.

With the objective of guarantying the same sensor-to-sink packet delay bound for all the nodes in the network, communication of alarms to the sink can be made in just one frame. However the size of the single frame is desirably the lowest possible. In this paper, several TDMA scheduling algorithms are simulated with the objectives of achieving low single frame sizes and low latencies.

This paper presents the related work in Section 2. In Section 3, the Cascading Minimum Single Frame Size Problem is defined. Section 4 presents the simulation model, the set of slot allocation algorithms, and their results in terms of achieving low single frame sizes, and low node-to-sink slot distances. Section 5 presents the simulations results obtained for a pre-determined frame size, in terms of node-to-sink slot distances. And finally, Section 6 presents simulation results, in terms of the actually required non-single frame sizes, and the respective worst-case delays of the communication of alarms to the sink.

RELATED WORK

While not being a pure TDMA protocol, the Data-gathering MAC (D-MAC) protocol, presented in Lu et al. (2004), uses staggered synchronization so that a data packet received by a node at one level of the tree, is transmitted to the next level in the following time period (i.e., cascading the transmissions in the overall transmission period). The node is then allowed to sleep until the reception period for its level occurs again. D-MAC is still a CSMA/CA based protocol as nodes at the same level of the tree have to compete for timeslot access and may also interfere with nodes located in the same area. Support of several sinks in D-MAC is troublesome.

The use of TDMA for fast broadcast (the converse problem of convergecast) is a well-known subject, which has been studied in the context of multi-hop radio networks. Chlamtac & Kutten (1987) show that the problem of determining optimal channel allocation for fast broadcasting is NP-hard. Two algorithms for tree construction and slot assignment are presented, namely a centralized version, and its distributed version. The distributed algorithm begins at the source node, for which the first slot is granted, and builds a spanning tree, such that each node has a slot number higher than its parent slot, but with the smallest possible value, in order to cascade the broadcast. Tree construction and slot assignment are performed depth-first, by means of passing a token to one node at a time, and by exchanging appropriate protocol messages with the neighbor nodes, in order to obtain a TDMA schedule that meets some slot allocation rules, and that achieves conflict-free schedules. These protocols are also designed to achieve spatial reuse of the slots, with relatively small TDMA frame sizes.

Another protocol that was designed to achieve TDMA conflict-free schedules is the DRAND distributed slot assignment protocol, presented in Rhee et al. (2006). As the authors state, the problem of obtaining a minimum slot frame size is NP-hard. DRAND is not particularly suited for fast broadcast or fast convergecast, as slot assignment is random. DRAND assures that nodes in a 2-hop neighborhood do not use the same slot, and it can operate with limited frame sizes. DRAND is also proved to have a message exchanging complexity of $O(\delta)$, where δ is the neighborhood of each node.

(Annamalai et al., 2003; Upadhyayula et al., 2003), present two centralized algorithms, CTC-CAA and CCA, which were specially designed to achieve low latencies in the convergecast process, namely by the use of cascading. CTC-CAA proceeds with the tree construction and slot allocation processes in a breadth-first top-down manner, while CCA proceeds in a bottom-up manner from the leaves of the tree to the sink node. The two algorithms differ in the way they establish the neighborhood of each node for the purpose of avoiding conflict schedules. However, they present the drawback of being centralized and thus not adaptive to the irregular propagation characteristics of the environment.

Kulkarni & Arumugam (2005) present SS-TDMA, which is a TDMA protocol designed for convergecast/broadcast applications. Its basic assumption is that the interference range is different from the communication range, and that the quotient y between them gives an estimation of the number of nodes within the interference range that can't have the same slot number. In the slot assignment process, each node receives messages from the neighbors with their assigned slots. The receiving node knows the direction of an incoming message, and adds fixed values to the neighbor's slot number, in order to determine its own slot number. Those values depend on the direction of the message, the y value, and the type of the grid, namely square and hexagonal. Although being a distributed algorithm, it needs a location service and topological knowledge about the networks, which limits its practical applicability. SS-TDMA follows a 2-hop neighborhood interference avoidance criterion, but it allocates different slots for the purposes of broadcast, and convergecast.

The problems of building routing trees, and minimizing convergecast latency in ZigBee networks, were studied by Pan & Tseng (2008). The authors prove that the problem of obtaining a conflict-free slot assignment that minimizes the convergecast latency is also NP-hard. The distributed version algorithm is essentially a breadth-first tree construction and slot allocation protocol that is based on HELLO messages transmitted by the relay nodes. The main contributions of this protocol are the slot reassignment rules: the nodes that have more interfering neighbors, that stay closer to the sink, or that have a lower ID (identification number), have priority to choose a given slot that minimizes the latency. The interference avoidance procedure of this protocol is also based on a 2-hop neighborhood criterion.

Bryan et al. (2007) present a centralized algorithm and two distributed algorithms (namely, the DSA-AGGR – Distributed CCH for Data Aggregation). All the three algorithms are designed to achieve low latency by means of cascading timeslots. DSA-AGGR begins to allocate slots from the sink to the leaves of the network tree, but each node is only eligible to allocate a slot if the following expression results in a value higher than 0.25 for the *color_score*:

$$color\_score = \frac{2.ColoredOneHop + ColoredTwoHop}{2.NumberOneHops + NumberTwoHops} \quad (1)$$

This means that this heuristic gives priority to the nodes that have a sufficient number of neighbors that are already colored. This heuristic is claimed to obtain low frame sizes. Interference avoidance is also based on a 2-hop criterion. However, convergence of this algorithm is not always guaranteed, as some *color_score* thresholds can't be attained in some topologies.

Gandham et al. (2008) present a TDMA scheme that allows the implementation of convergecast and the cascading of the timeslots, which is based on the reduction of the networks, namely tree networks, to multi-line networks. In a given linear branch each node knows its hop distance to the sink, and can schedule itself to be in one of three states: T for transmit, R for receive, and I for idle. These states are rotated in time and arranged in such a way that the packets flow in a propagation wave towards the sink. For more complex topologies, the networks can be transformed into multi-line networks, but the nodes need to acquire knowledge about the global topology in order to know their turns to transmit upstream. This scheme has a major drawback, as the complexity of the basic scheme increases significantly when it is progressively extended to deal with more realistic scenarios.

Lu & Krishnamachari (2007) present a set of joint routing and slot assignment algorithms that aim to achieve low latencies. However, the procedures have the drawbacks of being computational intensive, while the tree building process is centralized.

Finally, in Mao et al. (2007) centralized algorithms are presented, which can be used to optimize the energy or the latency of the data collection process. These algorithms are hybrid, being based on genetic algorithms and particle swarm optimization. The centralized nature of these algorithms also limits their potential use.

THE CASCADING MINIMUM SINGLE FRAME SIZE PROBLEM

Some of the works mentioned above are concerned with achieving short frame sizes, and therefore high throughput. As critical alarm-driven WSN applications should only report sporadic events, they do not need to periodically transfer bulk data to the sink. Therefore, achieving high throughput is not a specific design requirement and consequently the problem addressed in this paper is not to obtain the smallest possible frame sizes (i.e. maximum slot reutilization). The latter depends basically on the maximum degree (i.e., maximum number of neighbors) of the networks. Moreover, small frames sizes also lead to high duty-cycles, and also imply that the transmissions from nodes that are placed away from the sink will potentially have to span several frames. This can originate different delay bounds for nodes located at different levels of the network tree. In fact, the aim is to have similar delay bounds for the alarms transmitted by all the network nodes. This requirement can be accomplished by always transmitting the data in a single TDMA frame, whatever the location of the node. However, since different scheduling algorithms may lead to different single TDMA frame sizes, the objective is to find algorithms that lead to the smallest possible single TDMA frame size in such a way that it is able to accommodate transmissions from all the network nodes, spanning from the deepest leaf nodes to the sink node. We call this problem the Cascading Minimum Single Frame Size (CMSFS) problem.

Differently from some protocols and algorithms presented here and generally in the literature, in this paper, it is assumed that the slot assignment procedure takes place during the network setup phase, but only after the routing tree construction. The latter is assumed to make use of an energy-efficient contention MAC protocol like B-MAC, as presented by Polastre et al. (2004).

SCHEDULING ALGORITHMS AND RESULTS FOR THE MINIMUM SINGLE FRAME SIZE PROBLEM

Centralized slot allocation algorithms are potentially more optimal and more predictable in terms of convergence, but they require that the nodes communicate their local topology (e.g., their neighborhoods, parents, etc.) to the processing node (usually the sink node), which is a slow and communication intensive procedure. The distributed slot allocation algorithms can be particularly interesting because they do not require the sink node to know the network topology. Therefore, they are more scalable, flexible and adaptive, even if their convergence is less predictable and slower.

In this paper, for the matter of comparison, we did not consider the centralized strategies referred in the literature, or those that build the tree simultaneously with the allocation process.

The following slot allocation strategies were firstly considered: depth-first (DF) and breadth-first (BF) (see, Cormen et al., 2000), RANDOM, DSA-AGGR, and SS-TDMA. Note that DF and BF can be easily distributed (see, Chalamtac & Kutten, 1987, for the case of distributed DF). In this paper, DF and BF are implemented in such a way that their distributed behavior is emulated with maximum possible fidelity, albeit without implementing the particular protocol details. The same design option was also adopted for all the remaining allocation algorithms. All of the implemented strategies perform greedy cascading slot allocation by each node.

The Degree Heuristic, presented in West (2001), which allocates the nodes ordered by the size of their neighborhoods, the similar Minimum Neighborhood First (MNF), and the Progressive Minimum Neighborhood First (PMNF) of Ramanathan (1999), were not considered because they are centralized, and also because their performance is expected to be similar to DSA-AGGR of Bryan et al. (2007).

The RANDOM strategy consists in selecting randomly and allocating any node whose parent node has already allocated a slot. Therefore, the RANDOM strategy can descend the tree in several ways that fall between DF, and BF.

For the SS-TDMA slot allocation protocol, y was set to 2, meaning that the interference range was twice the communication range. In this way, if a node receives a message from its Northern closest neighbor (its parent), it allocates a slot number equal to its neighbor's minus one; if it receives a message from its Western closest neighbor, it allocates a slot number equal to its neighbor's minus (y+1), or, in the case, minus 3. The simulations showed that SS-TDMA was able to allocate all slot numbers of a given frame without any unused slots, also achieving spatial reutilization of the slots, while implementing a 2-hop neighborhood interference avoidance criterion. However, the SS-TDMA allocation assumes that the slots are unidirectional, and separates the slots that are destined to convergecast, from the slots that are destined to broadcast. The slots that were allocated by SS-TDMA cannot be made bi-directional, or otherwise the 2-hop interference avoidance criterion is not valid anymore.

Simulations were carried out considering a 100-node square grid physical topology, where the sink node was placed at the upper-left corner. A logical tree topology was assumed, with the sink being the root node, and each node communicating with a random neighbor, selected among those that stay closer to the sink. In this way, each node was allowed to choose as parent either the node that is closest to it in the West direction, or in the North direction.

A simple free space propagation model was used, with the path loss exponent set to 2, and radio propagation irregularity was not considered (see, Rappaport, 2002) for the sake of simplicity. This model was selected because it was found to be the most common in the related literature.

The dimension of the grid square edges was set to the approximate value of the communication range. The interference range was set to twice the communication range. Therefore, each node had a maximum of twelve 2-hop neighbors. The interference graph was built based on a 2-hop neighborhood criterion, as this is also a customary assumption in the related publications. Since each slot was considered bi-directional, the links that were considered as potentially interfering with the parent-child communication were all the links established by the 2-hop neighbors of both the parent node and of the child node, using the same timeslot.

The dimension of each slot was configured to offer three transmission opportunities, in order to maximize the probability that successful packet transmission is still achieved in a single TDMA frame.

As the square grid had 100 nodes, and the nodes located at the top edge of the square have always its closest Western neighbor as parent, while the nodes located at the left edge of the square have always its closest Northern neighbor as parent, 81 nodes can choose one node as parent, among its closest Western neighbor and its closest Northern neighbor. This means that 2^{81} different topologies can be generated, or 2.42×10^{24} topologies. For each of these topologies, there are also a huge number of different slot schedules that can be done by each slot allocation algorithm. For instance, DF can descend the tree visiting the branches in different sequences. These observations show that the number of possible slot schedules is a very huge number, being impossible to find the optimal TDMA single frame size based on an exhaustive search strategy. Therefore, for each different slot allocation algorithm, we ran 10,000,000 simulations, each having as input one different random logical tree, and resulting in one different random slot schedule. Although this number is small in comparison with the number of all possible combinations of topologies and slot schedules, it was considered to be sufficient to assess how the different slot allocation algorithms behave. The histograms of the frequencies for the respective TDMA frame sizes were then built. One reason that supports the decision of making only 10,000,000 simulation runs is that the obtained histograms are quite smooth in shape, resembling normal distributions. Probably even a smaller number of simulation runs would also be sufficient to obtain mean values with high accuracy, as high stability of the results was observed for different sets of simulation runs, but the problem is that some slot allocations, with slot counts approaching the edges of the distributions, would not appear. Since the RANDOM strategy can descend the tree in a much larger number of ways than the other disciplines, including, for instance, the breadth-first and the depth-first allocations as special cases, the number of simulations for this algorithm was raised to 100,000,000.

Figure 1 shows the histograms for the single frame size, obtained for the first set of slot allocation algorithms.

Figure 1. Single TDMA frame size histograms for the first set of slot allocation algorithms. The histograms are based on 10,000,000 simulation runs for each algorithm, with the exception of the 100,000,000 for the RANDOM algorithm.

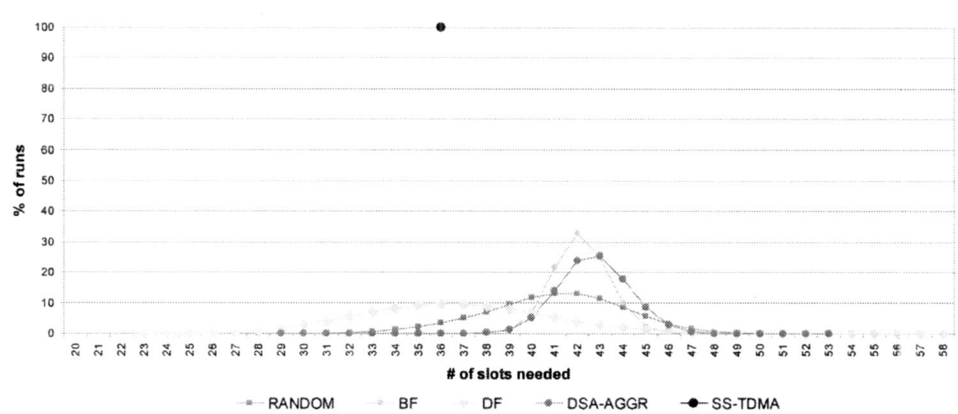

Table 1. Single frame size statistics for each allocation algorithm

Algorithm	*Average (slots)*	*Minimum (slots)*	*Maximum (slots)*
RANDOM	41	26	57
BF	42.1	29	49
DF	36.7	23	58
DSA-AGGR	42.7	29	53
SS-TDMA	36	36	36
DF-LPF	27.9	22	43
CENT-LPF	25.6	22	34
CENT-LDF	24.9	21	33
DIST-LDF	24.9	21	33

Those results show that BF has a lower variance than DF, but DF achieves a lower average value, respectively 36.7 slots for DF, and 42.1 for BF (see the five first rows of Table 1). DSA-AGGR does not behave as being a compromise between BF, and DF, with respect to finding low single TDMA frame sizes. However, the RANDOM algorithm behaves as expected: its histogram falls between those of BF, and DF.

It is interesting to note that SS-TDMA achieves always the same number of slots, i.e., 36, whatever the logical topology. The reason for this behavior is simple: the distances (in terms of slot numbers) between a given node and its neighbors are constant, and they depend only on the directions of the neighbors. As the nodes of all simulated grids have always the same coordinates, the frames have always the same size. However, the number of slots that is achieved with SS-TDMA is not particularly promising, as it is close to the average value that is obtained by DF, and substantially higher than the minimum values of DF.

DF can achieve a lower number of slots because it can descend first on larger branches. The cascading of timeslots in those larger branches results in a set of consecutive slots that can be reused in other adjacent upper and smaller branches of tree network. Inversely, if smaller branches are allocated first, those allocated slots cannot be used in larger adjacent branches, resulting in larger sensor-to-sink slot distances, and therefore larger TDMA frames. Note, however, that the slots that are used in the allocation of the longest branch first are not generally sufficient to color all the other network nodes, and that in most of the times there is a need for some extra slots: for instance, DF achieved a minimum of 23 slots, while the longest path had always a length of 18 hops. On the other hand, BF achieves higher single TDMA frame sizes, because it allocates the nodes of all the branches at some level in the same round, independently of their lengths. Therefore, in the allocation of the longest branches, the respective nodes find more slots that were already spent in previous slot assignments. Consequently, the slot distances between a parent node and its children are generally longer in those branches, resulting in increased single TDMA frame sizes.

DSA-AGGR seems to achieve the worst single TDMA frame sizes, as it tends to create hotspots of allocated nodes. Therefore, in the allocation of a given branch (namely the longest) the nodes tend to find more slots that were already assigned (sometimes to nodes placed deeper in the tree), resulting in larger frame sizes.

The remarks made above suggest that a longest-path-first strategy will lead to smaller TDMA frame sizes. In order to confirm this hypothesis, the next set of simulations considered a slot allocation algorithm that descends the tree in a longest-path-first scheme, when it has to make a decision of which path it chooses first, while allocating the other branches by backtracking in the same order of the depth-first strategy. This algorithm was designated depth-first-with-longest-path-first (DF-LPF). Longest-path first has been also used in the past in some similar contexts, namely for the wavelength division multiplexing (WDM) problem of high bandwidth optical WANs (see, for instance, Chlamtac et al., 1992).

Two centralized strategies were also investigated with the objective of achieving even smaller TDMA frame sizes.

The first centralized strategy (CENT-LPF, centralized longest-paths-first) allocates the branches in the descending order of their lengths, whatever their positions in the network, but always beginning in a node whose parent was already assigned a slot. When the paths have equal lengths, ties are broken giving priority to the paths that are situated deeper in the tree, and randomly if this rule is not enough to decide. The rationale for the first breaking ties rule is the same of the previously described DF-LPF.

The other strategy (CENT-LDF, centralized largest-distances-first) is similar to CENT-LDF, but allocates the branches in the descending order of the distances to the sink that the branches can reach, independently of their sizes. Breaking ties rules are the same as for CENT-LPF. In fact, largest-distance-first is not a novel concept, since sometimes - though rarely -, it appears in the literature.

A distributed version of CENT-LDF was developed, which is designated by DIST-LDF, whose pseudo-code is listed in Figure 2, where Δt represents the expected time needed to allocate a slot and c is a configurable constant value.

DIST-LDF descends the tree, allocating the slots, and when it has several different branches to allocate, it firstly (and immediately) initiates the distributed allocation of the branch that presents the longest path, while the other branches have to wait an amount of time before also starting to allocate slots for their nodes. The respective amounts of time are proportional to the difference between the lengths of their paths and the length of the longest path. In this way, longest paths are allocated in advance, reserving slots for them before the smaller branches. Therefore, nodes that are in branches that feature distances that are more distant from the sink are allocated first than nodes that are in branches that feature distances that are closer to the sink. Intuitively, in order for the allocations of the larger distance branches not to be disturbed by the allocations of shorter branches, the former have to be scheduled sufficiently in advance. It is worth to note that the DIST-LDF code is extraordinarily simple and easy to implement in a real distributed environment.

Figure 2. Basic pseudo-code of the distributed largest-distances-first (DIST-LDF) algorithm

```
At any node that has already a slot:
    If there is more than one child:
        For the child with the longest path to a leaf:
            Allocate a slot to it, within the time interval Δt;
        For each other child i, not in the longest path:
            Allocate a slot, within the time interval = c * (size-longest-
            path - size-path(i)) * Δt;
    If there is only one child:
        Allocate a slot to it, within the time interval Δt;
```

(Note: if there are two or more children with the same path lengths, they are scheduled randomly, with a separation interval of Δt between them)

Figure 3. Single TDMA frame size histograms for the second set of slot allocation algorithms, showing improvements over simple blind depth-first (DF).

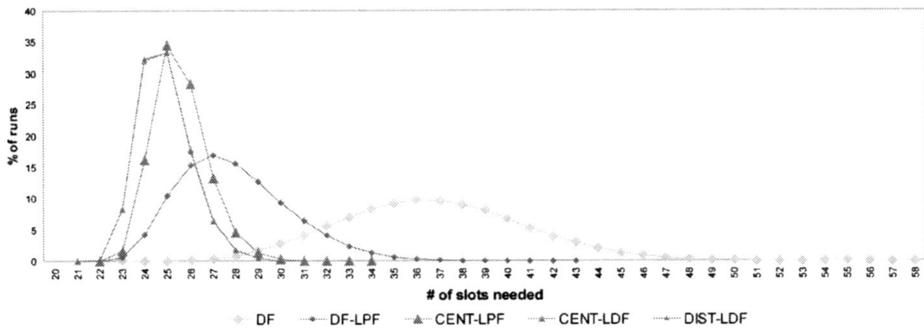

Figure 3 presents the single frame size histograms, showing that there is a systematic improvement on the single frame sizes, when DF-LPF, CENT-LPF, CENT-LDF, and DIST-LDF (whose results are the same as those of CENT-LDF) are successively considered. Average values for the single frame size were respectively 27.9, 25.6, 24.9, and 24.9, for these four slot allocation algorithms, compared with the 36.7 of blind DF, as it can be seen in the last rows of Table 1. The LDF algorithms also produce the smallest range of values, among all the considered slot allocation algorithms. CENT-LDF, and DIST-LDF, present single frame sizes that range from 21 to 33 slots, while blind depth-first (DF) presented a much broader range, from 23 to 58 slots.

Figure 4 and 5 are useful to understand the different behaviors of the BF and LDF disciplines, with respect to obtaining low single frame sizes. They show the allocations made by BF and DIST-LDF respectively, for some generated topology. In these figures, only a subset of 36 nodes of a total of 100 nodes in the network is shown. To the left of each node there is a pair of numbers: the first represents the order of the node in the resulting overall allocation sequence of the net-

Figure 4. Example of one BF strategy allocation. BF generates larger single frame sizes

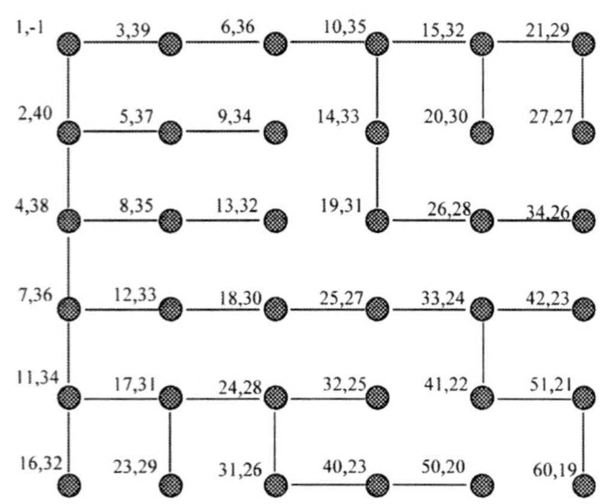

Figure 5. Example of one DIST-LDF strategy allocation. DIST-LDF generates smaller single frame sizes.

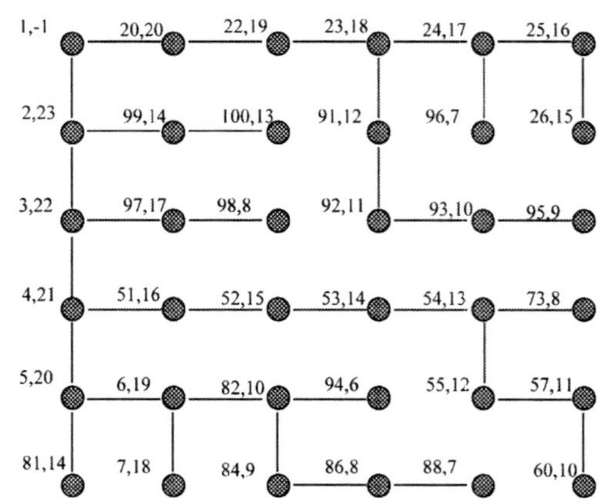

work, while the second represents the slot that was assigned to it.

What is clear from these two figures is that BF allocates the consecutive nodes of any branch with slots that have always leaps between them, because it considers the allocation in a breadth manner, assigning slots at the sides of any branch, which cannot be reused by the nodes in the branch. On the other hand, DIST-LDF can allocate the largest distance and longest branches without leaps in the slot numbers assigned to consecutive nodes. Consider the first path (therefore the longest path in the network) that was allocated by the DIST-LDF algorithm (see Figure 5). When the allocation process arrives to the second node of the lowest row, the slot distance to the sink is only 6 slots (23-18+1), while for BF (see Figure4) the distance is already equal to 12 (40-29+1). The same phenomenon occurs in the second path allocated by DIST-LDF, namely the branch situated at the topside of the network. These different behaviors explain the better performance presented by DIST-LDF over BF, with respect to obtaining smaller single TDMA frames.

Regarding again the DIST-LDF code of Figure 2, constant *c* was shown to generate the same results as the centralized algorithm when it takes values equal or greater than 5. For lower values, DIST-LDF performance degraded progressively into higher values for the TDMA frame size. In this case, some shorter distance branches do not wait for a sufficiently large time interval, and begin to allocate slots simultaneously with the longer distance branches. These concurrent actions are not separated enough in space and therefore the shorter distance branches can allocate slots that can't be used by longer distance branches, resulting in larger TDMA frames. In other words, longer distance branches are not scheduled sufficiently in advance. On the other hand, if the value of *c* is increased much beyond 5, the behavior of the DIST-LDF algorithm is not improved further, because the allocation of longer distance branches is already being performed at such distances that they cannot be affected by the allocations of shorter distance branches. Optimization of constant *c* for specific networks is, however, a subject for future work.

The slot distances from the nodes to the sink were also measured, and they are depicted in Figure 6. CENT-LPF, CENT-LDF, and DIST-LDF achieved the lowest average slot distances (and also the lowest maximum slot distances), respectively with the average values of 14.8, 14.7, and 14.7 slots, against 20.4 for RANDOM and BF, 18.7 for DF, 20.5 for DSA-AGGR, 18.2 for SS-TDMA, and 15.6 for DF-LPF, meaning that they can also attain the lowest delays in the communication to the sink. DIST-LDF still presents the best performance.

Figure 6. Maximum and average slot distances to the sink for the single frame scenario

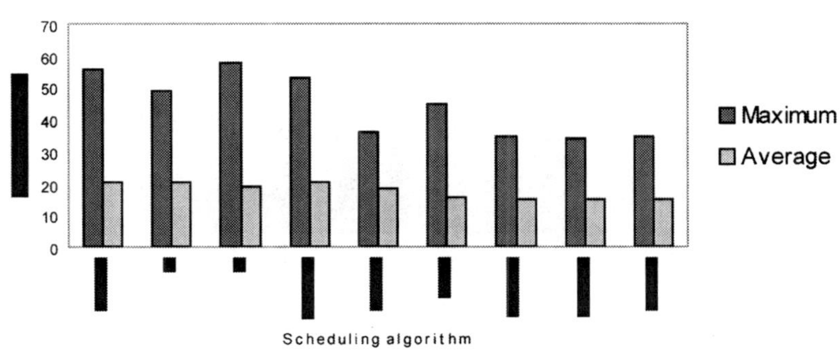

DIST-LDF presents the best results among the considered slot allocation strategies, but it requires each node to know the length of the branches that are rooted at each of its children. This may represent a significant disadvantage, since this information has to be propagated in the network tree, from the leaves to the sink, after the tree construction process. Such procedure may result in a significant overhead. Note, however, that the complexity of DIST-LDF in terms of the number of visited nodes is intuitively smaller than that of DF, since DIST-LDF does not need to backtrack in the tree structure, when it completely allocates a sub-tree. Therefore, it is also expected that DIST-LDF can achieve smaller execution times than DF. However, these two last observations need to be confirmed in future work.

SIMULATIONS RESULTS FOR A FIXED PRE-DETERMINED FRAME SIZE

Simulations have also been carried out to investigate the behavior of the scheduling algorithms, when the frame size is set to a fixed value, allowing that the sensor-to-sink communication can be completed in more than one frame.

Figure 7 shows the results of the nine different scheduling algorithms, for a fixed frame size of 30 slots, and for a grid network of 100 nodes. This frame size of 30 slots was found to be enough to avoid having nodes that could not be assigned a slot in any of the simulation runs. All simulation procedures and parameters were kept the same as in the previous section.

These results are very similar to those that were obtained for the single frame scenario: the proposed disciplines DF-LDF, CENT-LPF, and CENT-LDF, and DIST-LDF, featured a successive decrease on the slot distances. The explanation for this behavior follows the overall rationale that stands behind their design: when the smaller branches are allocated first, the longer branches will feature longer slot distances due to the increased difficulty of conflict-free cascaded allocation. Therefore, it is more advantageous to perform an allocation that results in larger slot distances on smaller branches, than to place larger slot distances in longer braches. This is achieved by allocating longer branches first and results in a more uniform branch delay bound distribution.

Similar results were also obtained with larger networks of 400 nodes. These results suggest that CENT-LPF, CENT-LDF, and DIST-LDF disciplines can also attain lower communication delays when more than one fixed sized frame is required to transport data from leaf nodes to the sink.

Figure 7. Maximum and average slot distances for a fixed frame size of 30 slots

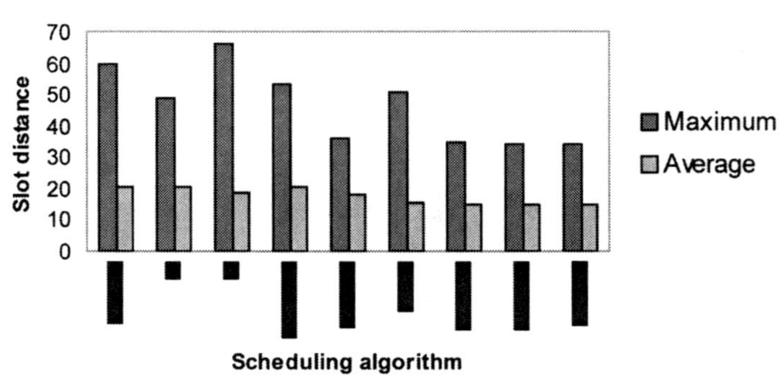

SIMULATION RESULTS FOR THE ACTUALLY REQUIRED FRAME SIZES IN A NON-SINGLE FRAME SCENARIO

Two questions can now be made: how do the different scheduling algorithms behave with respect to the minimum non-single frame sizes that they actually require? And how they behave with respect to slot distances that they achieve, when they use the frame lengths that they actually require?

In order to answer to these questions, a third set of simulations was carried out. The algorithm that was used to determine the required minimum frame sizes was the following: since each node in the grid has a maximum of 12 neighbors, the initial frame size is set to 9 (this number resulted to be adequate, since the obtained simulation runs always required 10 or more slots.). When a node tries to allocate a slot, it firstly tries to find a slot that is right next to its parent's slot, otherwise it tries each successive slot in the frame (eventually returning to the beginning of the frame), until a conflict-free slot is found or until it reaches its parent's slot. If no conflict-free slot is found, the number of slots is increased by one and the node tries to allocate the new slot. It may happen that a node will find the new slot already occupied by one of its neighbors in which case it will have to increase the frame size again and retry. In fact, the frame size is kept as a local variable, since in a distributed algorithm it is not feasible to communicate local frame size changes to distant nodes. At the end of the slot allocation process, the frame size is equal to the maximum slot number that was needed by all the nodes of the network.

Figure 8 shows the histograms of the required frame sizes for random topologies and allocations. And Figure 9 shows the respective worst-case delays in terms of slot-distances to the sink. The worst-case delays take into account a possible delay bias of one frame, for the case of a packet being generated just after the respective slot has elapsed.

Similar results were also obtained for larger networks of 400 nodes, but they are not shown for lack of space.

BF is known to be an efficient discipline with respect of requiring small frames, which is confirmed by the simulation results, which take into account cascading of TDMA slots.

BF achieved the lowest average value (12.0 slots) for the frame size, while the disciplines that descend the tree in-depth (DF, DF-LPF, CENT-LPF, CENT-LDF, and DIST-LDF) presented the highest values, being close to each other, (14.1,

Figure 8. Histograms of the minimum non-single TDMA frame sizes actually required by the nine allocation algorithms

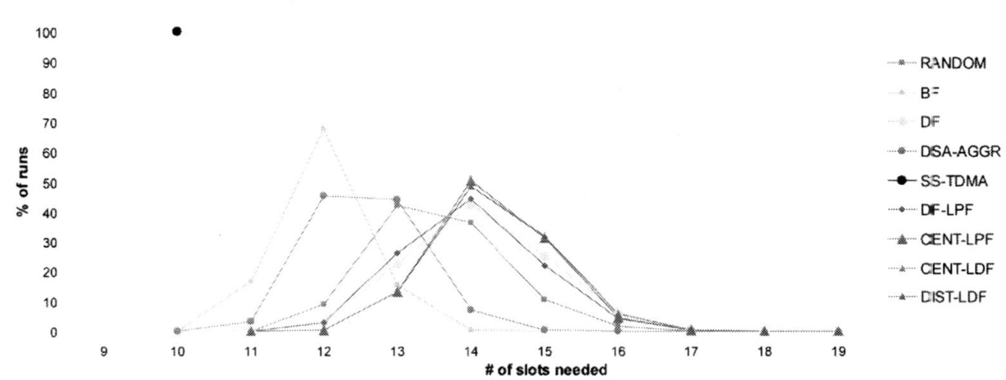

Figure 9. Maximum and average worst-case delays in terms of slot distance, for the non-single frame scenario

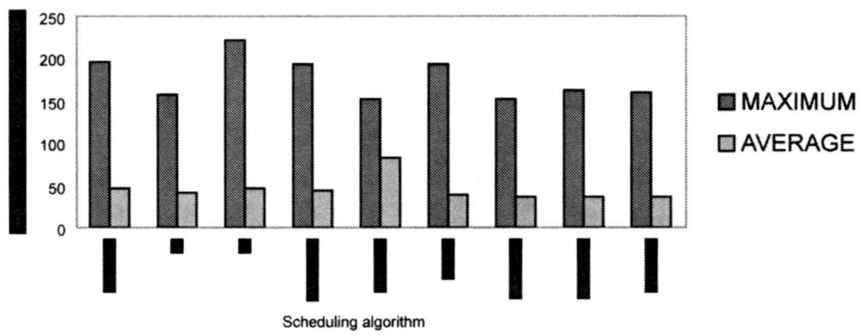

14.0, 14.3, 14.3, and 14.3 respectively). DSA-AGGR (12.6) is closer to BF than to DF. The RANDOM discipline (13.5) is closer to DF than to BF. SS-TDMA presented always a fixed value of 10 slots but, as already noted, its slots are not bi-directional, separating broadcast slots from convergecast slots. Therefore, it requires fewer slots. However, SS-TDMA performance, in terms of worst-case delays, does not benefit from its smaller frame sizes.

It is worth to point out that BF behaves quite differently in the slot allocation process when compared with the disciplines that descend the tree in depth, requiring fewer slots. The reason for its lower frame sizes rests in the fact that when BF allocates a slot for a given node, it finds only about one half of its interfering neighbors with a slot already assigned. These are all the neighbors located at the upper levels of the tree and some of the neighbors located at the same level. In contrast with this, in the disciplines that perform the allocation in-depth, when a node attempts to allocate its slot, it may find that almost all of its neighbors have already allocated one slot. These neighbors can be located in any adjacent and previously allocated branch, since after descending some branch in-depth, these disciplines can backtrack to any other branches, including of course those that have nodes in the neighborhood of the node that is trying to allocate a slot.

The same reasoning can also be used to explain the poorer results of RANDOM and DSA-AGGR when compared with BF. Note that the RANDOM and the DSA-AGGR disciplines may sometimes descend the tree in a more in-depth fashion.

Comparing these results with those of Section 4, it is interesting to note that the scheduling algorithms behave in opposite ways when cascading allocation is employed. BF is the best discipline to achieve minimum non-single frame sizes in comparison with in-depth disciplines, while the opposite happens regarding the minimum single frame sizes.

Finally, regarding worst-case delays, it should be noted that Figure 9 confirms the tendency of the longest-paths first and largest-distances first disciplines to present the lowest values, even if they need slightly larger frame sizes.

CONCLUSION

In most alarm-driven WSN applications, traffic can be characterized as very sporadic, but the generation of an alarm report demands an immediate response to the event. Low latencies and low duty-cycles can be simultaneously accomplished by using TDMA protocols. Cascading of slots may result in low latencies, while TDMA protocols can achieve low duty-cycles because each node only needs to be active in its own slots. Since this kind of applications does not need high throughputs, and it is also desirable to have the same delay bound for all the nodes in the network, we defined and investigated a new problem, which is that of allocating slots for all the nodes of the networks, such data can be always transmitted in a single TDMA frame, whatever is the place of the node in the network. However, such unique TDMA must also have the smallest possible size (a problem that we have designated the Cascading Minimum Single Frame Size – CMSFS – problem).

Several TDMA slot allocation strategies were comparatively evaluated with respect to the goal of minimizing the single TDMA frame size. The simulation results have shown that a breadth-first slot allocation strategy behaves poorly than depth-first, and that an informed depth-first strategy that visits the longest-path first, improves significantly the results when compared with blind depth-first. It was also shown that a largest-distances-first slot allocation algorithm would produce the smallest single TDMA frame sizes, and the smallest range of values, among all the scheduling algorithms that were considered. A distributed version (DIST-LDF) of this algorithm was implemented, which was able to obtain the same results as its centralized counterpart. This discipline is surprisingly simple, and easy to implement in real distributed slot allocation scenarios.

Simulations were also carried out to assess the behavior of the disciplines with respect to the slot distances and worst-case delays of the communication to the sink, for three scenarios: single frame sizes, non-single frame with a fixed frame size, and the non-single frame sizes that the algorithms actually require. In all these scenarios, the depth-first with the longest-path first, the longest-paths first and the largest-distances first disciplines, and its distributed version, presented successively lower values. This happened even if the in-depth disciplines required slightly larger non-single frame sizes than breadth-first. A comprehensive explanation for these last results was also provided in the paper.

As a main conclusion, it can be said that the largest-distances first discipline, and namely its distributed counterpart (DIST-LDF), can be a promising slot allocation strategy in order to obtain low single frame sizes and simultaneously obtain low delays in the communication to the sink, in a convergecast scenario.

ACKNOWLEDGMENT

The work described in this paper is based on results of IST FP6 project UbiSec&Sens. UbiSec&Sens receives research funding from the European Community's Sixth Framework Programme. Apart from this, the European Commission has no responsibility for the content of this paper. The information in this document is provided as is and no guarantee or warranty is given that the information is fit for any particular purpose. The user thereof uses the information at its sole risk and liability.

REFERENCES

Annamalai, V., Gupta, S. K. S., & Schwiebert, L. (2003). On Tree-Based Convergecasting in Wireless Sensor Network. *Proceedings of the IEEE Wireless Communications and Networking (WCNC 2003)*, New Orleans, LA, USA.

Bryan, K. L., Ren, T., DiPippo, L., Henry, T., & Fay-Wolfe, V. (2007). *Towards Optimal TDMA Frame Size in Wireless Sensor Networks (Technical Report, TR-xxx)*. University of Rhode Island.

Chlamtac, I., Ganz, A., & Karmi, G. (1992). Lightpath communications: an approach to high bandwidth optical WAN's. *IEEE Transactions on Communications*, *40*(7), 1171–1182. doi:10.1109/26.153361

Chlamtac, I., & Kutten, S. (1987). Tree-based Broadcasting in Multihop Radio Networks. *IEEE Transactions on Computers*, *C-36*(10). doi:10.1109/TC.1987.1676861

Cormen, T. H., Leiserson, C. E., & Rivest, R. L. (2000). *Introduction to Algorithms*. The MIT Press.

Gandham, S., Zhang, Y., & Huang, Q. (2008). Distributed Time-optimal Scheduling for Convergecast in Wireless Sensor Networks. [Elsevier.]. *Computer Networks*, *52*(3), 610–629.

Kulkarni, S. S., & Arumugam, M. U. (2005). *SS-TDMA: A Self-Stabilizing MAC for Sensor Networks. Sensor Network Operations*. IEEE Press.

Lu, G., & Krishnamachari, B. (2007). Minimum latency joint scheduling and routing in wireless sensor networks. [Elsevier.]. *Ad Hoc Networks*, *5*(6), 832–843. doi:10.1016/j.adhoc.2007.03.002

Lu, G., Krishnamachari, B., & Raghavendra, C. S. (2004). An Adaptive Energy-Efficient and Low-Latency MAC for Data Gathering in Wireless Sensor Networks. *Proceedings of the 18th International Parallel and Distributed Processing Symposium (IPDPS 2004)*, Santa Fe, NM, USA.

Mao, J., Wu, Z., & Wu, X. (2007). A TDMA scheduling scheme for many-to-one communications in wireless sensor networks. [Elsevier.]. *Computer Communications*, *30*(4), 863–872. doi:10.1016/j.comcom.2006.10.006

Pan, M. S., & Tseng, Y.-C. (2008). Quick convergecast in ZigBee beacon-enabled tree-basewireless sensor networks. [Elsevier.]. *Computer Communications*, *31*(5), 999–1011. doi:10.1016/j.comcom.2007.12.015

Polastre, J., Hill, J., & Culler, D. (2004). Versatile Low Power Media Access for Wireless Sensor Networks. *Proceedings of the 2nd ACM SenSys Conference*, (pp. 95-107), Baltimore, MD, USA.

Ramanathan, S. (1999). A unified Framework and Algorithm for Channel Assignment in Wireless Ad Hoc Networks. *Wireless Networks*, *5*(2), 81–94. doi:10.1023/A:1019126406181

Rappaport, T. (2002). *Wireless Communications: Principles and Practice* (2nd ed.). Prentice Hall.

Rhee, I. Warrier, A., Min, J., & Xu, L. (2006). DRAND: Distributed Randomized TDMA Scheduling for Wireless Ad-hoc Networks. *Proceedings the 7th ACM International Symposium on Mobile Ad Hoc Networking and Computing (MobiHoc'2006)*, Florence, Italy.

Upadhyayula, S., Annamalai, V., & Gupta, S. K. S. (2003). A low-latency and energy-efficient algorithm for convergecast in wireless sensor networks. *Proceedings of IEEE Global Telecommunications Conference (GLOBECOM '03)*, San Francisco, CA, USA.

West, D. B. (2001). *Introduction to Graph Theory* (2nd ed.). Prentice Hall.

This work was previously published in International Journal of Business Data Communications and Networking Volume 5, Issue 2, edited by V. Sridhar and D. Saha, pp. 68-83, copyright 2009 by IGI Publishing (an imprint of IGI Global).

Chapter 9
Shared Transport for Different Radio Broadband Mobile Technologies

Xi Li
University of Bremen, Germany

Thushara Weerawardane
University of Bremen, Germany

Yasir Zaki
University of Bremen, Germany

Carmelita Görg
University of Bremen, Germany

Andreas Timm-Giel
Hamburg University of Technology, Germany

ABSTRACT

This chapter introduces traffic separation technique and presents several traffic separation approaches to transmit HSPA (HSDPA/HSUPA) traffic and UMTS Release 99 (R99) traffic over a shared access transport network. The traffic separation technique enables QoS differentiations of HSPA and R99 traffic, while aiming to achieve a maximum utilization of the transport resources. In this chapter, two transport networks are studied for UMTS access network: ATM (Asynchronous Transfer Mode) based transport network and IP based transport network with DSL (Digital Subscriber Line) technology. In the ATM based transport network, the authors suggest the traffic separation approaches by using separate ATM Virtual Paths (VPs) or Virtual Circuits (VCs) for transmitting R99 and HSPA traffic with different ATM QoS class. With the introduction of IP transport, the authors propose to transport the HSPA traffic over the DSL network while transmitting the R99 traffic with the legacy ATM network. The benefit of applying traffic separation and its impact on the performance of the transport network as well as the end users are studied in this article. The quantitative evaluations are provided by simulations. The results presented are obtained from own developed UMTS R99 and HSPA simulation models, which can generate HSDPA and HSUPA traffic as well as R99 traffic in the same UMTS network and transmit them with different transport technologies and traffic separation approaches. The presented results demonstrate that applying traffic separation between HSPA and R99 traffic can considerably improve the performance of both HSPA and R99 traffic, and as well bring significant gain on efficient bandwidth utilizations.

DOI: 10.4018/978-1-60960-589-6.ch009

INTRODUCTION

Universal Mobile Telecommunication Systems (UMTS) is a key standard of the third-generation (3G) WCDMA-based cellular network. Since the operation of UMTS Release 99 (named as R99 in this article) which is the first UMTS release, UMTS has been widely deployed all over the world and experiences an intensive growth in recent years following a rapidly increasing number of mobile subscribers and a dramatically growing data traffic. The services offered by UMTS have been extensively expanded from primarily voice telephony service to a variety of appealing data and multimedia-based data services such as web browsing, email, FTP upload/download, video conferencing, video streaming, high resolution video, and IPTV, etc. The growing data service constitutes a dominant traffic share in the mobile networks and the resultant amount of data traffic still continues rising. In order to improve the support for the data services with enhanced resource efficiency and service quality, High Speed Downlink Packet Access (HSDPA) (3GPP TR 25.855, 2001) and High Speed Uplink Packet Access (HSUPA) also named as Enhance Uplink (3GPP TS 25.309, 2006) are introduced by 3GPP Release 5 and Release 6 individually as the evolution of UMTS to enhance the transmission of data packet traffic on the downlink and uplink separately. They offer very high data rate (up to 14.4 Mbps in the downlink with HSDPA and 5.76 Mbps in the uplink with HSUPA), low latency, and increased system capacity for transmitting data services with the use of fast Hybrid Automatic Repeat Request (HARQ), fast Node B scheduling, and short Transmission Time Interval (TTI). To support these new features, HSDPA uses a new downlink transport channel called High-Speed Downlink Shared Channel (HS-DSCH) that is shared by all HSDPA UEs in the cell and HSUPA uses a new uplink transport channel called E-DCH (Enhanced Dedicated Channel) for each HSUPA UE to provide high-speed data traffic transmission. HSDPA and HSUPA are jointly referred to as High Speed Packet Access (HSPA) (Dahlman, Parkvall, Sköld, & Beming, 2007).

In the current deployment of HSPA and R99 in the UMTS Terrestrial Radio Access Network (UTRAN), HSPA can co-exist with existing UMTS R99 technology by sharing the same access transport network, as illustrated in Figure 1. It is seen that one UMTS cell supports (1) normal UMTS R99 users like traditional voice users; (2) HSDPA users who require HSDPA service for high-speed data transfer on the downlink, e.g. Internet access; (3) HSUPA users who only uses HSUPA service for uplink data transmissions, e.g. FTP upload; (4) or HSPA users who use HSUPA on the uplink and HSDPA on the downlink simultaneously. In

Figure 1. UMTS network transmitting R99 and HSPA traffic

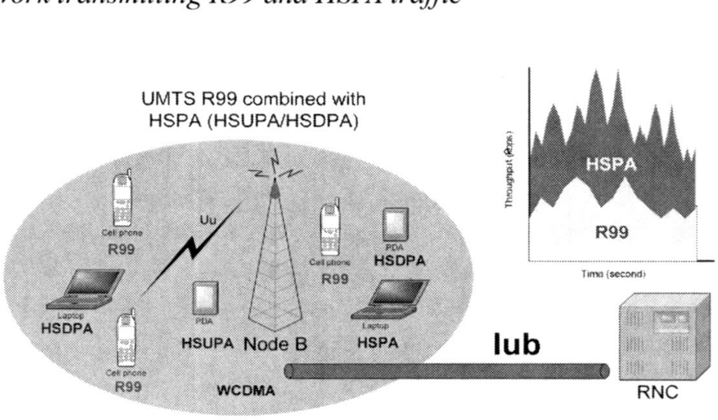

such network deployments, HSPA technology is integrated directly into the existing UMTS R99 nodes via software or hardware upgrades, i.e. the integrated Node B and RNC support both R99 and HSPA radio technologies. In this case, the Iub interface, which is the logical interface between the RNC and the Node B, carries both HSPA traffic and R99 traffic at the same time.

For such deployments, we need to properly design the transport network of the UTRAN, especially the Iub interface, to transmit the HSPA and R99 traffic over a shared transport in a cost-efficient way. We need to take two important considerations. Firstly, the UMTS R99 and HSPA services have rather different Quality of Service (QoS) requirements: R99 mainly carries delay sensitive traffic like conventional voice or real time video services; whereas HSPA traffic is primarily interactive and background traffic which is insensitive to the delay. That means, the R99 traffic has much higher delay requirements on the access network than the HSPA traffic. Furthermore, in R99 the WCDMA radio control functions such as soft-handover, power control, scheduling, and radio channel allocation (which are exchanged between the RNC to the Node B), etc., also impose strict delay and delay variation requirements on the UTRAN Iub interface in order to guarantee optimal radio resource utilization. The second consideration is that we should achieve a cost-efficient transport for transmitting the R99 and HSPA traffic. HSPA traffic is characterized by high peak data rate and high burstiness, which in turn requires considerably high transport capacity for the provisioning of high-speed transmission of HSPA packet data. In order to save the overall transport cost, the transport bandwidths for the Iub interface should be properly dimensioned such that the allocated transport bandwidths are the minimum while the required QoS is fulfilled. Thus, how to efficiently transport the R99 and HSPA traffic over a shared UTRAN transport network by guarantying their individual QoS requirements is a big challenge for designing the UMTS access transport networks.

For solving this problem, this article presents several traffic separation approaches to transmit both HSPA and R99 traffic in the same UMTS access transport networks, providing a differentiated QoS support for each type of traffic according to its individual QoS requirements. In this article, two different transport networks are studied: ATM (Asynchronous Transfer Mode) based transport network and IP based transport network with DSL (Digital Subscriber Line) technology. ATM was selected as the first transport technology for the UTRAN transport network in UMTS R99 by the 3GPP, due to its ubiquitous nature for the heterogeneous traffic types, quality of service guarantee and its widespread deployment in public networks. But its main problem is the expensive ATM lines. Adding ATM capacity by leasing additional E1/T1 lines leads to a rapid increase of the transport expenses, especially for data traffic. Therefore, IP is introduced by 3GPP Release 5 (3GPP TR 25.933, 2002) to replace the ATM in the UTRAN to transport data traffic. IP is now becoming the most cost-efficient and compatible transport technologies due to its considerably low cost and high flexibility. The use of IP transport can bring much more cost savings in the transport, and it is also an essential step towards one "All-IP" network merging the fix and the mobile networks. In the ATM based transport network, we suggest the traffic separation approaches by using separate ATM Virtual Paths (VPs) or Virtual Circuits (VCs) for transmitting R99 and HSPA traffic with different ATM QoS class. With the introduction of IP transport, we propose to transport the HSPA traffic over the DSL network while transmitting the R99 traffic with the legacy ATM network.

The contribution of this article is twofold: (1) To investigate how much performance gain can be achieved by applying traffic separation in the transport network on the user throughput, packet losses, and link layer transport efficiency, and in addition what will be the impact on the dimensioning of the transport network, especially the Iub interface that is between the RNC and the Node B. The presented results demonstrate that applying traffic separation to HSPA and R99 traffic can considerably improve the performance of HSPA and R99 traffic, and as well bring significant gain on efficient bandwidth utilization. (2) To achieve a cost-efficient dimensioning for the transport network, an optimum transport configuration is desired. So in this article, to transmit HSDPA, HSUPA and R99 traffic in the same transport network, different traffic separation configurations are investigated and compared with each other.

The article is organized as follows: in section 2 a general concept of traffic separation is described. Section 3 explains in detail the traffic separation in the ATM based transport and introduces various traffic separation configurations for transmitting R99, HSDPA and HSUPA traffic. In section 4, we present the approach of transporting the HSPA traffic over a DSL transport network. In section 5, we introduce the developed simulation models. Section 6 presents simulation results of ATM transport, giving a detailed analysis of the impact of applying traffic separation and comparisons of different traffic separation configurations. Section 7 analyzes the impact of using DSL transport on HSPA performance. Section 8 concludes the study and discusses the future work.

TRAFFIC SEPARATION FOR SHARED TRANSPORT OF HSPA AND R99 TRAFFIC

As mentioned above, R99 and HSPA traffic define different delay requirements on the UTRAN transport network. There is an extremely strict delay constraint on the Iub interface for DCH channels of R99, not only due to the delay requirements of the user traffic itself but also because of the requirements derived from supporting radio control functions such as outer-loop power control and soft handover. The excessively delayed Frame Protocol packets (their delay is larger than predefined delay boundaries) will be discarded at the Node B as they become too late to be sent over the air interface for the allocated time slot. However, HSPA traffic has significantly lower delay requirement on the Iub interface. Because for both HSDPA and HSUPA a fast scheduling is introduced at the Node B which reserves the time slot on the air interface replacing the scheduling at RNC in R99, and furthermore there is buffering in the Node B which supports fast HARQ. Thus, the delay requirements for HSPA are essentially only due to the service itself, which are mainly delay-tolerant best effort services that have loose constraints on the delay and delay variations. Thanks to the R99 traffic having a much more stringent delay requirement on the Iub interface, the R99 traffic is usually given a higher priority to transmit over the HSPA traffic.

In the case without using traffic separation, the R99 traffic and HSPA traffic are carried using a common transport path with a guaranteed bandwidth established between the RNC and the Node B. So the intermediate switches should not discard any packets (neither of R99 nor of HSPA) as long as the RNC and the Node B comply with the transport Service Level Agreement (SLA). In this case, the potential multiplexing gain is not considered and thus the required bandwidth is overspecified. This causes unnecessary high costs in terms of leased transport bandwidths.

In order to save the transport costs for the fixed lines, applying traffic separation techniques is a very popular and cost-efficient solution for the network operators or service providers.

In the ATM based transport network, the traffic separation technique is based on using separate ATM VPs or VCs for transmitting different types of traffic each with a different ATM QoS class. We suggest to use cheap UBR (Unspecified Bit Rate) or UBR+ (UBR with a minimum guaranteed rate) VPs to separately transport HSPA traffic whereas only use the CBR VPs to transmit R99 traffic, instead of paying for overestimated CBR (Constant Bit Rate) VPs to transmit all traffic types. The detailed traffic separation in the ATM transport will be explained in section 3. Furthermore, with the introduction of IP transport we propose to transport the HSPA traffic over the DSL (Digital Subscriber Line) network while transmitting the R99 traffic with the legacy ATM network. This part will be explained in section 4.

TRAFFIC SEPARATION IN THE ATM-BASED TRANSPORT

In this section the ATM transport is introduced. This section explains in detail how to apply the traffic separation to transmit HSPA and R99 traffic with the ATM technology and the required configurations.

ATM-Based Transport in UTRAN

ATM was introduced in UMTS R99 as the first transport technology. ATM is designed as a cell switching and multiplexing technology to combine the benefits of circuit switching and packet switching techniques. Circuit switching provides constant transmission delay and guaranteed capacity whereas packet switching provides high flexibility and a bandwidth efficient way of transmission. It uses very short fixed-length (53 bytes) packets, called cells, to transfer data traffic. Due to the short fixed length cells transmitted over the network, it can be used for the traffic integration of all services including voice, video and data.

ATM defines two levels of virtual connections: Virtual Paths (VP) and Virtual Channels (VC). The connection is identified by two values in the cell header: the Virtual Path Identifier (VPI) and the Virtual Channel Identifier (VCI). A virtual path (VP) is a bundle of virtual channels, all of which are switched transparently across the ATM network based on the common VPI.

ATM provides several types of logical connections between two end users, which can be set up statically or dynamically. The Permanent Virtual Circuits (PVCs) or Permanent Virtual Paths (PVPs) are usually created long before it is used and remains in place until the connection is deprovisioned. Bandwidth is allocated for them whether it is used or not. In this way they are similar to leased lines. The Switched Virtual Circuits (SVCs) in contrary are dynamic connections. They are established and released on demand and remain in use only as long as data is being transferred.

ATM technology is intended to support a wide variety of services and applications. The control of ATM network traffic is fundamentally related to the ability of the network to provide appropriately differentiated Quality of Service (QoS) for network applications.

The ATM forum defines five different service categories. Each class is designed to accommodate data bursts according to customer needs and provide the appropriate quality of service (QoS) for each service class. For each service class, a set of QoS parameters is defined to describe both the required traffic characteristic and the QoS that is required from the network, such as Cell Transfer Delay (CTD), Cell Delay Variation (CDV), and Cell Loss Ratio (CLR). The six service categories are listed as follows.

- Constant Bit Rate (CBR),
- Real time Variable Bit Rate (rt-VBR),
- Non-Real time Variable Bit Rate (nrt-VBR),
- Unspecified Bit Rate (UBR),
- Available Bit Rate (ABR),
- Guaranteed Frame Rate (GFR).

Service categories are distinguished as being real time or non-real time. The categories belonging to real time are CBR and rt-VBR support real time application services depending on the traffic descriptor specifications such as peak cell rate (PCR) or sustainable cell rate (SCR). The other four services categorize the support of non-real time services under the requirements of the traffic descriptor parameters. All service categories except GFR apply to both VPCs and VCCs. GFR is a frame aware service category which can only be applied to VC connections since frame delineation is usually not visible at the virtual path level.

Besides, on top of ATM the ATM Adaptation Layer type 2 (AAL2) (ITU-T Recommendation Q.2630.2, 2000) is used to transmit the user data at the user plane of the Iub interface. AAL2 was chosen because of its ability to perform efficient multiplexing of several data flows over a single ATM virtual channel and is designed for efficient transmission of low-bit-rate services with stringent delay requirements such as voice traffic.

Traffic Separation in the ATM-Based Transport

The basic idea of traffic separation technique is to apply separate ATM Virtual Paths (VPs) or Virtual Circuits (VCs) with different ATM QoS categories to transmit different traffic types. One example of using traffic separation to transmit R99, HSDPA and HSUPA traffic at the Iub interface is depicted in Figure 2. In this example, each traffic type is carried by one individual ATM VP. R99 traffic is transported with ATM CBR (Constant Bit Rate) (Ferguson & Huston, 1998) service category. It is defined as high priority traffic class, where bandwidth is reserved up to requested Peak Cell Rate (PCR) with guaranteed cell loss ratio and cell transfer delay. This also means high transport costs. However, the transport of HSDPA and HSUPA traffic uses ATM traffic class UBR (Unspecified Bit Rate) (Ferguson & Huston, 1998) or UBR+ ("Addendum to Traffic Management V4.1", 2000). UBR is a best effort service and is the lowest class of service in ATM. It is defined as low priority traffic class, which utilizes all bandwidth unused by the high priority traffic. Therefore it does not provide any guarantees for bandwidth, cell loss ratio and cell transfer delay. This traffic class has much lower transport costs. UBR+ is similar to UBR, but bandwidth is guaranteed up to a minimum rate MDCR (Minimum Desired Cell Rate). With UBR+, the HSPA traffic can be guaranteed up to MDCR.

With the use of traffic separation technique to differentiate the HSPA and R99 traffic over different paths with different priorities, the transport of R99 traffic is separated from the HSPA traffic and its stringent delay requirements can be guaranteed by using the CBR VP. On the other hand, using the UBR+ VPs for the transport of HSPA traffic can result in cell discards or long delays on the HSPA traffic if the MDCR is exceeded. But since the QoS requirements of the HSPA traffic is usually lower than the R99 traffic, some decreased performance of HSPA by using a separated UBR or UBR+ VP can be tolerated. In this way, compared to not using any traffic sepa-

Figure 2. Traffic separation in the ATM-based transport

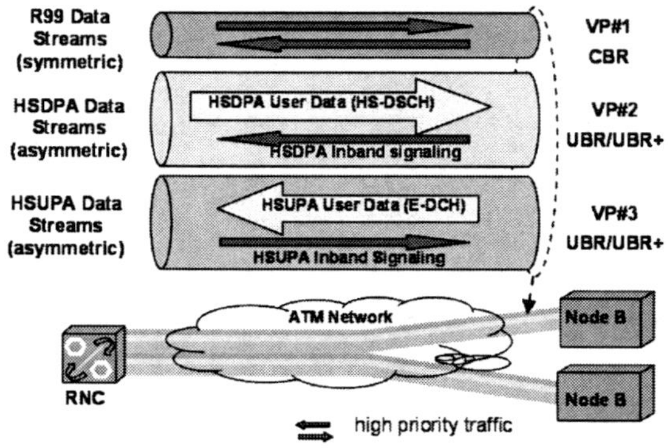

ration, the overall network cost is reduced at expenses of possible degraded HSPA performance.

Traffic Separation Configurations for the ATM-Based Transport

For setting up a traffic separation scenario, the following ATM parameters need to be configured:

- PCR (Peak Cell Rate) is the upper limit of the traffic that can be submitted to the link;
- MDCR (Minimum Desired Cell Rate) defines a minimum guaranteed cell rate on UBR VC. It is optionally configured on either a VC or VP connection.

PCR is required to configure for both ATM CBR and UBR/UBR+ service categories. Maximum allowed bandwidth can be set different for the uplink and downlink by means of an asymmetric PCR configuration of VPs and VCs. MDCR is only configurable for UBR+ VP/VCs.

As the R99 traffic consists of a considerable amount of symmetric voice traffic, CBR traffic class is elected for providing high QoS for the real time services and also a symmetric PCR is configured for both directions. On the HSDPA and HSUPA path, it allows an asymmetric configuration of UBR/UBR+ VPs or VCs, e.g. asymmetric PCR or MDCR settings, to support the asymmetric traffic property of HSPA traffic, i.e. HSDPA user data is only transmitted on the downlink and there is a small amount of inband signaling traffic on the uplink, and HSUPA user data is only transmitted on the uplink with a small amount of inband signaling on the downlink.

To transport HSDPA, HSUPA and R99 traffic simultaneously in the UTRAN transport network, there are mainly four possible scenarios to be considered:

1. 3 VPs: 1 CBR VP for R99, 1 UBR/UBR+ VP for HSDPA, 1 UBR/UBR+ VP for HSUPA;
2. 2 VPs: 1 CBR VP for R99, 1 UBR/UBR+ VP for HSPA with separated VCs to transmit HSDPA and HSUPA;
3. 2 VPs: 1 CBR VP for R99, 1 UBR/UBR+ VP for HSPA without separated VCs to transmit HSDPA and HSUPA;
4. 1 VP: 1 Common CBR VP or VC to carry all traffic types.

Scenario 1 applies three VPs each transferring one traffic type. Scenario 2 and 3 uses two VPs: 1 VP is assigned for R99 and the other one for the HSPA traffic. For these two cases, the HSDPA data traffic will be mixed with HSUPA inband signaling traffic and the HSUPA data traffic will be mixed with HSDPA inband signaling traffic. The difference of scenario 2 and 3 is whether to use separate VCs for transmitting HSDPA and HSUPA traffic. With separated VCs, each UBR/UBR+ VC can be configured with different PCR or MDCR for HSDPA and HSUPA individually. Moreover, in order to protect the HSPA inband signaling traffic which has high priority, Cell Loss Priority bit (CLP) that is defined in the ATM cell header can be used to select which cell to discard in case of congestion: CLP=1: for low priority traffic, cell may be discarded by ATM network in case of congestion; CLP=0: for high priority traffic, cell should not be discarded by ATM network. So we can set different CLP value for the separated VCs to differentiate the inband signaling traffic and HSPA traffic so that the HSPA inband signaling traffic can be protected. In scenario 4, all R99, HSUPA and HSDPA traffic share one common CBR VP/VC, i.e. there is no traffic separation in this case. For scenario 1, 2 and 3, the transport of HSPA traffic can either be on a UBR or UBR+ VP. If UBR+ VP is used, there is a guaranteed minimum bandwidth for transmitting the HSPA traffic, with which a minimum QoS is assured for the requested HSPA services.

HSPA TRAFFIC OVER DSL TRANSPORT

The Digital Subscriber Line (DSL) technology is widely deployed for fast data transmission in industry, business and home usage. The great advantage of DSL is that it can use the existing telephone network by utilizing free frequencies above the voice telephony spectrum. This technology is one of the cheapest ways for fast data transmission which can be effectively deployed for data access with many other transport technologies such as IP and ATM.

The idea of the Mobile Network Operators (MNOs) is to use such a cheap fast transmission technology for their transport network in UTRAN inside the HSPA network (mainly for data access). This replaces the costly bundled E1 (ATM). Due to factors like Bit Error Rate (BER), delay and jitter that depend on the line or cable distance, the performance over the DSL based transport networks has not yet been investigated for HSPA networks. Therefore in this work, the performance of the DSL based transport network is investigated.

Before the detailed modeling of the DSL line in UTRAN, a general introduction of DSL network is given in the following section 4.1.

Digital Subscriber Line (DSL) Technology

The Digital Subscriber Line (DSL) technology is one of the key technologies used in modern data communications. It uses existing twisted-pair telephone lines (Plain Old Telephony System, POTS) to transmit high bandwidth data. Commonly it is also known as xDSL where "x" distinguishes variants of DSL. Mainly there are two types: Symmetric DSL (also called Single pair High bit-rate Digital Subscriber Loop, SHDSL) (ITU-T Recommendation G.992.2, 1998) or Asymmetric DSL (ADSL) (ITU-T Recommendation G.992.1, 1999; ITU-T Recommendation G.992.2, 2002). The latter has different data rates in downstream and upstream whereas SDSL supports equal data rates in both directions. Both ADSL and SHDSL services provide dedicated, point to point, public network access using existing telecommunication infrastructure on the local loop which is commonly known as "last mile", that is the connection between the network service provider's switching center and the customer's site.

The twisted-pair copper wire supports frequency spectrums up to 1MHz for the transmission. However the normal voice telephony signal utilizes the spectrum below 4 kHz whereas the rest is not used. Therefore, DSL technology uses this leftover spectrum almost completely for high-speed data transmissions.

When the performance of the DSL technology is considered, there are a number of factors that can affect the frequencies in the DSL spectrum which reduce the effectiveness of the available bandwidth.

- Attenuation: depends on the length and the gauge of the line. The higher the length is, the higher the attenuation becomes, and also narrowing the gauge increases the attenuation as well.
- Reflection and noise: When bridging the lines, reflection and noise can be introduced.
- Cross-talk: bundling results in cross-talk which also depends on the relative position of the line.
- Radio frequency interference: RF interference can occur from external sources, for example any nearby station which transmits Radio Frequency (RF) waves.

ADSL uses the DMT (Discrete Multi-Tone) (ITU-T Recommendation G.992.1, 1999 and 2002) modulation scheme at the transmitter so that it can effectively cope with the above mentioned impairment conditions. DMT splits the available frequency spectrum into several sub-bands. There are 224 subcarriers in the downstream, each occupying a 4 kHz portion of the spectrum.

Protocol Architecture for DSL based UTRAN

As discussed above, DSL can be deployed to combine ATM or IP based transport technologies. Since IP over DSL is the most commonly used approach, it is implemented in this work for the UTRAN as a transport technology. The proposed protocol architecture is shown in Figure 3.

The above architecture shows that the Node-B is connected using Cell Site Gateway (CSG) and the RNC is connected using RNC Site Gateway (RSG). The RSG and CSG use Pseudo Wire Encapsulation (PWE) technology to adapt the ATM layer to the IP layer. The PWE technique is standardized by the Internet Engineering Task Force (IETF) Pseudo Wire Emulation Edge-to-Edge (PWE3) working group, which defines various types of Pseudo-Wires to emulate traditional and emerging services such as ATM over Packet Switched Network (PSN) (RFC 3916, 2004; RFC 3985, 2005; RFC 4717, 2006). The DSL line is connected between the xDSL modem and the Broadband Remote Access Server (BRAS). The traffic aggregation occurs at BRAS which connects the broadband network to the access network and vice versa. Further it performs the Point-to-Point Protocol (PPP) termination and tunneling. Layer 2 Tunneling Protocol (L2TP) network server (LNS) which is located between BRAS and RSG provides the layer 2 tunneling.

The CSG and the RSG gateways are connected to the Node-B and RNC respectively through the E1 based ATM links. To connect the ATM based network to the IP based Packet Switch Network (PSN), the PWE technology is deployed. In this work, PWE is used to couple the two transport technologies, ATM and Ethernet via an IP backbone. Thus, the ATM cells coming from the Node B or the RNC will be encapsulated into Ethernet packets within the CSG and the RSG gateways, and then the Ethernet packets are carried over the emulated Ethernet circuit, i.e. the Ethernet Pseudo-Wire. After the Ethernet packets arrive at the egress of the Ethernet network, they are decapsulated to the ATM cells and forwarded to their destination.

The peer-to-peer protocols for the CSG and the RSG are shown in Figure 4.

SIMULATION MODELS

This section introduces the HSPA and R99 simulation models which are developed in this work. In section 5.1, the HSPA and R99 Simulation Model based on the ATM transport is described.

Figure 3. IP over DSL protocol architecture for the UTRAN

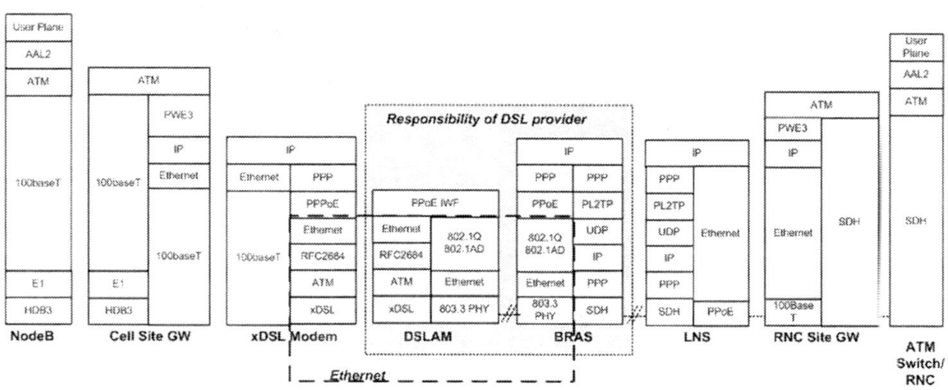

Figure 4. Peer-to-peer protocols in CSG and RSG entities

Afterwards, section 5.2 introduces the modeling of DSL link for the HSPA simulator.

HSPA and R99 Simulation Model using the ATM Transport

The ATM-based HSPA simulation model and R99 simulation model are developed using the OPNET simulation environment ("OPNET Modeler", n.d.). The main purpose of this model is to perform UTRAN transport network feature analysis. In addition the model is also designed to support individual performance analysis on the protocols like RLC, TCP/UDP and application layers on a per user level.

The modeling of R99 network is according to the 3GPP R99 specifications, including the basic UMTS R99 radio functions, channels, protocol stacks, and transport technologies with the main focus on the Iub interface. The Iub transport network is fully modeled with the resource management functions like admission control, traffic differentiation, priority scheduling and ATM QoS mechanisms. A more detailed introduction of the modeling of R99 is given in (Weerawardane, Li, Timm-Giel, & Görg, 2006).

The HSPA simulation model consists of both HSDPA and HSUPA, as shown in Figure 5. The red part is the implemented HSUPA protocols whereas the yellow one represents the HSDPA protocols. In addition, the HSDPA and E-DCH

Figure 5. HSPA simulation model with ATM transport

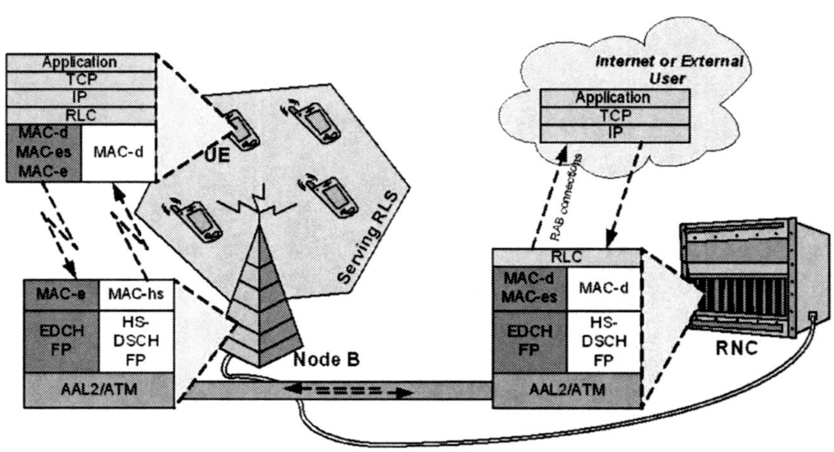

scheduler, Soft Handover (SHO), HARQ, NSRLS (Non Serving Radio Link Set) traffic, HSDPA/HSUPA congestion control (CC), and HSDPA flow control were implemented in the simulation model. More detailed introductions on the modeling of HSPA can be found in (Weerawardane, Li, Timm-Giel, & Görg, 2006; Weerawardane, Timm-Giel, Görg, & Reim, 2006; Weerawardane, Timm-Giel, Malafronte, et al., 2008; Zaki, Weerawardane, Li, Timm-Giel, et al., 2008).

In the network configuration, there are three different traffic types carried in the Iub transport network: UMTS R99, HSDPA and HSUPA. The simulation model can easily enable or disable the traffic separation for the transport network, and as well it can allow various configurations to transport HSDPA, HSUPA and R99 traffic as introduced in section 3.

DSL Line Emulation in the HSPA Simulator

The DSL line of the HSPA simulator is emulated using a real trace file taken at the Ethernet layer which records the absolute value and the variation of the delay for each packet arriving at the receiver. It also includes details about lost packets. The real traces are often short in length. Therefore, in order to use these traces in the network simulator, the raw data of approx. 100 traces are analyzed and a probability distribution function (CDF) is created. The delay and delay variation characteristics over the DSL line are simulated according to the derived distribution.

To include bit errors in the DSL transmission, the following approach is deployed. In general, for the DSL network, bit errors can be experienced in two forms, random bit errors and bursty bit errors. The burst errors which cannot be eliminated even with interleaving are due to the impulse noise and impairments of neighboring CPEs. Therefore these packet errors should be included within the model of the DSL line along with the delay and the delay variation distributions. To perform this, the following simple approach is used to emulate the packet errors in the system level simulator which provides sufficient accuracy for the end user performance analysis.

The main assumption of this method is that the total number of lost packets (or burst of lost packet events) is uniformly distributed over the given length of the trace file. First, all numbers of bit errors which create a corrupted packet and bursty errors which cause the loss of more than one packet are extracted from trace files based on their lengths. Then from these extracted details, the average packet and burst error rates are calculated. According to these average error rates, the packets are discarded when sending data over the DSL network. At the end, RLC or TCP has to recover those lost packets by resending them over the network. In order to analyze the end user performance, the average packet and burst loss rates are important and also provide sufficient accuracy for HSDPA and HSUPA networks.

The impact of the DSL transport technology on the HSPA end user performance is analyzed and presented in section 7.

RESULTS OF ATM-BASED TRANSPORT NETWORK

This section presents the simulation results for the ATM based transport network. Firstly, section 6.1 evaluates the impact of applying traffic separation on the network and end user performances as well as on the dimensioning. Section 6.2 compares different traffic separation configurations for transmitting HSDPA, HSUPA and R99 traffic simultaneously over the same ATM transport network. Section 6.3 summarizes the main findings from the simulation results.

Impact of Applying Traffic Separation

This section investigates the impact of using traffic separation for transmitting the HSPA and R99 traffic. The HSDPA and R99 traffic scenario is investigated as an example here for the investigations. In the following, the results of applying traffic separation to transport both HSDPA and R99 at the Iub interface is presented and compared to the scenario without traffic separation. The parameter settings for the traffic separation and its impact on the dimensioning will be also discussed. Additionally one example of the Iub dimensioning with traffic separation and without traffic separation is given and their transmission efficiency is compared.

Simulation Scenarios

The simulation scenario consists of one Node B and one RNC. In the HSDPA model, a Round Robin air interface scheduler is used in the simulations. In addition, in order to protect the congestion on the Iub link, flow control and congestion control schemes are applied on the Iub. The HSDPA traffic is modeled with 20 Internet users browsing the web. The web traffic model is defined by ETSI standards (3GPP TR 25.848, 2001), the traffic model parameters are given in Table 1. Each user requests multiple pages where the inactive time between pages follows the geometric distribution. The same traffic model is used for generating the R99 traffic where multiple Packet Switched (PS) Radio Access Bearers (RABs) are available for transmitting data.

When no traffic separation is applied in the Iub interface, R99 and HSDPA traffic are sharing one common ATM CBR VP, where the AAL2 priority is applied which assigns higher priority to R99 traffic over the HSPA traffic. While in the

Table 1. ETSI traffic model parameters

Page Interarrival Time (Reading Time)	Geometric distribution mean interarrival time = 5 seconds
Page size	Pareto distribution parameters: Shape=1.1, location=4.5 Kbyte max page size = 2 Mbyte mean page size = 25 Kbyte

case of using traffic separation, two ATM VPs are established: the transport of R99 traffic is over one ATM CBR VP and the transport of HSDPA traffic is on an ATM UBR+ VP. Here UBR+ VP is set to low priority.

The following metrics are used for performance evaluation:

- **Application Throughput**: the average throughput of transferring a web page at the application layer, excluding the reading time period. Throughput indicates the transaction speed, i.e. how long it takes to transfer a certain amount of data. It is directly related to the application delay and the volume of corresponding data transaction. The normalized application throughput is given in simulation results defined as the ratio of the application throughput under certain Iub link bandwidth to the maximum application throughput under an ideal Iub capacity.
- **Cell Discard Ratio**: in case of congestion of the Iub link, the ATM cells are discarded. The packet discard ratio is measured as the ratio of discarded ATM cells to the total ATM cells sent to the Iub link.
- **TCP Retransmission Counts**: the total number of TCP retransmissions.
- **Link Utilization**: the Iub link throughput over the given Iub link bandwidth. The link throughput includes transport network overheads as well as all TCP/RLC retransmissions.

Impact of Traffic Separation

In this section, the influence of traffic separation (TS) is investigated by comparing to the scenario without traffic separation technique in use in the transport network. In this example, there is in average 815.9kbps HSDPA traffic and 968.7kbps R99 PS traffic on the Iub link. In both with and without traffic separation cases, the offered HS-

DPA and R99 traffic is fixed while the common Iub link rate is step by step increased. For the configuration with traffic separation, the PCR of CBR VP for transport of the R99 traffic is set to 1600kbps, whereas the MDCR of UBR+ VP for transmitting the HSDPA traffic is increased from 0kbps up to 1400kbps which results in the increase of the total Iub link bandwidth.

Figure 6 compares the performance difference of using and not using traffic separation. It shows that with the usage of traffic separation technique, the end user application throughput is improved while the cell losses and resultant TCP retransmissions are reduced significantly. The major reason is that traffic separation provides a minimum bandwidth guarantee for HSDPA traffic, thus the HSDPA traffic will get less influence from the R99 traffic. Though the link utilization is similar in both scenarios, there is more link load contributed by RLC and TCP retransmissions in the case of no traffic separation.

From these results, we can conclude that to achieve the same application throughput or cell discard ratio target, using traffic separation needs less bandwidth on the Iub link, which means a more efficient utilization of the transport resources. For example, to achieve 90% normalized application throughput, applying traffic separation requires 2800kbps while no traffic separation requires 3300kbps on the Iub link. The obtained bandwidth saving is 15%. To guarantee less than 1% cell discard ratio, using traffic separation requires minimum 2100kbps bandwidth while no traffic separation requires minimum 2500kbps on the Iub link. The obtained bandwidth saving is around 16%.

Impact of MDCR Settings for UBR+ VP/VC

This part discuses the influence of MDCR settings of ATM UBR+ VP/VC on the overall performance, based on the results of the traffic separation scenario in the above example shown in Figure 6. As the PCR of CBR VP for transport of the R99 traffic is fixed to 1600kbps, the MDCR of UBR+

Figure 6. Performance comparisons: with TS and without TS

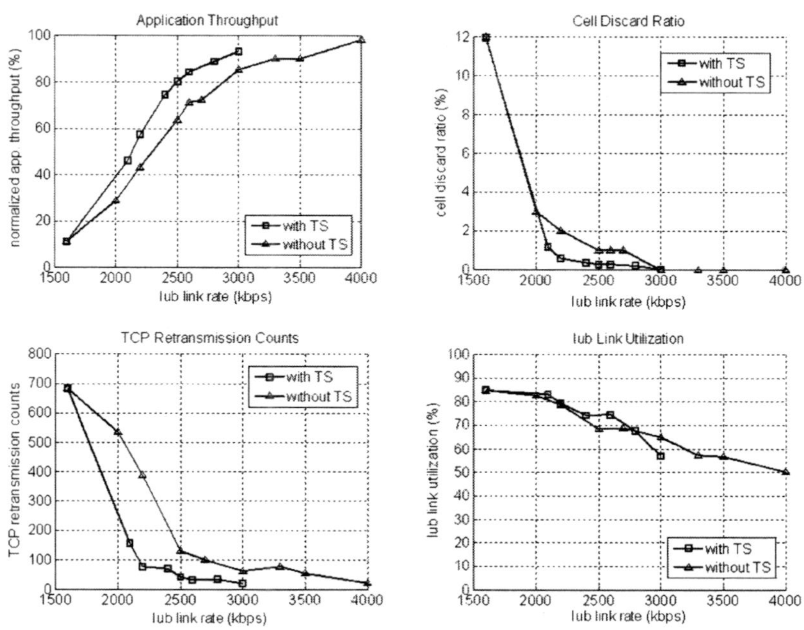

VP (MDCR = the total Iub link rate subtracted by the allocated bandwidth on R99 path) for transmitting the HSDPA traffic varies from 0kbps up to 1400kbps. It can be observed from Figure 6 that with the increased MDCR rates from 0kbps up to 1400kbps the total Iub link rate is increased from 1600kbps to 3000kbps, the end user application performance is improved considerably: the normalized application throughput is increased from 11% to 95%. Because with a higher MDCR rate, there is more bandwidth reserved for HSDPA traffic, and therefore the performance is better. Besides the improvement of application performance with a higher MDCR setting, the network performance is also enhanced. It is observed that RLC delays, cell discard ratio, number of TCP retransmissions are all decreased when MDCR increases. But on the other hand, the link utilization decreases due to a higher Iub link bandwidth caused by larger MDCR rates is configured to transfer the same offered traffic. Therefore, MDCR should be chosen as a compromise of the system performance and the Iub link utilization. That means, MDCR rate should be set properly to achieve the maximum link utilization while stratifying the QoS target.

Moreover it is observed that the application performance is much more sensitive to the MDCR setting than transport network performances. When MDCR is larger than 500kbps (i.e. Iub link rate = 2100kbps), the transport network performance such as cell discard ratio, TCP retransmissions, has been improved drastically. And afterwards, with further increased MDCR rate, the pace of the improvement is reduced and becomes more stable. But the application throughput is still quite low with 500kbps MDCR rate: only 46% of normalized application throughput is achieved. In order to achieve more than 90% of the application throughput, the MDCR need to be set higher than 1200kbps. So it is basically a choice of network operation to decide the MDCR rate based on its predefined QoS target. If the transport network performance is more important, then a smaller MDCR is adequate. If the end user application performance is the main target of dimensioning, the MDCR rate needs to be configured to a relative higher value.

Dimensioning Results

This section presents the results of dimensioning of the Iub link, which transmits the HSDPA and R99 traffic either with or without Traffic Separation (TS) technique. In the following example, R99 traffic contains 50% web traffic (web traffic model is defined in Table 1) and 50% voice traffic with AMR codec. The voice model consists of a series of ON and OFF periods with a service rate of 12.2kbps with Adaptive Multi-Rate (AMR) codec specified by 3GPP. ON and OFF states are exponentially distributed with a mean duration of 3 seconds (Türke, et al., 2002). HSDPA consists of purely web traffic (Table 1). In the following results, we fix the R99 traffic load and gradually increase the offered HSDPA traffic to the Iub link, and investigate the bandwidth demand for transferring the combined HSDPA and R99 traffic satisfying the predefined QoS targets of both traffic types. In this example, the QoS target for R99 traffic is 1% packet discard ratio and for HSDPA 95% normalized application throughput.

Figure 7 shows the required Iub link bandwidth over different offered UTRAN traffic loads in kbps. The offered UTRAN traffic is the total sum of traffic entering UTRAN network including HSDPA and R99. It shows that with the increased traffic demand, the required Iub bandwidth to achieve the predefined QoS targets is increasing. It can be also obviously seen that, the required Iub bandwidth for the traffic separation scenario is much lower than that for the case without traffic separation. Therefore it is concluded that applying the traffic separation technique brings a significant bandwidth saving for the Iub dimensioning, which reduces the transport cost.

The required capacity can be also expressed in terms of "Over-provisioning factor", β, which relates the capacity in the link (C) to the aggre-

Figure 7. Required Iub Bandwidth for different UTRAN load

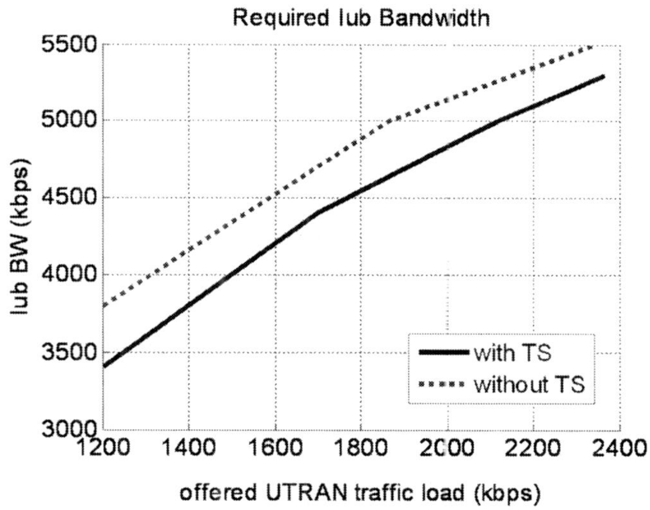

gated mean bit rate R_{iub} on the Iub link as given in the formula: $C = \beta R_{iub}$ (Galeana, Ferrus, & Olmos, 2007). This parameter indicates in addition to the mean traffic load on the Iub link how much extra bandwidth is needed in order to fulfill the QoS requirements. Figure 8 shows the obtained over-provisioning factor in percentage of the mean Iub traffic. As observed from Figure 8, the degree of over-provisioning decreases for higher traffic load on the Iub link in both with TS and without TS scenario. That means, with a larger traffic load a higher multiplexing gain is achieved which results in decreased over-provisioning factor. Furthermore, with traffic separation technique less extra bandwidth is required for transmitting the same amount of the traffic on the Iub link. And moreover, at the lower traffic load range, the over-provisioning factor of without traffic separation is much higher than that of the traffic separation scenario, and with the increase of the aggregated traffic load their gap is slowly reduced. This implies the traffic separation is able to achieve more bandwidth savings (compared to without traffic separation) at a lower mean Iub traffic load, where the room for the potential multiplexing gain is more.

Comparing Different Configurations for Traffic Separation

This section presents the simulation results of transmitting all HSDPA, HSUPA and R99 traffic simultaneously in the UTRAN transport network. Two different traffic separation solutions are compared: (1) 2 VPs among which one CBR VP for R99 and one UBR+ VP for both HSDPA and HSUPA without using separated VCs; (2) 3 VPs where each VP carries a separate traffic type: one CBR VP for R99, one UBR+ VP for HSDPA and one UBR+ VP for HSUPA. The goal of the comparisons is to find out the performance differences of the two traffic separation configurations, and discuss a more suitable configuration to apply in the UTRAN ATM transport network, which can achieve high bandwidth utilization while guaranteeing the desired QoS of each traffic type.

Simulation Scenarios

The scenario consists of one Node B and one RNC. In the simulated scenario, there are six HSPA users in the cell: 6 HSDPA in downlink and 6 HSUPA in uplink. For both HSUPA and HSDPA, a FTP

Figure 8. Over-provisioning factor over Iub link throughput

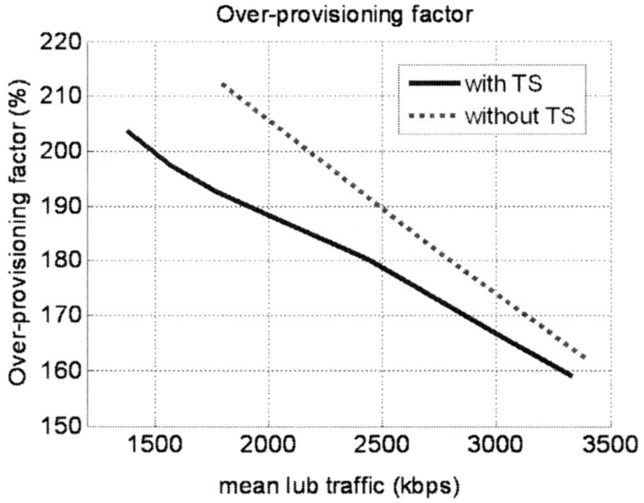

traffic model is used which generates heavy traffic on both uplink and downlink. The FTP traffic model parameters are defined in Table 2. With this FTP traffic model, all users are downloading or uploading very large files continuously. That means, all users have always data to be transmitted at any time and demanding the network resources constantly. Such a traffic model can be considered as a worst case traffic scenario from the network point of view. In addition there is a 2 Mbps R99 traffic transmitted on the Iub as well. In this setup, the Iub is heavily loaded, which is considered to be in a worst case scenario. In such overloaded traffic scenario, the investigated gain of applying traffic separation will be more significant.

The simulations scenarios are classified into two categories depending on the configurations of the transport network (2 VPs or 3 VPs). The transport network is configured with a line rate of 4 Mbps. The configurations of each VP are given in Table 3. In both scenarios, the 2 Mbps R99 traffic is fully utilizing its allocated 2 Mbps capacity.

Table 2. FTP traffic model for HSPA

File size	Constant Distribution Mean file size = 5 Mbyte
Inter-arrival time	~ 0.0 seconds which means immediately after the first file downloading of the second file is started

All simulation scenarios run for 1000 seconds. Following the main parameter configurations for the simulations are given:

- Radio configuration
    - TTI = 2 ms HSDPA, 10 ms HSUPA
    - Noise Rise = 6dB (for HSUPA)
    - Others-to-own interference factor = 0.6 (for HSUPA)
- Number of HARQ processes per user flow: 4
- RLC protocol: Operate in RLC AM mode
- TCP protocol: TCP New Reno version

Performances Analysis

In the following the comparison results are presented. Special attentions are paid to two main different performance aspects: transport network performance and the end user performance. In the transport network, the network-specific QoS

Table 3. Simulation scenarios

2 VPs	1 CBR VP for R99: PCR = 2 Mbps; 1 UBR+ VP for HSDPA/HSUPA with their inband signaling traffic, PCR = 4 Mbps, MDCR = 0.4 Mbps.
3 VPs	1 CBR VP for R99: PCR = 2 Mbps; 1 UBR+ VP for HSDPA and HSUPA inband signaling traffic, PCR = 4 Mbps, MDCR = 0.4 Mbps; 1 UBR+ VP for HSUPA and HSDPA inband signaling traffic, PCR = 4 Mbps, MDCR = 0.4 Mbps.

measures such as packet transport delay and packet losses need to be controlled to meet the agreed quality of a network, which is targeted to low delay and low loss. On the other hand, the end user performances like application throughput also need to be guaranteed to satisfy the user's particular QoS requirements. In the following presented results, the total discarded ATM cells and the link throughput represent the transport network performance. The user application throughput and the experienced TCP retransmissions indicate the end user performances.

Figure 9 shows the total discarded ATM cells on the downlink (left diagram) and the uplink (right diagram) individually. The ATM cell discard is due to the ATM buffer overflow, which is caused by the link overload. In Figure 9 the ATM cell discards only happen to HSDPA traffic on the downlink and HSUPA traffic on the uplink. The main reason of cell losses of HSPA traffic is because that with both 2 and 3 VPs configurations, the HSPA traffic is assigned to cheap UBR+ VP. In the heavy loaded situations, the intermediate ATM switches may drop HSPA cells if the MDCR rate has been exceeded by the amount of the injected HSPA traffic. In this case, the traffic which is above the MDCR has no bandwidth guarantee at all and therefore can be discarded in favor of R99 traffic. While on the R99 path, there is no cell loss on both directions as the R99 traffic is using a separate CBR VP which has a guaranteed bandwidth between the Node B and RNC.

When comparing the cell losses of using 2 and 3 VPs solution, it can be found that the 2 VPs scenario has more cell losses than the 3 VPs scenario on both uplink and downlink directions. This is because, in the configured 2 VPs scenario the HSDPA and HSUPA traffic are sharing one UBR+ VP but without separated VCs, i.e., HSDPA data and HSUPA signaling traffic are mixed on the downlink while HSUPA user data and HSDPA signaling traffic are combined on the uplink. Thus, there is no way to distinguish the signaling traffic from the user data traffic and therefore the signaling traffic can be dropped irrespective of its importance. Usually the signaling traffic consists of control messages sent by HSPA flow control or congestion control function, which are used to adapt the user data rate according to the available bandwidth in order to protect the Iub interface from heavy congestions. If the signaling traffic is dropped, the system will not react

Figure 9. Total ATM cell discards on downlink and uplink

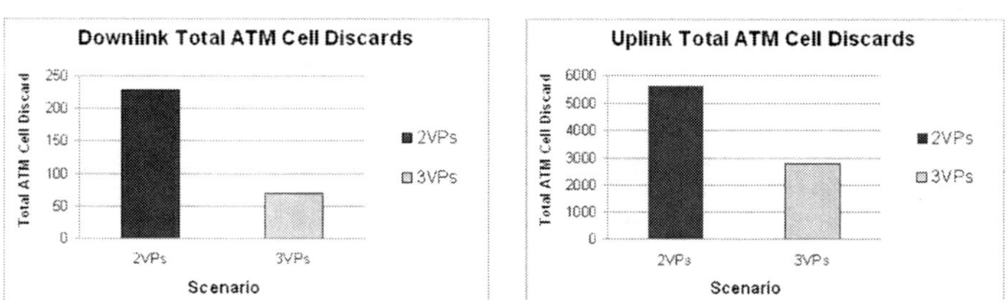

properly on the congestion situations and therefore the transport path keeps overloaded. But with 3 VPs solution, the signaling traffic is protected from the user data traffic as different VPs are used for transmitting HSDPA and HSUPA traffic separately. For each individual UBR+ VP to transmit either HSDPA or HSUPA traffic, the configured MDCR of will guarantee the bandwidth for the signaling traffic. Therefore, the signaling traffic is with a high probability not dropped in this case and hence the system can respond more correctly to the congestion situations. As a consequence, the resultant losses within the transport network in the 3 VPs scenario are less due to a better controlled transport network.

The cell discard on the ATM layer can further have influence on the TCP transport layer of the users. When the discarded cells are not able to be recovered by the RLC retransmissions, they will cause TCP retransmissions. It is seen in Figure 10 that resultant TCP retransmissions on HSDPA and HSUPA traffic is higher in the 2 VPs scenario than the 3 VPs scenario due to a higher cell losses. The TCP retransmissions will reduce the corresponding TCP congestion window and as a consequence decrease the user throughput.

Figure 11 presents the HSDPA per user application throughput (left figure) and the average application throughput of all HSDPA users (right figure). It can be obviously seen that the achieved average HSDPA user application throughput is slightly lower in the 2 VPs scenario. This is due to more TCP retransmissions, as explained above. Moreover, since there are more cell losses in the transport network as shown in Figure 9, the Iub congestion control function of HSDPA will be triggered more often to reduce the data rate of the users on the downlink direction. Similar is for the HSUPA traffic.

Figure 12 compares the resultant ATM link throughput on the downlink (left diagram) and the uplink (right diagram) individually. It shows that the link throughput obtained in the 3 VPs

Figure 10. HSDPA and HSUPA TCP retransmissions

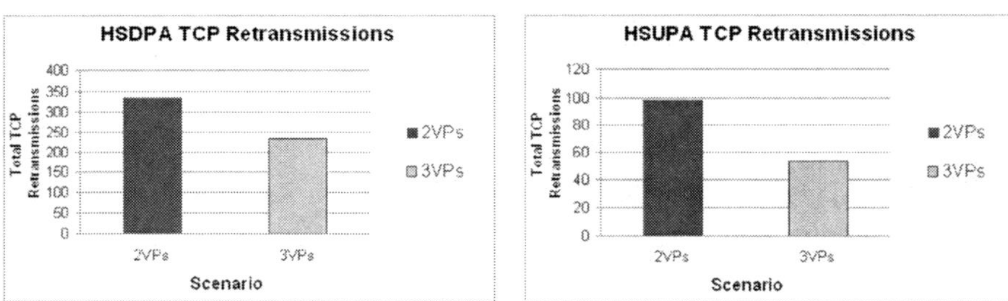

Figure 11. HSDPA application throughput

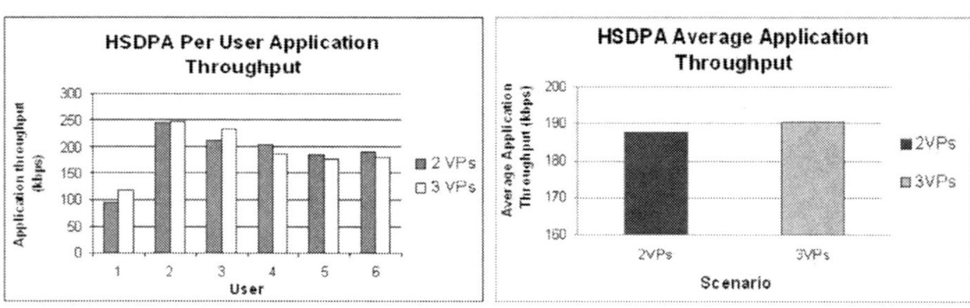

Figure 12. Downlink and uplink ATM link throughput

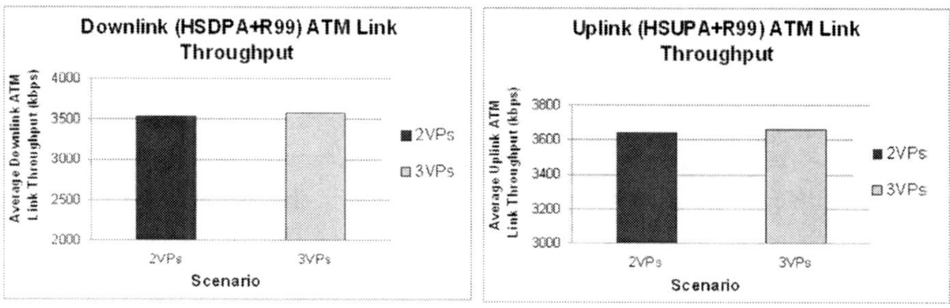

scenario is a little higher than the 2 VPs one, but the gap is not significant.

As a summary, through comparing the network and user performances of using 2 and 3 VPs as the transport solutions, it can be concluded the obtained user and transport network performances of using 3 VPs is slightly better than the 2 VPs due to a better protection for the signaling traffic at the expenses of one additional VP. From the performance perspective, the 3 VP setup is the best transport solution as it provides a clear traffic separation and QoS differentiation. However, it also requires additional cost for buying a separate VP. By taking considerations of both QoS and network costs, actually both 2 and 3 VPs solutions can be considered for the network operators or service providers to design a UMTS access network to transport both R99 and HSPA traffic.

SUMMARY

By investigating the different MDCR settings, it is concluded that MDCR should be chosen as a compromise of the system performance and the Iub link utilization, and also dependent on the QoS target defined by the network operator. Another contribution of this article is to investigate and compare different traffic separation configurations, i.e., 2 VPs or 3 VPs, to transmit HSDPA, HSUPA and R99 traffic within the same transport network. The presented simulation results demonstrate that both 2 VPs and 3 VPs configurations can be applied as the traffic separation solutions for the transport of HSPA and R99 traffic in the radio access network. Through comparing the network and user performances of the two traffic separation configurations, using 3 VPs is slightly better than the 2 VPs due to a better protection for the signaling traffic but at the expenses of paying for one additional VP. Therefore by taking considerations of both QoS and network costs, both 2 and 3 VPs solutions can be considered for the network operators or service providers to plan a UMTS access network to transport both R99 and HSPA traffic.

DSL-BASED TRANSPORT EFFECT ON HSPA PERFORMANCE

Several investigations and analyses have been performed to study the impact of the DSL based UTRAN on the HSPA performance. Two main investigations are presented in this section:

- Effects of the DSL transport delay and delay variation on the HSPA performance
- Effects of the DSL mode operation on the HSPA performance

In this work we study two DSL mode operations: DSL default mode and DSL fast mode. The DSL default mode uses random errors without considering the effects of impulse noise and impairment due to neighboring CPEs (Customer Premises Equipments), whereas the DSL fast mode considers the burst errors due to the impact of impulse noise and impairment. In section 7.1, the DSL default mode is investigated for different DSL delay and bandwidth configurations. In section 7.2, the DSL fast mode operation simulation results are compared with the results of the DSL default mode. In the downlink two different traffic models are used, i.e. ETSI based web traffic and FTP traffic. Whereas in the uplink a moderate FTP traffic model has been used instead that is taken from the standard 3GPP specification (3GPP TS 25.319, 2007). The detailed traffic model parameters are defined in Table 4 and Table 5.

Effects of DSL Transport Delay and Delay Variation on the HSPA Performance

The following analyses are carried out to characterize the effect of the DSL transport delay and delay variation on the HSPA performance. A trace file, taken from a DSL default mode operation (as described in section 5.2) in which the bit errors are randomly distributed along the DSL data stream, is deployed for this analysis. Three different simulation configurations are selected to analyze the performance, as given in Table 6.

All three configurations are configured with 6 FTP users in the uplink and 14 FTP users in the downlink that are transferring large FTP files all the time. This traffic model is considered as a worst case scenario. In addition to the HSPA users, there are R99 users in both downlink and the uplink. The R99 traffic is transported over ATM links with 2 Mbps capacity in each direction.

According to the above configurations, no effect is expected for configuration 1 and configuration 2 on the downlink due to sufficient capacity in the transport network which can fulfill the radio BW requirements. However, configuration 3 which has a limited uplink may block higher layer downlink signaling which can reduce the achievable throughput for the end users. The simulation scenarios (configurations) use the following naming: configurations 1, 2 and 3 are respectively called "ideal", "delay_ideal" and "delay_limited".

Table 4. ETSI Web traffic model parameters

Page Interarrival Time (Reading Time)	Geometric distribution mean interarrival time = 5 seconds
Page size	Pareto distribution parameters: Shape=1.1, location=4.5 Kbyte max page size = 2 Mbyte mean page size = 25 Kbyte

Table 5. FTP traffic model parameters

File Size	Constant Distribution μ_{MFS} = 12 Mbytes
3GPP FTP Traffic Model Parameters	
File Size	Truncated Lognormal Distribution Mean: 2 Mbytes, Max: 5 Mbytes Standard Deviation: 0.722 Mbytes
Inter Arrival Time	Exponential Distribution Mean: 180 sec

Table 6. Configuration of simulation scenarios

Configuration 1: Idle DSL transport and idle DSL delay	A high BW is allocated to the transport network which simulates a negligible delay and delay variation in the transport network
Configuration 2: Idle DSL BW and limited DSL delay	The idle BW is configured for the DSL link but static DSL delay and delay variation is emulated using the given ADSL trace file
Configuration 3: Limited DSL BW and Limited DSL delay	The DSL uplink BW is configured to 1.23 Mbps and DSL downlink BW is configured to 13 Mbps

Figure 13. Uplink per-user application throughputs

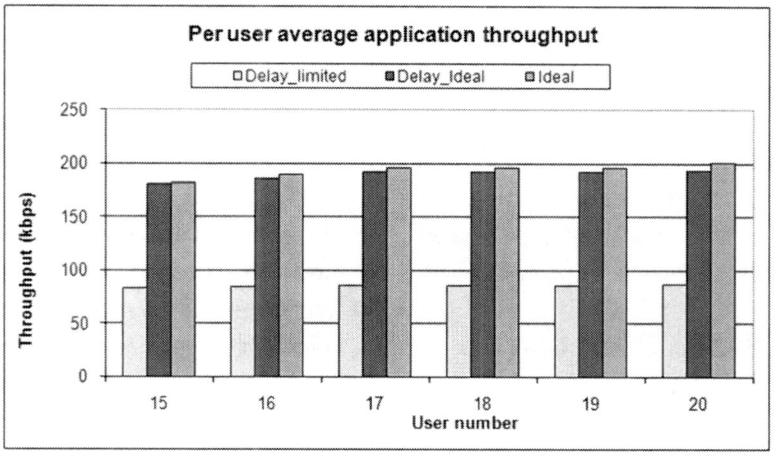

Figure 13 shows the uplink per user application throughput. The throughput is measured at the application layer; therefore all the effects of the lower layer are included. The idle scenario (config 1) has the best throughputs whereas the delay limited scenario has the lowest throughputs which are due to the limited uplink capacity. The fairness between the end user throughputs is excellent for the delay limited simulation scenario. The results of config 2 show that there is no significant impact of small static delays over the DSL network.

The average DSL uplink throughput between the DSL Modem and the BRAS are shown in Figure 14. The average throughput is approximately 1.15 Mbps for simulation config 3. This corresponds to the maximum BW value of 1.23 Mbps over the DSL link. Config 2 and 1 show higher throughput over the DSL link and it is corresponding to the radio interface capacity.

From the simulation results, we also observe that the uplink DSL delay varies in a small range between 26 ms and 27 ms for config 2 and 3 which is a very small jitter; and the uplink DSL delay

Figure 14. Uplink throughputs between DSL Modem and BRAS

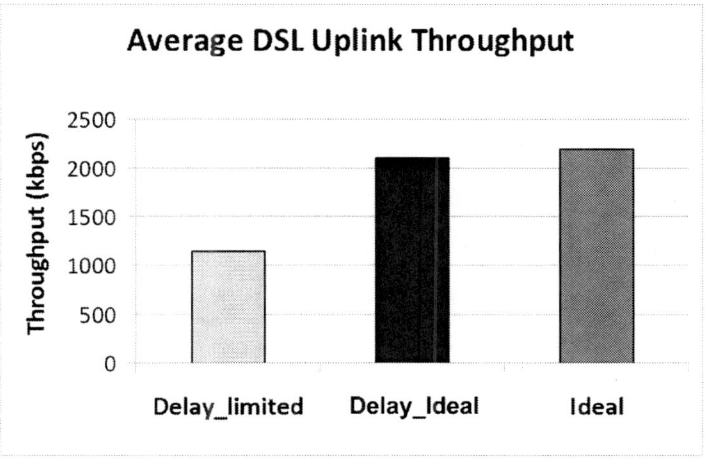

for config 1 is about 1 ms with also very small variation. This confirms that those small DSL jitter and static delays do not impact the HSPA end-to-end performance.

Figure 15 shows the total downlink application throughput. All simulation configurations show a similar HSPA performance for end user throughput and hence it can be concluded that having a limited uplink does not affect to the HSPA performance.

Effects of DSL Mode Operation on the HSPA Performance

This investigation is performed to compare the effects of the DSL default mode and the DSL fast mode and as well analyze the effect of burst errors on the performance of the HSPA networks. A trace taken from the DSL fast mode operation is deployed in the simulator as described in section 5.2. Both "default" and "fast mode" simulation scenarios are configured with the configuration 3 defined in Table 6 and use the same FTP traffic model.

The per-user throughputs for the uplink and the downlink are shown in the following left and right side of Figure 16 respectively. From the figure, it can be observed that there is no significant change in the per-user throughputs for both configurations. The results show that DSL burst errors do not have a significant impact on the HSPA performance.

Furthermore, there is also no significant effect on the DSL link throughputs for uplink and downlink for the default and fast mode scenarios. A bit higher uplink utilization is shown by the DSL fast mode compared to default mode. The small difference is due to more retransmissions at RLC layer for the fast mode simulation which has some bursty losses at the DSL transport network.

The uplink and the downlink DSL end-to-end delays for both simulation configurations are shown in Figure 17. The fast mode based configuration has a very low end-to-end delay compared to default configuration for both, the uplink and the downlink. This is mainly due to that the default configuration includes the interleaving delay where as the fast mode configuration does not include it. However, delay variation is same for both configurations.

Figure 15. DL overall throughput with and without delay and throughput constraints

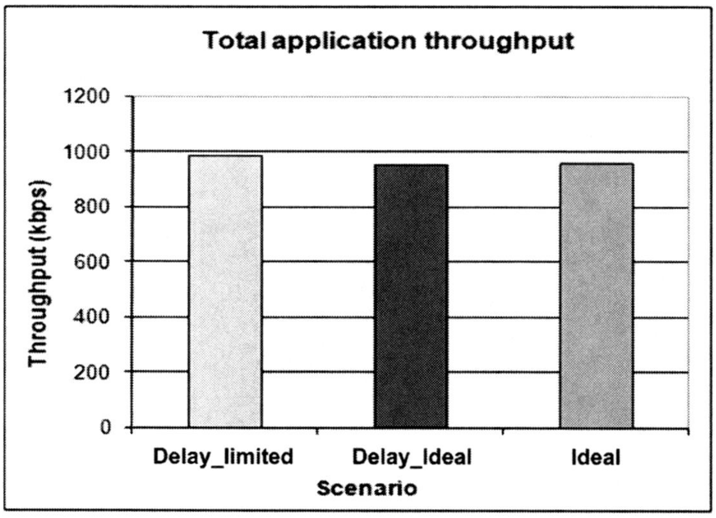

Figure 16. Uplink and downlink per-user application throughputs for the fast-mode DSL

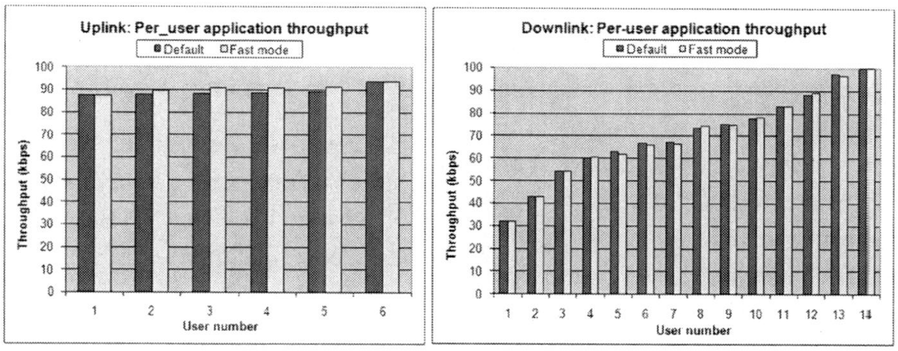

Figure 17. Uplink and downlink total DSL end-to-end delays for the default and fast mode

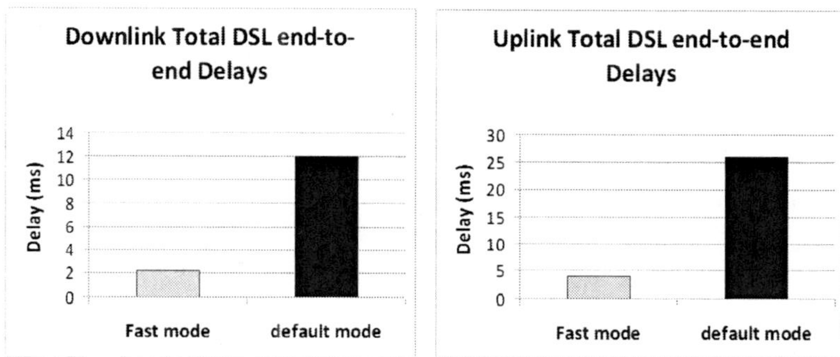

SUMMARY

During this investigation, effects of the DSL transport for HSPA end user performance have been studied. Different QoS parameters such as BER, delay and jitter have been analyzed for the two modes of DSL operations: DSL default mode and fast mode. The DSL bit errors causes the arrival of invalid packets whereas the DSL burst errors result in complete packet losses. All these DSL based packet errors can easily be recovered by the RLC layer without having impact on the TCP performance. Therefore, simulation results confirm that neither DSL random bit errors nor burst errors have a significant impact on the HSPA performance. Further analysis also confirms that the DSL delay and the jitter not affect the overall HSPA performance. In summary, the limited DSL network based simulations show similar performance which was achieved by the ATM based transport based simulations and therefore the expensive ATM transport links can be replaced by the cheap DSL links without having a severe impact on HSPA performance.

CONCLUSION

This article proposes using a traffic separation approach to transmit HSPA (HSDPA/HSUPA) and R99 traffic over a shared transport in the UMTS access network. In this work, we investigate the impact of using traffic separation, and explore its advantage compared to the cases without the traffic separation. The simulation results show that using traffic separation technique greatly improves the end user performance as well as the transport network performance, which in turn saves the transport bandwidth on the Iub interface while achieving the same QoS level. Therefore, we can conclude that deploying traffic separation between HSPA and R99 traffic can bring high cost efficiency in the transport resources in the UTRAN. Furthermore, we studied the ATM and IP based transport network for transmitting the HSPA traffic. From the investigations we find that the limited DSL network based simulations show similar performance which was achieved by the ATM based transport based simulations, and therefore the expensive ATM transport links can be replaced by the cheap DSL links without having a severe impact on the HSPA performance.

The future work of this study is to investigate the traffic separation for combing Long Term Evolution (LTE) technology with the current HSPA and R99 radio technologies in the same mobile backhaul network. Different radio networks can share the same IP transport.

ACKNOWLEDGMENT

This work is carried out with cooperation with Nokia Siemens Networks GmbH & Co. KG. Special thanks to Gennaro Ciro Malafronte, Stephan Hauth and Richard Schelb, who have contributed a lot of ideas into this work.

REFERENCES

Dahlman, E., Parkvall, S., Sköld, J., & Beming, P. (2007). *3G Evolution: HSPA and LTE for Mobile Broadband* (1st ed.). New York: Elsevier Ltd.

ETSI, *Universal Mobile Telecommunications System (UMTS)* (1998). *Selection procedures for the choice of radio transmission technologies of the UMTS (UMTS 30.03 version 3.2.0).*

Ferguson, P., & Huston, G. (1998). *Quality of Service: Delivering QoS on the Internet and in Corporate Networks* (1st ed.). New York: John Wiley & Sons, Inc.

Galeana, H., Ferrus, R., & Olmos, J. (2007, March). Transport Capacity Estimations for Over-Provisioned UTRAN IP-Based Networks. *Wireless Communications and Networking Conference (WCNC)* (pp. 4295-4300). Washington, DC:IEEE

ITU-T Recommendation G.992.2 (1998). *High bit rate Digital Subscriber Line (HDSL) transceivers.*

ITU-T Recommendation G.992.1 (1999). *Asymmetric digital subscriber line (ADSL) transceivers,* G.992.2 (1999). *Splitterless asymmetric digital subscriber line (ADSL) transceivers,* G.992.2 (2002). *Asymmetric digital subscriber line transceivers 2 (ADSL2),* G.992.2 (2003). *Asymmetric Digital Subscriber Line (ADSL) transceivers - Extended bandwidth ADSL2 (ADSL2plus).*

ITU-T Recommendation Q.2630.2. (2000). *AAL Type 2 Signaling Protocol (Capability Set 2).*

Li, X., Zeng, Y. Z., Kracker, B., Schelb, R., Görg, C., & Timm-Giel, A. (2008). Carrier Ethernet for Transport in UMTS Radio Access Network: Ethernet Backhaul Evolution. *Vehicular Technology Conference, 2008. VTC Spring 2008* (pp. 2537 – 2541). Washington, DC:IEEE

OPNET technologies Inc.(n.d.). *OPNET Modeler Accelerating Networks R&D.* Retrieved 12 4, 2008, from http://www.opnet.com

RFC 3916: *Requirements for Pseudo-Wire Emulation Edge-to-Edge (PWE3).* (2004).:IETF

RFC 3985: *Pseudo Wire Emulation Edge-to-Edge (PWE3) Architecture.* (2005).: IETF

RFC 4717: *Encapsulation Methods for Transport of Asynchronous Transfer Mode (ATM) over MPLS Networks* (2006): IETF

The, A. T. M. Forum (2000, July). *Addendum to Traffic Management V4.1 for an Optional Minimum Desired Cell Rate Indication for UBR.* Retrieved January 8, 2009, from http://www.ipmplsforum.org/ ftp/pub/approved-specs/ af-tm-0150.000.pdf

The 3GPP (2001). *Physical layer aspects of UTRA High Speed Downlink Packet Access* (3GPP TR 25.848 version 4.0.0).: 3GPP Technical Specification Group RAN.

The 3GPP (2002). *IP Transport in UTRAN Work Task Technical Report.* (3GPP TR 25.933 version 2.0.0).: 3GPP Technical Specification Group RAN

The 3GPP (2006). *FDD Enhanced Uplink; Overall Description; Stage 2* (3GPP TS 25.309 version 6.6.0).: 3GPP Technical Specification Group RAN.

The 3GPP (2007). *Enhanced uplink; Overall description; Stage 2.* (3GPP TR 25.319 version 7.2.0).: 3GPP Technical Specification Group RAN

The 3GPP(2001). *High Speed Downlink Packet Access (HSDPA); Overall UTRAN description.* (3GPP TR 25.855 version 5.0.0).: 3GPP Technical Specification Group RAN.

Türke, U., Winter, T., Perera, R., Lamers, E., Meijerink, E., Fledderus, E. R., et al. (2002). *Comparison of different simulation approaches for cell performance evaluation.* Retrieved January 5, 2009, from http://momentum.zib.de/ paper/ momentum-d22.pdf

Weerawardane, T. L., Li, X., Timm-Giel, A., & Görg, C. (2006). *Modeling and Simulation of UMTS HSDPA in OPNET.* Paper presented at OPNETWORK 2006, Washington DC.

Weerawardane, T. L., Timm-Giel, A., Görg, C., & Reim, T. (2006). *Impact of the Transport Network Layer Flow Control for HSDPA Performance.* Paper presented at IEE conference 2006, Sri Lanka.

Weerawardane, T. L., Timm-Giel, A., Malafronte, G. C., Gianluca, D., Hauth, S., & Görg, C. (2008). Preventive and Reactive Based TNL Congestion Control Impact on the HSDPA Performance. *Vehicular Technology Conference, 2008. VTC Spring 2008* (pp. 2296 - 2300).:IEEE

Zaki, Y., Weerawardane, T. L., Li, X., Timm-Giel, A., Malafronte, G. C., & Görg, C. (2008). Effect of the RLC and TNL Congestion Control on the HSUPA Network Performance. *Mosharaka International Conference on Communications, Computers and Applications, MIC-CCA 2008* (pp. 1-7).

Chapter 10
Strategic Scenarios for Fixed-Mobile Convergence:
An Integrated Operator Case

Jarmo Harno
Aalto University School of Science and Technology, Finland

K. R. Renjish Kumar
Aalto University School of Science and Technology, Finland

Mikko V. J. Heikkinen
Aalto University School of Science and Technology, Finland

Mario Kind
Deutsche Telekom, Germany

Thomas Monath
Deutsche Telekom, Germany

Dirk von Hugo
Deutsche Telekom, Germany

ABSTRACT

This study demonstrates that an integrated operator can benefit from cost savings, customer retention and prevention of revenue erosion by having a fixed-mobile convergence (FMC) migration strategy including introduction of advanced service packages. This development is driven by increasing importance of mobile network capabilities and services, as well as the decreasing gap between fixed and mobile systems, in terms of technological models and prices, making FMC both requested by the market and commercially feasible to provide. FMC is expected to offer benefits for network and service operators as well as businesses and consumers. The authors have analyzed the operator's dilemma on proper migration strategy in exploiting the benefits of cost savings and generating new revenues, but exposing oneself to the risk of substitution effects among its fixed and mobile products. They provide quantitative comparison of some strategic scenarios utilizing techno-economic case study methodology in modeling an integrated operator business in a Western European context.

DOI: 10.4018/978-1-60960-589-6.ch010

INTRODUCTION

Next Generation Services delivered via Fixed-Mobile Convergence (FMC) networks have been under discussion and consideration for several years. Though concepts and experimental implementations are widespread, truly operational experience is still lacking. Various players in the telecommunications value network follow different definitions of the FMC concept. Convergence may start either with common terminal equipment providing access to both fixed and mobile networks, or with a common billing and customer care center offering the user one-stop shopping. Various stages of convergence may be achieved on access and core network technology, management, service enabling, and content and application levels with individual or shared platforms.

In this chapter, we broadly define FMC as the end-to-end provisioning of unified services accessible by an end user independent of the underlying access and core network technologies. To enable an efficient realization of such an ecosystem, convergence has to occur at multiple levels, namely at the network, service, terminal and commercial level. One major enabler to achieve a seamless interconnection between all relevant entities is the use of a common underlying protocol infrastructure, which still is the Internet Protocol (IP). An overarching control platform for services, and underlying resources and transmission capacity, is the IP Multimedia Subsystem (IMS) as standardized and agreed upon in standardization organizations from both fixed and mobile industries.

Our study elaborates on a migration concept for an integrated operator from current separated traditional fixed and mobile networks towards FMC and IMS at different levels of service provisioning. The model investigates the impact on the overall profitability of the integrated operator. Our investigation considers different players in the FMC ecosystem, namely, the operators of access and core networks, service and content providers, hardware and software manufacturers and vendors, and legal authorities. Key drivers for industry development, for technology evolution and market demand are taken into account. The work is based on European CELTIC co-operation project ECOSYS (ECOSYS, 2004-2007) with partners from operators, universities, vendors and SMEs.

The chapter is structured as follows: After description of the players and the drivers in the FMC ecosystem, operators' motivations as well as strategic considerations for migration to FMC are compared. FMC framework and required investments assumed for the study are introduced, and considerations on OPEX (operational expenditure) are presented. Afterwards, the composition of FMC services offered is described and an elaboration of the common underlying geographical-economic model is given. The results for an integrated operator with and without FMC service provisioning are analyzed. The chapter concludes with an outlook on the next generation operator's future strategic decisions against the latest market development.

PLAYERS IN THE FMC SERVICE GAME

The highly complex FMC environment comprises the following players, who all are eager to gain a share of the value generated by FMC:

- Access and core network operators
- Transport network operators
- Service delivery platform developers
- Content providers that may produce content applicable for FMC devices
- Service operators offering value-added services to the end-users
- Software manufacturers providing client software for efficient switching between multiple radio technologies, terminal op-

erating systems, and desktop software vendors
- Handset manufacturers producing multi-radio handsets or smart cards
- Network vendors
- National regulatory and competition authorities

Some of the key drivers for FMC are described below:

- Industry drivers: From the industry's viewpoint some of the main drivers are: cost and investment savings, by eliminating redundancies and harmonizing the network and service management; faster time-to-market by having the ability to flexibly provide and deploy new and advanced services over multiple end-user devices simultaneously; and the ability to retain customers' loyalty by making access to services easy and convenient.
- Technology drivers: The main drivers include beside others: the progress in research, development, standardization and implementation of enabling technology platforms such as IP Multimedia Subsystem (IMS) (3GPP TS 23.228, 2008); availability of a growing number of multi-radio handsets.
- Market drivers: Some of the main drivers are: the customer demand for ubiquitous access of advanced content and services; single authentication, authorization and accounting (AAA) capabilities; ease of use; affordability of new services.

IMS is considered as the key platform required for migration to a fully functional FMC. However, operators have already taken the first step towards FMC with the deployment of transition technologies such as UMA/GAN (Unlicensed Mobile Access/Generic Access Networks) (3GPP TS 43.318, 2009). Examples of some such services are: BT's Fusion, NTT DoCoMo's PASSAGE DUPLE, and TDC's Duet (OECD, 2007), Vodafone One Net (Vodafone, 2010). The National Regulatory Authorities (NRAs) are following these developments and working towards amending existing policies and creating new ones that can enable the industry players in creating value, which will also benefit the end-users. A survey on the current situation in 18 EC countries stated that "along with the convergence of platforms, there is a clear trend towards bundled services, where operators offer a variety of services for a single global price often to the benefit of consumers. In this sense, end-users can benefit from fixed telephone and mobile access, IPTV and broadband (BB) internet connection through one operator's network" (ERG, 2009). Bundling mobile access with WLAN hot spot connectivity at an attractive tariff is a business case for fixed-mobile operators to use WLAN to offload expensive heavy data traffic from the cellular network (Intellinet, 2009).

Considering all these aspects of FMC, the chapter investigates the case of an integrated operator in a hypothetical Western European market setting, in its quest for migration to FMC.

MOTIVATION AND STRATEGIC CONSIDERATIONS OF AN INTEGRATED OPERATOR

As already mentioned, an integrated operator is characterized here as the owner of both fixed and mobile networks in a market, initially having a clear separation of the fixed and mobile operations' business units, i.e., each business unit having its own customer care, marketing, subscription management, network and service provisioning units. Traditionally, an integrated operator has been one of the incumbents in the market, often having a leading market share. Figure 1 illustrates some of the major reasons for an integrated operator's interest in migration to FMC from its existing situation.

Figure 1. Issues in integrated operator's business

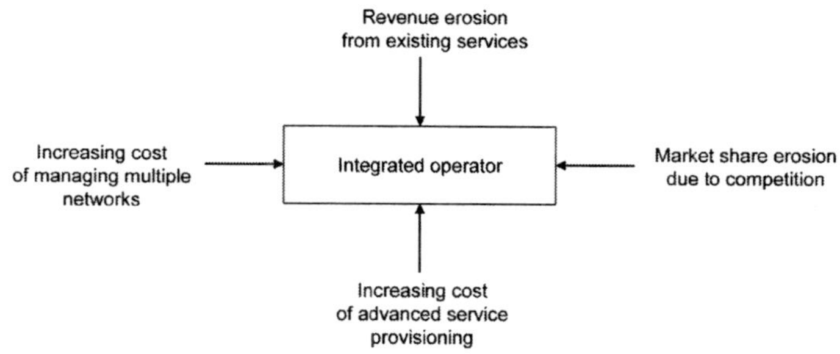

In order to solve the issues (as mentioned in Figure 1), integrated operators are motivated to deploy FMC, which will enable provisioning of advanced multimedia services easily and efficiently with a shorter time-to-market, across multiple access networks (with a common IP-based core network). Such services are expected to retain the loyalty of customers (i.e., reduce churn), and to generate new sources of revenue.

However the provision of a full FMC service portfolio also introduces new risks and challenges to a traditional network operator. New FMC services may substitute existing profitable products and thus cannibalize an operator's business case. As an alternative to caring for both high-quality connectivity through access network capacity, core network transport, service deployment, and provisioning, an operator may decide to concentrate on connectivity, together with network services as mobility, QoS, 4AC (i.e. AAA including Auditing and Charging), location information etc.. The following sections elaborate the study of an integrated operator's migration to FMC with and without providing full service control. The study investigates the impact on profitability, and discusses the related issues and possible solutions.

FRAMEWORK AND INVESTMENTS

An integrated operator, with ownership of both fixed and mobile networks (access, transport and core) is considered in this study. Such an operator may offer ubiquitous mobile services, which are purely based on cellular networks (GSM, 3G) in rural areas, complemented by Wireless Local Area Network (WLAN) hot spot coverage in popular public areas (mainly urban), whereas in indoor areas, the service offered may be a combination of home xDSL (Digital Subscriber Line), and WLAN, or even WiMAX (Wireless Interoperability for Microwave Access). The evolving cellular technology such as 3GPP's LTE (Long Term Evolution) and LTE-Advanced have not been included in the investigation due to time constraints. Nevertheless, with extended data rates and reduced transmission latencies that are in the order magnitude of the fixed access and the further progress towards "4G", the convergence between fixed and mobile continues: for example, the work plan for 3GPP's Release 11 discusses about "Management of Converged Networks" and clearly views IMS as the "Fixed-Mobile Convergence Core Network" to provide multimedia services (3GPP Workplan, 2010). Also the trend to move service and traffic to a pure packet domain with a common all-IP network in LTE prepares for fixed-mobile convergence (FMC), and enables new business models for triple-play and quadruple-play services (UMTS Forum White Paper, 2009). To achieve full FMC, both fixed and radio networks are based on IP. The IMS platform specifies interoperability and roaming, provides charging and security, thus making it easier to achieve service interoperability across different operators in terms of connectivity. The integrated operator will have a better potential compared to pure-play fixed or mobile operators, present only in either the fixed or the mobile business, to offer a single bill, a single number, as well as a set of services that may be delivered independent of, or specifically adjusted to a user's location, terminal capability and available access technologies.

In the modeling of integrated operator's migration to FMC, two key enablers, IMS and UMA/GAN were considered. IMS was considered as the de facto service control and execution architecture for next generation convergent all-IP networks by operators and standard bodies during the course of our analysis. Though it still is considered as the base framework for delivering convergence services, full-fledged IMS deployments are still missing. Instead, the discussion has been focused around adopting Service Delivery Platforms (SDP) vis-à-vis IMS and whether they are competing or complementary architectures (Current Analysis 2008). SDP is today widely adopted by operators worldwide as a unified platform for offering online as well as traditional services.

Similarly, adoption of femtocells (also known as Home NodeB in UMTS and Home eNodeB in LTE parlance) were still at an early stage of consideration by operators worldwide at the time of our analysis. WLAN penetration was rapidly growing and UMA/GAN was embraced by some of the operators and 3GPP. Hence, UMA/GAN was considered as our choice as the pre-IMS enabler. Today, femtocell adoptions have also picked up and major operators have already started deploying them in their networks (Internet Telephony 2010). Nevertheless, our analysis model is fully extensible in incorporating changes reflecting these new developments.

We first describe briefly the functionalities, dimensioning, and other assumptions of the different elements constituting IMS and UMA/GAN. IMS, as previously mentioned, is a platform for common control of integrated services including access and resource management of both wired and wireless technologies is seen as the prerequisite for achieving complete and true FMC. UMA/GAN is considered as a pre-IMS solution deployed with the primary goal of migrating existing customer base to FMC services and retaining the market share. Therefore, UMA/GAN's main contribution is in preventing the decline in existing revenues rather than generating new revenues.

The architectural components of UMA/GAN and IMS considered in our study are described as follows:

UMA/GAN Architecture and Cost Elements

UMA/GAN is a mechanism at the access network, which enables the seamless provisioning of circuit-switched (CS) services (such as GSM voice calls) over packet-switched (PS) unlicensed WLANs and broadband networks. The current 3GPP standard for UMA/GAN not only supports GSM and GPRS services (via the A/Gb interface) but includes also a mode to support UMTS/WCDMA (via the Iu interface) (3GPP TS 43.318, 2009). In the assumed scenario, UMA/GAN is deployed in the beginning of the modeling period as a provisional solution (IMS VCC was foreseen to be deployed in 2009[1]) to offer a user voice services across the CS GSM network at their homes via WLAN APs attached to a broadband access network. Such an overarching provisioning of voice calls enables operators to add value to the existing voice service of customers (by offering lower priced calls at home) and also allows for more efficient management of traffic load.

The UMA/GAN architecture considered for the model is shown in Figure 2. The key architectural elements considered in our study for UMA/GAN are as follows:

- Access Point (AP): An AP operates at the unlicensed 2.4 GHz and 5 GHz spectrums offering a local area network access to mobile handsets. These APs support IEEE 802.11 a, b and g standards. A single site

Figure 2. UMA/GAN network architecture

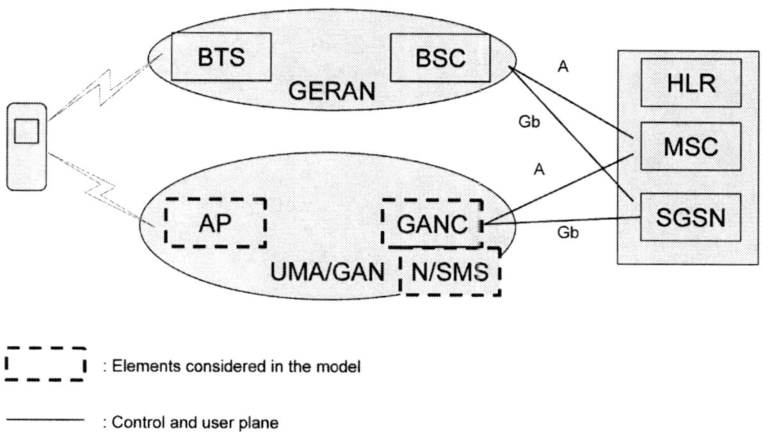

may have one or more APs depending on the capacity of each AP and the traffic demand.

- Generic Access Network Controller (GANC): A GANC acts as an aggregator of APs and performs Authorization, Authentication and Accounting (AAA) functionalities for the customers accessing services through APs. GANC is the equivalent of Base Station Controller (BSC) in GERAN.
- Network/Service Management System (N/SMS): An N/SMS element consists of network and service related management systems essential for managing the GANCs and services offered over APs.

These three cost elements are associated with UMA deployment in the business analysis with GANC contributing to 12%, APs to 2% and N/SMS to 1% of CAPEX during the study period.

IMS Architecture

IMS is generally accepted to deliver SIP-based session-oriented applications, including voice, dual-mode/FMC, unified messaging, presence, video sharing, enterprise integration (e.g., IP Centrex, Mobile PBX, Hosted Call Center), online and mobile games, group chat, and push-to-talk/push-to-video.

The IMS architecture used for the model is shown in Figure 3, where functionalities are mapped to physical components. The user equipment (UE) accesses IMS through an access network. The heart of the IMS system is the Call Session Controller (CSC). It is connected to a Home Subscriber Server (HSS) and two media controllers: a Media Resource Function Controller (MRFC) and a Media Gateway controller (MGC). AS refers to a generic application solution, which is basically a service delivery platform (SDP) offering both operator's own and third-party services. Support Functions (SF) are an estimate of necessary PS transfer (TSF) and business support system (BSS) upgrades due to IMS. VCC (3GPP TR 23.806, 2007) functions are provided for seamless voice call handover in different access networks both in CS and PS domains including Supplementary Services support like Line Identification, Call Forwarding, and Call Barring.

IMS Dimensioning

One of the main tasks of the IMS part of the FMC model is to calculate the annually required capacities for each IMS component, thereby determining

Figure 3. IMS architecture

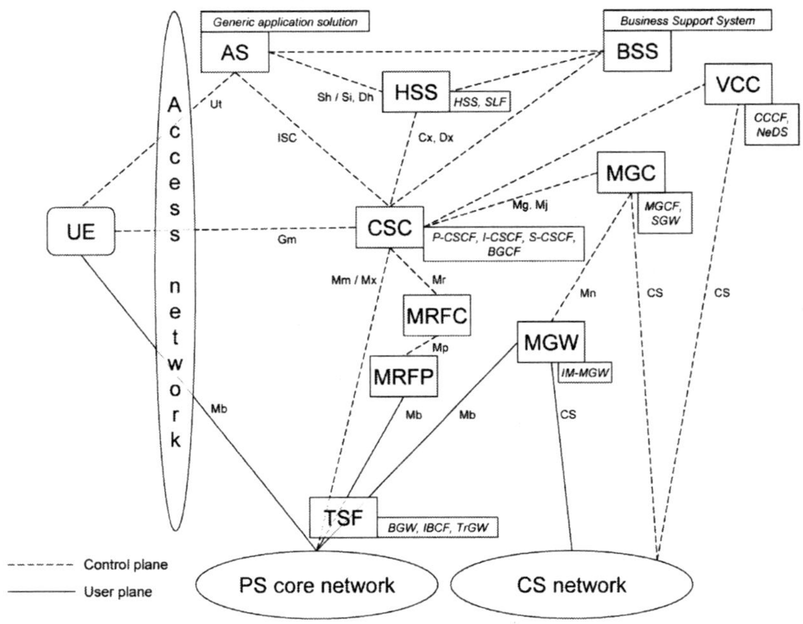

the number of each IMS component required for the rollout, i.e., the "shopping list", as shown in Figure 4. The main variables affecting the shopping list are service usage and the number of FMC subscribers. Usage translates into three technical dimensioning parameters. The number of subscribers is dependent on population, IMS compatible device penetration, IMS service penetration, and operator's market share.

Figure 4. IMS dimensioning general structure

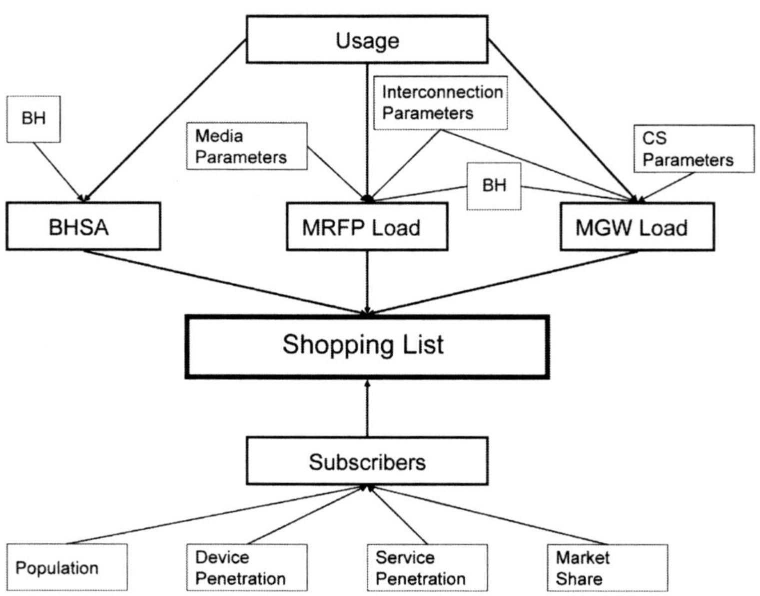

The input data used in dimensioning are based on insight within the ECOSYS consortium, six open semi-structured expert interviews conducted during November and December in 2006, analysis of IMS products from 11 vendors, and several academic, analyst and industry sources.

The dimensioning rules and the included functions of IMS components are presented in Table 1 based on expert opinions and IMS product analysis. The reference prices are for the year 2007, VCC being an exception with the reference year 2009. The components are subjected to price erosion defined later.

BHSA refers to Busy Hour Session Attempts. It is calculated from service usage by applying session attempt (SA) and a busy hour (BH) multipliers to the usage data. For a call, 1.0 SA is assumed. Other service dimensioning units are related to it: a message 0.5 SA; a presence update 0.2 SA; and a service setup 0.9 SA. BH usage is estimated to be 20% of total usage. For the VCC component only VCC related call usage is taken into account.

The ports number used with the MGW component refers to simultaneous CS calls processed. For calculating that, CS voice penetration and CS call interconnection rates are assumed. The final result is derived for the BH using Little's formula (Durrett, R., 2001), with a uniform distribution and average call lengths.

The MRFP component employs actual data transfer capacity. It is used for audio and video conferencing, network announcements, and voice mail. Estimates of shares of calls, average session lengths, data rates, and occurrence probabilities are used to yield data volumes. The end result for the BH is derived similarly as in the MGW case.

The IMS media and interconnection parameters, i.e., parameters related to MGW and MRFP, are presented in Table 2. They are derived based on expert opinions and a technical standard (3GPP TS 26.236, 2008). For both audio and video conferencing, 3.5 users per conference with a 39 kbps stream for audio and a 74 kbps for video per each user are assumed. Announcements and voice mail are played with a 39 kbps stream. The call interconnection departure rate is the number of calls departing to other networks out of the total number of calls. The call interconnection arrival rate is related to the number of calls arriving from other networks.

The number of subscribers is a multiplication of population, IMS device penetration rate, IMS service penetration rate, and operator market share. These estimations are handled in Chapter VII. Table 3 depicts other IMS time dependent parameters based on large set of analyst reports and

Table 1. IMS components' dimensioning

Comp.	Capacity	Unit	Included functions	Price (€)
AS	2 500 000	subscr.	Generic Appl. Solution	2 200 000
CSC	2 000 000	subscr.	P-CSCF, I-CSCF, S-CSCF, BGCF	3 000 000
	2 000 000	BHSA		
HSS	2 500 000	subscr.	HSS, SLF	1 100 000
MGC	2 000 000	subscr.	MGCF, SGW	800 000
MGW	25 000	ports	IM-MGW	2 300 000
MRFC	2 000 000	subscr.	MRFC	600 000
MRFP	1	Gbps	MRFP	2 400 000
SF	2 500 000	subscr.	BGW,IBCF,TrGW,BSS	900 000
VCC	2 000 000	BHSA	CCCF, NeDS	800 000

Table 2. IMS media and interconnection parameters

Voice conference share (% of calls)	0.1%
Voice conference average length (min)	30
Voice conference data rate (MB/min)	1.02
Video conference share (% of calls)	0.05%
Video conference average length (min)	30
Video conference data rate (MB/min)	1.94
Announcement probability	10%
Announcement average length (min)	0.2
Announcement data rate (MB/min)	0.29
Voice mail probability	5%
Voice mail average length (min)	0.5
Voice mail data rate (MB/min)	0.29
Call average length (min)	1.50
Call interconnection departure rate	50%
Call interconnection arrival rate	200%

studies, combined with in-project expert judgment (ARCEP, 2005; Bundesnetzagentur, 2009; Cap-Gemini, 2005; Informa, 2006; International Game Developers Association, 2005; Ofcom, 2006; RealNetworks, 2005; Strategy Analytics, 2006; UMTS Forum Rep. 33, 2003; Verkasalo, H., 2007),. The CS voice penetration rate is used in MGW traffic calculations. The VCC penetration rate is utilized in load calculations of the VCC component. The component price erosion depicts a common falling price curve for the IMS components. The usage figures are based on per user per month (PUPM) basis, and combine both downlink and uplink usage.

OPEX Considerations

The operational expenditures for an integrated operator can be divided in to seven general categories listed and described in the following:

- **Network related elements**: includes all the necessary costs for network operation; operation, maintenance, optimization and repair of the network elements, equipment and software licenses, rental of network resources, costs for site rental and electricity.
- **Interconnection and roaming costs**: termination fees for calling or completing a call or a session originated or terminated in another network.
- **Marketing and sales related elements**: costs including advertisement, customer acquisition, SLA (Service Level Agreement) negotiation and subsidization.
- **Customer service related elements**: costs associated to customer care and CRM (Customer Relation Management) operation.
- **Charging and billing**: includes traffic metering, accounting and controlling.
- **IT and general support related elements**: includes Business IT, management support, and costs regarding the purchase of licenses for content delivery.

Table 3. IMS time dependent parameters

	2007	2008	2009	2010	2011	2012	2013	2014
VCC penetration	0%	0%	50%	45%	40%	35%	30%	20%
Component price erosion		-10%	-5%	-2%	-2%	-2%	-2%	-2%
Calls (PUPM)	230	234	238	242	245	250	249	252
Messages (PUPM)	48	71	89	110	126	144	166	189
Presence updates (PUPM)	48	101	184	268	330	372	389	413
Service setups (PUPM)	6	10	19	32	49	71	91	107

- **Service development related elements:** includes new service related market research, design and development as well as integration to operational and business support systems for provisioning, management and monitoring

Costs within these categories have only been considered here in case they are affected by a potential operator decision for migration to FMC — in terms of savings and additional expenses affected by the FMC decision:

- Costs related to network elements and interconnection and roaming will generally be reduced for a convergent network with less elements and common management system.
- Marketing and sales, customer services, and IT and general support related elements are characterized by additional effort in the beginning, for example in terms of processes for and advertisement of the new FMC subscription, whereas towards the end of the study period savings thanks to unified products and a single interface towards the customer will occur in the FMC case.
- Marketing and development of services will become cheaper as the same services for both mobile and fixed systems have to be considered only once

The presented OPEX factors are estimated against the yearly development of the FMC migration, revealing that marketing and customer care categories will contribute the most to the overall cost reduction. A year by year breakdown of the induced cost changes concerning the OPEX categories will be given in the RESULTS section.

FMC SERVICE OFFERINGS

Modeling of the offered services and service bundles is an essential part of the scenario description.

The starting point for the FMC products is a converged voice product, which is enhanced over time with converged data services. The FMC integrated operator will initially offer four options for the customers: the FMC Bronze, Silver, Gold and Platinum packages (shown in Table 4).

Figure 5 illustrates the distribution of users that will use the FMC products, among the different offered packages. In addition to the components mentioned in Table 4, each of the above packages also includes some FMC services (e.g., free access to Wi-Fi spots, VoIP, instant messaging, etc.) as part of the bundle. The reasoning behind such free FMC services in the early stages of FMC introduction is to attract customers to these packages. The monthly revenue per FMC customer is calculated as the sum of the separate subscriptions (fixed voice, mobile voice, fixed BB and mobile data packet) minus a reduction of 10%. In addition to this, price of each FMC package, mentioned in

Table 4. FMC services

	Voice (mobile/ fixed unlimited national calls)	Fixed BB Internet	Mobile data quota per month	Initial tariff per person per month
Bronze	+	+	0	45 €
Silver	+	+	20 MB	50 €
Gold	+	+	200 MB	69 €
Platinum	+	+	2 GB	93 €

Table 4, as well as non-FMC service is assumed to decline by 5% annually. Such a price cut by as much as 15% in three years has been expected since some time (Exane BNP Paribas & Arthur D Little, 2006)

Upon these package based monthly fees, an FMC customer can subscribe to certain new services for additional fee. All these additional services are presented in Table 5. In our study, we have made a differentiation between FMC services being completely new services (such as IP/ Mobile TV, and Unified Communications), not fully available in non-FMC networks, and IMS-based services (such as instant messaging, and voice over IP) that potentially substitute the established service and thus cannibalize operator's revenues. Cannibalization means that customers adopting an instant messaging (IM) service partly discontinue the use of traditional SMS, and usage of Voice over IP (VoIP) will reduce the Circuit Switched (CS) voice revenues. However, for example SMS is still utilized to reach those users not yet migrated to IM. Also, usage of community-based services (groups, social networks, blogs etc.) will have some impact on SMS, with a raising effect on usage for those still participating through for example SMS-IM gateways. Unified Communication (i.e. offering all person-to-person communication services with single-sign on) will primarily be focused on business customers, adding value to the current person-to-person offerings (AT&T, 2008). Growing demand has been identified for location information by major market researchers (e.g., Gartner, 2006; Juniper, 2006; Canalys, 2006; Canalys, 2008). Location information, together with enhanced profile and presence information, increases the convergent service possibilities, both communication and content related. Improving interaction and communication between different electronic devices, over the air, increases further the benefits and attractiveness of the FMC offering.

Expected additional monthly revenue for the FMC operator is 10 € per subscriber of the new FMC services together (excluding Unified Communications) and 10 € for Unified Communications (intended mainly for the business customers). Unlike the FMC packages previously mentioned, no price erosion has been assumed for these new FMC services due to increasing customer value during the study period. We have assumed 30% operating profit margin for these services.

The new services will have an impact on transport costs, but assessing the current packet based network and radio technologies development, we estimate that the services can be provided feasibly in the framework of the mobile data quotas introduced in the packages above. The investments in core IMS are calculated through IMS dimensioning procedure presented in the

Figure 5. Services distribution

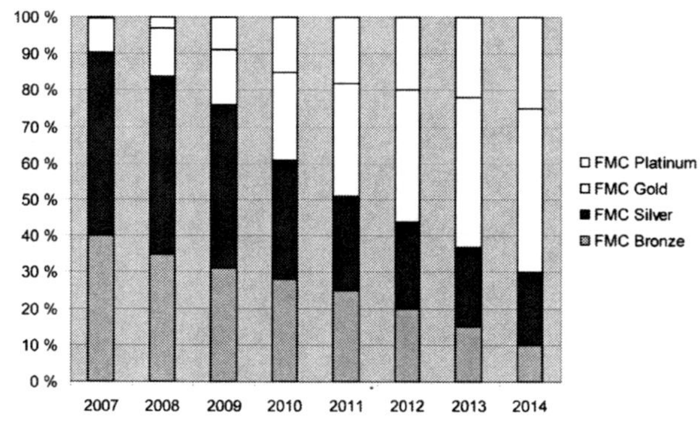

Table 5. New and substituting FMC services

FMC Services	FMC Service Enablers ("x" here represents the potential utilization for enhanced services)			% of customers adopting FMC services during study period		Service character	
	Location info	Presence/ Profile	QoS	2007	2014	New	Substitute
Instant messaging	X	x		2	30		x
Voice over IP	X	x	x	1	50		x
IP/mobile TV	X	x	x	0,5	15	x	
Network Gaming	X	x	x	0,5	20	x	
Video on demand			x	1	25	x	
Unified data access	X	x		1	50	x	x
Community- based services	X	x		1	50	x	x
Unified Communications	X	x	x	1	17	x	

previous chapter. In addition to that, investments for additional enablers, like application servers and support systems, are estimated to be 70 M€ per 2.5 million FMC subscribers during the study period.

GEOGRAPHICAL-ECONOMIC MODELING

The country type considered for the market modeling is a "Large Country" in Western Europe such as France, Germany, Italy, or the UK. The market's figures are calculated as averages of France, Germany, Italy and the UK. The total population is 65.2 Mio (with a yearly growth of 0.3%), the average household consists of 2.3 persons, while the surface area is 370,000 km^2 (European Communities, 2007). The overall surface area is not the sum of all the modeled areas (dense urban, urban, suburban, rural) because certain areas (lakes, mountaintops, etc.) are not covered. The study period is eight years, starting from year 2007[2]. The integrated operator has a strong market position in the beginning: 60% in fixed telephony, 50% in BB and 40% in mobile. Fifty percent market share of converged customers is assumed, if entering to the FMC path. If not, it is estimated that half of the potential FMC subscribers are lost. FMC offering can also lure new customers, but this has not been considered here. The basic assumption relating to competition has been that, if going for the FMC services the incumbent can keep its market share, but if not, the competing offerings will erode the incumbent's customer base, usage figures and revenues. Here we restrict ourselves to the situation of an operator within a saturated mobile market (such as North America or Western Europe) – the case may be quite different for emerging markets also studied within the framework of project ECOSYS (2004-2007). For the FMC case, we estimate the potential for integrated FMC customers from the penetration of IMS capable terminals and IMS service demand. Although the breakthrough is expected by UMTS Forum to take place only after 2012 (UMTS Forum Rep. 37, 2005), in our case where the operator strongly promotes FMC and IMS, we end up with more steady growth from the beginning (see Figure 6).

A report of EU regulators on FMC states that the lack of information on the number of FMC service customers (figures from only three countries were available) "seems to be related mostly to the fact that such services are in their initial stage of development" (ERG, 2009). On the other hand, the percentage of Europeans whose household buys two or more communication services as part of a bundle, range between 7%

Figure 6. Penetration of convergence customers for the FMC case

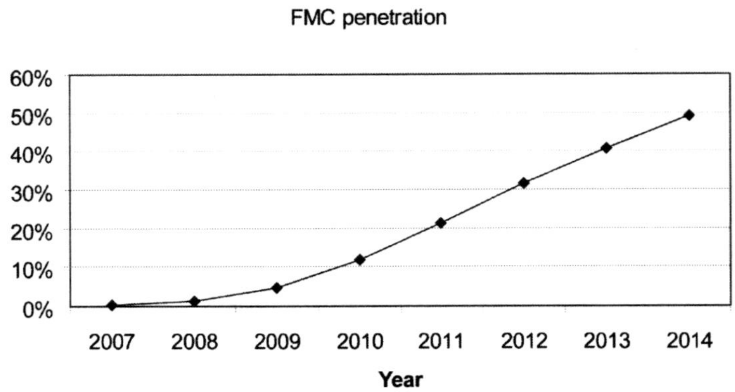

in Finland and 47% in Denmark with an average of 29%, i.e. an increase of 9% from 2007 to 2009, indicating the growing importance of FMC. Similarly, another study (Infocom, 2010) concludes that residential consumer customers seem to be more interested in simple product bundles offered at lower prices, rather than in truly integrated FMC services.

The analysis in this scenario focuses on calculating the effect of the integrated operator's decision, whether to migrate to FMC or not, on its overall business. This means that only the differences (i.e. delta) between the non-FMC and FMC (network and service) cases are considered. Revenues and costs that would exist regardless of FMC are not treated separately. In the beginning of the study period, the operator has two options: either to implement FMC, to begin the installation of the various IMS elements needed, and to start offering new services; or to continue with the existing condition, i.e., to continue operating two separate networks for fixed and mobile services.

RESULTS

The advantage of the FMC approach for an integrated operator is demonstrated as the delta difference between an FMC-driven strategy, and the expected development for a non-FMC scenario. This delta will show up in the operators' additional FMC related investments and additional operational expenditure (OPEX). The FMC approach affects also the market share and service demand figures, reflected in delta revenues related to both factors. In addition, the new convergence services add new profit streams to the business case. Cash flow comparison over the study period is provided in the Figure 7.

Application servers and the IMS related PS signaling and CS-PS gateways are the main sources of additional (delta) CAPEX for an integrated FMC operator. However, there are many sources of cost uncertainty dependent on the existing architectural structures of the operator. These include support function upgrades, service realization aspects and service usage. The extent of required business support system (BSS) updates due to IMS will be highly dependent on operator and service requirements. Operators with outdated legacy systems will face higher additional costs than operators with up-to-date systems.

Compared to the estimated rather small FMC-investments, i.e. 721 M€ over the whole study period, related to the FMC customers (without considering the network migration towards IP) there is a huge potential for FMC operator in OPEX savings.

Figure 7. FMC case cash flow effects for the integrated operator

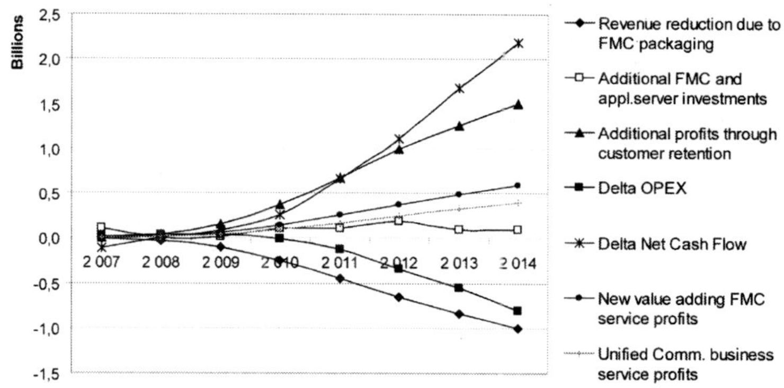

At the end of the study period, the yearly OPEX savings due to convergence are estimated as substantial, over 800 M€ in 2014, which is about 7% of the total yearly OPEX of the operator (see Figure 8). From the results, we can see that after some years, the operator will reduce its OPEX compared to the non-FMC situation. The trend in the OPEX savings is growing even after the study period, giving a considerable competitive advantage for the integrated operator selecting the FMC path.

Major contributors to overall OPEX over the entire study period are customer care (40%), network maintenance and management (19%), IT support (15%), and expenditures for marketing, sales, and customer acquisition (12%). Roaming regulation and expected interconnection models for IP networks are still uncertain and may not differ very much from the non-FMC situation, whereas figures for marketing will increase in the beginning as new products and services have to be established. The saving increases towards end of the study period because a concentration of converged products will occur. Similarly, the customer care effort will be reduced because of the reduction of separate subscriptions within the integrated converged network compared to two separate networks and multiple subscriptions per person.

If going for FMC, the operator is estimated to lose revenues, since the tariffs of the new products

Figure 8. FMC case OPEX (delta) breakdown (negative values show OPEX savings)

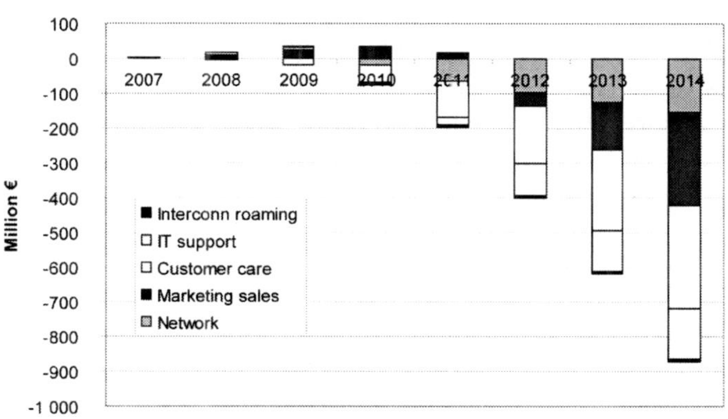

should have a discount compared to the sum of the mobile and fixed tariffs in the non–FMC case.

On the other hand, we estimate that if the integrated operator does not invest in FMC, it will start losing customers to other operators that are driving FMC. We estimate that in the non-FMC operator case the market share in all sectors, mobile, fixed voice and BB goes down due to competition and lower price bundles offered by the FMC operators. This is estimated to have a big positive effect in the long run, even 1500 M€ yearly impact in the end of the study period. The combined impact on profitability by these revenue streams sums to 500 M€ yearly delta net cash flows in 2014, as taking into account also the negative revenue "cannibalization" effect of the lucrative product bundles (see Figure 7).

As being in a strong position to provide the new services for its converged customers, the FMC operator is estimated to produce additional profits in the residential and business markets with total value growing up to almost 1000 M€ in the year 2014 as seen in Figure 7.

Combining these effects, the integrated operator selecting the non-FMC path is foreseen to be in a much weaker position in the end of the period, in this kind of competitive situation. As seen in Figure 7, the yearly net cash flow gap towards the FMC scenario may sum up to over 2000 M€ in the end, endangering the overall profitability of the non-FMC integrated operator.

Nowadays an incumbent integrated operator faces the challenges of growing competition by new entrants in the form of (Internet) service providers and meanwhile well established mobile operators with lower operational costs and generally higher flexibility to market dynamics. According to the results of this study, the negative effect of correspondingly resulting customer and revenue loss can be fought back by the introduction of FMC and the integration and cooperation of the fixed network operations with the mobile counterpart.

Only together with the use of common interoperable standards for both fixed and mobile networks and their integration, at least at the level of common service provisioning platforms (application plane), simplifying the deployment of intuitively usable converged services, with unified marketing (single bill), will the customer attractiveness and OPEX savings increase and stabilize profit and customer figures.

Further investigations are required to assess the impact of new integrated network and service platforms on the overall system performance, QoS, security, scalability and the potential for easy and fast deployment of new services (service modeling), including the converging front office and back office IT systems of the operator.

CONCLUSION AND DISCUSSION

Players in the ICT ecosystem are identifying strategies to maximize their benefits from convergence, driving new revenue streams, and reducing costs. Our study presents a techno-economic analysis of an integrated operator, demonstrates the impact of FMC approach on the operator's business, and compares FMC to sustaining with separate fixed and mobile business operations.

The results demonstrate increasing cost savings especially in terms of OPEX in later years of the study period. We expect that these savings can be sustained, thus giving the integrated operator a competitive advantage vis-à-vis operators without the convergence strategy. In addition to the estimated cost savings, the integrated operator gains better possibilities to retain its market share, which might have even wider impact on profitability than the potential cost savings. The third benefit is that the FMC strategy enables the operator to compensate the unavoidable revenue erosion from existing services by introducing packages with new FMC services for the advanced customer base. Despite our rather cautious assumptions, the results show clearly positive impact of the FMC strategy. Further work is required to investigate deeper the new service provisioning approaches, platforms, partnering and business models, and their effect on the overall business.

The suggested approach is a challenge for the operators, as it is not mainly a technological development step, but a strategic market approach and organizational change process. Although there has been plenty of technical supply from the equipment vendor side and most Western-European integrated operators have even had IMS trials starting from 2005 (for example, TIM in Italy and TeliaSonera in the Nordic countries), for most part of the industry the implementation of full fixed-mobile convergence has been delayed from the schedule we suggested.

The development in mobile technologies has been fast in recent years. A market disruption, where the mobile broadband through HSDPA is challenging the fixed DSL access, is at hand. In Finland, the number of DSL accesses started to decline in 2008, as customers migrated to mobile broadband, which reached a 30% share of all BB accesses in mid-2009 (FICORA 2009). This is not exactly what is meant with seamless service convergence, but one indication that through technology development the clear-cut separation between fixed and mobile access is vanishing, and the organizations and business architectures have to be adapted to that.

In recent years, many restructuring efforts have in fact been launched. In 2006, France Telecom rebranded its ISP operations (previously Wanadoo) under its mobile brand Orange and started to provide both mobile and fixed IP services together in various packages. The ordinary fixed telephony was still left as separated unit. The pressure towards restructuring has been high in France, causing even friction and problems among the personnel, which can be understood taking into account the long history of France Telecom as a state owned incumbent. Full integration of the fixed and mobile networks and services is still to be implemented, but IMS pilots for mobile multimedia services have started in 2009, with even interoperability tests between the major French mobile operators Orange, Bouygues Telecom, and SFR (Cellular-news 2009).

In spite of continuous development efforts and trials, Western-European markets have been generally lagging behind. In most cases, the fixed and mobile divisions of the operators are still in their own organizations, and even competing against each other in certain services like in the case of mobile broadband (HSDPA) and DSL access. In 2009, Deutsche Telekom announced that it will start merging its fixed and mobile divisions in Germany in the first quarter of 2010 using the One Company concept (Deutsche Telekom 2009). The designated CEO of the DT German operations has commented that T-Home's and T-Mobile's service and sales departments have already before that been 80 percent integrated and the marketing, technology and products will follow suit (Telecompaper 2009). In April 2010, Deutsche Telekom started to offer its consumers and business customers in Germany everything, whether fixed-network or mobile, DSL or mobile Internet, from a single source under the common corporate "T" brand.

The launch of the iPhone in 2007 marked a paradigm shift in the mobile communications ecosystem, as a device vendor managed to take a dominant role in the mobile service value network. Apple took a strong position in the offering of the whole packaged mobile solution to the customer. Other actors like Google (providing Android operating system for multiple vendors), Nokia (Symbian, Meego with Intel), RIM and Microsoft (Windows Phone) are following in row. Mobile phones as extremely personal items, contrary to home phones, and create increasing possibilities for market capture and customer retention through personalization. A mobile phone is even a personal style statement, just as a person's car is (Madeley 2007). People are increasingly using third-party mobile applications (such as email, Skype, Google

Voice, and Twitter) in their smartphones with data plans to replace circuit-switched voice and SMS.

The prices of many European operators for the premium mobile Internet packages are in line with the Platinum FMC package in our modeling, when only the mobile access part of the package is counted. In Finland, for example, the premium service charge per month for 1000 min voice, 1000 SMS, 100 MMS, 1GB data and Wi-Fi hot spot access in connection to the iPhone package is about 62€ per month in a 24-month contract (Sonera 2010). In the modeled Platinum package the price for 2 GB mobile data, unlimited voice minutes, Wi-Fi hot spot access and basic services like SMS and instant messaging was 93€/month (meaning under 80€ in 2010 with the price erosion), including additionally fixed BB and voice as part of the household deal. In Germany, Deutsche Telekom offers fully unlimited domestic mobile data, voice calls and 3000 SMS for the same price level as TeliaSonera in Finland with their iPhone premium package (Deutsche Telekom 2010a), and the unlimited mobile data applies even for the lower-level packages, whereas Sonera utilizes more limiting quotas for them. Noteworthy is also that in both cases the "Mobile Data Offload" tactics is utilized as combining the free hot spot Wi-Fi access into part of the package, thus providing high speed data access and at the same time reducing the mobile network load. Home DSL access is not yet combined in these packages; neither are the IMS based services such as Rich Communication Suite, including features such as enhanced phonebook and messaging, or enriched call with multimedia content (UMTS Forum White Paper, 2010).

However, some discounts are offered for customers combining the mobile and domestic fixed access packages (Deutsche Telekom 2010b). DT presented in 2010 also a packaged small office solution for converged fixed, mobile and IT communications, where a standardized user interface acts as a communication centre to integrate all communication channels such as telephony, e-mail, SMS, instant messaging and Internet platform for collaboration (Current Analysis 2010).

The development inside integrated operators confirms that despite the tough implementation task, the challenges and benefits modeled in this techno-economic delta analysis are so remarkable that the restructuring and fixed-mobile service alignment is largely seen inevitable and taking place. Though the eventual realization of FMC may be based on enablers different from those that we have considered (IMS and UMA/GAN) for our analysis, the philosophies they represent would remain the same. Most critical for the operators' successful FMC strategy is to identify compelling service use cases for the customers, which until now has been an elusive goal.

With the emergence of 4G/LTE technologies, offering fixed broadband-like bandwidths, the gap between a fixed and mobile access network is steadily narrowing, promising similar customer experiences across fixed and mobile networks. This can act as a key driver for enabling FMC, unlike in the case of 3G with lower access bandwidth. We therefore believe it is necessary for operators to revisit their FMC propositions and rebuild their business strategies reflecting the changes in available technologies and new market realities.

The realization of the new technologies and services combining IT applications, media and Internet require completely new business architecture and co-operation between the major forces in the market. Old organizational and technology-related constraints need to be removed to be competitive in the new communications business. This development, started in one form or another inside most operators, requires from them – to be able to maintain their position in the new ICT and media value network – in addition to keeping up with network and terminal evolution, also thorough service orientation, as the competitive added value is migrating from pure communications to network transparent experience of increasing choice of services, applications and content.

ACKNOWLEDGMENT

The authors would like to acknowledge the support, ideas and contributions of their colleagues from ECOSYS project and the project funding through the EUREKA/CELTIC program.

REFERENCES

ARCEP (2005). Annual Report.

AT&T. (2008). *Fixed-Mobile Convergence Designed to Lift Enterprises to a Higher Level of Performance*, White Paper. Retrieved January 2009, from http://www.business.att.com/enterprise/resource_item/Portfolio/enterprise-mobility-enterprise/Whitepaper/fixed_mobile_convergence_pov/

Bundesnetzagentur (2009). *Tätigkeitsbericht 2008/2009* (in German)

Canalys (2006). *Research release 2006/081*. Retrieved January 2009, from http://www.canalys.com/pr/2006/r2006081.pdf

Canalys (2008). *Global Mobile Navigation Device Shipments Hit 39 Million in 2007*. Retrieved January 2009, from http://www.canalys.com/pr/2008/r2008031.htm

CapGemini (2005). Growing Mobile Data Revenues, Opportunities in Infotainment. *Telecom & Media Insights,* (11), November 2005

Cellular-news. (2009). *Bouygues Telecom to Trial Alcatel-Lucent's IMS Platform*. Published 28 May 2009 in http://www.cellular-news.com/story/37708.php

Current Analysis. (2008). *IMS Status Report: A Protracted Adoption – Advisory Report*. Published 20 June 2008 in http://www.currentanalysis.com/m/ericsson/CurrentAnalysis-IMS.pdf

Current Analysis. (2010). *DT Offers Fresh Approach to Converged Fixed, Mobile and IT.* Published 8 Mar 2010 in http://www.currentanalysis.com/h/2010/DT-DeutschlandLAN.asp

Deutsche Telekom. (2009). *Invitation to the Extraordinary Shareholders' Meeting*. Bonn, October 2009

Deutsche Telekom. (2010a). Prices retrieved 9 November 2010, from http://www.t-mobile.de/iphone/tarife/0,18383,22271-_,00.html

Deutsche Telekom. (2010b). Retrieved 9 November 2010, from http://www.telekom.de/is-bin/INTERSHOP.static/WFS/EKI-PK-Site/EKI-PK/-/content/static_html/061102_290vorteil/index.html

Durrett, R. (2001). *Essentials of Stochastic Processes*. Harrisonburg, VA: Springer.

ECOSYS. (2004-2007). *ECOSYS project website*. Retrieved January 2009, from http://www.optcomm.di.uoa.gr/ecosys/deliverableslist.html

ERG (European Regulators Group). (2009). *Report on Fixed-Mobile Convergence: Implications on Competition and Regulatory Aspects*. Retrieved November 2010 from http://www.erg.eu.int/doc/publications/2009/erg_09_06_report_on_fixed_mobile_convergence.pdf

European Communities. (2007). *Europe in Figures: Eurostat Yearbook 2006-07*. Luxembourg

Exane BNP Paribas & Arthur D Little. (2006). *Facing Off on Convergence*. Retrieved January 2009, from http://www.cellular-news.com/story/16742.php

FICORA. (2009). *Market overview 2/2009 by FICORA (in Finnish)*. Retrieved from http://www.ficora.fi/attachments/suomimq/5jdCGapBJ/Markkinakatsaus_2_2009.pdf

Forum Rep, U. M. T. S. 33 (2003). *3G Offered Traffic Characteristics*.London: UMTS Forum.

Forum Rep, U. M. T. S. 37 (2005). *Magic Mobile Future 2010-2020*, London: UMTS Forum.

Forum White Paper, U. M. T. S. (2009). *Mobile Broadband Evolution: the roadmap from HSPA to LTE,* London.

Forum White Paper, U. M. T. S. (2010). *Recognising the Promise of Mobile Broadband.* London

Gartner (2006). *Forecast: GPS in Mobile Devices, Worldwide, 2004-2010.* Retrieved January, 2009, from http://gartner.com/DisplayDocument?id=498745

Global Insight, I. H. S. (2009). *Vodafone Spain Launches DSL/3G Bundle*. Published 12 Jun 09 in http://www.ihsglobalinsight.com/SDA/SDADetail17039.htm

3GPP TR 23.806 (2007). Voice Call Continuity between CS and IMS Study (Release 7)

3GPP TS 23.228 (2008). IP Multimedia Subsystem; Stage 2 (Release 8)

3GPP TS 26.236 (2008). Packet Switched Conversational Multimedia Applications; Transport protocols (Release 7)

3GPP TS 43.318 (2009). *Generic Access Network (GAN)*; Stage 2 (Rel. 9)

Infocom (2010). *Residential Fixed Mobile Substitution (FMS) and Convergence (FMC) – Showing Caution After the Hype*. Retrieved November 2010 from http://www.infocom-de.com/pressarchives/press_051010.html

(2006). *Informa* (4th ed.). Mobile Entertainment.

Intellinet (2009). *Mobile Data Offload for 3G Networks – A Whitepaper*. Retrieved November 2010 from http://www.intellinet-tech.com/Media/PagePDF/Data%20Offload.pdf

International Game Developers Association. (2005). *Mobile Games White Paper.*

Internet Telephony. (2010). *Revisiting the Femtocell: Carriers Move To Commercial Deployments, Equipment Prices Drop*. Published 01 October 2010 in http://www.tmcnet.com/voip/features/articles/114440-revisiting-femtocell-carriers-move-commercial-deployments-equipment-prices.htm

Juniper (2006). *Mobile Navigation Finds its Way*. Retrieved January, 2009, from http://www.wirelessweek.com/article.aspx?id=110108

Madeley, S. (2007). *Fixed-mobile Convergence - Who Owns The Customer? Connect-World, Europe II.* Retrieved November 2010 from http://www.connect-world. com/index.php/magazine/europe/item/1194-fixed-mobile-convergence-who-owns-the-customer

OECD (2007). *Fixed-Mobile Convergence: Market Developments and Policy Issues*. March 2007

Ofcom (2006). *The Communications Market.*

RealNetworks (2005). Annual Report

Sonera (2010). Prices retrieved 9 November 2010, from http://www.sonera.fi/puhelin+ja+liittyma/iphone+hinnat

Strategy Analytics (2006). Mobile Video Telephony: Growth Projection Upgrade and Key Success Factors

Telecompaper (2009). Retrieved 4 September 2009, from http://www.telecompaper.com/news/article.aspx?cid=689518

Verkasalo, H. (2007). *A Cross-Country Comparison of Mobile Service and Handset Usage*, Licentiate's Thesis Series, TKK Helsinki University of Technology

Vodafone (2010). *Vodafone One Net.*Retrieved from http://www.vodafone.com/start/about_vodafone/what_we_do/business_solutions/products/integrated_communications/one_net.html

3Workplan, G. P. P. (2010). *Overview of 3GPP Release 11.* Retrieved November 2010 from http://www.3gpp.org/ftp/Information/WORK_PLAN/Description_Releases/Rel-11_description_20100924.zip

ENDNOTES

1 Actually the majority of operators have still no plans for the launch of this convergent voice technology.

2 Note that the study period does not coincide with actual FMC deployments; rather it is to illustrate how an FMC deployment could have proceeded if it were launched in 2007.

Chapter 11
Subscription Policy Control Framework for IMS-Based Networks

Nidal Nasser
University of Guelph, Canada

Ming Shang
University of Guelph, Canada

ABSTRACT

The Policy and Charging Control (PCC) architecture was firstly introduced in the 3GPP's Release 7. However, the PCC has its problems. The main problems include the incapability of performing policy control with consideration of subscriber profiles and missing specification on how to organize and express the policy information. In addition, no policy control at application session establishment stage also contributes to its imperfectness. In this paper, the authors propose a subscription-based policy control framework that implements a subscription-centered approach for policy control and to enable flexible policy definitions based on the subscriber's profile at the application level. The framework also provides functionalities of organizing the subscription data, identifying the policy, regulating the policy control process, interpreting, managing and enforcing the corresponding policies. The main objective is to qualify the subscribers and thus, enhance the network customization through defining flexible policies based on policy control requirements for different subscribers.

INTRODUCTION

IP Multimedia Subsystem (IMS) (TS 22.228, 2007) is defined by 3rd Generation Partnership Project (3GPP) (Agrawal, 2008) as the All-IP based core network solution capable of providing real-time multimedia services. It is an access-independent and packet-based IP connectivity and service control architecture that enables various types of multimedia services for end-users using

DOI: 10.4018/978-1-60960-589-6.ch011

common Internet-based protocols (Camarillo & Garcia-Martin, 2006). It can also be viewed as the soft-platform combining the Internet protocols, multimedia applications and services (Mayer, Khartabil, Niemi & Poikselka, 2006). Figure 1 illustrates the conceptual composition of the IMS.

As a comprehensive core network solution plus its All-IP structure, it is not only regarded as the ideal core network of the Next Generation Networks (NGN), but is also seen as an important method for the implementation of Fixed-Mobile Convergence (FMC), Unified Communications (UC), Converged Networks (CN) as well as the key to achieve the universal control layer.

In contrast to traditional network architectures, IMS offers the network architecture through the use of standards, enable network elements like general purpose servers. With the 3GPP specifications, IMS enables many network functionalities to be reused and shared across multiple access networks, allowing for rapid service creation and delivery. To this end, IMS takes a layered network infrastructure which allows carriers to rapidly develop, deploy, and deliver a large number of new services, which in many cases will have been developed by a 3rd party content and service provider.

Among the IMS network entities, there are several key functional components defined in the 3GPP IMS as shown in Figure 2 (TS 23.002, 2007). The most important one is the Call Session Control Function (CSCF) (TS 23.228, 2007), which is the pivotal element to control the call session and is divided into three different entities: Serving CSCF (S-CSCF), Interrogating CSCF (I-CSCF) and Proxy CSCF (P-CSCF). The Home Subscriber Server (HSS) is the central data centre which stores subscriber's profile information and system provisioning data. The others include User Equipment (UE) and Application Server (AS).

Before illustrating the policy control within the IMS, we first present the concept of policy control. Policy control has been defined as a mechanism to control a network's resources through a device-independent approach (Yavatkar, Pendarakis & Guerin, 2000). It is achieved by using so called policies to express how a user may make use of the network resources. A policy is described as the combination of rules and services where rules define the criteria for resource access and usage. In policy control architecture (Bohm, 2003), a policy controller makes policy decisions and sends them to its policy clients. Policy clients are network devices that participate in the policy framework.

Figure 1. Conceptual composition of IMS

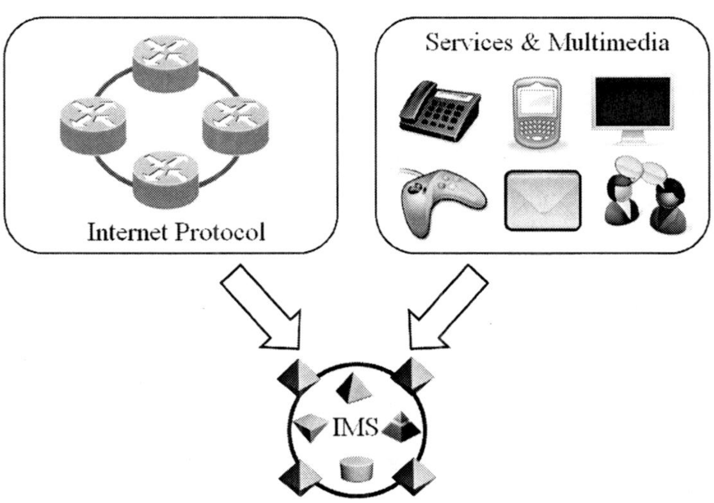

Figure 2. Core functional components of IMS

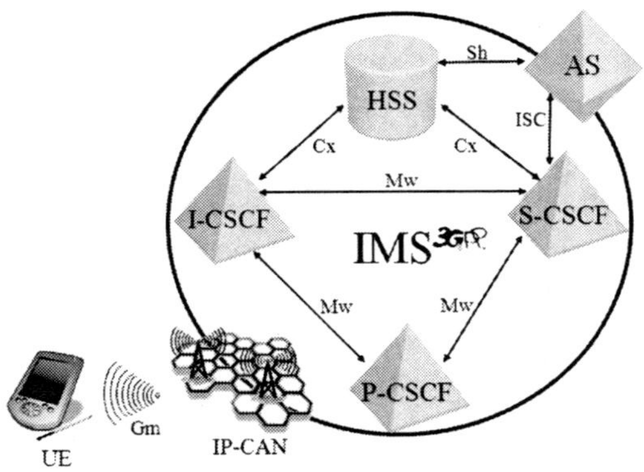

They initiate the communication and request policy decisions, which are sent back from the policy controller. The responsibility of the policy client is to carry out these decisions and to notify the policy controller if something happens that can affect the decisions made.

For the IMS, a new architecture named Policy and Charging Control (PCC) (TS 23.203, 2008) is introduced in 3GPP's Release 7, combining the functionalities of Policy and Charging Control. The aim of building such a merged architecture is to provide more efficient real time control of the service flows in the Gateways (GWs). As an important functional part of the whole IMS architecture, the PCC provides functionalities which encompass the high level functions for IP Connectivity Access Networks (IP-CANs) such as General Packet Radio Service (GPRS), Interworking WLAN (I-WLAN) and Fixed Broadband in the following aspects:

- Flow Based Charging, including charging control and online credit control (Kuhne, 2007).
- Policy control including gating control and QoS control (Rothenberg, 2008).

The PCC functionality is comprised by the functions of the Policy and Charging Enforcement Function (PCEF), the Policy and Charging Rules Function (PCRF), the Application Function (AF), the Online Charging System (OCS) (Mazzarella, 2006), the Offline Charging System (OFCS) and the Subscription Profile Repository (SPR). Within the PCC architecture, the PCRF is the policy entity that forms the linkage between the service and transport layers. The PCRF collates subscriber and application data, authorizes QoS resources, and instructs the transport plane on how to proceed with the underlying data traffic. The PCRF is connected on its Rx interface to the AF, an element residing on the service plane, which represents applications that require dynamic policy and QoS (TS 23.207, 2007) control over the traffic plane behavior. Within an IMS network, a P-CSCF would commonly fulfill the role of an AF. On the traffic plane, connected to the PCRF via the Gx (TS 29.212, 2008) interface, is the PCEF. The PCEF's role encompasses applicable traffic detection and resultant policy enforcement. This entity is typically located at a Gateway node implementing the IP access to the Public Data Network (PDN), which varies by transport layer (i.e. for GPRS, the GGSN, and for WLAN (TS 23.234, 2007), the PDG). A SPR node provides subscriber specific data to the PCRF, to assist in evaluating policy decisions. In addition, the PCEF is connected on its Gy interface to the OCS (TS 32.296, 2006), to allow online credit control for service data flow based charging. The functionalities required across the Gy reference point use the existing functionalities and mechanisms specified in RFC 4006 (Hakala, Mattila, Koskinen, Stura & Loughney, 2005). The PCEF is also connected on its Gz (TS 32.240, 2007) interface to the OFCS (TS 32.260, 2007), to enable transport of service data flow based on offline charging information. Figure 3 illustrates this architecture with relationships among different functional entities.

However, this new architecture has its problems. The main problems associated with the PCC include the incapability of performing policy control with consideration of subscriber profiles and missing specification on how to organize and express the policy information. In addition, no

Figure 3. PCC architecture in 3GPP's release 7

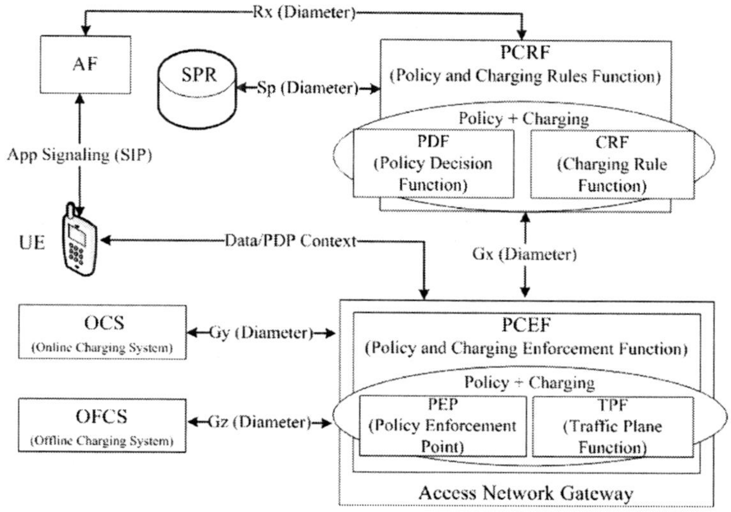

policy control at application session establishment stage also contributes to its imperfectness. Other problems include the missing reference point specification and no participation of core session control elements. To address the above shortcomings, we propose a Subscription-Based Policy Control Framework (SBPCF) that implements a subscription-centered approach for policy control and to enable flexible policy definitions based on the subscriber's profile. This framework is capable of organizing the subscription data, identifying the policy, regulating the policy control process, interpreting, managing and enforcing the corresponding policies within the current PCC architecture. Also, this framework works tightly with session control element such as S-CSCF to qualify the subscribers at application level as opposed to the PCC within which the policy control is taken place at network transport level.

RELATED WORK

In this section, we provide related work focusing on the area of policy controls which are helpful for the understanding of the materials to follow.

Common Open Policy Service

Although the IETF working groups have specified the RSVP protocol with which the QoS can be guaranteed, these protocols carry out merely resource-based admission control, which make admission decisions only based on available resources and may ignore factors that are also necessary for the decision making, like traffic type requirements, security considerations, identity of the users and applications. To overcome this problem, IETF has defined a new protocol, Common Open Policy Service (COPS) (Durham, Boyle, Cohen, Herzog, Rajan & Sastry, 2000). It is a protocol for device-independent network configuration using policies. The protocol is designed so that the messages are self-identifying policy objects, which contains policy elements. The policy elements are units of information necessary for the evaluation of policy rules. A single policy element may carry a user or application identification whereas another policy element may carry user credentials or other information. The policy elements themselves are expected to be independent of which QoS signaling protocol is used. The COPS message starts always with a common header. After the header, there are a number of policy elements following, which are used for transporting the information for decision-making. COPS protocol is used between a Policy Decision Point (PDP) and a Policy Enforcement Point (PEP) to transport policy-related information. The COPS protocol is a request/response protocol used between policy servers (PDPs) and their clients (PEPs). The clients send requests to their server, which responds back. Servers can also send unsolicited information to the clients.

Common Policy

With the purpose of building the framework for creating authorization policies, IETF defines Common Policy (Schulzrinne, Tschofenig, Morris, Cuellar, Polk & Rosenberg, 2007). As an authorization policy mark-up language, it can be used to describe the detailed access rights pertaining to the application-specific data by combining the common aspects of single authorization systems and location information, which previously had been developed separately. Although its origins lie in describing privacy settings that relate to geo-location information, in fact, any application that deals with access to a resource can reuse the underlying authorization mechanism and the basic structure of Common Policy and extend its permission statements to suit the application's special needs. The general framework defined in Common Policy is designed to be extended and enhanced by application-specific policies specified in any application domain. Common Policy defines an authorization policy as a set of rules called ruleset, which can be used to govern the access rights to information. Each rule grants permission based on certain matching criteria, like the identity of the user accessing the information, or the date and time of the day. These matching criteria are called conditions. Permission is further divided into an action and a transformation. The action part defines what happens when the conditions are met. Actions specify all remaining types of operations the policy controller is obliged to execute, i.e., all operations that are not of transformation type. The transformation part specifies operations that the policy controller must execute and modify the result that is returned to the client.

SUBSCRIPTION-BASED POLICY CONTROL FRAMEWORK

In this section, we present our proposed Subscription-Based Policy Control Framework (SBPCF) for 3GPP IMS and describe the behavior and functionalities of each component defined in this framework. The key concept behind the framework is to make the policy decision based on each subscriber's profile which is not provided in the current PCC architecture. Through this way, a customized policy definition is becoming feasible. In addition, it also enhances the differentiated service capability of 3GPP IMS since unqualified subscribers have been eliminated from certain service requirement and thus guarantee the service quality for those valued customers.

Problems in Policy and Charging Control Architecture

Figure 4 illustrates the IP-CAN session establishment process in the PCC. Among those steps, the Step-4 and Step-5 are designed to obtain the subscription-based information from the Subscription Profile Repository (SPR) using the reference point Sp. The subscription-based information obtained from the SPR will help the Policy and Charging Rules Function (PCRF) to make a proper policy decision on whether allow the establishment of the IP-CAN session or not. However, there are problems associated with the procedures such as:

- The detailed information associated with the Sp reference point is not specified in the 3GPP's Release 7.

Figure 4. IP-CAN session establishment

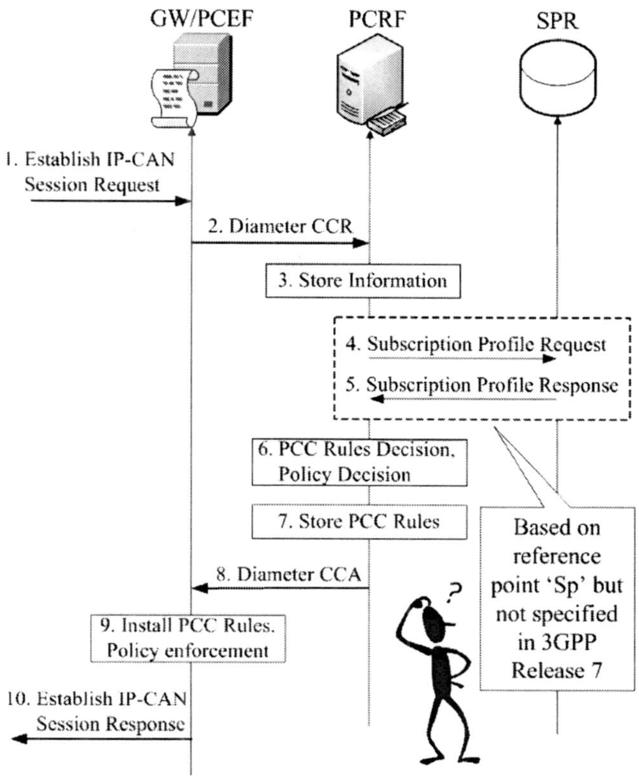

- Sp only obtains the subscription information related to IP-CAN transport-level policies from the SPR.
- There are no specifications on how to organize, define and enforce the subscription policy information.
- The participation of the session elements such as S-CSCF in the policy decision-making process is desirable but missing, which makes it impossible to discern sessions from different subscribers.

Figure 5 illustrates the AF (IMS) session establishment process (TS 29.213, 2008) in the PCC. We can see that it is possible and desirable to enforce policy control in S-CSCF at the Step-3 and Step-4 marked by the dotted rectangle. Due to the accessibility of the subscription data from the S-CSCF, subscription-based policy control is viable to be implemented here. This will solve the limitations that the PCC is designed to address the problems mostly at the network's transport level located at the edge of the networks without having knowledge of the application domain, which is necessary to implement the policy control and centralized management based on each subscriber. In addition, the policy control decision of the PCC is not made at the application-signaling stage, which would potentially introduce unnecessary signaling steps and is not efficient. Through adopting the policy control at AF session-establishment stage, requests from unqualified subscribers will be gated out and more accurate SDP derivation is able to be calculated before sending to the PCRF/PCEF.

Framework Design Objectives

The design objectives of the proposed framework are:

Figure 5. AF (IMS) session establishment

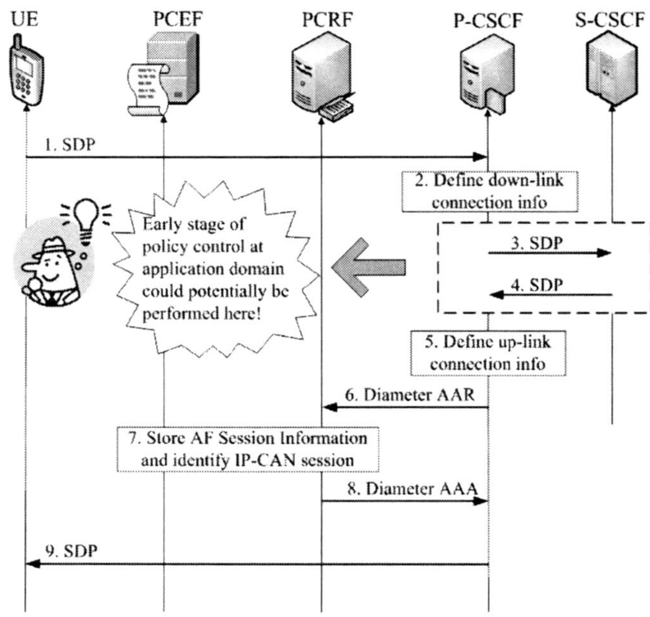

- Developing a subscription-based policy control framework, not service-based,
- Turning the concept of 'Framework = Markup Language + Codes' into the concrete model in the policy control area,
- Focusing on infrastructure and building block construction instead of designing complex policies in order to achieve flexibility,
- Using XML as policy expression vehicle to describe and organize the subscription-based policy information,
- Complementing to the current PCC architecture,
- Targeting on the application-level domain.

Framework System Architecture

The design purpose of the framework is to be part of the policy control at the application and provide complementary support in terms of subscription context at transport levels. The policy control could be carried out during the AF session establishment process. Through enforcing the subscription-based policies, it can qualify the subscribers and greatly reduce the unnecessary signaling messages defined in the PCC by terminating the unqualified users at this point. In addition, it also enhances the network customization based on the subscriber's profile and thus, improves the differentiating service capabilities of the network. The framework can also play a role at the transport level in the PCC architecture. It will provide crucial policy-related information based on subscription for the PCRF to make a wise policy decision. However, due to the volatile nature of this part designed in 3GPP and its tight relationship with these standards, emphasis of this framework design is focusing on the application domain. Figure 6 shows the proposed framework system architecture in 3GPP IMS network.

Framework Components

The SBPCF is comprised of four key components as shown in Figure 7.

The SBPCF Expressing Component (SEC) is designed to define the policy semantics and con-

Figure 6. Framework system architecture

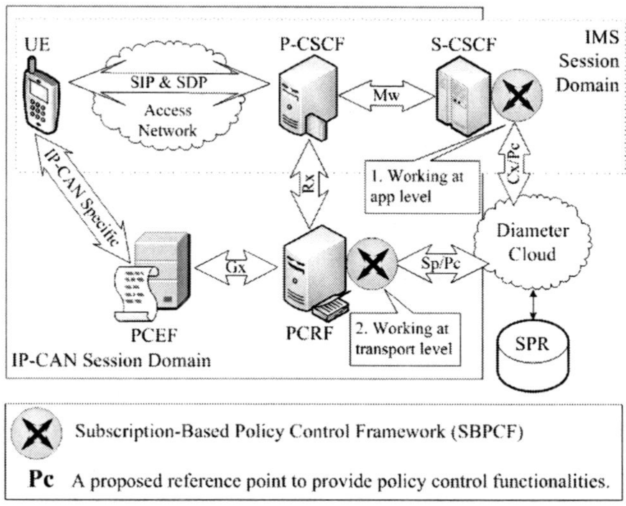

Figure 7. Framework composition

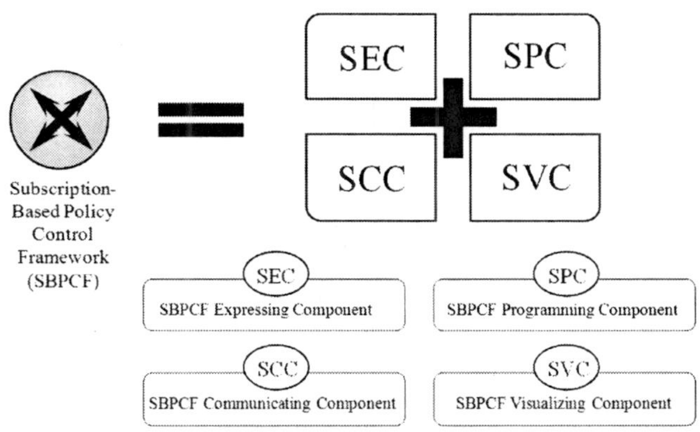

tent. In this part, the supported operation types, data types and other building-block entities such as property, functor and evaluator are defined. The SBPCF Programming Component (SPC) is a component library which contains collection of interfaces designed for the manipulation of the XML-based policy content through the object-oriented approach. It is designed to provide programming interfaces for developers to build their applications on top of the framework. The SBPCF Communicating Component (SCC) is designed for communications based on Diameter protocol. It includes the Pc reference point design, the messages and corresponding Attribute-Value-Pairs (AVPs). The SBPCF Visualizing Component (SVC) is the GUI part with the aim to provide visual approach to facilitate the management and design work.

Before we describe each component, we first present the operation steps of the framework. As indicated by Figure 8, when a subscriber tries to establish IMS session (Step-1), S-CSCF will solicit the framework to have the policy control based on the subscriber's profile. After receiving policy control request (fetched from the Request-Queue), the framework will check local cache for the corresponding subscriber's profile (Step-2). If not stored in the local cache, the framework will request the subscription policy information from the subscription repository (Step-3). The subscription policy information is stored in the subscription repository in the format specified by the SEC. After obtaining the corresponding subscription policy information, the framework will parse the subscription policy information (Step-4), enforce the policies (Step-5) and return the response to Response-Queue (Step-6). The pivotal function resides in the Step-5 at which the pre-defined policy definitions for a certain subscriber will be parsed and enforced by the framework. The response then goes to the CSCF (Step-7).

SBPCF Expressing Component

As the crucial part of the framework design, SEC is using the XML as the expression vehicle to specify the subscription information of the subscriber. The subscription node is defined as the root element to emphasize the subscription-based approach. In order to differentiate the various scenarios the framework is potentially applicable to, the application node is defined as multiple with each instance representing one applicable scenario. Within the application node, there are 3 sectors: profile, policies and gates. The profile sector

Figure 8. How the framework works

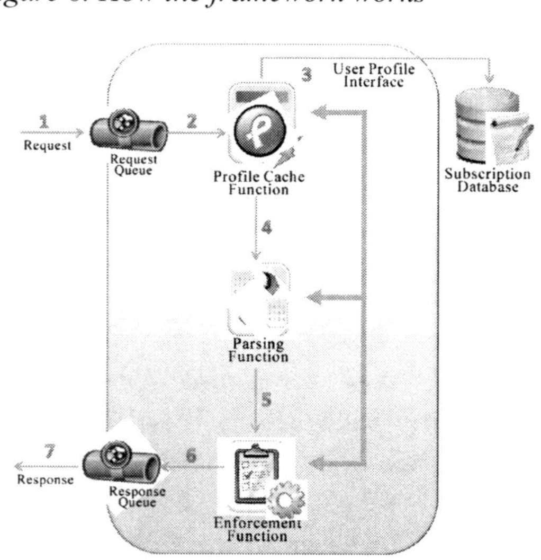

hosts the definitions of the basic information of the subscriber such as subscription identifier. In addition, property and functor related to this profile can also be defined here. The property behaves like the variable definition while the functor is for the purpose of defining functions. They act as building blocks to be referred by the other sectors. In the policies sector, the definitions of various policies are specified. They are referred and consumed by the various gates which will use the corresponding policy strategy to open or close the doors. The gates sector defines the gating functionalities of the current subscriber. They are categorized into two different groups: open gate and close gate. Any situation which does not match the open gate criteria or any situation which does match the close gate criteria will be disallowed. Each gating expression must be evaluated as Boolean result and be defined by the corresponding policy element in <policies> sector. The opengate and closegate are organized and evaluated as boolean expression in either Conjunctive Normal Form (CNF) or Disjunctive Normal Form (DNF). The absence value is CNF. A Boolean expression is said to be in CNF if it is expressed as a conjunction of disjunctions of literals (positive or negative atoms), i.e. as an AND of clauses, each of which is the OR of one of more atomic expressions. A Boolean expression is said to be in DNF if it is expressed as a disjunction of conjunctions of literals (positive or negative atoms), i.e. as an OR of clauses, each of which is the AND of one of more atomic expressions. Two extra attributes (group and negated) are also defined. The group attribute of the gate element allows the grouping of gates that will configure the sub-expressions inside a CNF or DNF expression. For instance, in the following CNF expression (A+B) (C+D), A+B and C+D correspond to different groups. In CNF, the group attribute identifies the ORed sets of gate instances within the same group. In DNF, the attribute group identifies the ANDed sets of gate instances within the same group. At least one group must be assigned for each gate.

Figure 9. Basic SEC structure

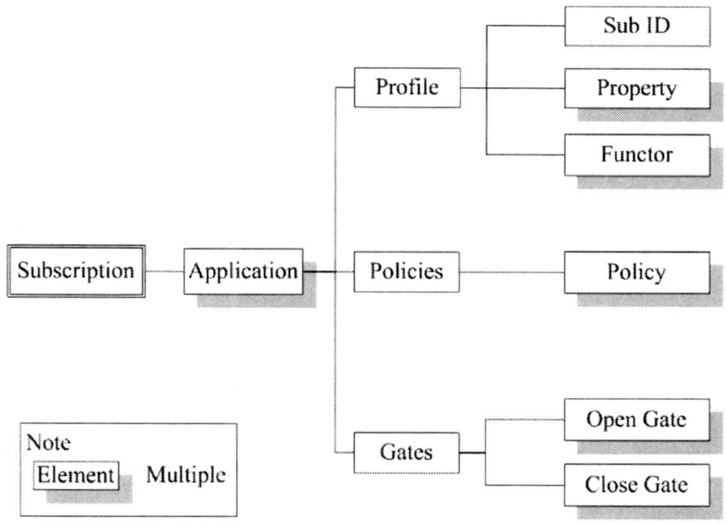

The negated attribute of the gate element defines whether the individual gate instance is negated (i.e. NOT logical expression). Figure 9 shows the tree-structure of the SEC.

Three categories of data types are defined in the framework: basic types, collection types and special types. Figure 10 lists each type defined in the framework.

Eight types of operations have been defined as illustrated by Figure 11. Each operation type takes two operands and is evaluated as Boolean type.

Three kinds of core element have been defined: property, functor and evaluator. They act as building blocks to be referred by the other sectors. The property behaves like the variable definition. The property element is defined by <property>. The property must be defined first before it can be referenced by ${property-id} in other places. In addition, property with input parameters need to be referred by <propertyref> before it can be referred by ${ref-property-id}. <propertyref> can be referred by ${ref-property-id} in other places since it has not parameters required. There are 3 types of properties: simple, persistent and reflective property. For the persistent property, there are three sub-categories, which are database, file and environment variables. Figure 12 illustrates the property types.

Figure 10. Data type definitions

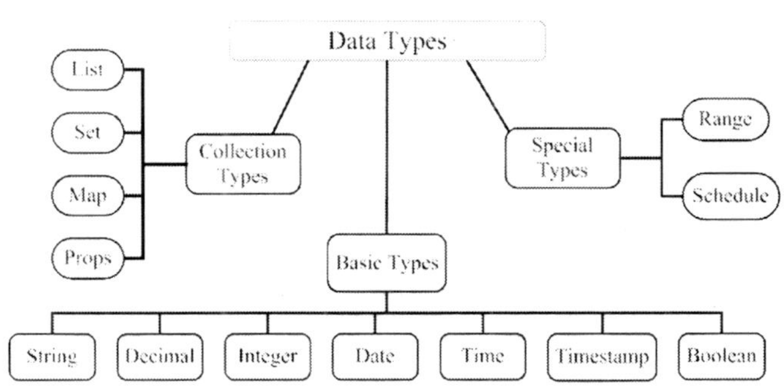

Figure 11. Operation type definitions

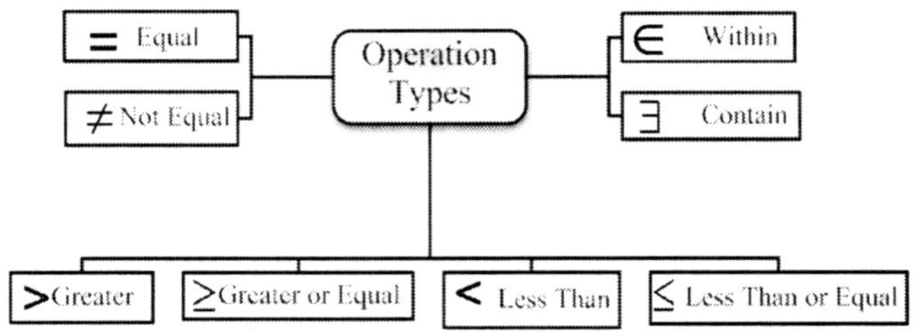

Figure 12. Property types

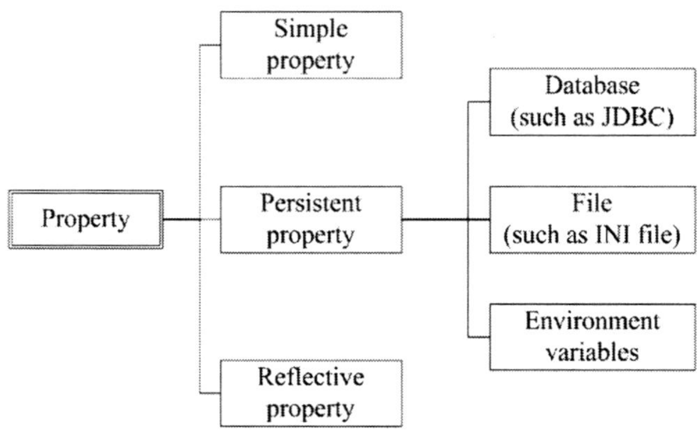

The functor behaves like functions with returned values. The functor must be defined first before it can be referenced by $[functor-id] in other places. In addition, functor with input parameters need to be referred by <functorref> before it can be referred by $[ref-functor-id]. <functorref> can be referred by $[ref-functor-id] in other places since it has not parameters required. The functor is defined by <functor>. There are 2 types of functors: Non-static and static functor. Within the non-static functor, there are two subcategories, which are constructor arguments and method arguments illustrated by Figure 13.

The evaluator behaves as Boolean operation with the result type that can be evaluated to TRUE or FALSE. The evaluator is defined by <eval>. The evaluator takes only 2 operands with one operator defined in operation types. Operand can

Figure 13. Functor types

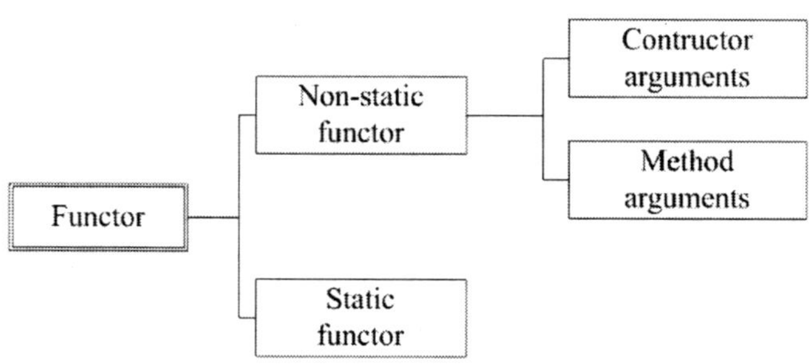

be basic-type, collection-type, special-type, property-reference and functor-reference.

SBPCF Programming Component

The SPC is designed to provide a set of component libraries with aim to programmably control the policies through the object-oriented approach. It is the reflection of the policy profile in terms of interfaces and objects. Figure 14 is a brief illustration of the structure of SBPCF Programming Component.

SBPCF Communicating Component

In order to exchange policy information with the subscription repository entities, Diameter is adopted as the underlying signaling protocol. Due to the unspecified status of the Sp reference point at this stage and the volatile nature this part may undergo, this part would be potentially under dramatic changes to fit into the future standards defined by 3GPP. Under such circumstance, the design work would focus on the minimum but essential elements this protocol needs to support rather than standardizing detailed information. The name of the reference point is tentatively termed as Pc (Policy control). The command code of the Pc reference point is 312, which is allocated by Internet Assigned Numbers Authority (IANA) exclusively for 3GPP but never used in any 3GPP's Diameter application as yet (TS 29.230, 2008). The designated command names for this function are Policy-Control-Request (PCR) and Policy-Control-Answer (PCA). The AVP codes will use the values ranging from 10,000 to 10,199 which are not conflicting with any 3GPP specific AVP code. The Pc application is defined as an IETF vendor specific Diameter application, where the vendor is 3GPP and the Application-ID for the Pc application is xxxxxxxx (tentative value is 16777255). The vendor identifier assigned by IANA to 3GPP is 10415. The application identification is included in the Auth-Application-Id AVP. With regard to the Pc reference point, the Policy Control Diameter Application (PCDA), which in our design is equal to SBPCF, acts as the Diameter client since it represents network element requesting the authorization of resources for a subscriber. The Subscription Repository Entity (SRE), which in our design is equal to SPR, acts as a Diameter server since it represents the network element that handles authorization requests for a particular realm.

The Policy-Control-Request (PCR) command, indicated by the Command-Code field set to 312 and the 'R' bit set in the Command Flags field, is sent by the PCDA to the SRE in order to request

Figure 14. Structure of SPC

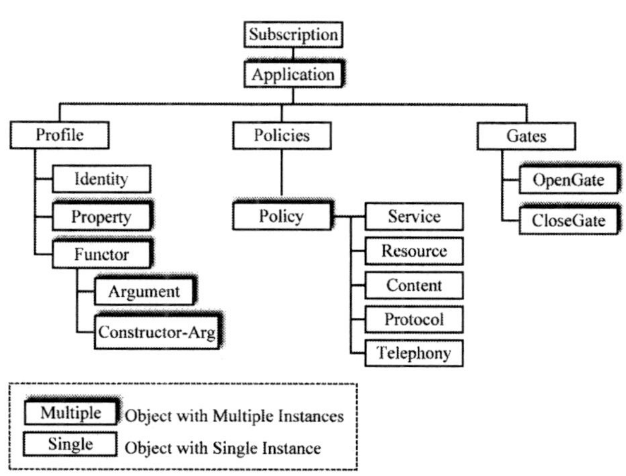

subscription data. The Policy-Control-Answer (PCA) command, indicated by the Command-Code field set to 312 and the 'R' bit cleared in the Command Flags field, is sent by the SRE to the PCDA in response to the PCR command. Several AVPs are also defined: Subscription-Id, Pc-Action and Subscription-Policy-Data. Subscription-Id is of type Grouped containing information which identifies the subscriber, to which the subscription request applies. The Pc-Action AVP is of type Enumerated, which identifies the request action. The Subscription-Policy-Data AVP is of type OctetString containing the subscription policy information and being encoded as XML with syntax and semantics specified in the previous sections.

SBPCF Visualizing Component

The SVC is designed to provide the graphical user interfaces to facilitate the management and design work. One of these interfaces is the policy editor with aim to provide the visual editing functionalities for policy design. In this design, we use the term 'Console' to refer to the SVC functional entity for this proposed framework. Each Console is able to provide certain functionalities in a certain aspect.

FRAMEWORK VALIDATION

In this section, we test our framework's capabilities to design subscription-oriented policy definitions according to different requirements, which would potentially enhance the IMS network's customization capabilities. To this end, two application scenarios will be simulated to demonstrate the functionality the framework can provide within the IMS network. For both scenarios, policy controls are going to take place at the stage of AF session establishment. The first application scenario is using content policy definition which defines whether the specific content is allowed to reach the present subscriber while the second one is using time-based service policy definition which defines whether the certain kind of service is reachable to the present subscriber according to the current time.

SCENARIO I

Malicious Content Prevention

Nowadays, most young people have their own mobiles which enable them to connect to the Internet for information. However, they are also exposed to malicious contents. Through defining the subscriber's content control policies, the framework is able to prevent the forbidden information from reaching subscribers under certain age. Figure 15 illustrates this process.

In this case, we are using indirect content mechanism specified by RFC 4483 (Burger, 2006) to cope with the scenarios where the session-related data does not directly reside on the Endpoint or User Agent. In such cases, the specification does allow the Session Initiation Protocol (SIP) (Rosenberg, Schulzrinne, Camarillo, Johnston, Peterson, Sparks, Handley & Schooler, 2002) message to contain an indirect reference to the desired content. The receiving party would then use this indirect reference to retrieve the content via a non-SIP transfer channel such as HTTP and FTP. The subscriber under the certain age (age information is stored in the subscriber's profile and in this case the subscriber's age is 16 and the threshold age value for content access restriction is 18) requests to access the certain URL using personal UE. The INVITE message is given as follows:

```
INVITE sip:sbpcf_demo@smile.com
SIP/2.0
Via: SIP/2.0/UDP 192.168.0.198:5040;
branch=z9hG4bKa4442d7fe9ed530bd71f766
d0dd39999;rport=5040
Call-ID: 9414d073f67818dd48dc9648ee-
```

Figure 15. Scenario of malicious content prevention

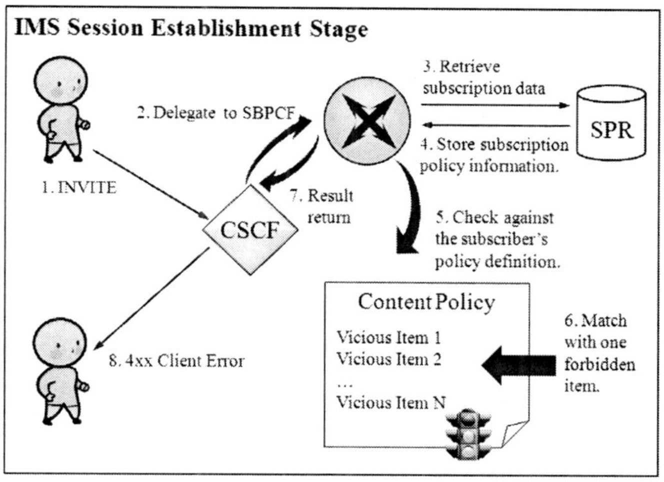


```
a397e0@192.168.0.198
CSeq: 1 INVITE
From: "Joe" <sip:Joe@smile.com:5040;t
ransport=udp>;tag=17605128
To: <sip:sbpcf_demo@smile.com>
Content-Type: message/external-
body;ACCESS-TYPE=URL;URL="www.sex123.
com"
Content-Length: 0
```

In the profile sector of the current subscriber policy, a functor called 'age' is defined as a reflection functor using a static method 'getAge' provided by the class '3gpp.ims.sbpcf.Utility'. This method is responsible for extracting the age information from the subscriber's profile. The following is the corresponding XML definition:

```
<functor id="age" type="Integer"
argc="0" static="1">
    <arg name="class">3gpp.ims.sbpcf.
Utility</arg>
    <arg name="method">getAge</arg>
</functor>
```

A second functor called 'uri' is defined as a reflection functor using a static method 'getURI' provided by the class '3gpp.ims.sbpcf.Utility' to extract the location of the external content specified by content indirection from the SIP message, which in this case has the value of 'www.sex123.com'. The following is the corresponding XML definition:

```
<functor id="uri" type="String"
argc="0" static="1">
      <arg name="class">3gpp.ims.sb-
pcf.Utility</arg>
      <arg name="method">getURI</arg>
</functor>
```

In the policies sector of the current subscriber policy, the policy with id value of 'policy.content.uri' is designed as the first operand referring to the 'uri' functor defined in the profile sector. The second operand is a Boolean type with 'True' value. The operator is defined as a regular expression to test whether the first operand matches the regular expression. The corresponding XML definition is as follows:

```
<policy id="policy.content.uri"
type="content">
    <eval>
        <operand refid="uri"
type="functor"/>
```

```
        <operator type="eq"
regex="1">sex|porn</operator>
        <operand
type="Boolean">True</operand>
    </eval>
</policy>
```

The second policy is defined with the first operand referring to the 'age' functor. The second operand is defined as an Integer type with the value of 18 in this case. The operator is defined as a less-than operation to test whether the first operand is less than the value of '18'. The corresponding XML definition is as follows:

```
<policy id="policy.content.age"
type="content">
    <eval>
        <operand refid="age"
type="functor"/>
        <operator type="lt"/>
        <operand type="Integer">18</
operand>
    </eval>
</policy>
```

In the gates sector of the current subscriber policy, two closegate entries are defined with reference to the policies 'policy.content.uri' and 'policy.content.age', respectively. The combination type is DNF. The following is the corresponding XML definition:

```
<closegate combination="dnf">
    <gate group="0" negated="0"
refid="policy.content.uri"/>
    <gate group="0" negated="0"
refid="policy.content.age"/>
</closegate>
```

We can see the URL requested by the subscriber is the inappropriate content for this subscriber. After receiving this request, the CSCF delegates the message to SBPCF for policy control checking as indicated by Step-2. When receiving the policy control request from the CSCF, the SBPCF will first check whether the subscriber's profile is already stored locally. If not, it would fetch the subscription information through sending request to SPR using PCR command with Pc-Action AVP value set to SUBSCRIPTION_POLICY_DATA_REQUEST as indicated by Step-3. In Step-4, the SPR will reply to the SBPCF using Policy-Control-Answer (PCA) command with Pc-Action AVP value set to SUBSCRIPTION_POLICY_DATA_ANSWER and Subscription-Policy-Data AVP value set to the XML content defined above. In Step-5 and Step-6, the requested URL value 'www.sex123.com' is checked against the inappropriate content items, which is a regular expression matching any content containing either 'sex' or 'porn'. As a result, both the 'policy.content.uri' policy and the 'policy.content.age' policy return True. Therefore, the closegate returns True, which means the gate is closed for this policy control checking. In Step-7, the forbidden result (415 Unsupported Media Type) is returned to the CSCF. Figure 16 demonstrates this simulation process.

SCENARIO II

Time-Based Services Control

Due to the capabilities of providing multimedia services, subscribers are able to use these services at any time during the day. However, some services such as FTP download consume a lot of bandwidth resources of the network. Therefore, it is desirable to have policy defined to set restrictions on services like this during the day time, especially for those downloading files with large size. Figure 17 illustrates this policy control process.

In this case, the service policy for the current subscriber is defined as FTP service is disabled each day at the time from 7:00am to 9:00pm to access any file with size greater than 1 MB. UE still use indirect content mechanism to request

Figure 16. Screenshot of malicious content prevention scenario

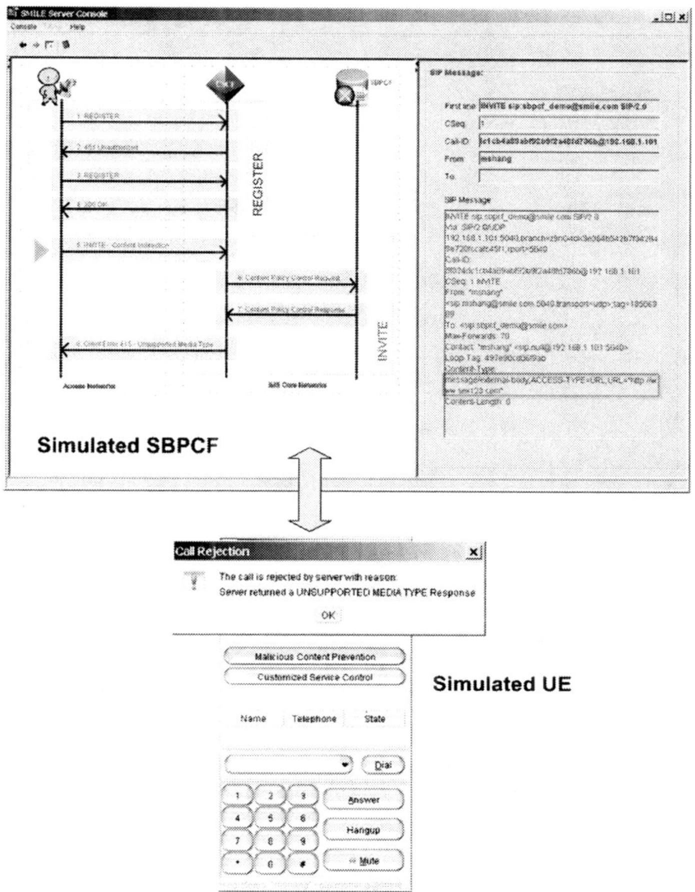

Figure 17. Scenario of time-based services control

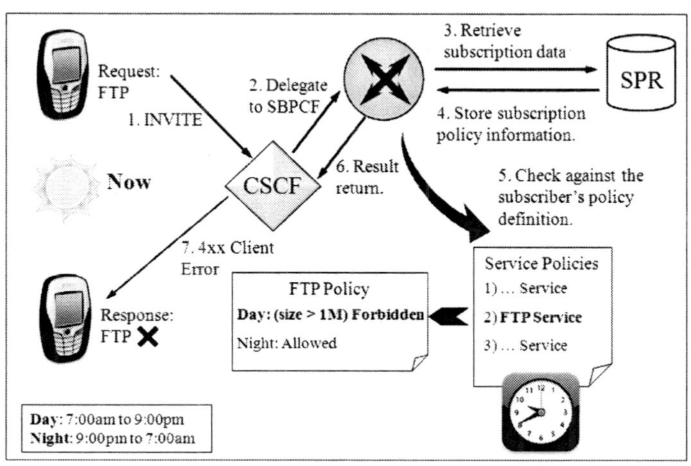

FTP services through the SIP INVITE message is given as follows:

```
INVITE sip:sbpcf_demo@smile.com
SIP/2.0
Call-ID: d8c13926a473a5d02be-
d800047afbab2@192.168.0.198
CSeq: 1 INVITE
From: "Joe" <sip:Joe@smile.com:5040;t
ransport=udp>;tag=33001201
To: <sip:sbpcf_demo@smile.com>
Via: SIP/2.0/UDP
192.168.0.198:5040;rport;branch=z9hG4
bK1331fa2911b0d5eb850abba71aaeaaf4
Contact: "Joe"
<sip:Joe@192.168.0.198:5040>
Content-Type: message/external-
body;ACCESS-TYPE=URL;URL="ftp://ftp.
songs.com/download/song.mp3"
Content-Length: 0
```

In the profile sector of the current subscriber policy, a functor called 'now' is defined as a reflection functor using a static method 'get-Time' provided by the class '3gpp.ims.sbpcf.Utility'. The corresponding XML definition is given as follows:

```
<functor id="now" type="Time"
argc="0" static="1">
    <arg name="class">3gpp.ims.sbpcf.
Utility</arg>
    <arg name="method">getTime</arg>
</functor>
```

A second functor called 'uri' is defined as a reflection functor using a static method 'getURI' provided by the class '3gpp.ims.sbpcf.Utility' to extract the FTP location from the INVITE message, which in this case has the value of 'ftp://ftp.songs.com/download/song.mp3'. The following is the corresponding XML definition:

```
<functor id="uri" type="String"
argc="0" static="1">
      <arg name="class">3gpp.ims.sb-
pcf.Utility</arg>
      <arg name="method">getURI</arg>
</functor>
```

A third functor called 'size' is defined as a reflection functor using a static method 'getSize' provided by the class '3gpp.ims.sbpcf.Utility' to obtain the file size of the requested URL, which in this case is greater than 1 MB. The following is the corresponding XML definition:

```
<functor id="size" type="Decimal"
argc="1" static="1">
    <arg name="class">3gpp.ims.sbpcf.
Utility</arg>
    <arg name="method">getFileSize</
arg>
    <method-args>
        <arg name="fileName"
type="String">$[uri]</arg>
    </method-args>
</functor>
```

In the policies sector of the current subscriber policy, the policy with id value of 'policy.service.time' is designed as the first operand referring to the 'now' functor defined in the profile sector. The second operand is a range type with 'from' and 'to' values of '07:00:00' and '20:59:59' respectively. The operator is defined as a 'within' operation to test the current time is within the time range. The corresponding XML definition is given as follows:

```
<policy id="policy.service.time"
type="service">
    <eval>
        <operand refid="now"
type="functor"/>
        <operator type="within"/>
        <operand type="range">
            <range type="Time" ne-
gated="0">
                <from
```

```
open="1">00:07:00</from>
                <to
open="1">20:59:59</to>
            </range>
        </operand>
    </eval>
</policy>
```

The second policy is designed with the first operand referred to the 'uri' functor. The second operand is a String value 'ftp'. The operator is defined as a regular expression to match whether the second group of the regular expression is equal to the 'ftp'. The corresponding XML definition is given as follows:

```
<policy id="policy.service.name"
type="service">
  <eval>
    <operand refid="uri"
type="functor"/>
    <operator type="eq" regex="1"
group="2">
        (([a-zA-Z][0-9a-zA-
Z+\\-\\.]*):/{0,2}([0-9a-zA-
Z;/?:@&=+$\\.\\-_!~*'()%]+))(#[0-9a-
zA-Z;/?:@&=+$\\.\\-_!~*'()%]+)?
    </operator>
    <operand type="String">ftp</op-
erand>
  </eval>
</policy>
```

The third policy is designed with the first operand referred to the 'size' functor. The second operand is a Decimal value '1'. The operator is defined as greater type to test if the size value is bigger than 1 MB. The corresponding XML definition is given as follows:

```
<policy id="policy.service.size"
type="service">
    <eval>
        <operand refid="size"
type="functor"/>
        <operator type=">" />
        <operand type="Decimal">1</
operand>
    </eval>
</policy>
```

In the gates sector of the current subscriber policy, three closegate entries are defined with reference to the policies 'policy.service.time', 'policy.service.name' and 'policy.service.size' respectively. The combination type is DNF. The corresponding XML definition is given as follows:

```
<closegate combination="dnf">
    <gate group="0" negated="0"
refid="policy.service.time"/>
    <gate group="0" negated="0"
refid="policy.service.name"/>
    <gate group="0" negated="0"
refid="policy.service.size"/>
</closegate>
```

We can see, in this case, all the three policies 'policy.service.time', 'policy.service.service' and 'policy.service.size' return True. Therefore, the closegate returns True, which means the gate is closed for this policy control checking. In Step-6, the forbidden result (415 Unsupported Media Type) returns to the CSCF. Figure 18 demonstrates this simulation process.

CONCLUSION

The Policy and Charging Control (PCC) is introduced to provide unified policy and charging control functionalities, which also reflect the great importance of the policy control functions playing within the IMS network. However, due to the initial stage of this effort, problems and limitations still exist. Under this circumstance, we proposed a complementary policy control framework tailored for 3GPP IMS with the em-

Figure 18. Screenshot of time-based services control scenario

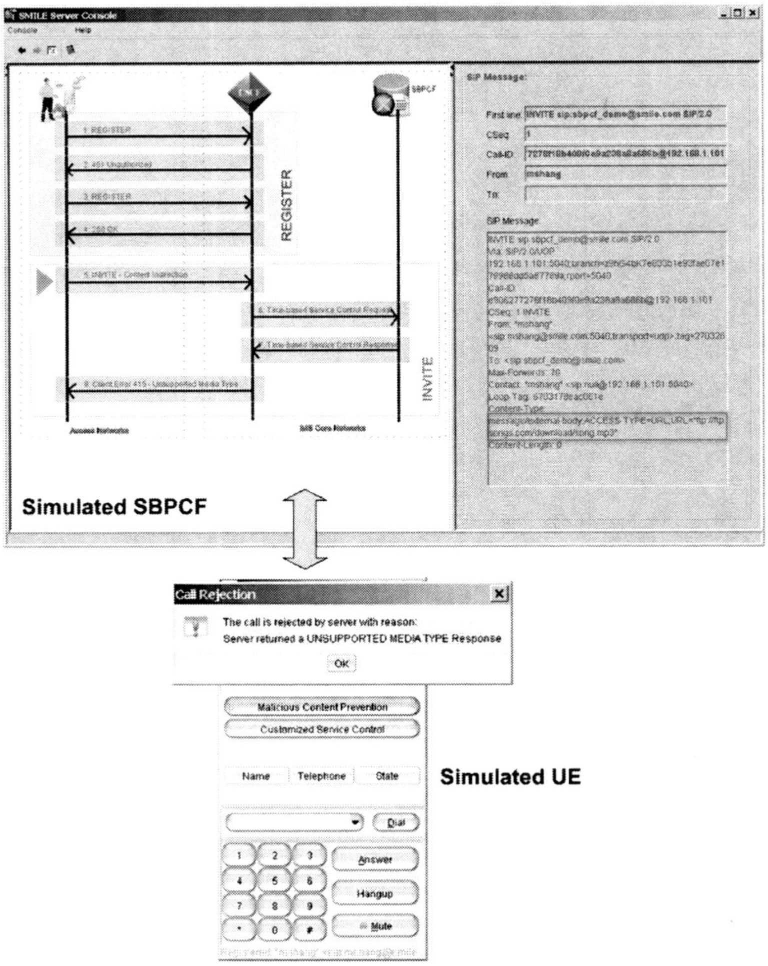

phasis on subscribers and the capability of defining flexible policies and building the policy control functions in a subscription-oriented approach. In this framework, we achieve:

- Turning the concept of 'Framework = Markup Language + Codes' into the concrete model in the policy control domain.
- Providing policy control functionalities during the application session establishment process.
- Providing infrastructure and building blocks capable of designing complex policies.
- Enforcing the subscription-based gating policies to qualify the subscribers using the pre-defined policy gates and deny those who cannot satisfy the gating policies.
- Participation of session level elements in the IMS such as S-CSCF in the policy decision-making process.
- Using XML as policy expression vehicle to describe and organize the subscription-based policy information.

However, there are still rooms for improvements. These aspects are listed below and may be studied in the future:

- Considering the maturity of Service-Oriented Architecture (SOA) and Web

Services, an enhanced programming interface with the capability of providing language-neutralized interfaces using service-oriented approach is desirable. In addition, as a bridge between the Internet application domain and Telecommunication domain, ParlayX is also specified in 3GPP IMS based on Web Services technology. Therefore, how to enhance this framework to be compatible with this standard is also a worthwhile endeavor.

- This framework is designed to work for 3GPP IMS at the present. However, there are other areas which also require policy control functionalities. Therefore, it is desirable to build a more generic policy framework to cope with the domains in addition to IMS.
- XML is a data self-expression language. However, it is not desirable to express complex logic sequence based on that. Therefore, to build a Domain-Specific Language (DSL) for policy control is also worth studying.

REFERENCES

Agrawal, P. (2008). IP Multimedia Subsystems in 3GPP and 3GPP2: Overview and scalability issues. *IEEE Communications Magazine*, *46*(1), 138–145. doi:10.1109/MCOM.2008.4427242

Bohm, W. (2003). *Policy based architecture for the umts multimedia domain. Proceedings second ieee international symposium on network computing and applications* (pp. 275–285). IEEE Comput. Soc.

Burger, E. (2006). *RFC 4483 A Mechanism for Content Indirection in Session Initiation Protocol (SIP) Messages*. IETF.

Camarillo, G., & Garcia-Martin, M. (2006). *The 3G IP Multimedia Subsystem: Merging the Internet and the Cellular worlds*. Hoboken, NJ: J. Wiley & Sons.

Charging architecture and principles (2007). *Stage 2*. TS 32.240, 3GPP.

Charging management: IP Multimedia Subsystem (IMS) charging (2007). TS 32.260, 3GPP.

Charging management; Online Charging System (OCS): Applications and interfaces (2006). TS 32.296, 3GPP.

Diameter applications: 3GPP specific codes and identifiers (2008). TS 29.230, 3GPP.

Durham, D., Boyle, J., Cohen, R., Herzog, S., Rajan, R., & Sastry, A. (2000). *RFC 2748 The COPS (Common Open Policy Service) Protocol*. IETF.

End-to-end Quality of Service concept and architecture (2007). TS 23.207, 3GPP.

3GPP system to Wireless Local Area Network interworking (2007). TS 23.234, 3GPP.

Hakala, H., Mattila, L., Koskinen, J.-P., Stura, M., & Loughney, J. (2005). *RFC 4006 Diameter Credit-Control Application*. IETF.

Kuhne, R. (2007). Charging in the ip multimedia subsystem: A tutorial. *IEEE Communications Magazine*, *45*(7), 92–99. doi:10.1109/MCOM.2007.382666

Mayer, G., Khartabil, H., Niemi, A., & Poikselka, M. (2006). *The IMS: IP Multimedia Concepts and Services*. Chichester, England; Hoboken, NJ: J. Wiley.

Mazzarella, NJ. (2006). Advantages of harmonized IMS-based charging architecture in different access technologies. *Bell Labs technical journal*, *10*(4), 109-116.

IP Multimedia Subsystem (2007). *Stage 2*. TS 23.228, 3GPP.

Network architecture (2007). *Stage 1.* TS 23.002, 3GPP.

Policy and charging control architecture (2008). *Stage 2.* TS 23.203, 3GPP.

Policy and Charging Control over Gx reference point (2008). *Stage 3.* TS 29.212, 3GPP.

Policy and Charging Control signaling flows and QoS parameter mapping (2008). TS 29.213, 3GPP.

Rosenberg, J., Schulzrinne, H., Camarillo, G., Johnston, A., Peterson, J., & Sparks, R. (2002). *RFC 3261 Session Initiation Protocol*. IETF.

Rothenberg, C. E. (2008). A review of policy-based resource and admission control functions in evolving access and next generation networks. *Journal of Network and Systems Management, 16*(1), 14–45. doi:10.1007/s10922-007-9096-3

Schulzrinne, H., Tschofenig, H., Morris, J., Cuellar, J., Polk, J., & Rosenberg, J. (2007). *RFC 4745 Common Policy: A Document Format for Expressing Privacy Preferences*. IETF.

Service requirements for the Internet Protocol (IP) multimedia core network subsystem (2007). Stage 1. TS 22.228, 3GPP.

Yavatkar, R., Pendarakis, D., & Guerin, R. (2000). *RFC 2753 A Framework for Policy Based Admission Control*. IETF.

This work was previously published in International Journal of Business Data Communications and Networking, Volume 5, Issue 3, edited by V. Sridhar and D. Saha, pp. 17-38, copyright 2009 by IGI Publishing (an imprint of IGI Global).

Chapter 12
What Happened to Preferences for Next Generation Internet?
A Survey of College Students in Taiwan

Wen-Lung Shiau
Ming Chuan University, Taiwan

Chen-Yao Chung
National Central University, Taiwan

Ping-Yu Hsu
National Central University, Taiwan

ABSTRACT

The growing popularity of the Internet has resulted in attracting many enterprises to do business transactions over the Internet. The current Internet protocol version 4 (IPv4) has been used for over 20 years. Even though IPv4 applications have been quite successful, it faces a problem of shortage in IP addresses, ineffective security mechanisms, and a lack of service quality management, etc. Scientists and engineers have devoted considerable effort to the development of next generation Internet protocol version 6 (IPv6), which is the core component of Next Generation Internet (NGI) to meet the future requirements of the Internet. Even though NGI is technically superior to the traditional Internet and is being established worldwide, few people have transmitted data through it. According to the Innovation Development Process in the Diffusion of Innovation theory, IPv6 is currently in a stage of technological diffusion. The research studies whether educating potential customers with more IPv6 knowledge created in the innovation process can increase their preference for the technology. With surveys collected from 596 undergraduate students, the results show that knowledge of the commercial applications of IPv6 in mobile communications and information appliances significantly contributes to a preference for the IPv6 technology.

DOI: 10.4018/978-1-60960-589-6.ch012

INTRODUCTION

The Internet has come into our lives and affects our ways of thinking, learning, working, communicating, entertaining, socializing, shopping, etc. Even though the Internet is viewed as a modern technology that has had an important impact on our daily life, the core component of current network is Internet protocol version 4 (IPv4), which has been used for over 20 years. IPv4 has gradually shown its age and is unable to cope with the demand of application trends. For example, IPv4 does not have enough IP addresses if all home appliances are to be connected, the security mechanisms that are needed for business transactions are not included in IPv4, and IPv4 cannot support service quality differentiation. These shortcomings greatly hinder the progress of application development for the Internet. Scientists and engineers of the Internet Engineering Task Force (IETF) therefore have devoted considerable effort to the development of the Next Generation Protocol (IPv6) in 1995. IPv6 can solve the above problems and provide a healthy internet platform for the development of new next generation applications (Bicknell, 2007; Everett, 2008; McLoughlin, 1999; Michael Mackay, 2003; Monteiro, 1998; Shiau, Chao, & Hsu, 2005; Shiau, Li, Chao, & Hsu, 2006; Weiser, 2001; Wright, 2007). However, with all its merits, Next Generation Internet (NGI) based on IPv6 has not been widely adopted (Hovav, Patnayakuni, & Schuff, 2004), which impedes the development of new applications. The lack of new applications in turn reduces the momentum of IPv6 adoption. The research studies whether educating potential customers with more IPv6 knowledge created during the Innovation Development Process can help with NGI diffusion. The Innovation Development Process consists of recognizing a problem or need, research, development, commercialization, diffusion, adoption and consequence (Roger, 1983). In the paper, a path analytical model is built to examine whether potential customers with knowledge created in former stages can have a better understanding of the knowledge in the latter stages and whether potential customers with NGI knowledge are more likely to prefer the technology. With survey results collected from 596 undergraduate students, we found that the knowledge of foreseeing the applications of IPv6 in mobile communication and information appliances has a direct and significant influence on the preference for NGI. The rest of the paper is organized as follows. Section 2 shows the research framework and hypotheses, which were derived from the Innovation Theory. Section 3 explains the data collection process. Section 4 shows the analysis results of reliability, correlation and analytic strategy for assessing the model. Section 5 shows the verified model and suggests ways to attract users' interest in NGI. Section 6 summarizes the findings of the research and any limitations it may have.

CONCEPTUAL FRAMEWORK AND RESEARCH HYPOTHESES

An innovation is an idea, practice, or object that is perceived of as new by an individual or other unit of adoption (Roger, 1983). Today we can see innovation taking place extensively in every field and industry, e.g. (Atuahene-Gima, 1996; Cabral, 1998; Hjalager, 1997; Johne, 1999; Keegan & Turner, 2002; Manu & Sriram, 1996; Silveira, 2001; Uzun, 2001) (Aa & Elfring, 2002; Atuahene-Gima, 1996; Evangelista, Perani, Rapiti, & Archibugi, 1997; Garcia & Calantone, 2002; Getz, Siegfried, & Anderson, 1997; Kano, 2000; Kuckartz, 2001; LaRose & Hoag, 1996; Mohamed, 1995; Sisaye., 1999; Vonortas & Xue, 1997). The innovation-development processes consists of recognizing a problem or need, research, development, commercialization, diffusion, adoption and consequence (Roger, 1983). In the first and second stages (Atuahene-Gima, 1996; Evangelista et al., 1997; Garcia & Calantone, 2002; Kano, 2000; LaRose & Hoag, 1996; Manu & Sriram,

1996; Roger, 1983; Silveira, 2001), scientists and engineers may perceive a forthcoming problem and initiate research activities to create an innovation to solve a problem or satisfy specific needs. In the third stage (Atuahene-Gima, 1996; Cabral, 1998; Evangelista et al., 1997; Garcia & Calantone, 2002; Hjalager, 1997; Johne, 1999; Kano, 2000; Kuckartz, 2001; LaRose & Hoag, 1996; Manu & Sriram, 1996; Roger, 1983; Silveira, 2001; Stock, Greis, & Fischer, 2002; Uzun, 2001), development is applied to tune the innovation to meet the needs of potential adopters. Production and service standardization are also created in this step to produce sellable products. In the fourth stage (Atuahene-Gima, 1996; Cabral, 1998; Evangelista et al., 1997; Garcia & Calantone, 2002; Johne, 1999; LaRose & Hoag, 1996; Mohamed, 1995; Roger, 1983; Silveira, 2001), development results are packaged as commercial products or services. These products along with knowledge are then pushed towards adopters during the diffusion stage. In the last stage (Aa & Elfring, 2002; Atuahene-Gima, 1996; Cabral, 1998; Evangelista et al., 1997; Garcia & Calantone, 2002; Getz et al., 1997; Hjalager, 1997; Johne, 1999; Kano, 2000; Keegan & Turner, 2002; Kuckartz, 2001; LaRose & Hoag, 1996; Manu & Sriram, 1996; Mohamed, 1995; Silveira, 2001; Sisaye., 1999; Uzun, 2001; Vonortas & Xue, 1997), namely, adoption and consequence, thus the technology is widely accepted(Dell, Kwong, & Ying, 2008; Hovav et al., 2004).

According to the Innovation Development Process (Roger, 1983), NGI is in the diffusion stage(Bicknell, 2007; Dell et al., 2008). Even though today's Internet applications have been quite successful, it faces the problem of a shortage of IP addresses, ineffective security mechanisms, and a lack of service quality management. Scientists and engineers have devoted considerable effort to the development of the next generation Internet protocol version 6 (IPv6), which is the core component of next generation Internet (NGI) to meet the future requirements of the Internet (Chen, Chao, & Kuo, 2006). A considerable number of enterprises have researched and developed many products to support NGI, such as routers and switches of Cisco Systems, Microsoft operating systems, Sun Microsystems and IBM(Clyman, 2007; Kuhl, 2007). Those companies not only want to commercialize the products but also need to diffuse the NGI into the market.

The research model proposed in the paper investigates whether potential adopters' knowledge of NGI innovation processes has a positive relationship with their preference for NGI and if potential adopters' knowledge in each stage has a positive relation to their understanding of the knowledge in the latter stages leading up to the diffusion. Figure 1 shows the proposed model. In the model, awareness of IPv4 limitation maps to the recognition of problems or needs. Appreciation of the technical merit of IPv6 maps to the appreciation of the research results. The knowledge of products supporting IPv6 maps to the knowledge of development result. Foreseeing the applications of IPv6 maps to the expectation of commercial products.

Awareness of IPv4 Limitation

Current network architecture based on IPv4 has three major constraints that need to be addressed for the further development of Internet applications. The first constraint is the lack of quality (bandwidth) guaranteed service. Applications such as video on demand and video conferencing which need to transmit images and voices simultaneously to provide meaningful presentations need guaranteed bandwidth. However, IPv4 offers no mechanism for such service. The second constraint is the lack of security and reliability measure needed to protect business transactions performed over Internet. Generic IPv4 does not provide end-to-end security. The third constraint is the insufficient number of IP addresses. The number of IP addresses defined in IPv4 is not enough to support the wiring of all home appli-

Figure 1. Proposed research model

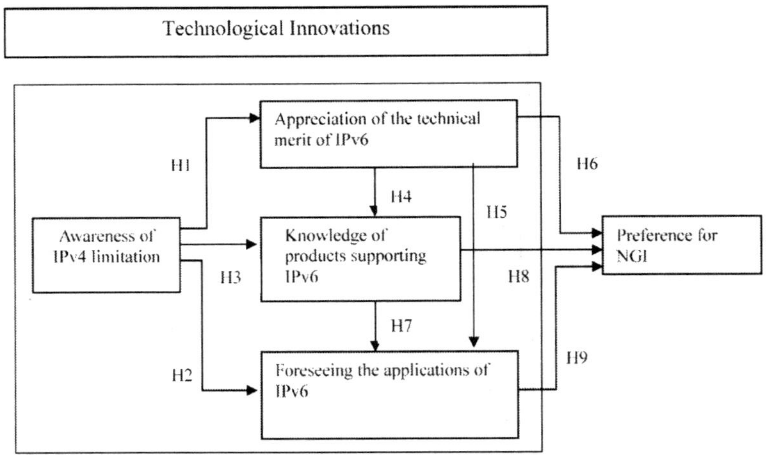

ances (Editorial, 2003; McLoughlin, 1999; Metz, 2003; Michael Mackay, 2003; Monteiro, 1998; Wright, 2007).

As Figure 1 indicates, the hypotheses and relations related to "Awareness of IPv4 limitation" are listed as follows:

Hypothesis 1(H1): Awareness of IPv4 limitation has a significantly positive relation with the appreciation of the technical merit of IPv6.

Hypothesis 2(H2): Awareness of IPv4 limitation has a significantly positive relation with the knowledge of products supporting IPv6.

Hypothesis 3(H3): Awareness of IPv4 limitation has a significantly positive relation with the foreseeing of the applications of IPv6 in mobile communication and information appliances.

There is no direct relationship between awareness of IPv4 issue and preference of NGI since the preference should be based on adopters understanding of the technology and its application.

Appreciation of the Technical Merit of IPv6

IPv6 adds considerable improvements to IPv4, including a vast new amount of IP addresses, guaranteed quality of service, more efficient transmission, better scalability and higher security (Alderman, 2003; Chen et al., 2006; Metz, 2003; Wright, 2007).

As shown in Figure 1, the hypotheses and relationships originating from "Appreciation of the technical merit of IPv6" are listed as follows:

Hypothesis 4(H4): Appreciation of the technical merit of IPv6 will have a significantly positive relation with the knowledge of products supporting IPv6.

Hypothesis 5(H5): Appreciation of the technical merit of IPv6 will have a significantly positive relation with the foreseeing of the applications of IPv6 in mobile communication and information appliances.

Hypothesis 6(H6): Appreciation of the technical merit of IPv6 will have a significantly positive relation with the preference for NGI.

Equipping with the Knowledge of Products Supporting IPv6

The global Internet consists of hosts, hubs, switches, routers, operating systems, and lots of applications. Today, the Internet is becoming a multi-service network. Significant changes to existing architectures, procedures and protocols are necessary in order to turn the Internet into a new multi-service network (Blefari-Melazzi, Sorte, & Reali, 2003). While the global Internet is migrating to next generation Internet, it is important to know what products support IPv6. These products can be desktop PCs, PDAs, and Notebooks(Wright, 2007), which support end-to end communication, real time video on demand, etc.

As shown in Figure 1, the hypotheses and relationships concerning "Knowing products supporting IPv6" are listed as follows:

Hypothesis 7(H7): Knowledge of products supporting IPv6 will have a significantly positive relation with the foreseeing of applications of IPv6 in mobile communication and information appliances.

Hypothesis 8(H8): Knowledge of products supporting IPv6 will have a significantly positive relation with the preference for NGI.

Foreseeing the Applications of IPv6

Mobile and wireless communication presents the largest commercial opportunity for NGI technology. Mobile and wireless technologies have advanced at an astonishing pace (Everett, 2008; Katz & Aspden, 1998; Kumar & Zahn, 2003). Over the past 10 years, wireless communications has been considered as the fastest growing segment of telecommunications market. In fact, mobile telephones have become a major communication tool for hundreds of millions of people. The demand for wireless access will exceed the number of fixed access lines by year 2010 (Ronald Beaubrun, 2001).

In retrospect, Internet and wireless communications have developed in parallel over the past decade, and converting these two popular technologies into one communication platform has been the target of many researches and projects. The latest developed 3G (Third Generation) wireless communication protocol promises to do so (DasBit & Mitra., 2003). In recent years, interest in home networking also has increased significantly due to technological innovations and market forces. To connect appliances into a network, each appliance will need an IP address. The new demand for IP addresses can only be met with IPv6.

As indicated in Figure 1, the hypothesis and relations which originated from the item is as follows:

Hypothesis 9(H9): Foreseeing the applications of IPv6 in mobile communications and information appliances, has a significantly positive relation with the preference for NGI.

Data Collections

The questionnaire included NGI background, Awareness of IPv4 limitations, Appreciation of the Technical Merit of IPv6, Knowledge of products supporting IPv6, Foreseeing the applications of IPv6 and Preference for NGI. All measures were rated on a 5-point Likert scale and were added to appendix A.

Awareness of IPv4 limitation: Three items were originally from the IPv4 drawbacks including IP shortage, lack of security and lack of quality service.

Appreciation of the Technical Merit of IPv6: Four items were originally from the IPv6 merits including numerous IP addresses, higher security, better quality service and faster speed.

Knowledge of products supporting IPv6: Three items were from the current IPv6 products including operating systems, network products and network applications.

Foreseeing the applications of IPv6: Three items were from the future IPv6 products including mobile systems, cell phone products and home applications.

Preference for NGI Six items were from the comparisons between IPv6 and IPv4 characteristics including more IP addresses, higher security, better quality service and faster speed, mobile features and home applications.

Since young people are more likely to appreciate new technology and about 80% of Internet users in Taiwan are college students (E. Li, 1999), therefore undergraduate students were the targets of the study. A pretest was performed with 20 representative undergraduate students in order to refine terms on the questionnaires. We also interviewed two internet experts to ensure content validity. In this study, 12 among 146 universities were chosen and questionnaires were distributed to the selected universities. A total of one thousand students from these universities were surveyed. Among them, six hundred and ten students responded and five hundred and ninety six returned valid questionnaires. Among valid responders, three hundred and thirty five students were male; two hundred and sixty one students were female.

RESULTS

Scale Reliability and Validity

The principal component analysis with Varimax rotation for the proposed constructs yielded five distinct factors: A: Awareness of IPv4 limitation, B: Appreciation of the Technical Merit of IPv6, C: Knowledge of products supporting IPv6, D: Foreseeing the applications of IPv6 and E: Preference for NGI, as shown in Figure 2.

Factor loading for all variables were greater than 0.6 with no cross-construct loading above 0.5, indicating good discriminant validity. The results demonstrated convergent validity with factor loading exceeding 0.50 for each

Figure 2. Scale reliability and validity

Scale items	Factor1 (A)	Factor2 (B)	Factor3 (C)	Factor4 (D)	Factor5 (E)
A1	.807				
A2	.817				
A3	.760				
B4		.873			
B5		.880			
B6		.875			
B7		.849			
C8			.858		
C9			.867		
C10			.802		
D11				.833	
D12				.859	
D13				.699	
E14					.926
E15					.930
E16					.938
E17					.911
E18					.939
E19					.939
Explaining percentage	11.96	19.36	13.78	11.24	31.20
Cumulative percentage	11.96	31.32	45.1	56.34	87.54

construct(Moon & Kim, 2001). The results of five factors accounted for 87.54 percent of the total variance. The reliability of all instruments was assessed by the Cronbach alpha reliability coefficient (Cronbach, 1951). The coefficient alpha (Cronbach alpha) for the total items was 0.93. The coefficient alphas for the awareness of IPv4 limitation, appreciation of technical merit of IPv6, knowledge of products supporting IPv6, and foreseeing the applications of IPv6 and Preference for NGI were 0.93, 0.96, 0.93, 0.83, and 0.97, respectively, which were above the conventional threshold level of 0.7 (Nunnally, 1978).

The bi-variant relationships indicated that all the variables were significantly correlated with each other as in Figure 3.

Analytic Strategy for Assessing the Model

The data were analyzed with the Path Analysis methodology (Kenny, 1979; Land, 1969; C. C. Li, 1975), which is a multivariate analytical methodology for empirically examining sets of relationships in the form of linear causal models (Duncan, 1966; C. C. Li, 1975). Path analysis has been applied to many disciplines, including sociology (Duncan, 1966) and management (Lederer, Maupin, Sena, & Zhuang, 2000; Lu & Yeh, 1998).

Unidirectional arrows linking two variables together represent the hypothetical causal relationships. Figure 1 depicts the proposed model of path diagram in this study. As shown in Figure 1, awareness of IPv4 limitation is assumed to have direct linear influences on appreciation of the technical merit of IPv6, knowledge of products supporting IPv6, and foreseeing the applications of IPv6 in mobile communication and information appliances. The appreciation of the technical merit of IPv6 is assumed to further influence knowledge of products supporting IPv6, foreseeing the applications of IPv6 in mobile communication and information appliances and Preference for NGI. The knowledge of products supporting IPv6 is assumed to have direct influence on both foreseeing the applications of IPv6 in mobile communication and information appliances and preference for NGI. The foreseeing of applications of IPv6 in mobile communication and information appliances is assumed to influence the preference for NGI.

The value of the path coefficient associated with each path represents the strength of the linear influence. The path coefficient has been shown to be identical to the standardized regression coefficients (C. C. Li, 1975; Lin & Lu, 2000). Although the path coefficients can be estimated in many ways (Kenny, 1979), multiple regression analyses (i.e., ordinary least-squares estimation) were used in this research, as in many other applications.

Figure 3. The First-order correlation coefficients among the variables

	A	B	C	D	E
A	1.000	.703*	.606*	.475*	.236*
B	.703*	1.000	.631*	.429*	.172*
C	.606*	.631*	1.000	.424*	.113*
D	.475*	.429*	.424*	1.000	.356*
E	.236*	.172*	.113*	.356*	1.000

** Significant level at 0.01 (two tails)*
A: Awareness of IPv4 limitation
B: Appreciation of the Technical Merit of IPv6
C: Knowledge of products supporting IPv6
D: Foreseeing the applications of IPv6
E: Preference for NGI

Evaluating the Hypothesized Model

Figure 4 presents the verified model with non-significant paths being removed. Multicollinearity is ruled out because the correlation between independent variables were all less than 0.8 (Emory, 1991). The scatter plots were examined to avoid problems of non-linear relationships.

Figure 4 shows that 49.3% of the variance in appreciation of the technical merit of IPv6 is

Figure 4. The verified research model

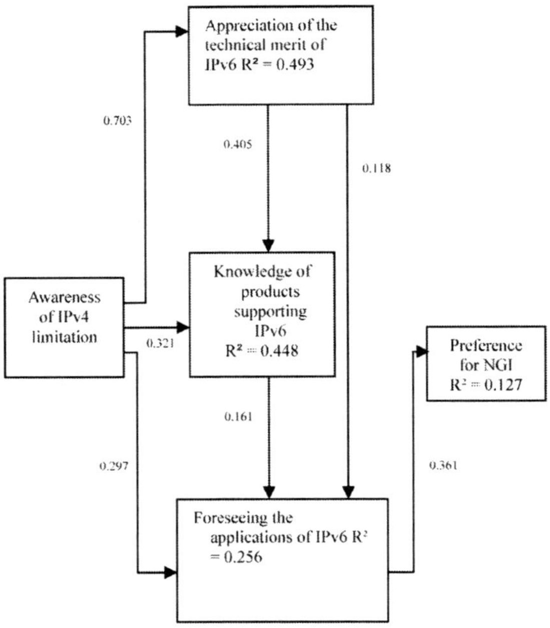

explained by an awareness of the limitation of IPv4. In addition, appreciation of the technical merit of IPv6 and awareness of IPv4 limitation accounted for 44.8% of the variance in knowledge of products supporting IPv6. Appreciation of the technical merit of IPv6, knowledge of products supporting IPv6 and awareness of IPv4 limitation accounted for 25.6% of the variance in foreseeing the applications of IPv6 in mobile communication and information appliances. Moreover, foreseeing the applications of IPv6 in mobile communication and information appliances accounted for 12.7% of the variance in the preference for NGI. The results are further articulated in following sections.

Awareness of IPv4 Limitation

All three hypothesized paths originated from the awareness of IPv4 limitation are supported by collected data. The paths linked to appreciation of the technical merit of IPv6, knowledge of products supporting IPv6, and foreseeing the applications of IPv6 in mobile communication and information appliances.

Hence, Hypothesis 1, Hypothesis 2, and Hypothesis 3 are supported.

The path coefficient (=0.703) from awareness of IPv4 limitation to appreciation of the technical merit of IPv6 was significant at $p< 0.001$. The path coefficient (=0.321) from awareness of IPv4 limitation to knowledge of products supporting IPv6 was significant at $p< 0.001$. The path coefficient (=0.297) from awareness of IPv4 limitation to foreseeing the applications of IPv6 in mobile communication and information appliance was significant at $p< 0.001$.

Appreciation of the Technical Merit of IPv6

The path coefficient (=0.405) from appreciation of the technical merit of IPv6 to knowledge of products supporting IPv6 was significant at $p< 0.001$. The path coefficient (=0.118) from appreciation of the technical merit of IPv6 to foreseeing the applications of IPv6 in mobile communication and information appliances was significant at $p< 0.05$. The path linked to the preference for NGI was removed.

Therefore, Hypothesis 4 and Hypothesis 5 are supported and Hypothesis 6 is not.

Knowledge of Products Supporting IPv6

Two hypothesized paths originated from knowledge of products supporting IPv6, namely the links to foreseeing the applications of IPv6 in mobile communication and information appliances and preference for NGI. Only the path to foreseeing the applications of IPv6 in mobile communication and information appliances was supported and the other path was removed. The path coefficient (=0.161) from knowledge of products supporting IPv6 tow foreseeing the applications of IPv6 in mobile communication and information appliances was significant at $p< 0.001$. The path from knowledge of products supporting IPv6 to preference for NGI was removed.

Hence, Hypothesis 7 is supported and Hypothesis 8 was not.

Foreseeing the Applications of IPv6 in Mobile Communication and Information Appliances

Only one hypothesized path from foreseeing the applications of IPv6 in mobile communication and information appliance was formulated, namely, to the item of preference for NGI. The path was supported with path coefficient =0.361 with significant level of 0.001. Therefore, Hypothesis 9 was supported at the 0.001 levels of significance.

DISCUSSIONS

Based on the result depicted in Figure 4, only two ways could significantly attract users' preference of NGI. The first route starts from educating potential adopters on the limitation of IPv4 and to help them understand the technical merit of IPv6 and then expose them to the products supporting IPv6 and show them the commercial applications of integrating IPv6 with mobile communication and information appliances. The second route starts from educating potential adopters on the limitation of IPv4 and then showing them the commercial applications of integrating IPv6 with mobile communications and information appliances. These people then have a greater possibility of preferring NGI to the current network. The finding concurs with other researches (Davis, 1989; Davis, Bagozzi, & Warshaw, 1989) and points out that usefulness plays an important role in raising interest to new technology.

Awareness of the limitation of IPv4 is a very important factor to raise the appreciation of the technical merit of IPv6, which is consistent with the results of previous studies (Amichai-Hamburger, 2002; Carpenter, 2003; Guice, 1998; Monteiro, 1998; Pospischil, 1998; Weiser, 2001). Appreciation of the technical merit of IPv6 plays a very important factor in understanding the products supporting IPv6. Appreciation of the technical merit of IPv6 does not directly contribute to preference for NGI. Knowledge of products supporting IPv6 increases the ability to foresee the applications of IPv6 in mobile communication and information appliances. Knowledge of products supporting IPv6 does not directly contribute to preference for NGI. Foreseeing the application of integrating IPv6 with mobile communication and information appliances has a significant effect on NGI preference.

CONCLUSION AND LIMITATIONS

Even though the Internet is viewed as a modern technology that has had an important impact on our daily life, the core component of the current network is Internet protocol version 4 (IPv4), which has been used for over 20 years. IPv4 has gradually shown its age and is unable to cope with the demand of application trends. Scientists and engineers of the Internet Engineering Task Force (IETF) therefore have devoted considerable effort to the development of Next Generation Protocol (IPv6) in 1995. IPv6 can solve the above problems and provide a healthy Internet platform for the development of new generation applications (McLoughlin, 1999; Michael Mackay, 2003; Monteiro, 1998; Weiser, 2001). However, with all the merits, the adoption of Next Generation Internet (NGI) based on IPv6 is surprisingly slow, which impedes the development of new applications. The lack of new applications in turn reduces the momentum of IPv6 adoption.

This study investigates the possible paths to increase potential adopters' preference of NGI based on the knowledge created in the Innovation Development Process (Roger, 1983), which states that the process consists of recognizing a problem or need, research, development, commercialization, diffusion, adoption and consequence (Roger, 1983). The study tries to find whether educating with knowledge created or identified during the process can increase potential adopters' interest in NGI and if gaining the knowledge of the former stages can help to understand the knowledge created in the latter stages. The knowledge of the first four stages is mapped to the awareness of IPv4 limitation, appreciation of the technical merit of IPv6, knowledge of products supporting IPv6, and foreseeing the applications of IPv6 in mobile communication and information appliances.

Results show that awareness of IPv4 limitation strongly contributes to appreciation of the technical merit of IPv6, knowledge of products supporting IPv6, and foreseeing the applications of IPv6 in mobile communication and information appliances. But among the knowledge created through technological innovation, only the foreseeing the applications of IPv6 in mobile communication and information appliances contributed significantly to the preference of NGI.

The research results of this study should be accepted with caution due to several inherent limitations. First, most of the undergraduate students have not had NGI experience. This study provides only the understanding and attitude of undergraduate students toward NGI before they actually use related products and services. Second, the study does not include real cost factor in the model because no Internet Service Providers have announced their pricing. However, the results can still offer insight into the facilitation of NGI diffusion, and provide an impetus for future research.

REFERENCES

Aa, W. d., & Elfring, T. (2002). Realizing innovation in services. *Scandinavian Journal of Management*, *18*, 155–171. doi:10.1016/S0956-5221(00)00040-3

Alderman, J. (2003). IPv6: Asia's Agent of Change. www.japaninc.com*; J@pan Inc* (pp. 24-27).

Amichai-Hamburger, Y. (2002). Internet and personality. *Computers in Human Behavior*, *18*, 1–10. doi:10.1016/S0747-5632(01)00034-6

Atuahene-Gima, K. (1996). Market orientation and innovation. *Journal of Business Research*, *35*, 93. doi:10.1016/0148-2963(95)00051-8

Bicknell, D. (2007). *Preparing your business for IPv6* (pp. 26–28). Computer Weekly.

Blefari-Melazzi, N., Sorte, D. D., & Reali, G. (2003). Accounting and pricing: a forecast of the scenario of the next generation Internet. *Computer Communications*, *26*, 2037–2051. doi:10.1016/S0140-3664(03)00146-4

Cabral, J. E. D. O. (1998). Survey on technological innovative behavior in the brazilian food industry. *Scientometrics*, *42*, 129–169. doi:10.1007/BF02458353

Carpenter, B. E. (2003). Future applications and middleware, and their impact on the infrastructure. *Future Generation Computer Systems*, *19*, 191–197. doi:10.1016/S0167-739X(02)00145-0

Chen, J.-L., Chao, H.-C., & Kuo, S.-Y. (2006). IPv6: More than protocol for next generation Internet. *Computer Communications*, *29*, 3011–3012. doi:10.1016/j.comcom.2005.11.001

Clyman, J. (2007). The Insider's Guide to Windows Vista. (cover story). *PC Magazine*, *26*, 74–88.

Cronbach, L. J. (1951). Coefficient alpha and the internal structure of tests. *Psychometrika*, *16*, 297–334. doi:10.1007/BF02310555

DasBit, S., & Mitra, S. (2003). Challenges of computing in mobile cellular environment - a survey. *Computer Communications*, *26*, 2090–2105. doi:10.1016/S0140-3664(03)00117-8

Davis, F. D. (1989). Perceived Usefulness, Perceived Ease Of Use, And User Accep. *Management Information Systems Quarterly*, *13*, 319. doi:10.2307/249008

Davis, F. D., Bagozzi, R. P., & Warshaw, P. R. (1989). User Acceptance Of Computer Technology: A Comparison Of Two. *Management Science*, *35*, 982. doi:10.1287/mnsc.35.8.982

Dell, P., Kwong, C., & Ying, L. (2008). Some reflections on IPv6 adoption in Australia. *Info*, *10*, 3–9. doi:10.1108/14636690810874034

Duncan, O. D. (1966). Path analysis: Sociological example. *American Journal of Sociology*, *72*, 1–16. doi:10.1086/224256

Editorial. (2003, July 21). IPv6It's time. *eweek*, 48.

Emory, C. W., & Cooper, D. R. (1991). *Business research method* (4th ed.). Homewood, IL: Irwin.

Evangelista, R., Perani, G., Rapiti, F., & Archibugi, D. (1997). Nature and impact of innovation in manufacturing industry: Some evidence from the Italian innovation survey. *Research Policy*, *26*, 521–536. doi:10.1016/S0048-7333(97)00028-0

Everett, C. (2008). An IP address for every machine. *Wireless Asia*, *11*, 28–28.

Garcia, R., & Calantone, R. (2002). A critical look at technological innovation typology and innovativeness terminology: A literature review. *Journal of Product Innovation Management*, *19*, 110–132. doi:10.1016/S0737-6782(01)00132-1

Getz, M., Siegfried, J. J., & Anderson, K. H. (1997). Adoption of innovations in higher education. *The Quarterly Review of Economics and Finance*, *37*, 605–631. doi:10.1016/S1062-9769(97)90013-2

Guice, J. (1998). Looking backward and forward at the Internet. *The Information Society*, *14*, 201–211. doi:10.1080/019722498128827

Hjalager, A.-M. (1997). Innovation patterns in sustainable tourism an analytical typology. *Tourism Management*, *18*, 35–41. doi:10.1016/S0261-5177(96)00096-9

Hovav, A., Patnayakuni, R., & Schuff, D. (2004). A model of Internet standards adoption: the case of IPv6. *Information Systems Journal*, *14*, 265–294. doi:10.1111/j.1365-2575.2004.00170.x

Johne, A. (1999). Using market vision to steer innovation. *Technovation*, *19*, 203–207. doi:10.1016/S0166-4972(98)00112-6

Kano, S. (2000). Technical innovations, standardization and regional comparison - A case study in mobile communications. *Telecommunications Policy*, *24*, 305–322. doi:10.1016/S0308-5961(00)00020-3

Katz, J. E., & Aspden, P. (1998). Theories, Data, and Potential Impacts of Mobile Communications - A Longitudinal Analysis of U.S. National Surveys. *Technological Forecasting and Social Change*, *57*, 133–156. doi:10.1016/S0040-1625(97)00089-9

Keegan, A., & Turner, J. R. (2002). The management of innovation in project-based firms. *Long Range Planning*, *35*, 367–388. doi:10.1016/S0024-6301(02)00069-9

Kenny, D. A. (1979). *Correlation and causality*. New York: Wiley.

Kuckartz, M. (2001). Innovation market - the economic exploitation of property rights in high-quality inventions. *World Patent Information*, *23*, 67–70. doi:10.1016/S0172-2190(00)00098-3

Kuhl, C. (2007). 'Vista' worries far off on cable's horizon: New OS supports IPv6, viewed as potential upgrade for online services. *CED*, *33*, 42–46.

Kumar, S., & Zahn, C. (2003). Mobile communications: Evolution and impact on business operations. *Technovation*, *23*, 515–520. doi:10.1016/S0166-4972(02)00120-7

Land, K. C. (1969). Principles of path analysis. *American Sociological Association 1.*

LaRose, R., & Hoag, A. (1996). Organizational adoptions of the internet and the clustering of innovations. *Telematics and Informatics, 13*, 49–61. doi:10.1016/0736-5853(96)00003-2

Lederer, A. L., Maupin, D. J., Sena, M. P., & Zhuang, Y. (2000). The technology acceptance model and the World Wide Web. *Decision Support Systems, 29*, 269–282. doi:10.1016/S0167-9236(00)00076-2

Li, C. C. (1975). Path analysis: A primer. Paci®c Grove

Li, E. (1999). *Report on Taiwan's network usage for 1998* (pp. 38–41). Advertising Magazine.

Lin, J. C.-C., & Lu, H. (2000). Towards an understanding of the behavioural intention to use a web site. *International Journal of Information Management, 20*, 197–208. doi:10.1016/S0268-4012(00)00005-0

Lu, H.-P., & Yeh, D.-C. (1998). Enterprises' perceptions on business process re-engineering: A path analytic model. *Omega, 26*, 17–27. doi:10.1016/S0305-0483(97)00029-7

Mackay, C. E., Dunmore, M., Chown, T., & Carvalho, G. (2003). A Scenario-Based Review of IPv6 Transition Tools. *IEEE Internet Computing, 7*, 27–35. doi:10.1109/MIC.2003.1200298

Manu, F. A., & Sriram, V. (1996). Innovation, marketing strategy, environment, and performance. *Journal of Business Research, 35*, 79. doi:10.1016/0148-2963(95)00056-9

McLoughlin, G. J. (1999). Next generation Internet and related initiatives. *Journal of Academic Librarianship, 25*, 226–229. doi:10.1016/S0099-1333(99)80204-3

Metz, C. (2003). Moving Toward an IPv6 Future. *IEEE Internet Computing, 7*, 25–26. doi:10.1109/MIC.2003.1200297

Mohamed, M. Z. (1995). Innovation implementations in Malaysian firms: Process, problems, critical success factors and working climate. *Technovation, 15*, 375. doi:10.1016/0166-4972(95)96598-N

Monteiro, E. (1998). Scaling information infrastructure: The case of next-generation IP in the Internet. *The Information Society, 14*, 229–245. doi:10.1080/019722498128845

Moon, J.-W., & Kim, Y.-G. (2001). Extending the TAM for a World-Wide-Web context. *Information & Management, 38*, 217–230. doi:10.1016/S0378-7206(00)00061-6

Nunnally, J. C. (1978). *Psychometric theory* (2nd ed.). New York: McGraw-Hill.

Pospischil, R. (1998). Fast Internet. *Telecommunications Policy, 22*, 745–755. doi:10.1016/S0308-5961(98)00053-6

Roger, E. M. (1983). *Diffusion of Innovation* (3rd ed.). New York: Free Press.

Ronald Beaubrun, S. P. (2001). Technological developments and socio-economic issues of wireless mobile communications. *Telematics and Informatics, 18*, 143–158. doi:10.1016/S0736-5853(00)00026-5

Shiau, W.-L., Chao, H.-C., & Hsu, P.-Y. (2005). Obstacles of enterprises moving towards the next generation internet. [IJIPT]. *International Journal of Internet Protocol Technology, 1*, 34–39. doi:10.1504/IJIPT.2005.007557

Shiau, W.-L., Li, Y.-F., Chao, H.-C., & Hsu, P.-Y. (2006). Evaluating IPv6 on a large-scale network. *Computer Communications, 29*, 3113–3121. doi:10.1016/j.comcom.2005.11.011

Silveira, G. D. (2001). Innovation diffusion: Research agenda for developing economies. *Technovation, 21*, 767–773. doi:10.1016/S0166-4972(01)00007-4

Sisaye, S. (1999). An organizational approach for the study of the diffusion of process innovation strategies in internal auditing and control systems. *International Journal of Applied Quality Management, 2*, 279–293. doi:10.1016/S1096-4738(99)80095-5

Stock, G. N., Greis, N. P., & Fischer, W. A. (2002). Firm size and dynamic technological innovation. *Technovation, 22*, 537–549. doi:10.1016/S0166-4972(01)00061-X

Uzun, A. (2001). Technological innovation activities in Turkey: The case of manufacturing industry, 1995-1997. *Technovation, 21*, 189–196. doi:10.1016/S0166-4972(00)00033-X

Vonortas, N. S., & Xue, L. (1997). Process innovation in small firms: Case studies on CNC machine tools. *Technovation, 17*, 427–438. doi:10.1016/S0166-4972(97)00016-3

Weiser, M. (2001). Whatever happened to the next-generation Internet? *Association for Computing Machinery. Communications of the ACM, 44*, 61–68. doi:10.1145/383694.383708

Wright, M. (2007). Unique IP addresses will prove main IPv6 benefit. *EDN, 52*, 14–14.

This work was previously published in International Journal of Business Data Communications and Networking, Volume 5, Issue 3, edited by V. Sridhar and D. Saha, pp. 39-52, copyright 2009 by IGI Publishing (an imprint of IGI Global).

APPENDIX

Awareness of IPv4 Limitation

A1 Are you aware of the IPv4 shortage problem?
A2 Are you aware that IPv4 lacks security?
A3 Are you aware that IPv4 lacks quality service?

Appreciation of the Technical Merit of IPv6

B4 Do you appreciate the numerous IP addresses of IPv6?
B5 Do you appreciate the security of IPv6?
B6 Do you appreciate the quality of service of IPv6?
B7 Do you appreciate the speed of IPv6?

Knowledge of Products Supporting IPv6

C8 Do you have the knowledge of operating system supporting IPv6?
C9 Do you have the knowledge of network products supporting IPv6?
C10 Do you have the knowledge of home applications supporting IPv6?

Foreseeing the Applications of IPv6

D11 Do you foresee that the mobile system supporting IPv6?
D12 Do you foresee the cell phone products supporting IPv6?
D13 Do you foresee the home applications supporting IPv6?

Preference for NGI

E14 Do you prefer NGI when IPv6 provides more IP address than IPv4?
E15 Do you prefer NGI when IPv6 provides more security than IPv4?
E16 Do you prefer NGI when IPv6 provides better quality service than IPv4?
E17 Do you prefer NGI when IPv6 provides better speed than IPv4?
E18 Do you prefer NGI when IPv6 provides better cell phone service than IPv4?
E19 Do you prefer NGI when IPv6 provides better home applications than IPv4?

Chapter 13
On Demand Bandwidth Reservation for Real-Time Traffic in Cellular IP Network using Particle Swarm Optimization

Mohammad Anbar
Jawaharlal Nehru University, India

D. P. Vidyarthi
Jawaharlal Nehru University, India

ABSTRACT

Cellular IP network deals with micro mobility of the mobile devices. An important challenge in wireless communication, especially in cellular IP based network, is to provide good Quality of Service (QoS) to the users in general and to the real-time users (users involved in the exchange of real-time packets) in particular. Reserving bandwidth for real time traffic to minimize the connection drop (an important parameter) is an activity often used in Cellular IP network. Particle Swarm Optimization (PSO) algorithm simulates the social behavior of a swarm or flock to optimize some characteristic parameter. PSO is effectively used to solve many hard optimization problems. The work, in this paper, proposes an on demand bandwidth reservation scheme to improve Connection Dropping Probability (CDP) in cellular IP network by employing PSO. The swarm, in the model, consists of the available bandwidth in the seven cells of the cellular IP network. The anytime bandwidth demand for real-time users is satisfied by the available bandwidth of the swarm. The algorithm, used in the model, searches for the availability of the bandwidth and reserves it in the central cell of the swarm. Eventually, it will allocate it on demand to the cell that requires it. Simulation experiments reveal the efficacy of the model.

DOI: 10.4018/978-1-60960-589-6.ch013

INTRODUCTION

Advancement in technology opens many challenges in wireless communication especially in cellular wireless communications. Many of these challenges are due to optimism to extract more from the technological growth. To maintain better Quality of Service (QoS), to the users in a cellular system, is quite challenging task. Cellular IP network, in wireless cellular systems, uses packet switching techniques in transferring data from one user to another and is based on one of the Internet Protocols IPV4 or IPV6. The problems are to be studied well as the flow of multimedia traffic, in cellular IP network, is growing enormously. Non-real time traffic e.g. e-mail, text data etc., though important in cellular IP network, is given less importance in comparison to the real-time traffic. Often, bandwidth reservation has been advocated to provide better QoS in cellular systems. The work, proposed here, uses bandwidth reservation method for better QoS giving priority to real-time traffic. Thus the packets have been given more importance for the flow in Cellular IP network. The proposed scheme uses bandwidth reservation algorithms similar to the Adaptive Resource Reservation schemes studying bandwidth reservation using Support Vector Machine and Particle Swarm Optimization (Chenn-Jung Huang, Yi-Ta Chuang, Wei Kuang Lai, Yu-Hang Sun,& Chih-Tai Guan., 2007). Also a Probabilistic Resource Estimation and Semi-reservation scheme have been considered which uses the resource semi-reservation approach rather than the conventional full reservation method. The semi-reservation approach consumes less time of the network (Geng-Sheng Kuo, Po-Chang Ko, & Min-Lian Kuo, 2000).

Another scheme that uses Bandwidth Reservation is Dynamic Grouping Bandwidth Reservation scheme for multimedia wireless networks which is based on probabilistic resource estimation. According to this scheme, when the Mobile Host (MH) requests a new connection flow or it handoffs to a new cell, it provides some important information e.g. the estimated switching time and the estimated staying time etc (Jau-Yang Chang, & Hsing-Lung Chen, 2003).

The proposed model is an attempt to provide better QoS in Cellular IP network for on demand bandwidth reservation using Particle Swarm Optimization (PSO) algorithm. On-demand refers to a service which addresses the user's need for instant and immediate use. It means that when a Mobile Host (MH) requests a new connection from a cell in a Cellular IP network, the base station will perform a processing based on swarm optimization algorithm to reserve the available free bandwidth in all the cells of the swarm. If it fails in doing so, the cell may demand the reservation of the bandwidth which is assigned to the non real time users. Experimental studies have been carried out and the results have been compared with the Probabilistic Resource Estimation and Semi-reservation scheme.

The rest of the paper is organized as follows. In the next section, we briefly describe cellular IP network and elaborate bandwidth reservation for real time packets in cellular IP network. Section 3, discusses Particle Swarm Optimization (PSO) algorithm. In section 4, the proposed model is elaborated. Section 5 contains the simulation experiments, the results obtained and the comparison of the results with the Probabilistic Resource Estimation and Semi-reservation scheme. In final section 6, some conclusions and observations, with the results obtained, have been made.

BANDWIDTH RESERVATION IN CELLULAR IP NETWORKS

It is imperative to brief cellular IP network before going through the bandwidth reservation in cellular IP network.

Cellular IP Network

Cellular IP is a type of cellular systems that inherits group of cellular network features and is based on the IP design principles.

Components of cellular IP are the Base Stations (BS) that works as a router for packets destined to a Mobile Host (MH). It works as a wireless access point as well. Gateway is the connector between the cellular IP and the internet Mobile Host. Figure 1 shows cellular IP access network enabled internet (A. G. Valkó, Javier Gomez, S. Kim,& Andrew T. Campbell, 1999).

In general, there are two types of traffic in any communication network and therefore in Cellular IP network. The class I which includes real time traffic like (video, voice, etc.) and the class II with non real time traffic such as (e-mail, paging, etc.). Due to delay sensitive nature in real time traffic it is to be served before non real time traffic. In all aspect, the class I traffic requires more attention than class II traffic. A better QoS demands the efficient service of class I traffic. At the same time, serving class I traffic without neglecting (dropping) class II traffic will be a good accomplishment. Thus any scheme to enhance QoS has to consider both the aspects while serving the users of the network.

Figure 1. Cellular IP network connected to internet

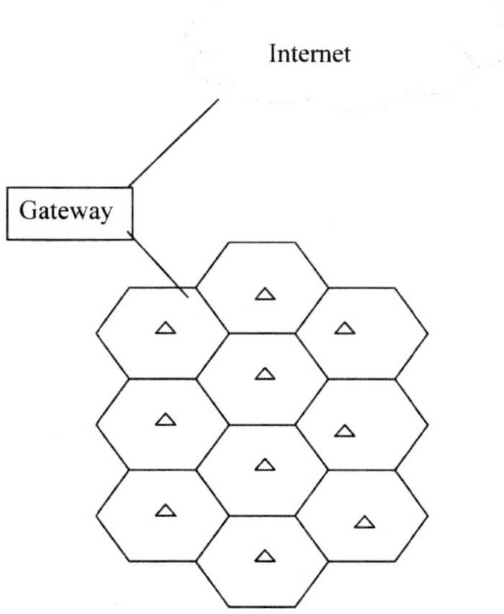

Bandwidth Reservation

Resource reservation in general and the bandwidth reservation in particular is the most important task in a Cellular IP network. Bandwidth reservation is a good approach through which the QoS in the network can be improved. It is proved (by Sunho Lim, Guohong Cao, & Chita R. Das, 2001) that reserving bandwidth reduces the Connection Dropping Probability (CDP) and thus leads to better QoS. Bandwidth Reservation implies that a part of assigned bandwidth to a cell in a Cellular IP network is reserved aside and is issued to high priority establishments such as real-time traffic connections (Roland Zander, & Johan M. Karlsson, 2005). There are two types of bandwidth reservation schemes; static and adaptive bandwidth reservation schemes (dynamic schemes). Static schemes allow a user, admitted to a cell, to reserve the required bandwidth only once and do not allow to reserve more if the user further puts the demand for more bandwidth. It results in connection drop. Adaptive bandwidth reservation schemes are more flexible and allow the user to reserve the bandwidth as and when the need arises. Adaptive schemes, performs based on the resources availability and according to the applications (user's requirements) (Xiang Chen &Yuguang Fang, 2003). The aim of both the bandwidth reservation schemes, static or dynamic, is improving the QoS in network by making the best utilization of the available bandwidth. It also results in the reduction of packet loss, delay and the jitter in the network (Manvi. S.S & Venkataram. P, 2002).

The proposed model uses dynamic bandwidth reservation that allows the user to reserve the available bandwidth from the neighboring cells of the same group of cells (swarm) based on some statistics and parameters and the user's requirements. It is an on-demand scheme as it caters to immediate requirement.

Bandwidth Reservation to Improve QoS

For the multimedia services in the wireless networks, a certain amount of bandwidth is necessary in order to meet the requirements of these services; therefore, bandwidth reservation is one of the best approaches to guarantee QoS for such services. It has been proved that resource reservation, especially bandwidth, can improve QoS in wireless networks. Ad-Hoc network is one example in which bandwidth reservation is applied and the results show the improvements in QoS (Chih-Shun Hsu, Jang-Ping Sheu, & Shen-Chien Tung, 2006).

QoS can be affected by many factors such as the limitations in bandwidth, transmission characteristics etc. These issues impose constraints on the amount of administrative and control information to be exchanged, as a result affecting the QoS. In order to satisfy the requirements to achieve good QoS for all the users in the network, there should be enough resources. Bandwidth is one of the prime resources. Availing bandwidth during the service time can be done depending on the bandwidth reservation schemes.

PARTICLE SWARM OPTIMIZATION

Swarm intelligence is an intelligent paradigm to solve the hard optimization problems. It is based on the behavior of the social insects such as bird flocks, fish school, ant colony etc., in which individual species change its position and velocity depending on its neighbor's movement. This is done by mimicking the behavior of the creatures within their swarms or colonies.

Particle Swarm Optimization (PSO) is an algorithm based on the swarm intelligence. It is a population based optimization tool. As an algorithm, the main strength of the PSO is its fast convergence in which due to its well organized logic and procedures the optimal solution for a specific problem can be attained very fast. PSO shares many common points with Genetic Algorithm (GA). Both algorithms start with a group of randomly generated populations and both have fitness function to evaluate the population. Also, both update the population and search for the optimal solution with random techniques (Nadia Nedjah, & Luiza de Macedo Mourelle, 2006). Any problem to be solved (e.g. bandwidth reservation problem) can be transformed to the function optimization problem.

PSO model is a swarm of individuals called particles. Particles are initialized with the random solutions. These particles move through many iterations to search a new and better solution for the problem. Each particle is represented by two factors; one the position (x), in which each particle has a specific position and at the beginning initialized by the initial position. The other factor is the velocity (v), where each particle moves in the space according to a specific velocity. During the iteration time (t), the particles update their position and their velocity (x and v) (Nadia Nedjah, & Luiza de Macedo Mourelle, 2006).

PSO simulates the behavior of the bird flocking. Consider the following scenario. A flock of birds are flying randomly and searching for food in an area and there is only one piece of food in the area being searched now. All the birds in the flock have learned that there is food in this area but none of them know where the food is? The best strategy to find the food is to follow the nearest bird to the food (Jang-Ho Seo, Chang-Hwan Im, Chang-Geun Heo, Jae-Kwang Kim, Hyun-Kyo Jung, & Cheol-Gyun Lee, 2006).

In PSO algorithm, there are two types of best values: one is (Pbest) which is the local best position for each particle in the swarm and must be updated depending on the fitness value for each particle. The second best value is (Gbest) which is the global best value for the swarm in general.

This value must be checked with each iteration and be changed by (Pbest) if the (Pbest) for the current iteration is better than (Pbest) for the last iteration.

The pseudo-code of the PSO algorithm is as follows.

PSO ()
{
Initialize the swarm by giving initial and random values to each particle.
For each particle do
{Calculate the fitness function
If the value of the fitness function is better than the best fitness value (pbest) in history then,
Set the current value as the new best.
Choose the particle with the best fitness value of all the particles as the gbest.

Update the velocity of each particle as

$$V_j^{k+1} = w.V_j^k + c1.r1.(P_{best_j}^{\ k} - X_j^k) + c2.r2.(G_{best}^{\ k} - X_j^k)$$

Update the position of each particle:

$$X_j^{k+1} = X_j^k + V_j^k.\Delta t$$

}

Until the solution converges.
}

In the above algorithm,

V_j^k: is the velocity of particle j in iteration k

Pbest: is the best solution it has achieved so far for each individual

Gbest: is the global best value for the swarm

X_j^k: is the current position of particle j in iteration k

w: is inertia weight and is varied from 0.9 till 0.4

r1, r2: are random numbers between 0 and 1

c1, c2: are acceleration factors that determine the relative pull for each particle toward Pbest and Gbest and usually c1, c2 = 2.

Δt: is a time step

THE MODEL

The model, proposed in this work, uses the PSO algorithm for the optimization of the Connection Dropping Probability (CDP) making bandwidth utilization in Cellular IP. The model and the algorithm have been described as follows.

Model Description

Particle Swarm Optimization (PSO) algorithm is employed to reserve the bandwidth for the real-time users that exist in a cell within a swarm. The swarm consists of seven cells, one centre cell surrounded by six neighbor cells. The cell announces that there are some real-time users and their connections may be dropped in case there is not enough bandwidth. Reservation scheme is done according to the swarm shown in Figure 2.

The central cell of the swarm (Figure 2) is cell A. Bandwidth reservation for the swarm i.e. asking for help is done in the cell A, in which the bandwidth available in any of the cells of the swarm will be transferred. Eventually, the bandwidth may be given to the cell asking for it. This is because cells are away from each other but are close to the central cell to which the bandwidth is transferred. Bandwidth reservation is done through two-step reservation procedure. In the first step the swarm will search, executing PSO algorithm, for any free bandwidth available in all

Figure 2. The swarm of seven cells

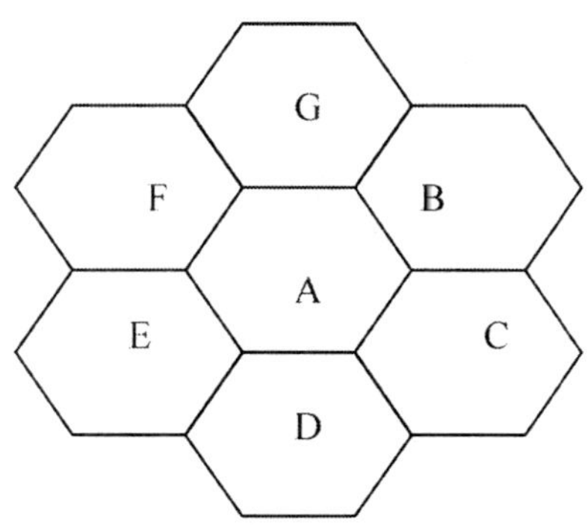

cells of the current swarm. If there is some available bandwidth, designated as free bandwidth, it will be reserved in the cell A. The swarm will adjust accordingly. It will then calculate the fitness value (CDP) of the cell asking for the bandwidth. If the swarm could satisfy the requirement of the seeking cell, it is served. Otherwise the swarm will search for another solution in terms of reserving the available bandwidth in each cell and assigned to the on going non real-time users in the swarm. It is done keeping in mind that the real-time traffic has higher priority than the non real-time traffic. Though it is to be handled carefully and should not affect the ongoing non real time sessions in any of the cells of the swarm. After that the swarm will execute PSO algorithm and adjust it according to the new reserved bandwidth. This time the PSO is to be executed with many iterations and different numbers of users with different sizes of packets (real-time as well as non real-time packets).

THE ALGORITHM

PSO algorithm, used in this work, has been applied in two-steps. In each step, the same PSO algorithm is applied with some modifications in bandwidth reservation. Algorithm is described as follows.

1. Distribute the available amount of bandwidth randomly, among the cells in the swarm, where each cell will have a part of this distributed bandwidth assigned to real-time traffic in the cell. The other part of this bandwidth is assigned to the non real-time traffic in the same cell. Also some part of this bandwidth is addressed as free bandwidth. Each type of the distributed bandwidth (real-time, non-real-time, and free) has been generated (for experimental purposes) and distributed randomly among the cells. Also the proportion of bandwidth assigned to each cell is not fixed or static.
2. Users start arriving to any cell of the swarm. They start their real-time or non real-time sessions and therefore start generating random packets. Assign 1 for real-time packets and 2 for non real-time packets. Also for each session that is going on assign the session number.
3. Calculate the number of users and the required amount of bandwidth (to continue the connection) in each cell for each user. Take the decision according to these statistics.
4. Execute PSO algorithm for the cell asking for more bandwidth in the swarm. In first step (PSO1) by reserving the free bandwidth in each cell. This free bandwidth will be transferred to the cell A (central cell) and then to the cell asking for the bandwidth. Compute the fitness function (CDPbest) in every iteration with the new value of CDP and if it is less, then set CDPbest = CDP.
5. Repeat step 4 for a given number of iterations.
6. If after the step 4 and 5 we get CDP for the asking cell >0.5 then the swarm starts second step (PSO2) of the bandwidth reservation.
7. Reserve the available bandwidth for non real-time traffic in each cell by transferring it to the cell A (central cell) and eventually to the cell asking for it. Take into account that none of the on-going non real-time sessions are dropped.
8. Repeat steps 6 and 7 for different numbers of iterations or until CDP<0.5.

SIMULATION EXPERIMENTS

The model is simulated and many experiments have been conducted to prove the efficacy of the model in decreasing Connection Dropping Probability (CDP) and improving the QoS.

Experiment 1: To Study the Effect Of Available Bandwidth on CDP

This experiment considers that the cell asking for bandwidth contains four real-time users and four non real-time users. Size of every packet is 10 Kbytes. The CDP obtained are shown by graph in Figure 3.

Experiment 2: To Study the Effect of Number of Users Asking for Bandwidth (Real-Time) on CDP

Experiment considers size of every packet as 10 Kbytes, available amount of bandwidth as 40 (Mb/sec) and the number of packets in each session as 10. The CDP obtained are shown by graph in Figure 4.

Experiment 3:To Study the Effect of Real-Time Packets in The Real-Time Sessions on CDP

It is when the number of packets in the on-going session changes due to the data exchanged between the users. This experiment is carried out for 4 real-time users and 4 non real time users in the cell asking for bandwidth. Size of each packet is 10 Kbytes and the available amount of bandwidth is 40 (Mb/sec). Figure 5 shows the effect of number of packets on the CDP for real time traffic.

Experiment 4: To Study the Effect of Number of Iterations on CDP with PSO1

This experiment is carried out with the following input data: available bandwidth is 40(Mb/sec),

Figure 3. Effect of available bandwidth on CDP

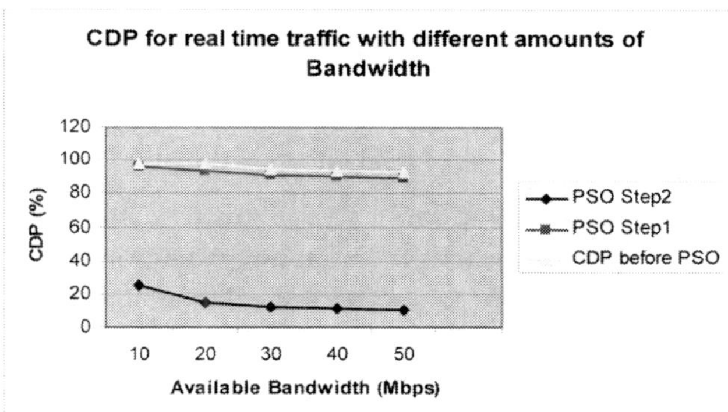

Figure 4. Effect of number of users on CDP

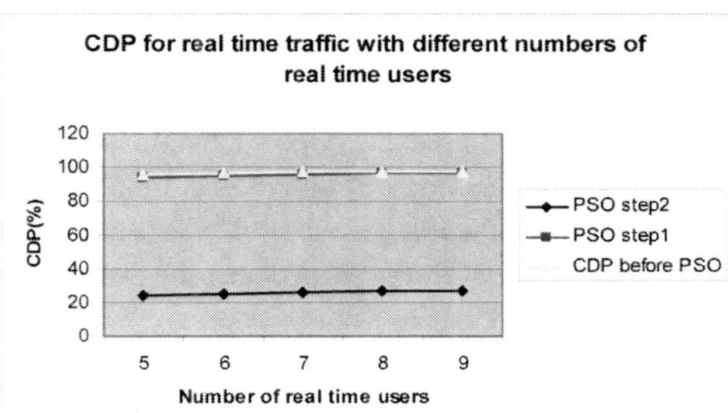

Figure 5. Effect of number of packets on CDP

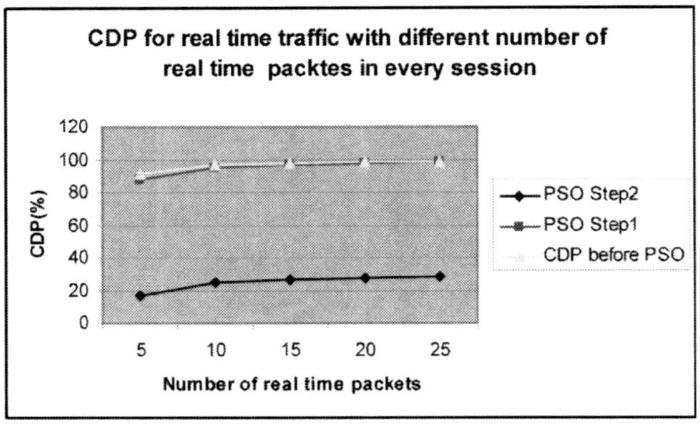

Figure 6. Effect of number of iterations on CDP in step 1

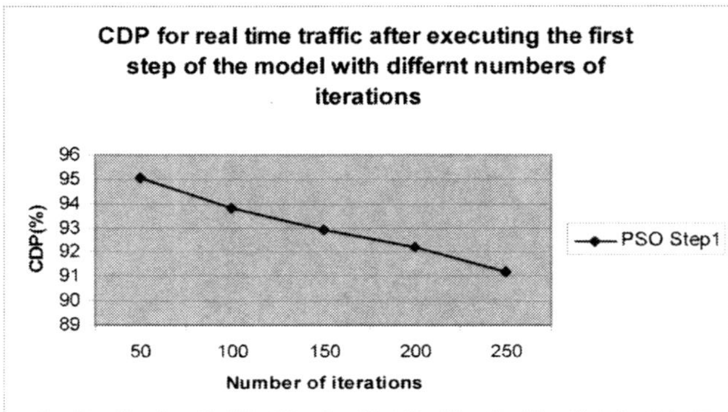

number of real time users in the cell asking for bandwidth is 4, the number of non real time users is 4, number of packets in each session is 10, size of each packet is 10 Kbytes. The effect of number of iterations is shown in Figure 6.

Experiment 5:To Study the Effect of Number of Iterations on CDP with PSO2

Again an experiment is carried out for this purpose with the following input data: available bandwidth is 40 (Mb/sec), number of real time as well as non real time users in the cell asking for bandwidth is 4, number of packets per session is 10, size of each packet is 10 Kbytes. After PSO1, if CDP is still greater than 0.5 then the swarm searches for another solution with different numbers of iterations. Figure 7 exhibit the result.

Experiment 6:To Compare the Proposed Model with the Probabilistic Resource Estimation Semi-Reservation (PS) Model

The comparison is done for the following input data; number of real time users in the cell asking for the bandwidth is 4, number of non real time users is 4, number of packets per session is 10, size of each packet is 10 Kbytes. The comparison between the two models is shown in Figure 8.

Figure 7. Effect of number of iterations on CDP in step 2

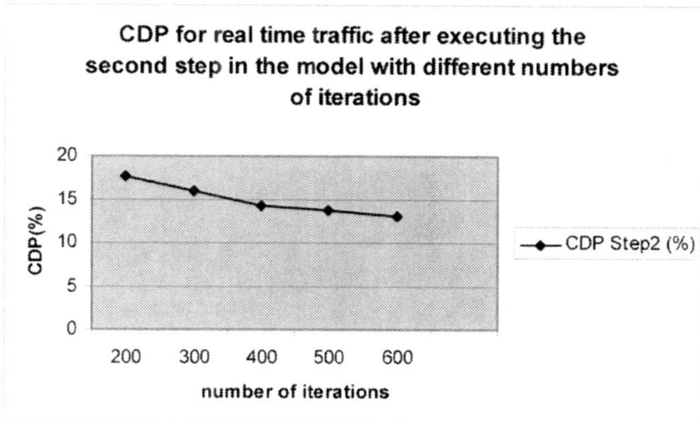

Figure 8. Comparative study with the PS model

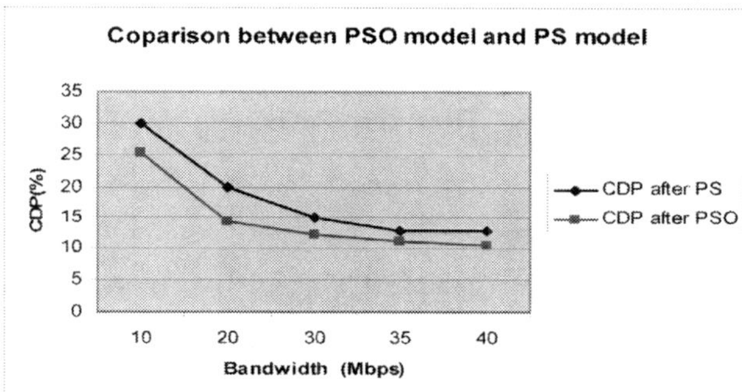

Experiment 7: To Study the Change in CDP on Real-Time and Non Real-Time Traffic

This experiment is conducted with the following input; the available amount of bandwidth is 40 (Mb/sec), number of real time users and non real time users in the cell asking for the bandwidth is 4, number of packets per session is 10, size of each packet is 10 Kbytes. The observation in the CDP is done for real time and non real time traffic.

Figure 9 shows CDP for both real-time and non real-time traffic with specified number of users with non real-time packets. It is conspicuous that CDP for real-time traffic in the cell demanding the bandwidth is decreasing when the swarm is reserving the bandwidth assigned to the non real-time users. Also, the Connection Dropping Probability is increasing for these non-real-time users, but without reaching to the value that drops this connection.

Experiment 8: To Study the Effect of Packet Generation Rate on CDP

Packet Generation Rate is an important factor that plays an important role in CDP. Study of the effect of packet generation rate is done in this experiment by choosing 5 real-time users, 2 minutes as a session time and the available bandwidth as 40 (Mb/sec). Results obtained from this experiment are shown in Figure 10.

Experiment 9: To Study the Effect of Time on CDP

The time of the session has been changed every time with 5 real-time users, 40 (Mb/sec) as

Figure 9. Bandwidth assigned to non real time users reserved without connection drop

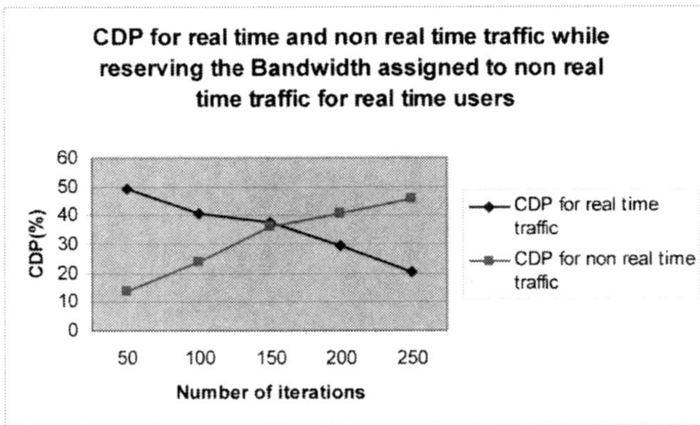

Figure 10. Effect of packet generation rate on CDP

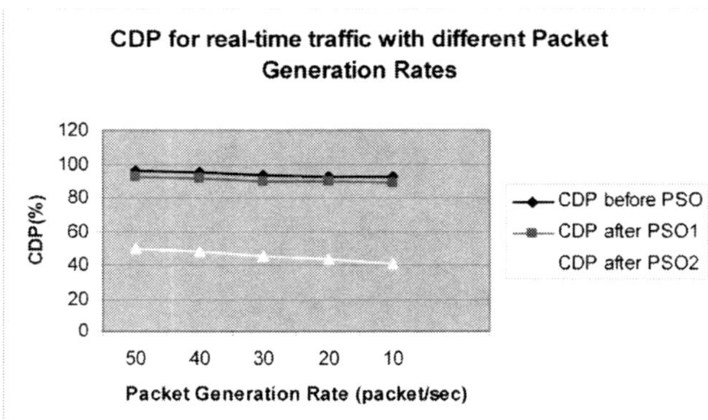

available bandwidth and 30 packet/sec. as packet generation rate. Effect of the time change is shown in Figure 11.

OBSERVATIONS AND CONCLUSION

The following observations are drawn from the experiment conducted in the previous section.

In all the experiments, the bandwidth availability has been taken care of as the bandwidth is a limited resource.

Each cell in the swarm gets different amount of bandwidth. Sometimes free bandwidth in some cells is zero or very small due to random distribution; therefore CDP by reserving the free bandwidth is not improved i.e. it is almost similar to the CDP before using the PSO algorithm. After using PSO we observe the improvement in terms of decreased CDP. It is obvious that the CDP will be less when the available free bandwidth is higher.

When real-time traffic increases, the demand for the bandwidth will also increase and consequently the consumed bandwidth. Applying PSO reduces CDP when the traffic increases. Using PSO for bandwidth reservation, the swarm can accept more users (real-time users) without dropping their connections. Even the available amount of bandwidth for the swarm remains the same. This is an important achievement of the model.

The ability of the swarm to accept more real-time users is applicable for non real-time users as

Figure 11. Effect of session time on CDP

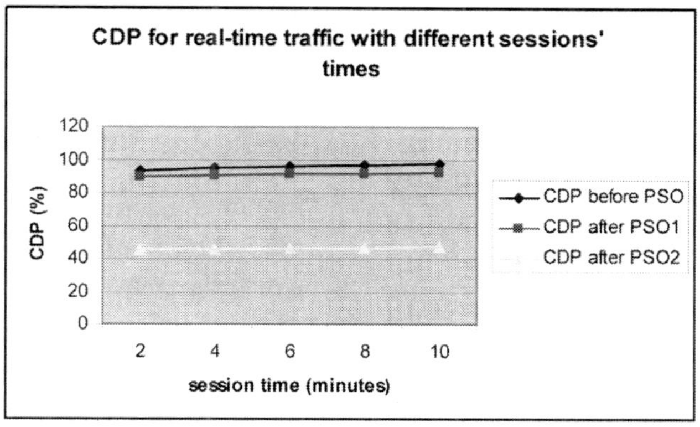

well. It is because, as mentioned before, the model takes into account even the connection drop for non real-time users.

It is obvious that CDP increases when the number of packets per session increases. But using PSO we can increase the number of exchanged packets per session without fast reaching to the dropping probability value (in the model value greater than 0.5) at which the connection will be dropped. This issue is important in saving the time of the users (by sending more packets per session) where a user can send all the packets in one or two sessions instead of four or five sessions, for instance, and this eventually will save the time of the network as well.

Though in all the experiments, we conducted, it is found that the cell demanding bandwidth from the swarm for its real-time users is not satisfied by the available free bandwidth but it doesn't mean that it is impossible to satisfy so by the available free bandwidth. The bandwidth demand of the cell may be satisfied by the available free bandwidth when the swarm is not highly loaded with the real time traffic. In that case most of the allocated bandwidth to the cells and therefore to the swarm will be free that can satisfy the demand of the other cells containing real time traffic.

By increasing the number of iterations it is observed that the CDP is decreasing. Thus after certain number of iterations the swarm searches a good solution and the CDP is decreased to a small value. Though there is a tradeoff in terms of time as the decision is to be taken as quickly as possible.

The experiment of iterations (experiment 4 and 5) is conducted exclusively for different numbers of iterations for PSO1 and PSO2. It is because in PSO1 the amount of bandwidth is less and therefore tolerance to reserve more bandwidth is less. The tolerance in case of PSO2 is more due to more availability of bandwidth reserved from non real-time traffic. Increasing number of iterations is important in PSO2 because it leads to good results.

In comparison with Probabilistic Resource Estimation semi-Reservation scheme, it is observed that the Probabilistic Resource Estimation model doesn't take care of non real-time traffic while the proposed model takes care of real-time traffic along with ongoing non real-time session. Apart form this; CDP for real-time users for the same number of cells and the same amount of bandwidth is also less in the proposed model.

Overhead in this model is to be seen in terms of number of dropped non-real-time users if there is no available bandwidth for real-time users. The connection for non real time users will be dropped in order to avail the bandwidth to the real-time users which is more important.

To summarize, in this model a method is proposed for bandwidth reservation based on the Particle Swarm Optimization algorithm. This model considers two types of traffic; real time and non real time. A swarm is constituted consisting of seven cells. This model is on-demand model, in which the cell demands for bandwidth. The swarm will execute PSO in every base station of the swarm. The model improves CDP for real-time traffic and at the same time takes care of the non real time traffic. In comparison with the Probabilistic Resource Estimation semi-Reservation model the proposed model performs better in terms of reduction of CDP for the same number of cells and the same amount of available bandwidth. The model can easily be incorporated for actual implementation in cellular IP network.

The authors would like to accord their sincere thanks to the anonymous reviewers of this paper, whose suggestions have resulted in the quality improvement of this paper.

REFERENCES

Al-Karaki, J. N., & Chang, J. M. (2004). A simple distributed access control scheme for supporting QoS in IEEE 802.11 wireless LANs. *Wireless Communications and Networking Conference* (Vol. 1) (pp. 213–218). *IEEE.*

Bader, A.-M., Nidal, N., Abu, A. N., & Ho, H. (2008). Efficient bandwidth management in Broadband Wireless Access Systems using CAC-based dynamic pricing. *33rd IEEE Conference on Local Computer Networks, (pp 484 – 491).*

Chang, J.-Y., & Chen, H.-L. (2003). Dynamic-Grouping bandwidth reservation scheme for multimedia wireless networks. *IEEE Journal on Selected Areas in Communications*, *21*(10), 1566–1574. doi:10.1109/JSAC.2003.814863

Chen, X., & Fang, Y. (2003). An Adaptive Bandwidth Reservation Scheme in Multimedia Wireless Networks. *Global Telecommunications Conference, GLOBECOM '03. IEEE, 5,* 2830–2834.

Chung, Y.-L., & Tsa, Z. (2006). Modeling and Analysis of Dynamic Spectrum Allocation of Two Wireless Communication Systems. *IEEE 17th International Symposium on Personal, Indoor and Mobile Radio Communications* (pp. 1–5).

Francois, J. M., & Leduc, G. (2005). Mobility prediction's influence on QoS in wireless networks: a study on a call admission algorithm. *Third International Symposium on Modeling and Optimization in Mobile, Ad Hoc, and Wireless Networks* (pp. 238-247).

Georgantas, N. V., Papadoukakis, N. K., Liveris, A. D., & Sykas, E. D. (1999). Modelling of a bandwidth management scheme for DECT wireless access to a local area network. *Conference on Local Computer Networks* (pp. 43–48).

Grewal, J., & DeDourek, J. (2004). Provision of QoS in wireless networks. *Second Annual Conference on Communication Networks and Services Research (pp. 337–340).*

Guo, H.-B., & Kuo, G.-S. (2005). A dynamic and adaptive bandwidth management scheme for QoS support in wireless multimedia networks. *IEEE 61st Vehicular Technology Conference* (Vol. 3) (pp. 2081–2085).

Hac, A. (1991). Bandwidth management in the switch with various traffic burstiness. *Third IEE Conference on Telecommunications* (pp. 343–347).

Hsu, C.-S., Sheu, J.-P., & Tung, S.-C. (2006). An On-Demand Bandwidth Reservation QoS Routing Protocol for Mobile Ad Hoc Networks. *IEEE International Conference on Sensor Networks, Ubiquitous, and Trustworthy Computing* (Vol. 2) (pp. 198–207).

Huang, C.-J., Chuang, Y.-T., Lai, W. K., Sun, Y.-H., & Guan, C.-T. (2007). Adaptive resource reservation schemes for proportional DiffServ enabled fourth-generation mobile communications system. *Computer Communications Journal*, *I*(7), 1613–1623. doi:10.1016/j.comcom.2007.01.015

Hwang, W.-S., & Tseng, P.-C. (2005). A QoS-aware residential gateway with bandwidth management. *IEEE Transactions on Consumer Electronics*, *51*(3), 840–848. doi:10.1109/TCE.2005.1510493

Kuo, G.-S., Ko, P.-C., & Kuo, M.-L. (2000). A probabilistic resource estimation and semi-reservation scheme for flow-oriented multimedia wireless networks. *IEEE Wireless Communications and Networking Conference, (WCNC 2000)* (Vol. 3) (pp. 1046 – 1051).

Lim, S., Cao, G., & Das, C. R. (2001). *A Differential Bandwidth Reservation Policy for Multimedia Wireless Networks* (pp. 447–452). Parallel Processing Workshops.

Liu, K., Petr, D. W., Frost, V. S., Zhu, H., Braun, C., & Edwards, W. L. (1997). Design and analysis of a bandwidth management framework for ATM-based broadband ISDN. *Communications Magazine, IEEE*, *35*(5), 138–145. doi:10.1109/35.592108

Lorenz, P. (2006). QoS in Wireless Networks. *The IET Workshop on Practical Applications for Wireless Technology* (pp. 1–22).

Manvi, S. S., & Venkataram, P. (2002). Adaptive bandwidth reservation scheme for multimedia traffic using mobile agents. *High Speed Networks and Multimedia Communications 5th IEEE International Conference, ISBN: 0-7803-7600-5* (pp. 370- 374).

Nagarajanvelmurugan, K. S., & Dananjayan, P. (2007). Resource Allocation Management for Multimedia QoS in Wireless Networks. *International Conference on Signal Processing, Communications and Networking* (pp. 122–127).

Nedjah, N., & de Macedo Mourelle, L. (2006). *Swarm Intelligent Systems*. Berlin, Heidelberg: Springer-Verlag. doi:10.1007/978-3-540-33869-7

Padilla, J. J., Paradells, J., & Rodriguez, A. (2006). Supporting QoS over IPv6 wireless networks with IntServ6. *IEEE 17th International Symposium on Personal, Indoor and Mobile Radio Communications* (pp. 1–6).

Samad, M., & Herman, S. H. (2005). Quality of service for mobile IP services in wireless network. *Asia-Pacific Conference on Applied Electromagnetic* (p. 4).

Saraireh, M., Saatchi, R., Al-khayatt, S., & Strachan, R. (2006). Development and Evaluation of a Fuzzy Inference Engine System to Incorporate Quality of Service in IEEE 802.11 Medium Access Protocol. *Conference on Wireless and Mobile Communications International* (pp. 29–29).

Seo, J.-H., Im, C.-H., Heo, C.-G., Kim, J.-K., Jung, H.-K., & Lee, C.-G. (2006). IEEE Transactions on Multimodal function optimization based on particle swarm optimization. *Magnetics*, *42*(4), 1095–1098. doi:10.1109/TMAG.2006.871568

Shan, T., & Yang, O. W. W. (2007). Bandwidth Management for Supporting Differentiated Service Aware Traffic Engineering. *IEEE Transactions on Parallel and Distributed Systems*, *18*(9), 1320–1331. doi:10.1109/TPDS.2007.1052

Valkó, A. G., Javier Gomez, K. K., & Campbell, A. T. (1999). On the Analysis of Cellular IP Access Networks. *Proceedings of the IFIP TC6 WG6.1 & WG6.4 / IEEE ComSoc TC on Gigabit Networking Sixth International Workshop on Protocols for High Speed Networks VI, 158,* 205–224.

Zander, R., & Karlsson, J. M. (2005). Combining Bandwidth Borrowing and Reservation in Cellular Networks. *International Journal of Wireless Information Networks*, *12*(3), 187–201. doi:10.1007/s10776-005-0004-z

APPENDIX

Table 1. Effect of available bandwidth on CDP

CDP after PSO2 (%)	CDP after PSO1 (%)	CDP before PSO (%)	Bandwidth (Mb/sec)
25.33	95.73	97.8	10
14.34	93.09	97.72	20
12.23	90.95	94.26	30
11.22	89.51	93.12	40
10.45	89	92.76	50

Table 2. Effect of number of users on CDP

CDP after PSO2 (%)	CDP after PSO1 (%)	CDP before PSO (%)	No. of real-time users in the cell asking for bandwidth
23.96	93.82	95.93	5
24.902	95.04	96.73	6
25.64	95.78	97.22	7
26.404	96.307	97.56	8
26.702	96.77	97.87	9

Table 3. Effect of number of packets on CDP

CDP after PSO2(%)	CDP after PSO1 (%)	CDP before PSO (%)	Number of packets per session
16.944	87.77	91.94	5
24.9	95.04	96.73	10
26.88	96.86	97.93	15
27.87	97.8	98.55	20
28.34	98.33	98.9	25

Table 4. Effect of number of iterations on CDP in step 1

PSO Step1	No. of iterations
0.9504	50
0.9382	100
0.9291	150
0.9223	200
0.9174	250

Table 5. Effect of number of iterations on CDP in step 2

CDP Step2 (%)	No. of Iterations
17.56	200
15.85	300
14.27	400
13.65	500
13.03	600

Table 6. Effect of packet generation rate

Packet generation rate (packet per second)	CDP before PSO (%)	CDP after PSO1 (%)	CDP after PSO2 (%)
50	95.93	92.44	49.73
40	94.72	91.23	47.56
30	93.57	89.78	45.33
20	92.60	89.60	43.12
10	92.77	88.77	41.10

Table 7. Effect of session time

Session Time (minutes)	CDP before PSO (%)	CDP after PSO1 (%)	CDP after PSO2 (%)
2	93.57	89.78.	45.33
4	95.44	90.31	45.45
6	96.31	91.57	46.01
8	97.21	91.88	46.33
10	97.56	92.12	47.13

This work was previously published in International Journal of Business Data Communications and Networking Volume 5, Issue 3, edited by V. Sridhar and D. Saha, pp. 53-66, copyright 2009 by IGI Publishing (an imprint of IGI Global).

Chapter 14
Mobile Division Query Processing Incorporating Multiple Non-Collaborative Servers

Say Ying Lim
Monash University, Malaysia

David Taniar
Monash University, Australia

Bala Srinivasan
Monash University, Australia

ABSTRACT

The demand of mobile technology is growing and mobile information services including the gathering of information and manipulating them are also becoming more critical in the world of mobile technology. These allow mobile users to download useful data, possibly from multiple sources. In this article, we introduce two basic mobile query processing, namely: (i) mobile device side processing (MDSP), and (ii) server side processing (SSP). Our focus in this article is on performing relational division and multiple division operations using both MDSP and SSP techniques. In addition, walkthrough examples are also illustrated. A series of performance evaluation of our proposed techniques are presented and analyzed.

INTRODUCTION

Mobile technology has been increasingly in demand and is widely used to allow people to be connected wirelessly without having to worry about the distance barrier (Waluyo, Srinivasan, & Taniar, 2005c,d; Bohl, Manouchehri, & Winand, 2007). This new technology enables users to access information anytime, anywhere. Mobile devices are capable of processing and retrieving data from multiple remote databases (Lee, Zhu, & Hu, 2005). This allows mobile users who wish to collect data from different remote databases by sending queries to the servers and then be able

DOI: 10.4018/978-1-60960-589-6.ch014

to process the multiple information gathered from these sources locally on the mobile devices (Lim, Taniar, & Srinivasan, 2005). By processing the data locally, mobile users would have more control of what they actually want as the final results of the query. They can therefore choose to query information from different servers and join them to be processed locally according to their requirements. Also, being able to obtain specific information over several different sites would help to optimize results to mobile users' queries, as different sites may give different insights into a particular issue.

In order to derive and assemble data from several servers in a mobile device, a new research domain known as *Mobile Query Processing incorporating Multiple Non-collaborative Servers* has arisen (Lim, Taniar, & Srinivasan, 2005). Since servers may be just independent service providers, often these servers are specialized within the domain. Hence, the information they are providing, such as information on restaurants disseminated by a server, normally just focuses on the restaurants information with perhaps additional limited supporting information which might include how to get there (e.g. information about transportations). However, this is only supporting information since it does not exactly show the route that the user should take from his present location. Therefore, there is still a need to obtain full information from multiple servers, in this case, the servers that deal with restaurants and transportation separately.

Example: An international tourist, while traveling in a foreign country, does not know the whereabouts of the available vegetarian restaurants. He looks for restaurants recommended by both the Tourist Office and Vegetarian Community. First, using his wireless PDA, he would download information broadcast by the Tourist Office (Gulliver, Ghinea, & Patel, 2007). Then, he would download the information provided by the second organization mentioned above. Once he has obtained the two lists from the two information providers that may not correspond to each other, he may perform an operation on his mobile device that joins the contents from the two related providers.

This example illustrates the importance of being able to collate, in a mobile device, information obtained from various non-collaborative sources. In addition, not all service providers are supported by the use of a mediator. Therefore, information obtained from other independent non-related service providers needs to be processed individually. It should not be assumed that every service provider is linked through a mediator. Hence, in our research we focus on non-collaborative service providers and we refer to them as independent servers. Thus, it is vital to gather information from independent servers because it is often not enough to obtain data from just a single server.

This article focuses on a query type, called *division*, which is common in relational databases (Elmasri & Navathe, 1994). Division is even more common in mobile query processing involving multiple non-collaborative servers. In this article, we particularly focus on two types of division queries, namely: (*i*) relational division, and (*ii*) multiple group division. While relational division is already common traditionally in relational databases, multiple group division is a novel query type (Taniar & Rahayu, 2002a, 2002b). In this article, we proposed a number of algorithms to process these query types.

Division query processing is a powerful concept for querying databases, which enables a database system to evaluate complex "for-all" predicates (Elmasri & Navathe, 1994). This is also known as the universal quantification query and has been neglected in past research even though the power of universal quantification and relational division enables the analysis of many-to-many relationships and set valued attributes. The same goes for queries within the mobile environment. This type of division query processing appears to be useful and appropriate in certain circumstances and therefore is vital to explore its potential use in the mobile environment.

The rest of this article is structured as follows. Firstly, we review mobile database environment and the basic of mobile query processing involving multiple non-collaborative servers. Following this, we will give an overview of the two division query types mentioned above. Then we present our contributions, which are algorithms to process both query types, followed by some experimental results.

BACKGROUND

As the background of this article, we explain briefly a mobile database environment, and some basic mobile query processing techniques incorporating multiple non-collaborative servers.

Mobile Database Environment

Generally, mobile devices are defined as electronic equipment which operate without cables for the purposes of communication, data processing and exchange, which can be carried by their users and can receive, send or transmit information anywhere, anytime due to their mobility and portability (Waluyo, Srinivasan, & Taniar, 2005c). In particular, mobile devices include mobile phones, Personal Digital Assistants (PDAs), laptops that can be connected to a network, and PDA-mobile phones that add mobile phone functionalities to a PDA.

Mobile users, with their mobile devices and servers that store data, are involved in a typical mobile environment. Each of these mobile users communicates with a single or multiple servers that may be independent of one another (Jayaputera & Taniar, 2005). Communication between mobile users and servers is required in order to carry out any transactions and information retrieval. Basically, the servers are normally static, whereas mobile users can move from one place to another and are therefore dynamic. Nevertheless, mobile users have to be within a specific region so as to be able to receive signals in order to connect to the servers.

Figure 1 illustrates a scenario of a mobile database environment, which involves mobile users moving from one location to another location (Jayaputera & Taniar, 2005; Lim, Taniar, & Srinivasan, 2005). Whenever mobile users are within a specific region or cell, they can access information provided by the servers within that region. And the mobile users will obtain the desired data and downloaded into their mobile device. When they move to a different location, the information that they access may have been changed due to the change of the region. So in a new region, mobile users can download other data. Assuming is *C*3 currently in Location *B* and is accessing *S*2. Once he receives the desired data, he moves to Location *C* and now accesses *S*3 that provides different data. Having these two individual data sources, the mobile device can manipulate the data locally.

Example: A tourist is visiting an entertainment complex which locates many cinemas. The tourist using his mobile device downloads a list of top box office movies from the local film association. While visiting the entertainment complex, the tourist also downloads a list of movies shown in each cinema in that complex, and would like to find out which cinema that currently shows all top box office movies.

The above example is a typical example of a division query, whereby there is an element of universal quantifier. In this case, query condition it that the cinemas must show ALL movies listed in the top box office. This example also shows how mobile device can be used to obtain information from various independent sources or servers; in this case, they are local film association, and the cinema/entertainment complex. The user uses his/her mobile device to acquire information from these multiple non-collaborative sources and

Figure 1. A mobile database environment

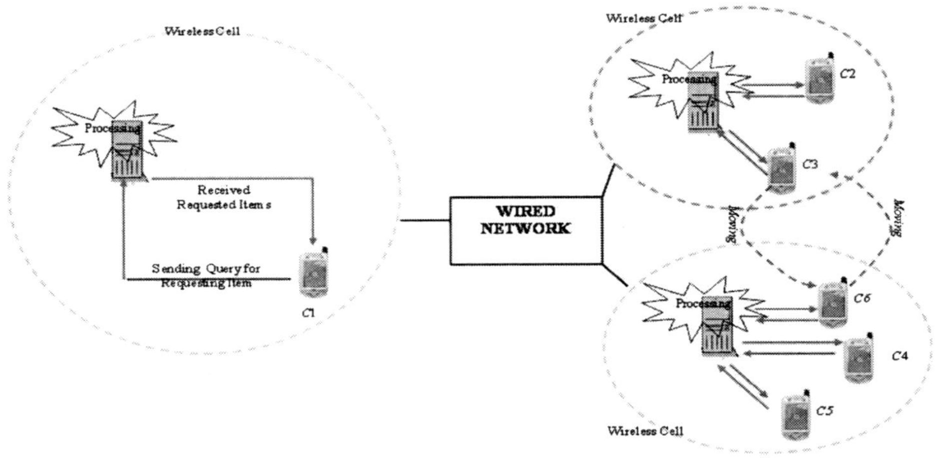

then assemble the required information in his/her mobile device.

Basic Mobile Query Processing incorporating Multiple Non-Collaborative Servers

Our previous work has identified two different ways of processing mobile queries involving multiple non-collaborative servers. They are: (*i*) Mobile Device Side Processing (MDSP) and (*ii*) Server Side Processing (SSP).

i. **MSDP Processing**
   In general, mobile device side processing deals with acquiring information the servers to the mobile device to be processed locally on the mobile device. Hence, the mobile device will process locally in determining the matching items. There are many variations of the MDSP method: (*a*) Download everything from both servers and then perform a local processing (MDSP1) (see Figure 2). (*b*) Download everything from one server, but only download the primary key from the other server. Then a local processing is carried out (MDSP2), and (*c*) Download only primary keys from both servers, and then perform the process locally (MDSP3).
   The performance results vary depending of various factors. Basically, there is a trade-off between full download in the beginning (MDSP1&2) and more communication round trips (MDSP3).
ii. **SSP Processing**
   Base on the possible large amount of data to be downloaded from both servers that may not be able to be supported by the limited memory and processing capability of mobile device, alternative techniques that can help to minimize the amount of transfer data taking into account the mobile device limitations need to be used. This alternative moves the processing of the downloaded lists in determine the qualified data from locally processing on the mobile device side to processing on the server side. Hence, with this technique it only requires one list from one of the server to be downloaded and then the mobile device act as a mediator to send the primary key to the other server to be process to get the qualified match.

There are two variants of server side processing: (*a*) Download everything from one server

Figure 2. Mobile device side processing (MDSP)

```
Algorithm: Mobile Device Side Processing
(MDSP)
// Step 1:
1. Send Query-1 to Server-1:
   Select * Into R
   From Table_R
   Order By PK;
// Step 2:
2. Send Query-2 to Server-2
   Select * Into S
   From Table_S
   Order By PK
// Step 3: On-mobile merging
3. i=1; j=1;
4. Let Query result Qr = {}
5. Do
6.   If i is EOF(R) or j is EOF(S) Then
7.      Exit Loop
8.   End If
9.   If Ri.PK = Sj.PK Then
10.     Qr = Ri = Sj;
11.     i++; j++;
12.  Else
13.     If Ri.PK > Sj.PK Then
14.        j++
15.     Else
16.        i++
17.     End If
18.  End If
19.End Do
```

and then send a query to the other server using the data downloaded from the previous server and performing a server processing (SSP1) (see Figure 3), and (*b*) Download the primary key from one server and then send a query to the other server using the primary key from the previous server and do a server processing (SSP2). If an index is used, SSP2 method may accept the index of primary keys which will then be used in the server processing (Waluyo, Srinivasan, & Taniar, 2008; Taniar & Rahayu, 2002, 2004).

Overview of Relational Division Queries

Division query is often known as the *universal quantification* and is not supported directly in most traditional database systems (Elmasri & Navathe, 1994). Despite this concept being a powerful means of querying sets and databases that adds significant power to the query processing and inference capabilities, the universal quantification algorithms have been omitted and have not been explored for large databases. Basically, it is a query request to find an attribute in one relation that satisfies all of a given list of criteria from another relation. Figure 4 illustrates a division query between two tables: table *R* and table *S* based on attribute *attr*2 of table *R* and attribute *attr*1 of table *S* whereby there exists a record in the relation matching tables *R* and *S*.

Apart from the relational division query mentioned above, there is a novel division query, which we call, *multiple group division*.

The difference between relational division and multiple group division is that in the multiple group division, it processes a division query where both sources of data are groups (Taniar & Rahayu, 2002a,b). Comparing it with the relational division, notice that in Figure 4 the divisor contains

Figure 3. Server side processing (SSP)

```
Algorithm: Server Side Processing (SSP)
1. Let query result Qr = {}
2. Send Query-1 to Server-1:
   Select * Into R
   From Table_R
   Order By PK;
3. Send Query-2 to Server-2:
   Select * Into S
   From Table_S
   Where Table_S.PK IN {R.PK.item1,R.PK.item2...R.PK.itemn};
4. Qr = S
```

Figure 4. Relational division operation

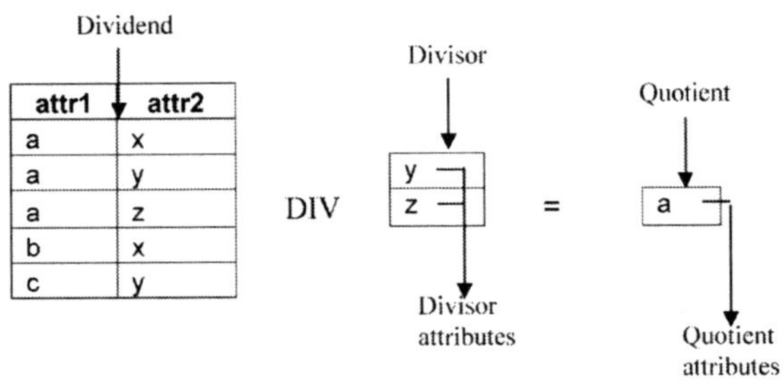

only individual attributes that are of single value without any grouping. However, in the multiple group division, the divisor would be groups. Figure 5 illustrates a multiple group division query between two tables: table *R* and table *S* based on attribute *attr*2 of table *R* and attribute *attr*1 of table *S* the divisor table has multiple groups of divisors as indicated by *attr2* in the divisor table.

PROPOSED METHODS 1

Relational Division Techniques

In this section, we proposed techniques for relational division processing on mobile device based on two basic techniques mentioned earlier, called *MDSP* (*Mobile Device Side Processing*) and *SSP* (*Server Side Processing*) (Lim, Taniar, & Srinivasan, 2005).

Generally, *Mobile Device Side Processing* (MDSP) with division operation deals with acquiring information from one server in terms of blocks and sending it to the mobile device to be processed locally in executing the division operation. We extend MDSP to relational division by having a full list downloaded from one server to the mobile device and another list in blocks from another server and, and with the two lists currently in the mobile device, the division operation will be executed to determine the qualified match.

The second method is based on *Server Side Processing* (SSP). This technique utilizes the di-

Figure 5. Multiple group division operation

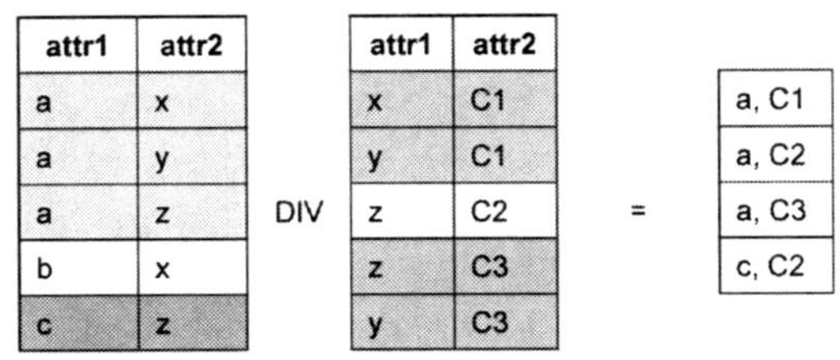

vision operation to transfer the processing to the servers instead of having it done locally on the mobile device. Therefore, for instance, we would now download in data blocks from Server 1 and send it to Server 2 for carrying out the division operation. And after the qualified match is determined by Server 2, it would return the results to the mobile device.

Both MDSP and SSP based Relational Division Techniques are based on count-based or aggregate-based. The details of each method are elaborated in the following sections.

Count-Based Relational Division Technique with MDSP

Now, we would like to demonstrate how the relational division works in a mobile environment. Figure 6 shows two tables that have been obtained from two different servers that contain different tables holding different types of records. For instance, Server 1 represents the table recording the movies in a year that are showing in respective cinemas while Server 2 represent the table that lists all movies that received awards. We will be using the following tables as a running example. Note that the data in Figure 6 is not necessarily sorted on the basis of any attribute. The sample data is used only to assist readers to understand the example.

In this example, our aim is to find cinemas, which show all the movies in the awards list. In other words, we need cinemas that have all movies in the awards list. We know that this means that each cinema must have at least the same count

Figure 6. Tables from two different servers

Table 1 in Server 1	
Cinema_ ID	**Movie_ ID**
A	1
A	2
A	3
A	4
A	5
A	6
A	8
B	2
B	1
C	3
C	10
C	8
C	9
C	1

Table 2 in Server 2
Movie_ID
1
2
3
6
8

as, or greater count than, the movies in the awards list. Therefore, in this example, our first step is to obtain the count of the awards list, which means we will download the count of the divisor, which is from Server 2 into the mobile device. Then, we will send a query to obtain the count of the dividend for each block of cinemas using the Group-By cinemas function. Since there are three different cinemas, there are three blocks of counts that are to be downloaded from Server 1 into the mobile device.

Now that we have the two lists of counts on the mobile device as in Figure 7, we need to locally compare them to check which cinema contains Count (Movie_ID) the same as, or greater than, the Count (Movie_ID) obtained from the divisor which is the awards list.

Figure 7 shows that only Cinemas '*A*' and '*C*' might contain all the movies that exist in the movie awards list. Thus, we will now discard cinema '*B*' because it can never be a potential query result.

Next, we need the divisor. So we will now send a query to the Server, which is the awards list to obtain the details of the awards which in this case is the actual data of the Movie_ID itself as in Figure 8b. We do the same thing for the dividend; we will now send a query to the Server 1 to obtain the full details record for the potential qualifying match which in this case only involves Cinemas '*A*' and '*C*' because we know Cinema '*B*' can never satisfy the division condition based on the previous count matching comparison we done as in Figure 8a. This shows that we have made it possible to eliminate unnecessary downloads to the mobile device.

The comparison between the dividend and the divisor data is to make a final comparison based on the Movie_ID to ascertain which cinema actually has all Movie_ID that exist in the Movie_ID in the awards list. In Figure 8a we can see that Cinema_ID '*A*' contains 7 movies 5 of which exist in the awards list as in Figure 8b. This shows that Cinema_ID '*A*' has satisfied the division condition and therefore will be our qualified match. On the other hand, Cinema_ID '*C*' has 5 movies but only 3 of these exist in the awards list. This makes it a non-qualifying match because we are interested only in a cinema that shows all the movies in the awards list. Therefore, the results for the query asking which cinema shows all

Figure 7. Counts from both dividend and divisor

Cinema_ID	**Count(Movie_ID)**
A	7
B	2
C	5

Count(Movie_ID)
5

Figure 8. a) Dividend b) divisor c) quotient

a	
Cinema_ID	Movie_ID
A	1
A	2
A	3
A	4
A	5
A	6
A	8
C	3
C	10
C	8
C	9
C	1

b
Movie_ID
1
2
3
6
8

c
Cinema_ID
A

movies that received awards are as shown in Figure 8c.

The count-based relational division algorithm is shown in Figure 9. Note that the algorithm is divided into four phases.

- In Phase 1, the count request is sent to the respective servers (Dividend and Divisor Servers).
- In Phase 2, counts from the respective servers are compared. This is needed in order to eliminate unqualified quotient groups.

Figure 9. Count-based relational division based on MDSP algorithm

```
Algorithm: Count-based Relational Division MDSP version
Input: Dividend from server R and Divisor from server S
Output: Quotient at the mobile device

// Phase 1: Count Request:
1. Send a count request to Server S (Divisor)
   Select Count(*)
   From Table_S;
2. Receive the divisor count and store it to CountS
3. Send a group count request to Server R (Dividend)
   Select QuotientAttr, Count(*)
   From Table_R
   Group By QuotientAttr;
4. Receive the group count and store it to CountR temp table

// Phase 2: Count comparison
5. For each group in CountR table
6.   If group count in CountR table NOT >= CountS Then
7.     Delete record from CountR table
8.   End If
9. End For

// Phase 3: Qualifying group download
10.Send a request to Server S (Divisor) to download the data
11.Send CountR temp table to Server R and request to send the
   actual data according to CountR
// Phase 4: Record comparison
12.For each QuotientAttr
13.  Read a record
14.  Perform merging between QuotientAttr record and Divisor
15.  If all divisor records have match Then
16.    Store the QuotientAttr into the Quotient Table
17.  End If
18.End For
```

- In Phase 3, the actual records are requested from the respective servers. Since we are going to use a merging approach, the divisor data is requested to be sorted (Taniar & Rahayu, 2002c), and the dividend data is sorted based on the quotient attribute (primary sort) and then for each quotient attribute, the divisor attribute is sorted (secondary sort).
- In Phase 4, the local merging is carried out. Quotient table will be produced to store the division results.

Count-Based Relational Division Technique with SSP

The only difference between the SSP-Division and MDSP-Division is where the processing takes place. Therefore, instead of having the division operation being done on the mobile device locally, we will have it done on the server. Our aim is still the same which is to get the results of a query for the cinema which shows all the movies that received awards based on the award list.

In the SSP, we could maximize our memory efficiency slightly because we would distribute some processing to the server instead of having everything done locally on the mobile device. Firstly, we would obtain the count of the divisor and then send it to Server 1 which is the dividend for processing to compare the count between the dividend and the divisor so that only the count of the dividend that is same as, or greater than, the count of the divisor will be kept as a potential qualify match. This is where the difference lies, since in the MDSP base, we would have to download both count lists but in the SSP, we download only one count list.

This process can be illustrated as in Figure 10. In other words, count comparison is done in the dividend server. This ends the first phase. The next phase is the actual data comparison.

By processing and obtaining the potential qualifying match based on the count comparison, the result will now be returned to the mobile device together with its full details which are the Cinema_ID and Movie_ID. With the result on the mobile device, we now send it to Server 2 which is the divisor so that a final processing at the server to obtain the qualify match can be done. This is where the server will process the list obtained from mobile device and compare it with its award list to check all Movie_ID from the mobile device that exist in the Movie_ID in the server. Once the server has finished processing, the final output will be returned to the mobile device as the qualify match. This process is illustrated in Figure 11.

There is another option for processing the above division query whereby, instead of sending the dividend to the server for processing, the divisor is sent for processing. This avoids the need for block processing, which is required when sending the dividend. This is because, by sending

Figure 10. Count-comparison done in the dividend server

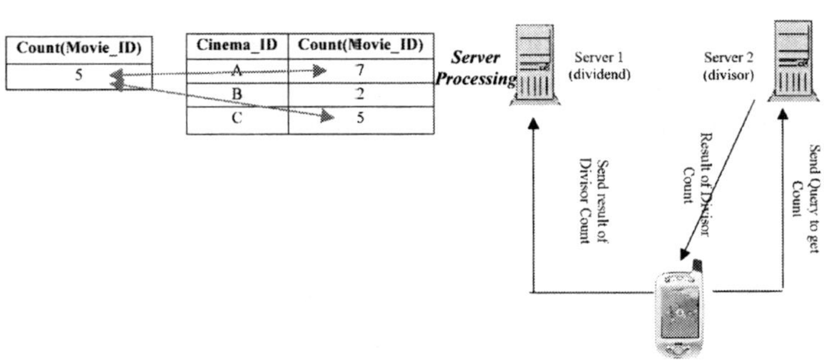

Figure 11. Data-comparison done in the divisor server

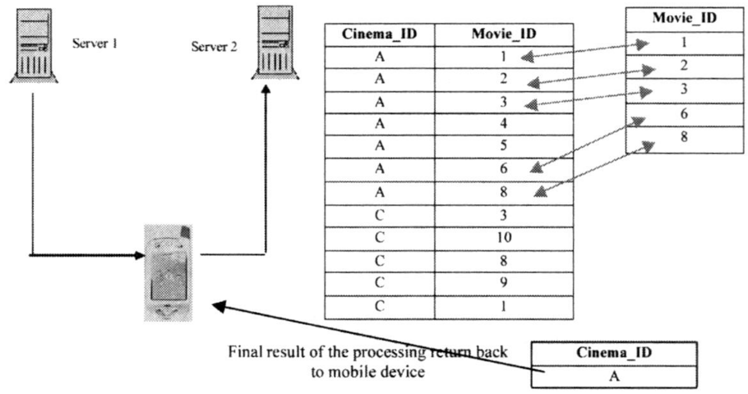

the divisor, the whole list will be sent at once rather than needing to be broken down into blocks as in the case of the dividend.

The choice of selecting whether to send the dividend or the divisor for processing in the server side will depend greatly on circumstances. For instance, if the dividend does not contain a long list that needs to be broken down into blocks, then it may be a wise choice to send the dividend. However, if the list of the divisor appears to be carrying a large amount of data, then downloading the divisor to be sent to the server for processing may not be the better choice.

In summary, the Relational Division uses the following steps:

- **Step 1:** Obtain the count from the divisor server.
- **Step 2:** Send the count of the divisor to the dividend in the server to compare the count of the dividend server and discard those that contain a count less than the count of the divisor. The server will count for each of the groups and compare the count received from the mobile device. It will process only those counts that are the same or greater and then it will send the result back to the mobile device.
- **Step 3a:** Option 1: Using MDSP: Upon obtaining the counts that are either the same or greater between the divisor and dividend and sending them to the mobile device, a query will then be sent to the divisor server to obtain a full list of details so that a local comparison matching can be done to see whether or not they qualify. This is done in blocks within the dividend.
- **Step 3b:** Option 2: Using SSP: Obtain the qualify match of the result in Step 2 together with its full details and send it to the divisor server for final comparison of the qualify match. This means that the results of the count that are the same or greater obtained from Step 2 will be downloaded to the mobile device and included with the other attributes records. This is done in blocks within the dividend.
- **Step 4:** Repeat the same process until all blocks that qualify based on the count within the dividend have been compared and processed.

As shown above, count-based relational division can be either integrated with the mobile device side processing (MDSP) or server side processing (SSP). There is only a slight difference between using the MDSP or SSP as outlined in Steps 3a and 3b, where MDSP does the main processing in the mobile, whereas SSP does it in the divisor server.

Figure 12. Count-based relational division based on SSP algorithm

```
Algorithm: Count-based Relational Division SSP version
Input: Dividend from server R and Divisor from server S
Output: Quotient at the mobile device

// Phase 1: Count Request
1. Send a count request to Server S (Divisor)
   Select Count(*)
   From Table_S;
2. Receive the divisor count and store it to CountS
3. Send the divisor count to Server Dividend to process

// Phase 2: Count comparison in the Server Dividend
4. The Server Dividend groups the quotient attribute
5. Count each group
6. Compare each group count with CountS received from mobile

// Phase 3: Qualifying group download from Server Dividend
7. Qualified dividend records are downloaded to mobile device

// Phase 4: Record comparison in Server Divisor
8. Qualified divident records are sent to Server Divisor for
   comparison of mathces.
9. Quotient tables as the result of the comparison are sent
   to mobile
```

PROPOSED METHODS 2

Multiple Group Division Techniques

In this section, we propose two different methods for tackling group division. The proposed techniques are called (*i*) Sort Merge Multiple Group Division, and (*ii*) Aggregate Multiple Group Division. These techniques aim to answer the query of "Listing the cinemas that show ALL movies in at least one category". This query will use the following tables as in Figure 13 to illustrate the process of obtaining the results.

Note the data for the multiple group division query has an additional group attribute in the Divisor. In the above example, the additional attribute is the Category attribute which acts as a group attribute for the multiple division. This attribute will also appear in the Quotient table. Therefore, the Quotient table will have two attributes: the original quotient attribute from the dividend table and this new group attribute from the divisor table. The former is now called the first quotient attribute, whereas the latter is now called the second quotient attribute, producing a Quotient table with two attributes.

Sort-Merge Multiple Group Division

As in the example shown in Figure 14, using the Sort-Merge Multiple Group Division, we would first request to download all movies from all cat-

Figure 13. Sample data for multiple group division

Cinema_ID	Movie_ID
A	1
A	3
A	4
A	5
B	1
B	2
B	6
C	2
C	3
C	6

Movie_ID	Category
1	Drama
2	Drama
2	Action
3	Action
6	Action
3	Horror
4	Horror

egories in the divisor which contain the Movie_ID and Category records. When issuing the query to download the records, ensure that movies from each category are sorted in ascending order.

The next step is to download the dividend in a block manner. The block is defined for each cinema (using the above example). In other words, each separate cinema, regardless of how many movies it holds, is considered as a block. The same as for Step 1, we need to ensure that the Movie_ID for each cinema is sorted as in Figure 15.

Step 3 is the local processing itself. With two lists of records which are the divisor and a block of the dividend that are downloaded one at a time in the mobile device, the merging process will be carried out. The process will merge one block of the dividend with all blocks of the divisor.

In the process of merging the first block of the dividend with the whole divisor, we will first compare the Cinema '*A*' with Category '*Drama*' to check each element. This means to merge *A*(1,3,4,5) and *Drama*(1,2). After merging in order to ascertain whether or not they match, move on to the next group of the divisor, which is comparing Cinema '*A*' with Category '*Action*'. This process is repeated until all categories in the divisor have been matched against the block of the dividend that has been downloaded. This can be seen in Figure 16.

After a comparison of the one block of the dividend with all divisor groups has been completed, Step 4 is to download the subsequent block of the dividend, which in this case, is Cinema '*B*' and repeat the same process of comparing them with the whole divisor.

This process will finish once all blocks from the dividend have been downloaded and compared with the divisor.

The following is the summary of the steps for Sort Merge Multiple Group Division:

Figure 14. Download of the Divisor Table (divisor attribute values within each quotient attribute are sorted)

Movie_ID C	ategory
1	Drama
2	Drama
2	Action
3	Action
6	Action
3	Horror
4	Horror

Figure 15. Download of the Dividend Table (Dividend attribue values are sorted for each quotient attribute value)

Cinema_ID	Movie_ID	
A	1	Block 1
A	3	
A	4	
A	5	
B	1	Block 2
B	2	
B	6	
C	2	Block 3
C	3	
C	6	

Figure 16. Merging first block of Dividend with the Divisor

Cinema_ID	Movie_ID
A	1
A	3
A	4
A	5

Movie_ID	Category
1	Drama
2	Drama
2	Action
3	Action
6	Action
3	Horror
4	Horror

- **Step 1:** Download the divisor that is large enough to fit into the main memory of the mobile device. If the main memory fits the whole divisor, then download the whole thing. Otherwise, if the main memory fits only 2 blocks of divisor (for example, the 'Drama' and 'Action' categories) then download only these 2 blocks.
- **Step 2:** Download a block of the dividend (e.g. Cinema '*A*').
- **Step 3:** Process the comparison of the single block of the dividend with whatever divisor is in the main memory.

Now we address the issue of limited memory in the mobile device. Often it may not have enough memory space to hold the whole block of the divisor that has been downloaded from the server to the local memory. Therefore, to solve this problem, there are two options in Step 4 from which to select in order to save memory.

- **Step 4a:** Download another block of the divisor if in Step 1 it has only partly been downloaded (For example, if not all categories from the divisor can be fitted into main memory).
- **Step 4b:** The other option is to process by downloading all dividend blocks first instead of downloading al the divisor first.

As a summary, the two options of processing the multiple group division query using the proposed sort merge have a trade-off between them. In Option 1, we download all divisors for each block of dividend; whereas in Option 2, we download all dividend blocks for each block in divisor. Option 1 would appear to be more efficient if the divisor is smaller than the dividend; whereas Option 2 would be more desirable if the dividend is smaller than than divisor. Thus, when selecting which option to use for processing the group division queries, ascertain whether the divisor or the dividend is smaller and then select the right option accordingly for the processing.

Aggregate Multiple Group Division Technique

In this proposed technique, we make use of the count for each group using the aggregate count function. Therefore, first we obtain the count for each block in the dividend and the divisor. This is done by sending a query to the respective servers to obtain the two lists of data in sorted order based on the value of the count.

Figure 17. Sort-merge multiple group division algorithm

```
Algorithm: Sort-Merge Multiple Group Division
Input: Dividend from server R and Divisor from server S
Output: Quotient at the mobile device

// Phase 1: Download Divisor
1. Send a request to Server S (Divisor) to download divisor
   Select *
   From Table_S
   Group By Table_S.quotient
   Order By Table_S.divisor;

// Phase 2: Download one dividend block at a time and process
2. For each block in Dividend
3.   Send a request to Server R (Divident) to download one
     divident block, sorted by divisor attribute
4.   Compare with the divisor
5.   If matched Then
6.     Store the match in the Quotient
7.   End If
8. End For
```

Figure 18. Complete the dividend for each divisor block

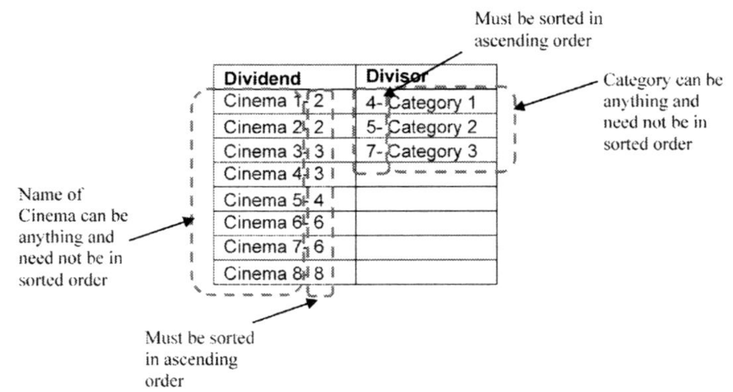

For instance, Figure 18 illustrates the list of data from the dividend and divisor that are to be downloaded to the mobile device.

Once the mobile device has the information of the count, the mobile can then execute the pruning phase. Based on the count information, we can actually prune the comparison. In this pruning phase, the comparison between the count of the dividend and divisor is performed. The condition is that the count of the group in the dividend has to be at least the same as the count of the group in the divisor. Thus, only the count of the dividend that is same or greater than the divisor qualifies. Figure 19 indicates how the comparison is being done.

In the pruning phase as illustrated in Figure 19, category 1 in the divisor shows a count of 4 that needs to be compared with the count of the cinema which is the dividend starting with count 4 moving downwards (e.g. 4, 6, 6, and 8) as indicated by the arrow, whereas category 2 with count of 5 needs to be compared only with cinemas with counts 6,6 and 8. As for Category 3 with count of 7, this needs to be compared only with cinemas with counts of 8. Therefore with this scheme, cinemas with counts 2, 2, 3, 3 are eliminated right away.

Once the pairs have been formed, the mobile device can request the details from the dividend and divisor to download the details in sorted form to be merged. This process is then similar to what we did in sort-merge multiple group division technique. Therefore, as can be seen, this technique is an enhanced version of the previous technique. It appears to be more efficient since, even though we need to get all divisors downloaded, not all the dividends will be needed. Furthermore, the divisor with a large count needs to be compared only with the dividend with a large count. This is not a pure nested loop approach but rather, a pruning approach. This is because a high divisor count such as '7' will be compared only with the count of a dividend that is the same or higher which is '8' in the example given above.

To differentiate the pruning approach and the pure nested loop approach, Figure 20(a) and (b) show the difference between the pure nested loop approach and pruning approach whereby in a pure nested loop, the divisor '7' has to be compared

Figure 19. Pruning phase

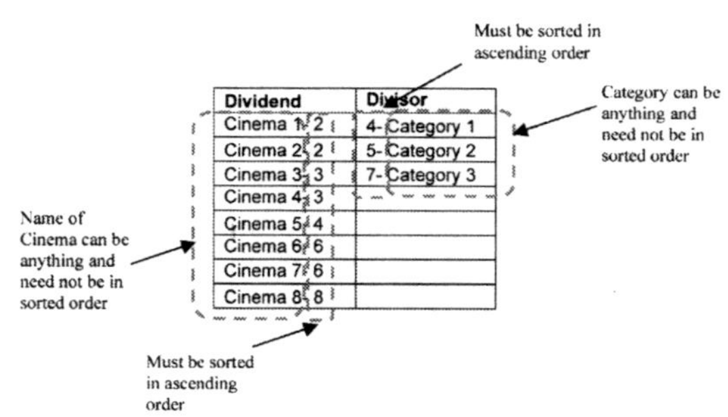

with every single record in the dividend which is 4, 6, 6, 8, 10 and 12; whereas in the pruning, divisor '7' is compared only with counts of the dividend of 8, 10 and 12. The others such as 4, 6, 6 are automatically eliminated.

Figure 21 shows the algorithm for the aggregate-based multiple group division. In the first phase, the counts from both dividend and divisors are obtained. In the second phase, the pruning is in action. The result of this phase is the candidate pairs. In the third phase, each of the candidate pairs is sent to the respective servers to obtain the full data. The comparison will then be carried out in the mobile device.

PERFORMANCE EVALUATION

We studied the two areas of relational division and multiple group division. The relational division queries are quite common in relational databases, but we have additionally proposed multiple group

Figure 20. Pure Nested loop vs. pruning phase

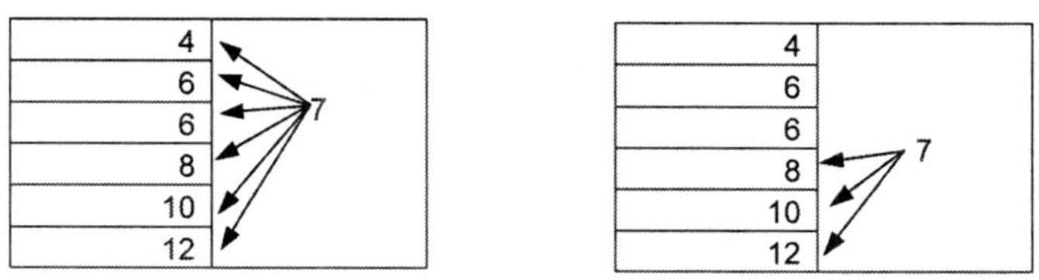

(a) Pure nested-loop **(b) Pruning phase**

Figure 21. Aggregate-based multiple group division algorithm

```
Algorithm: Aggregate-based Multiple Group Division
Input: Dividend from server R and Divisor from server S
Output: Quotient at the mobile device
// Phase 1: Count Request:
1. Send a count request to Server S (Divisor)
   Select QuotientAttr2, Count(*)
   From Table_S
   Group By QuotientAttr2
   Order By Count(*);
2. Receive the divisor count and store it to CountS
3. Send a group count request to Server R (Dividend)
   Select QuotientAttr1, Count(*)
   From Table_R
   Group By QuotientAttr1
   Order By Count(*);
4. Receive the group count and store it to CountR temp table
// Phase 2: Pruning phase
5. For each record in CountS
6.    For each record in CountR
7.       If CountR.count >= CountS.count Then
8.          Put both QuotientAttr from CountR and CountS into
            Candidate Pairs
9.       End If
10.   End For
11.End For
//  Phase 3: Comparison
12.For each pair in Candidate Pairs
13.  Send QuotientAttr1 to Server R (Dividend) and
     Download the records
14.  Send QuotientAttr2 to Server S (Divisor) and
15.  Download the records
16.  Compare the two quotients
17.  If matched then
18.    Store both quotients into the Quotient Table.
19.  End If
20.End For
```

division for mobile query processing. We did a simulation study using a network simulator, *Planimate*, to simulate the mobile environment to process division queries. Some of our experimental results are presented in the following sections.

PERFORMANCE RESULTS FOR MOBILE RELATIONAL DIVISION QUERIES

In the experimentations, we have focused particularly on the impact of dividend size (number of records in the dividend table) and the impact of selectivity ratio, on both transfer and processing costs.

In mobile relational division query processing, we have proposed three techniques, namely (*i*) MDSP relational division, (*ii*) SSP dividend, and (*iii*) SSP divisor. The difference between the two SSP techniques is where the division is carried out; that is, in the dividend server or in the divisor server. In the performance graphs, we compare the three techniques to examine the efficiency of each.

Figures 22(a)-(b) show the performance in terms of the data transfer costs. Figure 22(a) shows that SSP dividend gives the most efficient performance. In SSP dividend, the divisor table which is relatively smaller than the dividend, is sent to the dividend table to be processed. Because the divisor table is small, the transfer cost is then low. At the other extreme, SSP divisor is the most expensive. When using the SSP model, we need to choose the smaller table to be transferred to the other server.

The MDSP technique in Figure 22(a) shows a medium-level performance. In MDSP, the transfer cost will also be dominated by the dividend table, similar to the SSP divisor. The only difference is the final query results; when using the SSP version, the results have to be transferred from the server to the mobile, and this incurs additional transfer cost.

Figure 22. a) Impact of dividend size on transfer costs b) Impact of selectivity ratio on transfer costs

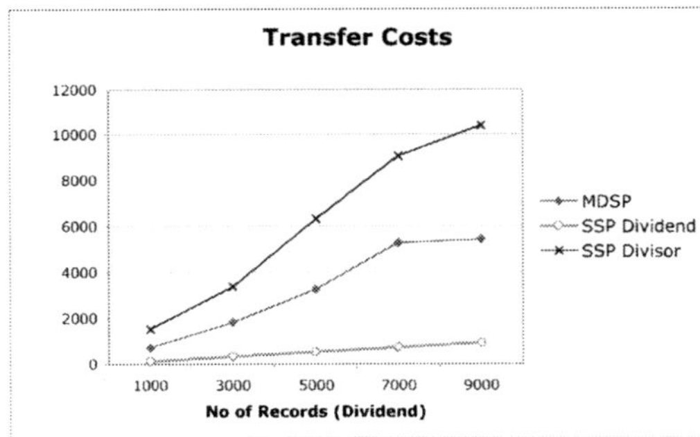

(a)

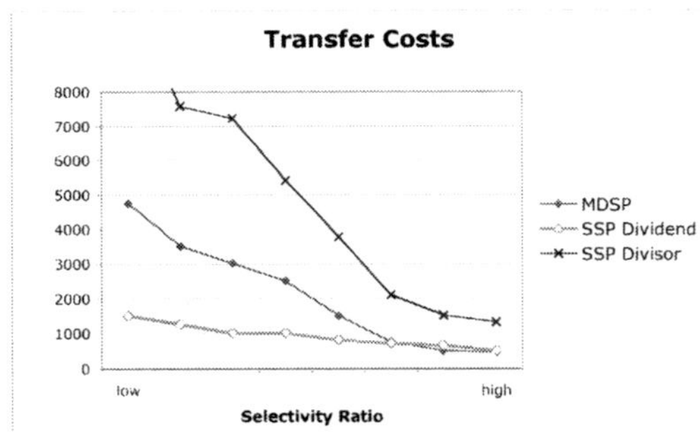

(b)

Figure 22(b) shows the impact of selectivity ratio on the transfer cost. Again, SSP dividend gives the best performance, even when the selectivity ratio is quite low. A low selectivity means that the selectivity pruning is low, implying that the results are large. The MDSP and SSP dividends are both quite good when the pruning is high. When the pruning is high, using MDSP, some of the dividend records are not downloaded because they are pruned away. Whereas with the SSP dividend, the low transfer cost is a consequence of the small number of query results. The SSP divisor is very expensive, again due to the data transfer from the dividend to the divisor server.

Figures 23(a)-(b) studied the processing costs based on the two variables: dividend size and selectivity pruning. Figure 23(a) shows that the SSP divisor's processing cost is the lowest. This is in contrast to the transfer cost, where the SSP divisor is the most expensive. The low processing cost of SSP divisor is because the some of the records from the dividend have already pruned by the group counting comparison done on the mobile side. Due to this, the division processing in the divisor is discounted. For the SSP dividend, there is no group elimination, because all dividend records need to be accessed from the local disk in the dividend server anyway. So the processing cost will not be as low as that for the SSP divisor. For MDSP, the processing cost is the most expensive, simply because of a limitation in resources in the mobile device. In the experimentations, the divisor table was very small, and therefore we did not encounter any main memory overflow. Nevertheless, because of the difference in processing speed between server and mobile, the poor performance of MDSP reflects this fact.

Figure 23(b) shows the processing costs due to the selectivity pruning. Again, the SSP divisor shows its superiority; whereas, the SSP dividend does not improve so much even when the pruning is high. This is because, when the division operation is done in the dividend table, the pruning of the dividend does not affect the entire processing. For MDSP, it is very similar due to the more expensive mobile processing.

Figure 23. a) Impact of dividend size on processing costs b) Impact of selectivity ratio on processing costs

(a)

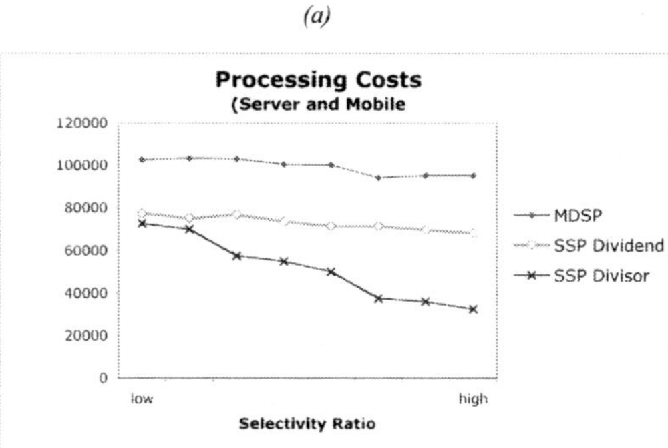

(b)

Overall, there is a contradictory result especially for the SSP divisor, which is high in transfer cost but quite efficient in the processing cost. However, because the transfer unit cost is much more expensive than the processing unit cost, SSP divisor overall performs poorly. We need to rely on the SSP dividend in processing mobile relational division query. For the MDSP approach, the performance is not too bad. MDSP is a good option, if the mobile environment does not permit any query to be processed by the server (if the server does a data broadcast only) (Waluyo, Srinivasan, & Taniar, 2005a,b; Waluyo, Srinivasan, & Taniar, 2007). MDSP performance is acceptable especially if the divisor table is quite small, since this is not likely to incur any main memory overflow in the mobile processing.

Performance Results for Mobile Multiple Group Division Queries

We have proposed two processing techniques: one is based on sort-merge, and the other on count aggregate. In the experimentations, we have examined the performance of these two techniques. Additionally, we have studied the effect of pruning in the aggregate-based of multiple group division, since pruning is one of the optimization techniques we developed for multiple group division query processing.

Figures 24(a)-(b) demonstrate the effect of pruning in the aggregate-based multiple group division query. We compare this with the other two methods (sort-merge and the non-pruning aggregate-based).

Figure 24(a) shows the impact of the pruning level on the transfer costs. Since the sort-merge

Figure 24. a) Impact of pruning level on transfer costs b) Impact of pruning level on processing costs

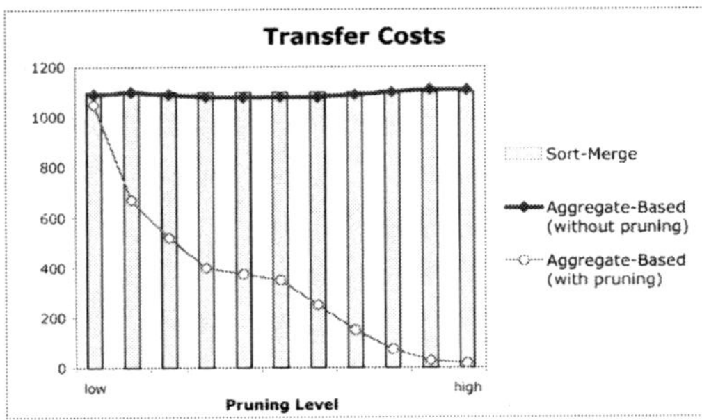

(a)

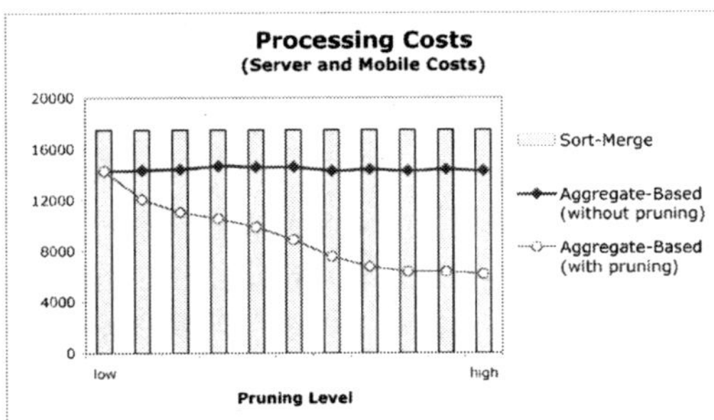

(b)

and the non-pruning aggregate-based are not affected by the pruning level, there is only one value in the performance graph. Figure 24(a) shows that as the pruning level increases (becomes high), the aggregate-based with pruning performance drops significantly. In contrast, the sort-merge and the aggregate without pruning are not affected by the pruning level, as they do not possess the pruning features. The pruning in the aggregate-based technique works by first requesting from the servers the group summary, which is then analyzed and compared in the mobile device. Only those groups, which will be likely to contribute to the final query results, will then be requested from the servers. As a result, if most of the groups have already been pruned away at this stage, fewer records need to be downloaded, resulting in a low data transfer cost, as happened with the aggregate-based with pruning as shown in Figure 24(a). On the other hand, the other two methods do not include any group comparisons to prune away non-qualified groups.

Figure 24(b) shows the impact of pruning on the processing costs covering server and mobile processing costs. Note that the trend is quite similar to transfer costs, only with a number of differences. One is that the drop in the processing cost of the aggregate-based with pruning technique is not as sharp as that of the transfer cost. This is due to the majority of the server cost having been absorbed by the retrieval and the grouping cost in the beginning. The other difference is that the aggregate-based without pruning performs slightly better than the sort-merge technique. This is because the sorting unit cost is slightly higher than the aggregate unit cost, resulting in only a slight improvement. Nevertheless, the aggregate-based technique should employ the pruning method in order to take full advantage of the aggregate-based.

Figures 25(a)-(b) show the impact of the size of dividend table on the transfer and processing costs. We can certainly expect that the cost will increase as the number of records in the dividend table grows. However, we would like to compare the three methods.

Figure 25(a) shows that the transfer costs of the sort-merge and aggregate without pruning are quite comparable. We run a simulation to generate the synthetic data randomly, and the results for these two techniques are quite comparable, although Figure 25(a) shows that the sort-merge incurs a higher transfer cost. However, the most important observation is that the aggregate-based with pruning performs the best. It is roughly around 60% of the other two techniques. The saving demonstrates the benefits of pruning at the initial grouping stage.

Figure 25(b) shows the processing cost. It is clear that the sort-merge processing cost is the worst of the three, and the aggregate-based with pruning saves between 20%-30% of the processing costs compared with the other two techniques.

Both graphs demonstrate that the performance of the aggregate-based with pruning is quite prom-

Figure 25. a) Impact of size of dividend on transfer costs b) Impact of size of dividend on processing costs

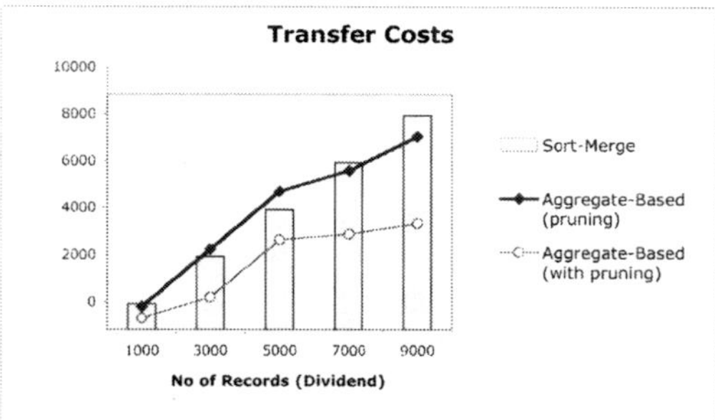

(a)

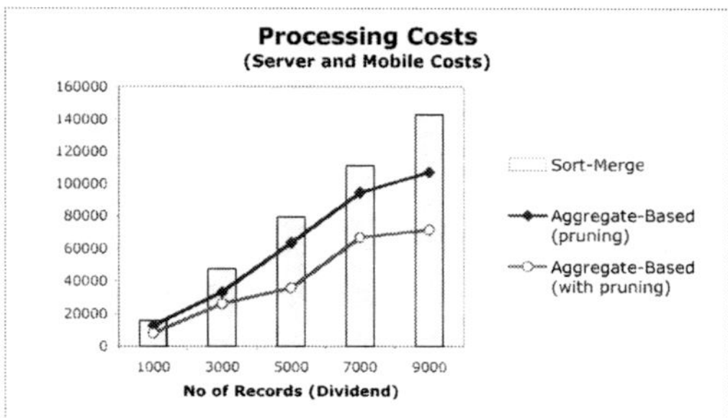

(b)

ising, although the effectiveness of the technique depends on how much pruning can be done. Nevertheless, even when the amount of pruning is small, some performance benefits are still produced, thereby indicating that it is beneficial to add a pruning stage to the aggregate-based method. On the contrary, the sort-merge performance is not attractive.

CONCLUSION AND FUTURE WORK

Traditionally, research in universal quantification has been lacking, especially in cases where large databases are involved. Thus, the existing techniques need to be studied so that they can be enhanced or modified for use in the mobile environment. Basically, it is a query request to find an attribute in one relation that satisfies all of a given list of criteria from another relation. We examine ways to overcome the shortcomings of the traditional division techniques and apply them to the mobile environment incorporating multiple non-collaborative servers. We also extended the traditional division query processing by considering multiple group division query processing, whereby the divisor is not a single group from a relation, but multiple groups from a single relation. This type of query in particular is common in multiple independent servers query processing in a mobile environment.

Our contributions are also twofold in this regard. Firstly, we focused on the relational division based on sort-merge and aggregate functions. Our second contribution is the main one, which is multiple group division. None of the existing work has considered multiple group division. In addition to this major contribution, we enhance the performance by introducing some pruning techniques so that unnecessary records will be pruned away, and with consequent gains in performance.

There are two important lessons drawn from the experimentation results: (*i*) For the *Relational Division* operation, since the processing itself is a heavy task, the SSP method is the best option. It is also important to choose the correct server to download, whether it be the divisor server or the dividend server. Because the transfer cost is a dominant factor, SSP dividend method delivers a better performance due to the lowest communication costs. (*ii*) For the *Multiple Group Division* operation, the *pruning* method used in the aggregate-based is very attractive. Even when the pruning level is low, the performance is still very good, in terms of transfer costs and processing costs.

In the future, we plan to extend this work to cover top-k join and restricted join (Ozakar, Morvan & Hameurlain, 2005), context-sensitive mobile operations (Aleksy, Butter & Schader, 2008), as well as spatial mobile navigation (Xuan, et al, 2008; Zhao, et al, 2008), all of which are extremely important in information processing on mobile devices, especially when information are provided from various non-collaborative servers.

ACKNOWLEDGMENT

This research has been partially funded by the Australian Research Council (ARC) Discovery Project (Project No: DP0987687).

REFERENCES

Aleksy, M., Butter, T., & Schader, M. (2008). Architecture for the development of context-sensitive mobile applications. *Mobile Information Systems*, *4*(2), 105–117.

Bohl, O., Manouchehri, S., & Winand, U. (2007). Mobile information systems for the private everyday life. *Mobile Information Systems*, *3*(3-4), 135–152.

Elmasri, R., & Navathe, S. B. (1994). *Fundamentals of Database Systems* (2nd ed.). Benjamin/Cummings.

Gulliver, S. R., Ghinea, G., Patel, M., & Serif, T. (2007). A context-aware Tour Guide: User implications. *Mobile Information Systems*, *3*(2), 71–88.

Jayaputera, J., & Taniar, D. (2005). Data retrieval for location-dependent queries in a multi-cell wireless environment. *Mobile Information Systems*, *1*(2), 91–108.

Lee, D. K., Zhu, M., & Hu, H. (2005). When location-based services meet databases. *Mobile Information Systems*, *1*(2), 81–90.

Lim, S. Y., Taniar, D., & Srinivasan, B. (2005). On-Mobile Query Processing Incorporating Multiple Non-Collaborative Servers. *Ingénierie des Systèmes d'Information*, *10*(5), 9–38. doi:10.3166/isi.10.5.9-38

Ozakar, B., Morvan, F., & Hameurlain, A. (2005). Mobile join operators for restricted sources. *Mobile Information Systems*, *1*(3), 167–184.

Taniar, D., & Rahayu, J. W. (2002a). Parallel sort-hash object-oriented collection join algorithms for shared-memory machines. *Parallel Algorithms Appl.*, *17*(2), 85–126.

Taniar, D., & Rahayu, J. W. (2002b). Parallel sort-merge object-oriented collection join algorithms. *Comput. Syst. Sci. Eng.*, *17*(3), 145–158.

Taniar, D., & Rahayu, J. W. (2002c). Parallel database sorting. *Inf. Sci.*, *146*(1-4), 171–219. doi:10.1016/S0020-0255(02)00196-2

Taniar, D., & Rahayu, J. W. (2002d). A Taxonomy of Indexing Schemes for Parallel Database Systems. *Distributed and Parallel Databases*, *12*(1), 73–106. doi:10.1023/A:1015682215394

Taniar, D., & Rahayu, J. W. (2004). Global parallel index for multi-processors database systems. *Inf. Sci.*, *165*(1-2), 103–127. doi:10.1016/j.ins.2003.09.019

Waluyo, A. B., Srinivasan, B., & Taniar, D. (2005a). Efficient Broadcast Indexing Scheme for Location-dependent Queries in Multi Channels Wireless Environment. *Journal of Interconnection Networks*, *6*(3), 303–322. doi:10.1142/S0219265905001447

Waluyo, A. B., Srinivasan, B., & Taniar, D. (2005b). Global Indexing Scheme for Location-Dependent Queries in Multi Channels Mobile Broadcast Environment. *Proceedings of the 19th International Conference on Advanced Information Networking and Applications (AINA 2005)* (pp. 1011-1016).

Waluyo, A. B., Srinivasan, B., & Taniar, D. (2005c). Research in mobile database query optimization and processing. *Mobile Information Systems*, *1*(4), 225–252.

Waluyo, A.B., Srinivasan, B., & Taniar, D. (2005d). Research on location-dependent queries in mobile databases. *Comput. Syst. Sci. Eng.*, *20*(2).

Waluyo, A. B., Srinivasan, B., & Taniar, D. (2008). Indexing schemes for multichannel data broadcasting in mobile databases. *International Journal of Wireless and Mobile Computing*, *3*(1-2), 79–90. doi:10.1504/IJWMC.2008.019720

Waluyo, A. B., Zhu, F., Taniar, D., Srinivasan, B., & Rahayu, J. W. (2007). Multiple Entity Types Wireless Broadcast Database System. *Proceedings of the 21st International Conference on Advanced Information Networking and Applications (AINA 2007)* (pp. 179-186).

Xuan, K., Zhao, G., Taniar, D., & Srinivasan, B. (2008). Continuous Range Search Query Processing in Mobile Navigation. In *Proceedings of the 14th International Conference on Parallel and Distributed Systems (ICPADS 2008)* (pp. 361-368).

Zhao, G., Xuan, K., Taniar, D., & Srinivasan, B. (2008). Incremental k-Nearest-Neighbor Search on Road Networks. *Journal of Interconnection Networks*, *9*(4), 455–470. doi:10.1142/S0219265908002382

This work was previously published in International Journal of Business Data Communications and Networking Volume 5, Issue 3, edited by V. Sridhar and D. Saha, pp. 67-90, copyright 2009 by IGI Publishing (an imprint of IGI Global).

Chapter 15
A Survey on Classical Teletraffic Models and Network Planning Issues for Cellular Telephony

Francisco Barcelo-Arroyo
Universitat Politècnica de Catalunya, Spain

Israel Martin-Escalona
Universitat Politècnica de Catalunya, Spain

ABSTRACT

Research on traffic characterization and analysis of cellular networks has been very active in the past decades. However, it is difficult for networks planners to incorporate new results to network engineering. On the one hand, some models are very complex and need advanced programming and skills to be properly computed. On the other hand, reliability on those models is poor because there is a general lack of published field studies to corroborate them. This paper proposes simple well-known teletraffic models for cellular networks compared to the conventional Erlang-B frequently applied when a first estimate is needed. To this purpose, the latest results on the characterization of the arrival process and service time in cellular systems are reviewed. According to the arrival and service characteristics, three models are proposed to obtain a first approach to the performance characteristics of the cellular system. The first two models are extremely simple, allowing direct computation of performance through closed formulas, while the third requires simple programming.

DOI: 10.4018/978-1-60960-589-6.ch015

Air pollution is an important environmental issue that has a direct effect on human health and ecological balance. Factories, power plants, vehicles, windblown dust and wildfires are some of the contributors to pollution. Reasonable simulation tools exist for evaluating large scale sensor networks; however, they fail to capture significant details of node operation or practical aspects of wireless communication. Real life testbeds capture the realism and bring out important aspects for further research. In this paper, we present an implementation of a wireless sensor network testbed for automatic and real-time monitoring of environmental pollution for the protection of public spaces. The paper describes the physical setup, the sensor node hardware and software architecture for "anytime, anywhere" monitoring and management of pollution data through a single, Web-based graphical user interface. The paper presents practical issues in the integration of sensors, actual power consumption rates and develops a practical hierarchical routing methodology.

INTRODUCTION

Motivation

In recent years, many papers have been published on models for personal communications in cellular networks. Mathematical analyses, simulations and field studies have been presented separately with little effort dedicated to the link between them and the engineering consequences of using a specific modeling tool. Most of these recent studies show that the classical Erlang-B theoretical hypotheses seldom occur in actual operative networks: arrivals might be not Poisson, channel-holding time not exponentially distributed, blocked handoff attempts not immediately lost if not attended, coverage overlapping areas, and so forth. Newer models try to overcome these issues through relaxing hypotheses and adding features to previous ones, but there is a general lack of comment on how the new hypotheses and features fit the real cellular world, what happens when these hypotheses are relaxed, and how to apply new models. The consequence is that the planner fails to take advantage of the huge amount of research work carried out in this field and often remains loyal to the conventional Erlang-B model.

Teltraffic modeling of cellular networks is a complex issue because there are many variables involved. Some of them are traffic related: the channel-holding time (CHT) is a fraction of the duration of the whole connection (i.e., unencumbered holding time), arrivals are not always Poisson (specifically handoff arrivals), and so forth. Others are mobility related: speed and direction of the Mobile Node (MN), different user patterns, and so forth. There are also radio constraints: propagation, multi-path, antenna radiation diagram, cell shape, and so forth. This is merely a very brief summary of related variables, since it is almost impossible to produce a comprehensive list of all of them.

In addition there are strong relations between variables. For instance, the cell shape obviously depends on all the variables related to the radio environment. The consequences of the traffic load on the cell shape are not so obvious, but the planner cannot overlook the fact that under heavy load, the MN is often connected to a base station (BS) that is not the nearest one. As explained below, MNs within overlapping areas may belong to different cells depending on the traffic circumstances.

Goals and Organization

This work deals with the first stage in network planning: applying simple teletraffic models to draw approximate figures of capacity-related performance figures. This first step is necessary before carrying out simulations, which consume a much greater amount of time and resources. The advantage of simple models is that they quickly produce a first approach to the system performance while allowing a high level of understanding about the impact of each single phenomenon. However, the estimates obtained through this analysis are always rough and never permit to elude the need for simulations that will provide more accurate results by allowing the designer to relax the necessarily simple hypotheses of the model.

The objective of this paper is to provide general considerations that help the planner to better understand:

- The reasons why a specific queueing model is applied. Simplicity and ease of use are major advantages for planners.
- The differences between the hypotheses in the model and the true network.
- The consequence of those differences on the achieved performance evaluation figures. An estimate of the level of accuracy provided by the model is also important.
- How to slightly change classical teletraffic models in order to apply them to a specific cellular network. Notice that most variables and hypotheses can significantly change between networks.

The paper is organized to highlight the main differences between cellular and non-cellular networks. We first present a review of the features of the arrival process, and then present the CHT. We follow with the QoS figures obtained from a teletraffic model and analyze them. We then highlight some aspects related to the use of the Erlang-B model and introduce other models that can be applied to cellular networks without loosing simplicity. We offer several hints to generalize these models.

THE ARRIVAL PROCESS

The most widely used arrival process in teletraffic analysis is the Poisson process generated by an infinite population. For practical purposes, a large population that independently generates low individual (i.e., per terminal) traffic is a Poisson traffic source. This clearly occurs in conventional telephony, where at least several thousands of subscribers are usually connected to the same switch. But in cellular networks two traffic flows must be distinguished according to their different features.

Fresh Traffic

In cellular networks, the flow due to new call establishments is commonly known as "fresh traffic," as opposite to the flow generated by handoffs. In general, this traffic is recognized to be Poisson (Bregni, Cioffi, & Décina, 2006), that is, generated by a large population under the cell coverage. The arrival rate is represented by λ. Only under certain circumstances this fresh traffic could differ from a Poisson flow.

Current trends in public networks lead to the deployment of microcells. Both the new assigned radio bands and the need to increase capacity suggest that more small cells perform better than few but larger cells. Microcells are more likely to have a traffic population that is not large enough to be considered infinite (i.e., Poisson). For most planning purposes this effect can be neglected because the difference between the true finite population traffic and the assumed Poisson is small unless the number of terminals within the cell is very low (i.e., low compared to the number of channels in the cell). In addition, Poisson traffic leads to worse QoS figures than the smoother[1] finite population traffic. For planning purposes, having worse-case figures that lead to moderately conservative designs is desirable.

Another deviation from the Poisson model is caused by the fact that cells receive overflow traffic from others (e.g., neighbor cells fully busy). As explained below, most of the territory is covered by more than one Base Station (BS). Hence, if a new call finds all channels busy on its first attempt, it will possibly try connection to another

Figure 1. The handoff traffic as aggregate of carried traffics in a) hexagonal layout b) highway layout

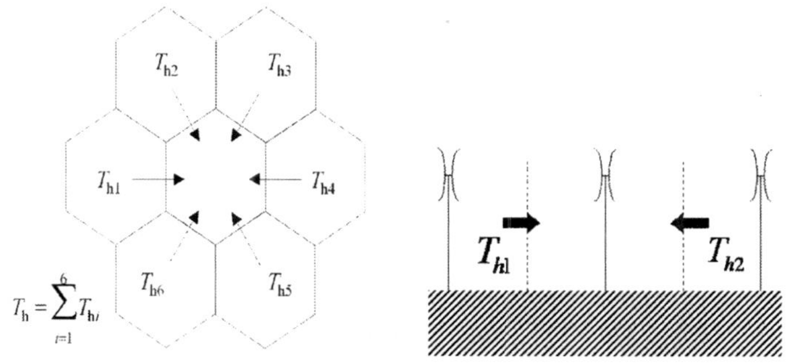

BS (e.g., with less but enough power received). This is overflow traffic, hence superexponential (in this work, distributions with coefficient lower than unity are called smooth or subexponential while superexponential is reserved to coefficients of variations larger than one). Overflow traffic is peak traffic showing a peakedness factor (Z) higher than unity, the consequence being a poorer performance than that expected for Poisson traffic. Under normal traffic working conditions and low congestion, the impact of this effect can be neglected. But under heavy traffic and overflow representing more than 5–10% of the offered traffic the Poisson model could lead to non-conservative designs.

Handoff Traffic

According to the Poisson nature of fresh traffic, handoff traffic should be smooth. This is because handoff traffic is carried traffic from a Poisson offered stream (i.e., traffic carried in the surrounding cells that have a finite number of channels each). Figure 1 shows how the handoff traffic to the central cell is the aggregate of 6 carried traffics for hexagonal layout and only two for highway layout. Rajaratnam and Takawira (2000) analytically proved that handoff traffic is smooth with peakedness factors between 0.85 and 1 (i.e., not very low), but due to the limitations imposed by the analysis, many factors such as radio propagation are not included in the study. Field tests carried out by Barceló and Sanchez (1999) weakly support the same hypothesis: show that the overall arrival process to a particular channel is smooth and conjecture that if the fresh traffic was Poisson (not proved in the field) the handoff traffic must be smooth.

The average number of handoffs per call is hereafter represented by α. Accordingly, the average number of cells visited during a call is $\alpha+1$. Every fresh call generates $\alpha+1$ arrivals (i.e., the fresh-star of the call plus α handoffs, hence the overall arrival rate (i.e., fresh plus handoff) is $\lambda(\alpha+1)$, being λ the arrival rate of fresh traffic.

CHANNEL HOLDING TIME

The CHT in cellular networks is only a fraction of the overall (i.e., unencumbered) call holding time: while the call holding time accounts for the whole call duration, the CHT only accounts for the time spent within a certain cell (e.g., since the call starts until the first handoff). The average CHT is represented by $1/\mu$. The negative exponential distribution used to model the call holding time is not the best choice in current systems. Field studies carried out by Bolotin (1994) showing that the call holding time in fix telephony is better modeled by a mix of lognormal distributions was corroborated by Chlebus (1997) for the unencumbered holding time in cellular telephony. However, there are many modulation stages between the call and the channel holding times in cellular networks: user mobility patterns including speed and direction (Thajchayapong & Peha, 2006), cell shapes including terrain class and shape, radio propagation conditions, and so forth. Accordingly, there is nothing that suggests that CHT should be modeled by an exponential distribution. Some discussion on this variable according to the methods recently used to characterize it follows.

Analysis

Many analytical studies have been published on the probability distribution of CHT in the last decade, most of them showing that it is not exponentially distributed. Modeling CHT through mathematical analysis can only capture the effect of a few of the related variables due to the need to simplify the problem in order to keep it tractable. Some recent studies include round (Fang, Chlamtac, & Lin, 1998), hexagonal (Christensen, Nielsen, & Iversen, 2001), Manhattan and highway cells (Cho & Sung, 1998). However, in real networks most BS are sectored and radiate three 120° cells. It is easy to understand why operators are reluctant to show their ratio of sectored cells to all BSs, but most planners agree that more than 95% of cells are sectored in urban environments. Analytical studies usually assume the BS to be in the middle of the cell (i.e., center of the circle or hexagon), while real BSs are in corners of 120° diagrams. Another mismatch between analytically obtained CHT and true networks comes from the fact that most cells are located in areas that are not flat and that have non-homogeneous characteristics (e.g., concrete, greens, etc.). Hence the actual cell shape depends on the terrain shape and features.

Another important constraint of analysis is that most of the territory is covered by more than two cells. Engineers commonly recognize the area in which the MN is able to connect to more than two cells to be greater than 90% in urban areas, while the area covered by more than one BS (i.e., the so-called handoff or overlapping area) is almost 100%. Analytical studies usually need to assume a clearly defined border between cells, which only occurs in true networks when, with light traffic, most MNs connect to the nearest BS (i.e., that which most receives power). However, this is difficult to accept under heavy traffic conditions when MNs connect wherever possible.

There are other real world patterns not included in analytical studies, but the two aforementioned ones are enough to prevent planners from trusting analytical studies. Of course these comments do not reduce the interest or usefulness of analytical tools, which must be the starting point of any serious research in this field and contributes greatly to the knowledge of the cause-effect relationships.

Simulations

The logical way to overcome analytical constraints is to use simulations. Most simulations capture effects that are very difficult or impossible to include in an analysis. In (Guerin, 1987; Zonoozi, Dassanayake, & Faulkner, 1995) several mobility patterns are included in simulations to obtain CHT samples: the negative exponential and the Gamma distribution are obtained respectively. More complex simulations include the effects of radio-path features and non-Poisson handoff arrivals (Spedalieri, Martin-Escalona, & Barceló, 2005) to obtain the lognormal distribution as the best fit. However, simulations seldom include sectorization and overlapping to the extent that occurs in real networks. Simulation has the advantage of allowing the researcher to feed it with realistic figures for the intended scenarios and of offering a certain degree of control over the cause-effect relationship, but many researchers and planners complain that the results remain unrealistic.

Field Studies

Once real systems are working, field studies are possible and necessary. Field studies cannot be criticized as unrealistic since they measure the real

Figure 2. Comparison of empirical histogram and two probability mass functions for the duration of the connection

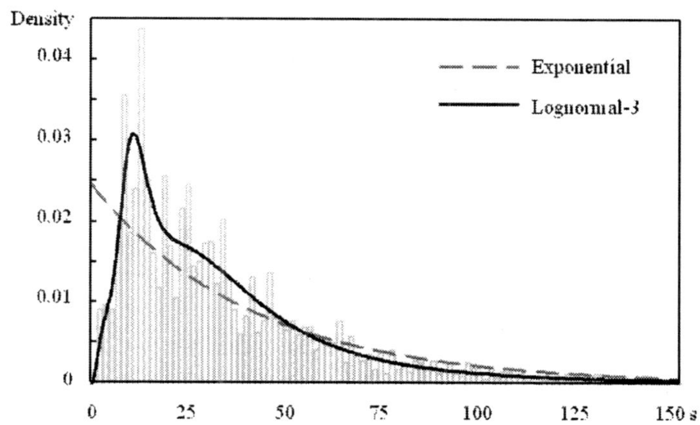

world. However, authors of such studies recognize that they never accurately know the "feeding figures" like radio conditions, mobility patterns, individual traffic per MN, cell shapes, and so forth. In addition, the results of the field study are biased by network settings. The lognormal and mix of lognormals distributions were identified as the best fit by Jedrzycki and Leung (1996) and Barceló and Jordán (2000); the reported squared coefficients of variation are between 1.7 and 4, showing the superexponential nature of CHT. An example of the empirical histogram and underlying candidate probability mass functions is shown in Figure 2. In both works a bias due to the minimum time to retry a handoff from the instant at which the last handoff occurred was detected. This minimum time is commonly set to a few seconds by the operator in order to avoid very frequent handoffs between adjacent cells for a mobile that remains on the border zone. The percentage of channel occupancies is artificially high at this point and this may differ in different networks.

One must also consider the consequences of using maximum likelihood estimation (MLE) to find the parameters of the conjectured distribution. MLE is more powerful than moment matching, as it leads to the best statistical significance; from this perspective it better serves the purpose of showing that the underlying distribution is the one conjectured. However, when using the parameters obtained by MLE, the engineer may find a slight mismatch between the average of the sample and the average of the proposed distribution. This mismatch is most inconvenient since the average CHT has a direct impact on the traffic, which is key for planning.

However, the main drawback of field studies for use by planning engineers is that it is impossible to generalize a field study or extrapolate the results to different scenarios. The field study is particular to the case in hand, and although the planner can reapply the modeling technique in the area of interest, he will be confronted with yet another particular case. This model will be more suited to the scenario in question, but not practical for obtaining data for conditions (e.g., traffic or mobility) different to those under study. This highlights the need for a link between modeling tools.

Is There a Link Between Different Tools?

In general modeling, the passing of time tends to make for a decrease in interest in analysis and simulations and an increase in interest in field studies that are only possible once true networks exist. The history of teletraffic results for conventional telephony followed these stages, which greatly increased the reliance on the proposed models. However, this did not happen with cellular networks, for two reasons. First, field studies in the open technical literature are scarce owing to the disproportionate degree of confidentiality maintained by operators: many consider CHT and other teletraffic data as highly sensitive. Second, mainly because of the lack of field studies, analytical and simulation results are used to bridge the gap without the necessary support of true data.

Unfortunately the link between results obtained through different tools is poor and this remains an unresolved issue. Few authors have carried out the necessary effort of trying to match results obtained through different tools in order to show that simulations or field studies fit the analysis (Orlik & Rappaport, 1998). Although questionable, since the hypotheses of the analysis are far removed from the reality in the field, such attempts are of great value. The consequence is that, in spite of the many published models, planners are reluctant to use them to plan the network due to the difficulties found to identify a specific model with the conditions found in the real planned scenario. After reading the literature, planners can at best be vaguely confident that a specific model is better suited to a specific environment, based on the hypotheses of the paper that supports it, or try to reapply the modeling methodology to the scenario in question.

QUALITY OF SERVICE

One key objective of network planning is to forecast the Quality of Service (QoS). Following ITU-T, the difference between QoS and GoS is that the figures belonging to the former are perceived by the user while the latter is a set of engineering figures that may not be clearly perceived at the user plane. The probability of a new call being rejected, the so-called probability of blocking (*PB*), is QoS because the user could measure it as the percentage of call attempts blocked. However, the probability of a handoff being dropped (*PD*), due to the lack of channels in the neighboring cell, cannot be measured by the user, who should remain unaware of the handoff process. This is a Grade of Service (GoS) figure, useful for engineering but not relevant at the user plane.

The consequence of a dropped handoff is the forced interruption of an ongoing call. The probability of a forced interruption (*PF*) is again a QoS figure since it is clearly perceived by the user (Grillo, Skoog, Chia, & Leung, 1998). The probability of forced termination due to handoff failure, taking into account that one call involves α handoffs on average, is calculated as:

$$PF = 1 - \left(1 - PD\right)^{\alpha} \approx \alpha PD \,. \qquad (1)$$

This latter approximation is valid as long as *PD* is kept very low and this should always be a design condition. From the above equation it is easy to understand why a handoff scheme is needed. If no handoff mechanism is used (i.e., handoffs are not prioritized) the obvious consequence of managing fresh calls and handoffs the same way is that *PB*=*PD*. Hence, according to Equation (1) the mobility $\alpha > 1$ due to possible high speed, small cells, and so forth makes interruptions occur more frequently than blocking. Since the subjective degradation (i.e., as perceived by the user) caused by the interruption of an established call is much worse than that of blocking a call establishment attempt this must be avoided by all means unless all QoS metrics are really very low.

In Figure 3 the main QoS are related to specific events during call. Blocking can take place at A and G. At B, D and H a handoff is requested while it is granted in C, E and I while still within the overlapping area.

Figure 3. Teletraffic events during the mobile call

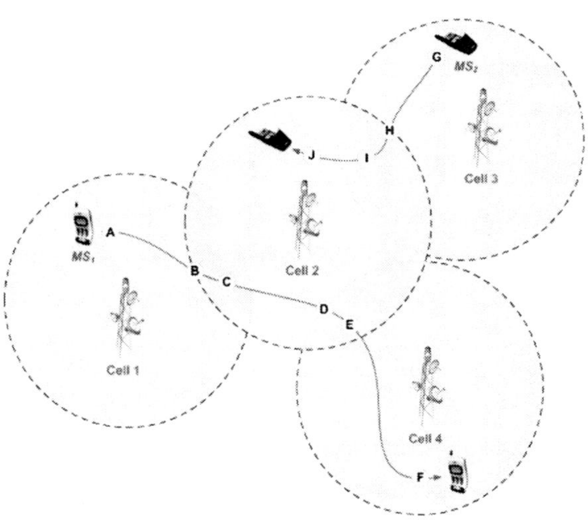

ON QUEUING MODELS

Blocked Calls Policy

It is important to determine how the system deals with each of the two arrival flows. For the fresh traffic, the Blocked Calls Lost (BCL) model has been widely applied. The assumption of this model is basically true for most systems: if a new call does not immediately find a free channel in the cell it is definitely rejected. But this immediateness can be questioned since operators tend to implement procedures to improve revenue: allowing new calls to wait until a channel in the cell is released is one of these procedures. The user must not perceive this wait. This can be achieved by appropriately sizing the number of channels in such a way that the wait is shorter than a few seconds (i.e., in the same time range involved in call establishment) in a high percentage of attempts.

Handoffs should always be modeled as queued arrivals. Handoffs occur within the overlapping area and therefore are allowed a certain time to reach the channel in the new cell in case it cannot be immediately provided. In most systems the handoff is retried with an extremely short retry period. This should be modeled as a virtual queue with Random Service Order (RSO) queueing discipline. Of course the handoff cannot retry forever, but only for the time that the MN remains within the overlapping area. The time within the overlapping area has been modeled as combination of memoryless stages (Spedalieri et al., 2005) and Gaussian (Ruggieri, Graziosi, & Santucci, 1998). The handoff dwell time (i.e., a fraction of the time within the overlapping area, only until the handoff is achieved) has been studied analytically (Ruggieri et al., 1998) and through simulation (Ruiz, Doumi, & Gardiner, 1998; Spedalieri et al., 2005); unfortunately, this author knows no field studies.

BCL Model and Erlang-B

Frequently a cell of a cellular system has been modeled as a BCL system with Poisson arrivals for both new and handoff attempts, and exponential CHT. According to this M/M/C/C queueing model, Erlang-B formula with offered traffic $A=(1+\alpha)\lambda\mu$ must be applied to determine QoS. It is a well-known result that the probability of blocking obtained by Erlang-B holds for the M/G/C/C queue. Hence, moments of the CHT higher than the average have no effect on the obtained network performance. Since the overall traffic tends to be slightly smooth as above explained, Erlang-B tends to slightly overestimate the blocking probability.

BCL is far from reality, mainly because assuming that handoffs are immediately rejected if all channels are busy is not realistic. New calls can have a queueing margin of a 0–10 seconds, depending on the operator's policy, but handoffs can reach average queueing margins of 50–100% CHT. This is more likely to happen with smaller cells and larger proportion of overlapping area which is the current design trend. While neglecting the possible short wait of fresh calls has a limited impact on the resulting QoS metrics, the impact of neglecting the wait of handoffs is noticeable and twofold. First the error incurred is obviously larger due the longer possible wait. Second, the wait of a handoff while insisting in seizing a channel has the same effect as a priority owned by the handoff in front of the fresh call: even if no specific handoff priority method is used, handoffs see less blocking than new calls since they are available to be served upon any channel release.

The Cell as a Delay System

If losses are kept very low as they should be, a pure Blocked Calls Delay (BCD) model can be used. A BCD M/G/C queue should be used instead of M/M/C. This is because the holding time is far from being exponentially distributed. However, it must be noticed that the waiting probability (*PW*) for the M/M/C is an excellent approximation for the M/G/C, so Erlang-C is appropriate to compute *PW* but not valid for other QoS metrics. The waiting time mean and distribution are sensitive to the general service time. Exact solutions for the M/G/C queue require programming whereas approximations are simple to compute and results have a degree of accuracy of around 3%, smaller than the uncertainty of inputs such as forecasted traffic or the expected rate of handoffs to new calls. The first and second moments represented by the average and coefficient of variation are enough to capture the effect of the general service time on the achieved performance. Further details on all approximations proposed in the remaining of this section can be found in the work by Kimura (1994).

According to the M/G/C BCD model, QoS should be approximately computed according to the following steps, all of them include approximations. First, the average waiting time is computed according to:

$$R_G = \frac{2kR_D}{2c^2 R_D + \left(1 - c^2\right)}, \tag{2}$$

where c_s^2 represents the squared coefficient of variation of the service time and *W(M/M/C)* the average waiting time in the *M/M/C* queue with the same offered traffic and channels (i.e., computed through Erlang-C). R_D represents the relative mean waiting time of the M/D/C queue and can be approximated as:

$$R_D = \frac{1}{2}\left\{1 + \left(1-\rho\right)\left(s-1\right)\frac{\sqrt{4+5s}-2}{16\rho s}\right\}, \tag{3}$$

where $\rho = \dfrac{\left(1+\alpha\right)\lambda\mu}{C}$ represents the channel load.

Now losses should be computed as the percentage of arrivals that exceed a certain time in the queue: the queueing margin for new calls (t_n) and the time in the overlapping area for handoffs (t_h). The waiting time can be approximated as exponentially distributed, hence the GoS can be obtained as:

$$\begin{aligned} PB &= \Pr\left(t > t_n\right) = \\ &PW \exp\left(-\frac{PW}{W\left(M/G/C\right)}t_n\right) \\ PD &= \Pr\left(t > t_h\right) = \\ &PW \exp\left(-\frac{PW}{W\left(M/G/C\right)}t_h\right) \end{aligned}. \tag{4}$$

Note that this approach overcomes the drawbacks mentioned, showing estimates of both *PB* and *PD* with *PD*<*PB* as long as $t_h > t_n$. It is conservative since it neglects the reduction of load due to traffic losses. The approach is accurate for normal working conditions (i.e., low losses) and too conservative for heavy load and congestion situations.

A More General Model

The model presented in this section is more sophisticated than the previous ones: it cannot

Figure 4. Markov B-D process modeling a cell

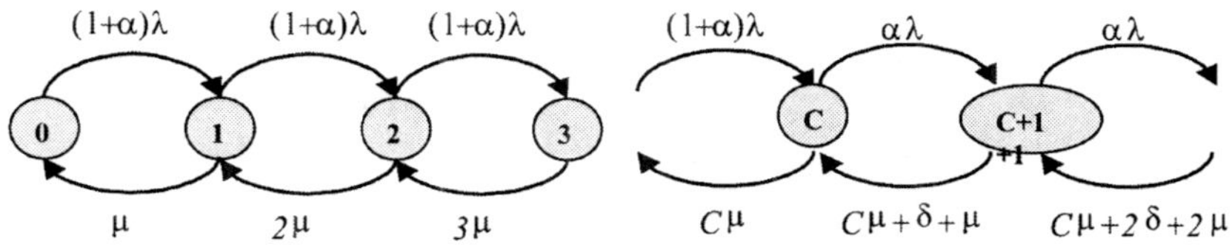

be computed through closed formulas but needs some simple programming. The proposed model is a mix loss-delay system with early departure. The birth-death (B-D) process is represented in Figure 4. Note that all the traffic $(1+\alpha)\lambda$ is offered to the C channels while only the handoff traffic $(\alpha\lambda)$ is allowed to queue. The departure rate for states $i<C$ represents the natural ending of the service. Departure rates larger than $C\mu$ for queueing states represent early departure: μ represents services completed while requesting a handoff while δ represents the drop of a handoff because the overlapping area is left without having obtained a channel (i.e., time within the handoff area is assumed exponential with mean $1/\delta$).

The solution is simple to compute (Akimaru & Kawashima, 1999), being the offered traffic $A=(1+\alpha)\lambda\mu$:

$$P_0 = \left[1 + \sum_{i=1}^{\infty} \frac{\lambda_0\lambda_1 \cdots \cdots \lambda_{i-1}}{\mu_1\mu_2 \cdots \cdots \mu_i}\right]^{-1} = \left[1 + \sum_{i=1}^{C} \frac{A^i}{i!} + \frac{A^C}{C!} \sum_{i=C+1}^{\infty} \frac{\left(\alpha\lambda\mu\right)^{i-C}}{\prod_{j=1}^{i-C}\left(i + j\gamma^{-1}\right)}\right]^{-1} \tag{5}$$

$$P_i = P_0 \frac{A^i}{i!} \quad \text{for} \quad 1 \le i \le C$$

$$P_i = P_0 \frac{A^C}{C!} \prod_{i=1}^{C} \frac{\left(\alpha\lambda\mu\right)^{i-C}}{\prod_{j=1}^{i-C}\left(i + j\gamma^{-1}\right)} \quad \text{for } C < i, \tag{6}$$

where $\gamma=\mu/\delta$ represents the proportion of CHT within the handoff area and $\alpha\lambda/\mu$ the traffic produced by handoffs only.

Now the GoS figures can be computed. Note that PD does not include early departure at rate μ since it does not contribute to handoff dropping.

$$PB = \sum_{i=C}^{\infty} P_i. \tag{7}$$

$$PD = \frac{\sum_{i=C+1}^{\infty} \left(i - C\right)\delta P_i}{\sum_{i=0}^{\infty} \alpha\lambda P_i} = \frac{1}{\alpha\lambda\mu\gamma} \sum_{i=C+1}^{\infty} \left(i - C\right) P_i \tag{8}$$

This model has obvious advantages compared to the simpler BCL and BCD. It allows including queueing for fresh traffic with a slight modification of the B-D process: an additional early departure must be included. It captures the effect of the lost traffic, not captured by the previous models, and is valid for all loads, including heavy traffic. Another advantage is that it easily allows the inclusion of handoffs schemes (Barceló, 2004; Candan & Salamah, 2007). The drawback is that it neglects the superexponential nature of the holding time (Xhafa & Tonguz, 2008).

FURTHER REMARKS

Macrodiversity

This feature must be considered with care when modeling: there is an additional load caused by the simultaneous connection of the MN to more than one BS in overlapping areas, that is, almost 100% of the territory (Jung, Hong, Chang, & Lie, 2009). However, this load has the lowest priority in front of handoff traffic and new calls, so neglecting this load is not a bad trade-off to compute GoS figures while maintaining simplicity.

Discontinuous Transmission

Current trend to exploit discontinuous transmission (DTX) must also be accounted for. Voice activity is between 30% and 40% of the time, depending on the voice activity detector (VAD). At this level, the channel occupancy time equals the holding time and there is no need to account for handoffs in the model since the average talkspurt is 1.5 seconds, much lower than the cell residence time.

When modeling at the talkspurt level, the QoS parameters presented in Section 4 loose their meaning: a talkspurt can neither wait more than a few miliseconds (e.g., at most 30 ms.) nor be fully blocked. The MN must transmit the talkspurt after a very short time, independently of whether there are free channels or not. Hence front-end clipping occurs if all channels are busy when the talkspurt stars. The probability of clipping a talkspurt must be investigated (Barceló & Ramón, 1997). Talkspurt arrivals for a single conversation have been found to follow a hyperexponential distribution, but assuming Poisson traffic for the aggregated traffic can lead to good estimates. For the service time, squared coefficients of variation of around 2.5 have been reported. Slight modifications on the BCD model can lead to good estimates while using the Blocked Calls Held (BCH) computed through the Molina distribution (so called Poisson) is more realistic.

Multiservice

If the network additionally supports data services on the same pool of resources used for voice, the analysis is complex. However, the fact that voice is delay sensitive while data can tolerate delays, lead most operators to prioritize voice. In such a case, and depending on the sharing and prioritization policies, the models proposed in this paper will give good performance estimates (Barceló, Paradells, & Setaki, 2003).

CONCLUSION

Nowadays, there is a gap between theory and practice in the use of teletraffic models for cellular systems. On the one hand, research is very active in this field and complex models have been presented; most of them are far from being simple and require advanced programming. On the other hand, there is a general lack of field studies that corroborate the models. This situation prevents the engineer from using them as long as reliability is not proved.

Along this paper, several hints have been provided to allow the planner to introduce some of the available research material into the first stage of network planning. First, a review of the arrival traffic and CHT features has been presented. According to this review, the arrival process can be approximated by Poisson without loosing much accuracy, at most moderately conservative performance results will be obtained. The service time cannot be approximated by the negative exponential distribution without losing accuracy since superexponentiallity has been widely showed.

The models proposed to evaluate performance are basically simple in order to be useful for a quick first approach to the problem before the simulation phase. The Erlang-B is the simplest model but shows obvious drawbacks. The delay model can also be computed through closed formulas and captures further effects: general service time, handoff dwell time and possible queueing of new calls. The more general model is more accurate and valid for heavy traffic, but requires simple programming and neglects the general service time. An important advantage of this latter is that it is easily adaptable to handoff schemes.

REFERENCES

Akimaru, H., & Kawashima, K. (1999). *Teletraffic, theory and applications*. Berlin, Germany: Springer-Verlag.

Barceló, F. (2004). Performance analysis of handoff resource allocation strategies through the state-dependent rejection scheme. *IEEE Transactions on Wireless Communications*, *3*(3), 900–909.

Barceló, F., & Jordán, J. (2000). Channel holding time distribution in public telephony systems (PAMR and PCS). *IEEE Transactions on Vehicular Technology*, *49*(5), 1615–1625.

Barceló, F., Paradells, J., & Setaki, F. (2003). Design and modelling of Internode: A provider provisioned mobile VPN. *Mobile Networks and Applications*, *8*(1), 51–60.

Barceló, F., & Ramón, A. (1997, December). PRMA performance evaluation based on queueing theory. In *Proceedings of the IEEE International Conference on Personal Wireless Communications-ICPWC'97,* Mumbai, India (pp. 455- 459). IEEE.

Barceló, F., & Sánchez, J. I. (1999, June). Probability distribution of the inter-arrival time to cellular telephony channels. In *Proceedings of the 49th IEEE Vehicular Technology Conference (VTC'99),* Houston, TX (pp. 762-766). IEEE.

Bolotin, V. A. (1994, June). Telephone circuit holding time distribution. In *Proceedings of the 14th International Teletraffic Congress,* Antibes Juan-les-Pins, France (pp. 125-134). Elsevier Science.

Bregni, S., Cioffi, R., & Dècina, M. (2006, November 27-December 1). *An empirical study on statistical properties of GSM telephone call arrivals*. Paper presented at the IEEE Global Telecommunications Conference (Globecom'06), San Francisco.

Candan, I., & Salamah, M. (2007). Analytical modeling of a time-threshold based bandwidth allocation scheme for cellular networks. *Computer Communications*, *30*, 1036–1043.

Chlebus, E. (1997, June). Empirical validation of call holding time distribution in cellular communications systems. In *Proceedings of the 15th International Teletraffic Congress,* Washington, DC (pp. 1179-1188). Elsevier Science.

Cho, H. S., & Sung, D. K. (1998). Channel holding time distribution in cross- and cigar-shaped urban microcells. *IEICE Transactions on Communications. E (Norwalk, Conn.)*, *81-B*(6), 1275–1279.

Christensen, T. K., Nielsen, B. F., & Iversen, V. B. (2001, December). Distribution of channel holding time in cellular communication networks. In *Proceedings of the 17th International Teletraffic Congress,* Salvador de Bahia, Brazil (pp. 471-480). Elsevier Science.

Fang, Y., Chlamtac, I., & Lin, Y. B. (1998). Channel occupancy times and handoff rate for mobile computing and PCS networks. *IEEE Transactions on Computers*, *47*(6), 679–692.

Grillo, D., Skoog, R. A., Chia, S., & Leung, K. K. (1998). Teletraffic engineering for mobile personal communications in ITU-T work: The need to match practice and theory. *IEEE Personal Communications*, *5*(6), 38–58.

Guerin, R. A. (1987). Channel occupancy time distribution in a cellular radio system. *IEEE Transactions on Vehicular Technology*, *35*(3), 89–99.

Jedrzycki, C., & Leung, V. C. M. (1996, April 28-May 1). Probability distribution of channel holding time in cellular telephony systems. In *Proceedings of the IEEE 46th Vehicular Technology Conference*, Vancouver, British Columbia, Canada (pp. 247-251). IEEE.

Jung, S. H., Hong, J. W., Chang, W., & Lie, C. H. (2009). Quality of service estimation for soft handoff region ratio and call admission control in CDMA cellular systems. *Computers & Industrial Engineering*, *56*, 1429–1441.

Kimura, T. (1994). Approximations for multi-server queues: System interpolations. *Queueing Systems*, *17*, 347–382.

Orlik, P. V., & Rappaport, S. S. (1998). A model for teletraffic performance and channel holding time characterization in wireless cellular communications with general session and dwell time distributions. *IEEE Journal on Selected Areas in Communications*, *16*(5), 788–803.

Rajaratnam, M., & Takawira, F. (2000). Nonclassical traffic modeling and performance analysis of cellular mobile networks with and without channel reservation. *IEEE Transactions on Vehicular Technology*, *49*(3), 817–834.

Ruggieri, M., Graziosi, F., & Santucci, F. (1998). Modeling of the handover dwell time in cellular mobile communications systems. *IEEE Transactions on Vehicular Technology*, *47*(2), 489–498.

Ruiz, G., Doumi, T. L., & Gardiner, J. (1998). Teletraffic analysis and simulation for nongeostationary mobile satellite systems. *IEEE Transactions on Vehicular Technology*, *47*(1), 311–320.

Spedalieri, A., Martin-Escalona, I., & Barceló, F. (2005, March 13-17). Simulation of teletraffic variables in UMTS networks: Impact of lognormal distributed call duration. In *Proceedings of the IEEE Wireless Communications and Networking Conference (WCNC),* New Orleans, LA (Vol. 4, pp. 2381-2386). IEEE.

Thajchayapong, S., & Peha, J. M. (2006). Mobility patterns in microcellular wireless networks. *IEEE Transactions on Mobile Computing*, *5*(1), 52–63.

Xhafa, A. E., & Tonguz, O. K. (2008). Handover performance of priority schemes in cellular networks. *IEEE Transactions on Vehicular Technology*, *57*(1), 565–577.

Zonoozi, M. M., Dassanayake, P., & Faulkner, M. (1995, November). Mobility modelling and channel holding time distribution in cellular mobile communication systems. In *Proceedings of the IEEE Globecom,* Singapore (pp. 12-16, 1995). IEEE.

This work was previously published in International Journal of Business Data Communications and Networking Volume 5, Issue 4, edited by V. Sridhar and D. Saha, pp. 1-15, copyright 2009 by IGI Publishing (an imprint of IGI Global).

Chapter 16
Testbed Implementation of a Pollution Monitoring System Using Wireless Sensor Network for the Protection of Public Spaces

Siuli Roy
Indian Institute of Management Calcutta, India

Anurag D
Indian Institute of Management Calcutta, India

Somprakash Bandyopadhyay
Indian Institute of Management Calcutta, India

ABSTRACT

Air pollution is an important environmental issue that has a direct effect on human health and ecological balance. Factories, power plants, vehicles, windblown dust and wildfires are some of the contributors to pollution. Reasonable simulation tools exist for evaluating large scale sensor networks; however, they fail to capture significant details of node operation or practical aspects of wireless communication. Real life testbeds capture the realism and bring out important aspects for further research. In this article, we present an implementation of a wireless sensor network testbed for automatic and real-time monitoring of environmental pollution for the protection of public spaces. The article describes the physical setup, the sensor node hardware and software architecture for "anytime, anywhere" monitoring and management of pollution data through a single, Web-based graphical user interface. The article presents practical issues in the integration of sensors, actual power consumption rates and develops a practical hierarchical routing methodology.

DOI: 10.4018/978-1-60960-589-6.ch016

INTRODUCTION

The availability of consistent, accurate and timely information on environmental conditions greatly improves the speed of planning and decision making (Hammons & Chisholm, 2006). Such information, although required for most businesses, is of particular importance in chemical factories, mines and other high investment industries. The availability of precise information enables businesses to respond in quick time and take preventive measures against emission of poisonous gases and other hazards, thus improving the safety of personnel and equipment.

Air pollution (Jung, Lee, Lee, Ryu, & Nittel, 2008; Ma, Richards, Ghanem, Guo, & Hassard, 2008) is an important environmental issue that has a direct effect on human health and ecological balance. The primary airborne pollutants covered by European legislations are: SO_2, NO_x (NO/NO_2), Benzene, Ozone, CO/CO_2 and particulate matter (PM10/PM2.5). Air pollution has diverse causes and sources. "Stationary sources" such as factories, power plants and smelters; "mobile sources" such as automobiles; and "natural sources" such as windblown dust and wildfires are primary contributors to air pollution. Due to the trans-boundary nature of airborne pollutants, it is difficult for any single organization to take responsibility for overall emission levels. Thus, the control of air pollution is entirely legislation driven. As such the passing of new legislation may only be effective if the pollution level in the specified compounds can be monitored accurately using sensors.

Environmentalists can use sensors to measure atmospheric pollution and monitor industrial emissions; safety monitors can use sensors to detect harmful chemical vapours and explosives in public spaces, government or military facilities and chemical processing plants.

Gas sensor technologies are still developing and have yet to reach their full potential in capabilities and usage (Jung et al., 2008). Some technologies are particularly accurate but prohibitively expensive for large-scale deployment. By using a sensor network, the problem of false positives could potentially be reduced. Multiple outputs can be compared for a more accurate analysis.

However, realizing sensor-based networks using wires are not feasible solutions for large scale deployment. Wireless sensor networks (Akyildiz & Wang, 2005) offer powerful new ways to monitor air quality and without the costs of major new installations or wire-runs that are typically associated with these types of projects. Wireless sensor networks comprises of:

- A multitude of bi-directional radio transceivers with sensors known as sensor nodes or tags. The sensors include a wide variety of pollution detectors and are connected together using the communicational capability of the wireless nodes to form the core of our proposed pollution monitoring system. Sensors capture pollution data and communicate them wirelessly over the network to a remote control station for further analysis and generation of alerts during critical events.
- The wireless sensor nodes are arranged in a networking topology called a "mesh." Mesh networking is a type of network where each node can communicate with multiple other nodes thus enabling better overall connectivity than in the traditional hub-and-spoke or star topologies. State of the art mesh networks often have some or all of the following characteristics:
    - They are self-forming. As nodes are powered on, they automatically enter the network, compute efficient routes for data forwarding through their neighbors and are ready for operation.
    - They are self-healing. As a node leaves the network, the remaining nodes automatically re-route their packets around the out-of-network

node to ensure a more reliable communication path.

- They support multi-hop routing. This means that data from a node can jump through multiple nodes before delivering its information to a host gateway or controller that may be monitoring the network thus extending the reach and range.

The self-forming and self healing coupled with battery operable attributes of a mesh sensor network make it ideal for environmental monitoring applications in a wide range of facilities.

This article presents a generic framework for "anytime anywhere" visibility of sensor data from a remote wireless sensor network using the Internet (Pujolle, 2006). Active RFID tags, combined with sensors and actuators, form the proposed wireless sensor network based on IEEE 802.15.4 (IEEE, 2003). The architecture also enables the control of the environment by triggering actuators. Lastly, we elaborate a test-bed implementation for monitoring environmental pollution for possible detection of toxic gases based on the above framework.

The article is structured as follows. We explore the related work on the testbed implementation of wireless sensor networks and present the proposed conceptual framework of a wireless sensor network for pollution monitoring. Detailed system architecture and different technology options for the wireless sensor network, components of the wireless sensor node, software architecture and setting up of testbed is then illustrated. We conclude the article highlighting the scope for further work.

RELATED WORK

In Jung et al. (2008), the design and implementation of an air pollution monitoring system based on a geo-sensor network has been discussed. In order to detect the status of air pollution, a context model which includes sensor data abstraction and air pollution prevention model has been designed. The sensed data is combined by the abstraction model after collecting the various kinds of geo-sensor data. The level is derived by the user defined rules using the values of the geo-sensors such as dust, carbon dioxide, ultraviolet, and so forth. The abstracted data also includes the boundary of the area, the sensors' properties, air pollution types and maximum levels. This local abstracted data is combined in the global area by the air pollution prevention model. The system provides an alarm and safety guideline for near future dangerous situations and prevents severe damage and mitigates recovery cost. It also supports the flexible sampling-interval change depending on the pollution conditions of the context model.

In order to monitor the pollutants and analyze their effects on the environment, a Mobile Discovery Net (MoDisNet) was developed in (Ma et al., 2008). It collects real time pollution data on key aspects of traffic, emissions, ambient pollutant concentration and human exposure. The purpose is to develop a capability to measure, model and predict a wide range of environmental pollutants and hazards using a grid of pervasive roadside and vehicle/person-mounted sensors. The major challenge, as pointed out, is integrating data from heterogeneous fixed and mobile environmental sensor grids in real time to provide dynamic estimates of pollutants and hazard concentrations. This should demonstrate how such data can be usefully correlated with a wide range of other complementary dynamic inputs, such as weather, health or traffic. The MoDisNet system is designed as a two-layer network architecture—the mobile sub-network formed by the Mobile Sensor Nodes installed in vehicles and the stationary sub-network organized by the Static Sensor Nodes. The mobile nodes may pre-process the raw data (such as the noise reduction, local data cleaning and fusion, etc.) and send the data to a nearest static node. The static nodes now in turn, account for the data reception, updation, storage and exchange. Cooperating with a grid architecture, the static nodes can realize the distributed data analysis and mining.

As pointed out in Werner-Aller, Swieskowski, and Welsh (2005), deploying a wireless sensor network into a realistic environment requires iterative programming of dozens of nodes, locating them throughout an area large enough to produce an interesting radio topology and extract debugging and performance data. The advent of networked interfaces for sensor nodes, such as the Crossbow MIB-600, makes remote reprogramming and monitoring of permanently-powered sensor network nodes possible. The MoteLab sensor network testbed platform addresses these challenges. MoteLab also allows users to interact with individual nodes directly during experiments. By providing a web interface to the testbed, MoteLab simplifies and accelerates deployment and evaluation of wireless sensor network applications. They have also instrumented a node in MoteLab with a network-connected digital multimeter, allowing the MoteLab to continuously monitor the energy usage of the node. Current consumption data is logged and returned with other data generated during the experiment.

The experience of implementation, deployment and operation of SensorScope, an indoor environmental monitoring network is reported in (Schmid, Dubois-Ferrière, & Vetterli, 2005). From the data gathered, it has been observed that the network performance is greatly improved by using MAC layer retransmissions.

The tesbed experimentation done so far on wireless sensor networks for pollution monitoring does not take care of the requirements of "anytime, anywhere" access of pollution data as well as control of environment, alert generation and sleep management through a single user interface. We propose to address these issues through development of a generic framework.

CONCEPTUAL FRAMEWORK

Small, autonomic wireless devices, cheaply priced and capable of forming communications networks are expected to be the next technological innovation to bring about a mass cultural change. This new concept is due to the emergence of Device Area Network where billions of devices are inter spread in the everyday activities of humans. The devices help organize, manage and control a host of equipment and services. This device area network can be further extended to offer "Anytime Anywhere" visibility of network data over the Internet. This initiates a new technology paradigm called the "Internet of Things" (International Telecommunications Union, 2005).

Such a network comprises of wireless devices typically, sensors and actuators which are capable of forming wireless networks among themselves and are designed to run on batteries for prolonged periods of time. These devices and such networks need to support a low data rate and the computational capability is restricted. Low data rate protocols have been designed specifically for these needs lately, a significant deviation from the trends of building high-bandwidth networks. The protocol ratified by IEEE is 804.15.4 in 2003 (IEEE, 2003) and updated in 2006 (Wheeler, 2007). The 802.15.4 standard provides specifications for the Physical and Medium Access Layer. The upper layers (Network to Application) have been developed by a consortium of wireless device manufacturers and enthusiasts called ZigBee (ZigBee Alliance, 2004).

In this article, we have developed an end to end architecture for remote monitoring of environment and remote actuation necessary for controlling the environment. The architecture comprises of three basic components: (a) the wireless mesh network, consisting of RFID tags connected to sensors and/or actuators (which may also be called as wireless sensor actuator network), (b) the gateway or the central server and (c) the Internet. The mesh networks collect local information (pollution level) which is communicated through a gateway to the central server. The gateway handles the messages to and from the mesh network. The gateway is expected to be A/C mains powered while the individual nodes (sensors/actuators) are battery driven. The information is then transported to a web end-point by the Central Server. The central server essentially does the job of protocol conversion from the Internet to the sensor network and vice-versa. The information is displayed in the requisite format and provides an interface for object management. The web end point application can be a simple hosted page viewed on a personal computer or a message transmitted to a mobile device. We could think of supporting both push based and pull based systems. A push based system updates the status of the devices automatically, either periodically or when a status changes. In a pull based system, the latest information is provided when the user asks for it. The conceptual framework is pictorially depicted in Figure 1.

DETAILED SYSTEM ARCHITECTURE

As depicted in Figure 1, the proposed architecture has three components: Wireless Sensor-based Mesh Network, Central Server or Control Station and any web-enabled device accessing the central server through the Internet. Each of these components is explained in greater detail below.

a. Wireless Mesh Network

A Wireless Mesh Network, as its name implies, is a type of network where each node in the network can communicate with multiple wireless nodes, thus enabling better overall connectivity. A wireless mesh network has two kinds of nodes: the mesh routers and the mesh clients. The mesh clients can be made to communicate with multiple routers/other clients, or arranged in a hierarchical fashion where every client has a single router as its parent. A major advantage of this kind of network, other than self-forming and self-healing capability, is multi-hop routing. This means that data from a wireless node can jump through other nodes before delivering its information to a remote host gateway or controller that collects the data from sensors for further processing. Low power mesh networking, sensing and active RFID-based real-time location tracking has enabled us to design systems for tracking, locating and monitoring of people and other assets including environmental conditions.

Passive RFID tags (RFID without a battery) are already driving shifts in supply chain and retail capabilities for automatic identification of objects. However, Active RFID has much broader potential in the enterprise. Firstly, active RFID can form a wireless mesh network, providing automatic, dynamic visibility into what is going on in and around the enterprise. Secondly, active RFID technology, if combined with sensors and

Figure 1. A conceptual framework for Monitoring and Actuation of environment through the Internet

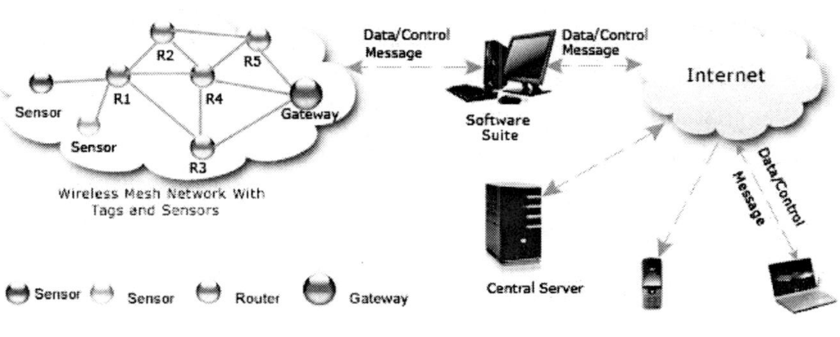

actuators in a networked environment, enables a spectrum of applications that can exponentially increase visibility and monitoring. Offering much more information than passive RFID, sensors can monitor and record conditions like temperature, humidity, pressure, wind direction and speed, illumination intensity, vibration intensity, sound intensity, chemical concentrations, pollutant levels and so on.

An important aspect of the wireless mesh network is to develop an efficient route to multi-hop the data packet from the sensor to the sink (or gateway). The de-facto standard for routing in mesh networks is AODV (Ad-Hoc On-demand Distance Vector) (Al-Karaki & Ahmed, 2004; Perkins & Royer, 1999) which develops route tables by broadcasting control packets. If we study the topology of the wireless mesh, the sensors and the gateway are static. The intermediate routers, which relay the data, are static as well. The AODV algorithm was designed for mobile nodes and its application in a static mesh network would be wasteful in terms of resources. A competing routing methodology is the hierarchical tree which labels every node such that no route tables are needed. Such labeling does not need broadcast control packets. A comparison (Cuomo, Luna, Monaco, & Melodia, 2007) with the hierarchical algorithm has shown AODV to incur a large number of control packets for developing and maintaining an active route thus expending a larger amount of energy and results in a slightly lower throughput. The hierarchical algorithm on the other hand suffers from its inability to support networks of large depths and fault tolerance (Wheeler, 2007). The ZigBee Stack (ZigBee Alliance, 2004) supports the hierarchical (denoted as C_{skip}) and AODV based routing and they have been compared in (Peng, Mao-heng, & You-min, 2006). Devices which join the network are given an address based on rules set by the network parameters. These parameters are predetermined and cannot be changed throughout the network lifetime. The Parameters are C_m, R_m and L_m; where C_m = Maximum Number of children that a full functional device (FFD) can have; R_m = Maximum number or children (out of C_m) which are FFDs; and L_m = Maximum depth of the network. The values of these parameters are stored in the NIB (Network Information Base) in each device (Ho-In & Yeonsoo, 2007).

An address of a requesting child is generated by the equation:

$$A_n = A_{parent} + C_{skip}(d)\, R_n + n \quad (1)$$

where A_n is the address which the new device will take. A_{parent} is address of the parent of the device that will assign the address. $C_{skip}(d)$ is determined as follows:

$$C_{skip}(d) = (1 + C_m - R_m - C_m * R_m C_m - d - 1) / (1- R_m) \quad (2)$$

$C_{skip}(d)$ determines the block of addresses which the parent device must skip before assigning the next address. The algorithm assumes the worst case scenario and makes provision for accommodation for all devices in the pre-determined network architecture. This assumption of a worst case scenario severely restricts the network depth. For example, values of $C_m = 6$, $R_m = 6$ makes L_m=7 for a 16 bit short address.

The addressing scheme based on C_{skip} develops a tree topology which makes possible the optimum routing. Routing in such networks is made by comparing the destination address with the C_{skip} allocation block. If the destination address is within the C_{skip} block of any of its children, the packet is forwarded to that child, else the packet is forwarded to its parent. It has been shown that such tree based routing provides the minimum latency (Peng et al., 2006). The address wastage problem in C_{skip} is pictorially shown in Figure 2.

Our test bed implementation used a variation of the hierarchical network where node ad-

Figure 2. The address wastage problem in C_{skip}

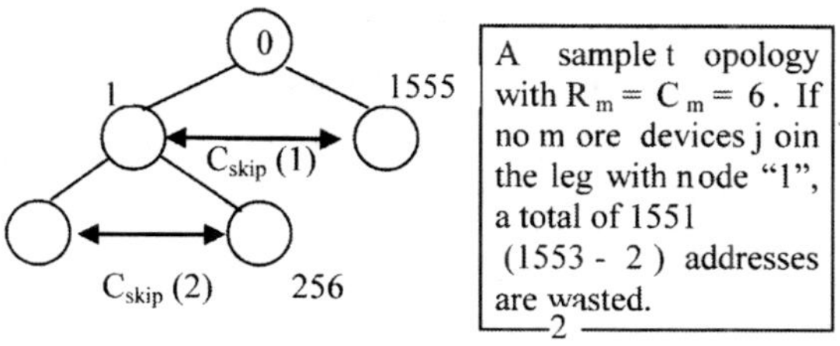

dresses are made once the network is in place. The addressing follows a depth first approach where the first address is given to the leaf of the network and works upwards toward the root. The addressing and routing in such a mechanism is show as algorithm 1 & 2.

The addressing algorithm is akin to the C_{skip} in the sense of addressing in the depth first approach. However, C_{skip} assumes the worst case and would earmark addresses for non-existent nodes. This leads to a huge amount of address wastage and this precisely is avoided in the static algorithm. The monitoring applications have a well defined path and this is known in advance. Thus, the need to maintain addresses for future nodes does not arise. Further, the address allocation can be suitably tweaked to include an address provision for future nodes as the need may be. This can be achieved, by including a larger value for E_n.

Every router maintains an address list, which has the addresses of its child routers and child end devices. The routing is done based on this and is shown as Algorithm 2.

The address of a node denotes the maximum address of all devices under it. A router assigns address to end devices as node_address – n. Where $1 <= n <= E_n$. At any given point, the property of the node addresses is maintained and thus routing is achieved. For example, consider an end device joins the network at node 35. It is assigned an address of 34. The routing decisions would be at each node and checks under which child, can the destination address exist and forward accordingly as shown in Figure 3.

Algorithm 1. Modified hierarchical address allocation

```
Determine location of FFDs and their parent/children
input: En, set R of Routers and their children
Initialise node ← PAN coordinator, addr ← 0
For all children of node from left to right
If node has no children
assign node address ← addr
Increment addr by En + 1
End if
Else
assign node ← node's children
End if
End For
```

Algorithm 2. Routing in modified hierarchical topology

```
input: set C, children address of Router R, Destination address, D, of received packet at R
For all children of node R from left to right
If children address <= D
assign next hop ← children address
break from loop
End if
End For
Return next hop
```

Figure 3. Addressing and routing in the modified hierarchical tree

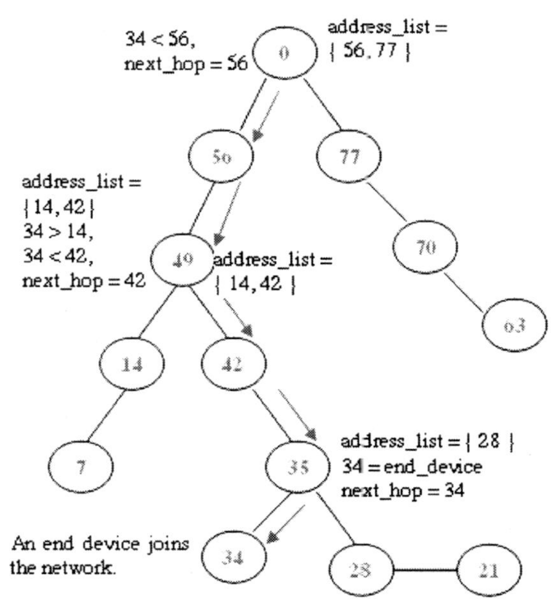

b. The Central Server

The data obtained from the wireless mesh network is primarily stored in different staging tables in a server database and subsequently processed and analyzed by the software to get a real-time view and status of the monitoring zone. The data may be classified into two categories:

- Sensor data from sensors
- Periodic health beacons from Routers indicating their presence and battery status

The software system consists of the following modules:

- **The Middleware:** It accepts the data through the serial port of the server and filters them for duplicates and is stored in an acceptable format in different tables depending on the type of data. It also accepts the control messages from the software system (central server or Internet), translates it into a proper format and is dispatched to the Gateway for the intended recipient.
- **System configurator:** is used to map the mesh network components into the software system. The Mesh network is essentially a sensing zone with routers and actuators placed at different strategic locations.
- **Network manager:** Allows the user to view status of the devices in the network (alive, dead, battery status, etc.)
- **Sensor monitor:** Allows the user to monitor the status of different sensor nodes in the sensing zone.
- **Response handler:** It enables the user to send context-sensitive control message/ commands to any device (actuators) in the network in response to some critical event.
- **Report generator:** Allows the user to generate customized reports based on the available sensor data.
- **Query manager:** This is a query-driven interface that allows the user to send a query to the mesh network asking for information about the status of sensor node(s).
- **Message handler:** Allows the user to send control message to the network for actuation or sleep control.

c. Accessing through Web-Enabled Devices

A Web-based Monitoring system offers the flexibility to monitor the environment through the Internet, thus providing "anytime anywhere" visibility of the sensor data. Remote users can monitor through the control station and automated control or manual decisions can be taken. Other features like, Sensor monitoring, Response handler, Report generator, Query manager and Message handler are also available in the web-based version of the software. The components of the central software are shown in Figure 4.

WIRELESS SENSOR NETWORK TESTBED

In this section we weigh the different technology options for a pollution monitoring testbed and provide a detailed account of the different aspects on the off the shelf sensors and the results of the implemented testbed.

Exploring different technology options for wireless sensor network

We have explored the different technology options for wireless sensor network like 802.11 and 802.15.4. It is identified that for real-time monitoring of pollution level in the environment, each sensor node needs to send its data (very low volume) at regular intervals to the central control station. Though it is possible to achieve the goal using 802.11 based technology, it is generally power-hungry and not suitable to operate in outdoor environments for a long period of

Figure 4. The system architecture

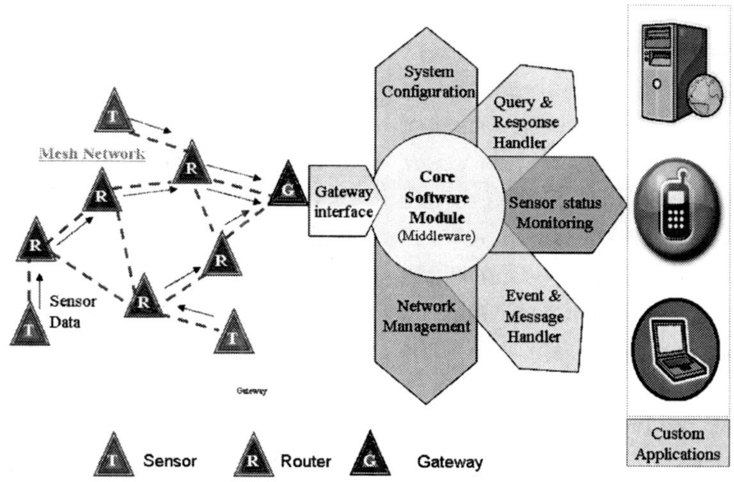

time. Moreover, our application does not require the high data rate provided by 802.11. Thus, we have decided to explore low-power, low data rate 802.15.4 based ZigBee and non-ZigBee platform.

Each wireless sensor node in our system would be an IEEE 802.15.4 based RF module having interfaces to connect different types of pollution sensors. Since sensors and RF modules will be mostly battery-powered, in order to extend the battery life of those sensors as well as RF devices, we are exploring different sleep mechanisms. The devices can be made active (from the sleep mode) either periodically or on user-defined conditions (such as when the sensed data is beyond acceptable range). We have primarily selected products from Maxstream for this development. XBee and XBee-Pro OEM RF Modules from Maxstream are Zigbee/IEEE 802.15.4 compliant solutions that satisfy the unique needs of low-cost, low-power wireless sensor networks. The modules are easy-to-use, require minimal power and provide reliable delivery of critical data between devices. XBee-Pro yields 1.5 km LoS (line of sight) data communication range whereas XBee offers 100 m LoS data communication range. Our target is to implement a mesh network with XBee and XBee-Pro (Figure 5) Modules and interfacing them with Sensors and Actuators.

Pollution Sensors

The Wireless Sensor network based air pollution monitoring is all about monitoring the amount of pollutants in air wirelessly. We identify the pollutants normally present in air to be CO (carbon monoxide), CO_2 (carbon di-oxide), CH_4 (methane)

Figure 5. XBee/ XBee-Pro Zigbee OEM wireless modules

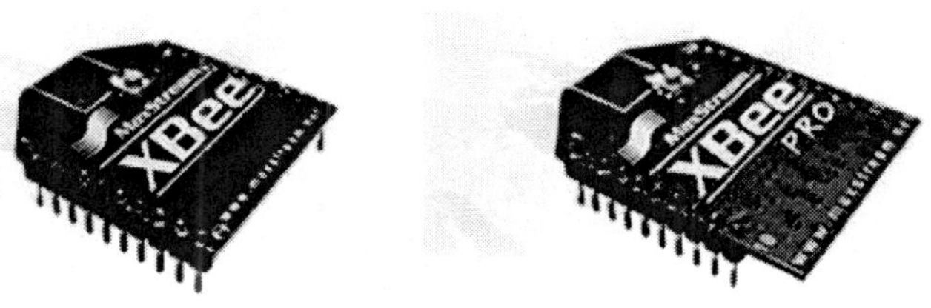

and volatile organic compounds (VOC). According to our proposed system, wireless devices with sensors attached will be put in different locations covering the entire area to be monitored. Sensor data will be captured by the respective sensors and will be sent wirelessly to the control station. If the control station is not within the radio range of the wireless sensor device, sufficient number of routers needs to be placed in between the sensor devices and the control station to transmit the data in multi-hop. On-line monitoring software at the control station will show the status of sensors in different areas so that the criticality in environment can be detected and necessary actions may be taken. The semiconductor sensors used are shown in Figure 6 and their properties in Table 1.

Wireless Sensor Network Testbed

We have used 20 wireless sensor nodes, spread across four different buildings within our institute campus. A snapshot of testbed scenario is shown in Figure 7. Two buildings are approximately 500 meters apart. Therefore, we have used multiple routers to collect sensor data from the buildings and send them to the control station which is approximately one km away from the deployment field. Each individual wireless sensor node can accommodate maximum of four different types of pollution detection sensors. For example, sensor "5678" can sense three parameters,: temperature, humidity and air contaminants, sensor "5679" consists VOC and solvent vapor sensors, "5681" measures CO and "5685" measures CH_4. We have also connected additional actuators like buzzers with some nodes in each location for generating alerts at the location in case of critical events. Our testbed is not only a typical wireless sensor network, but also serves as a wireless actuator network supporting two-way communication between the deployment and control station.

Each sensor node sends sensor data at a 10 second interval with its sleep cycle at 10 seconds. Once active (i.e., after completing a sleep cycle), the device waits for 3 seconds to warm up the heater coil (the sensors we used relies on heater coil to sense parameters) to enable the sensor reading to stabilize and then sends the observed sensor data over the network to the remote control station. Since the sensors draw significant power for each reading, therefore, even if proper sleep management technique is employed, node power drains out quickly. Therefore, for this experimentation, we have used mains to power the sensors, while the communication node is battery powered. To test the robustness of the IEEE 802.15.4 framework, the system was checked for the packet arrival variance for different sleep times of the sensors and is depicted in Figure 8.

Figure 6. Pollution sensors

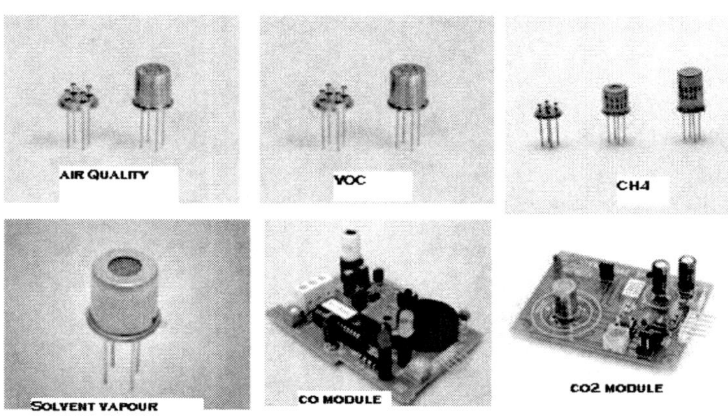

Figure 7. A snapshot of Testbed scenario

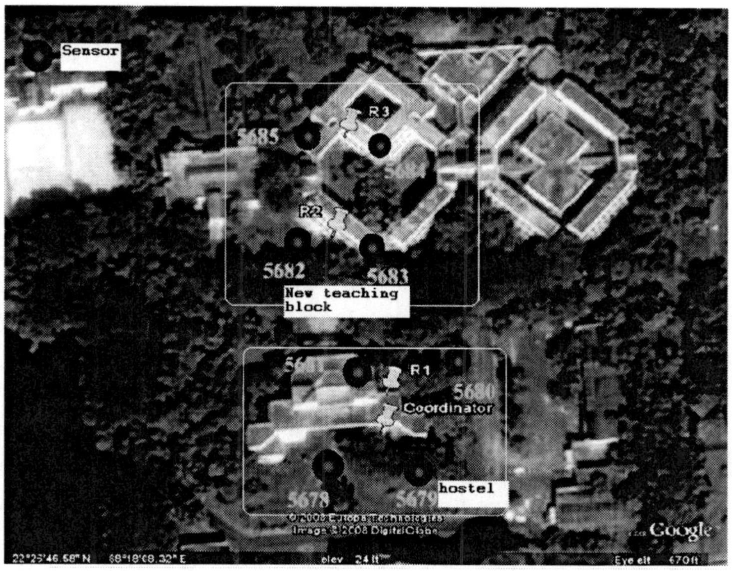

Figure 8. Variance in packet arrival rates

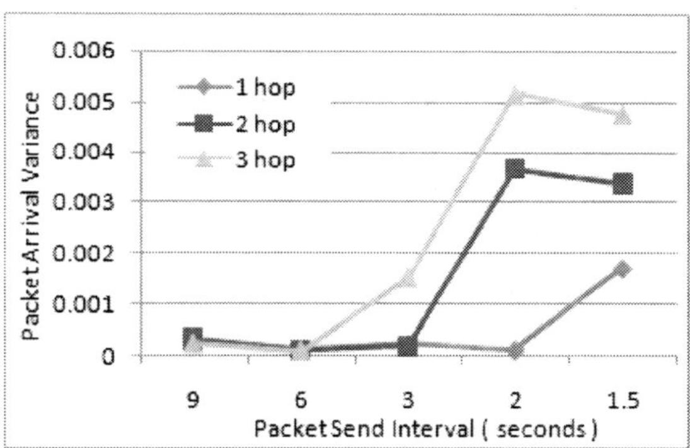

The results show stable data communication performance for multihop routing when the packet send interval and thus the sleep time is 10 seconds. However, the requirement to wait for stable sensor data means that the life time of the battery operated nodes is affected. The actual measured power consumption is shown in Table 2.

Accounting for the sleep duration of 10 seconds, an active duration of 3 seconds and Tx Current equals the Rx current, the average current consumption works to (55 *3/13 + 2*10^-3*10/13) = 12.66mA. With standard AA batteries of 1400 mAHr, the life of a node is 110 hours.

Our testbed enables the user to change the sleep time of the sensors from the central server depending on the criticality of the situation. It can be achieved by sending a command from the central control station to the specific sensor in the format <<sensor node-id>> <<sensor type>> <<new sleep time>>. Since the target sensor may be in the sleep mode while the command is issued, therefore, we have devised a strategy to hold the command at the nearest router in the network. Since the network is static, the nearest router for a sensor node is also known to the system. Each sensor node after the sleep cycle will broadcast an "awake beacon" and a router, on receiving it, will send the pending "change sleep time" information accordingly. If a sensor shows a reading beyond a preset threshold the software will wait for a predefined time to evaluate whether the criticality is real or due to a spurious reading. If real criticality is detected then a command will be automatically generated by the system to sound the buzzer at that location.

Figure 9 shows the web based graphical user interface to monitor pollution data "anytime anywhere" using any web-enabled machine. In the top panel of the left-most pane of the GUI list of all wireless sensors nodes deployed in the field are shown. Middle panel shows the current sensor reading for each sensor associated with a single wireless sensor node. Bottom panel shows the continuous graphical representation of the trend and variation of readings for a selected sensor type. Map-based view panel shows the location and current readings of all the deployed sensors on the map of the deployment site. If the reading of a sensor crosses a preset threshold, the reading will be shown in red for quick identification of critical conditions and the location of the event so that proper control action can be taken. If for any reason (i.e., due to any fault or if it runs out of battery), a sensor node stops sending data, that can be detected at once from the map-based view, where that particular sensor will be shown in grey implying that it is deactivated.

CONCLUSION AND FUTURE WORK

In this article, we have shown a generic architecture to achieve the "Internet of Things" paradigm and made a practical implementation using IEEE 802.15.4 based devices. We have provided detailed explanation of the various stages of the network setup and the aspects of selecting and configuring sensors for pollution monitoring. Contributions of our work include the generic architecture, the modified routing algorithm suited for the sensing system and the identification of the practical issues in the configuration and set-up of sensors for the mesh network. Specifically, research has to be directed at achieving low cost, low latency and low power sensors to achieve a truly usable pollution monitoring system. Till date, a large portion of the research has been directed at communicational aspects of the sensor network with little enhancements made to the actual sensor technology. The results confirm this proposition, where stable communication from the IEEE 802.15.4 based devices was achieved while the average lifetime deteriorated due to the latency requirement of the sensors.

There are a few aspects we would like to address in the future. The framework developed

Figure 9. The GUI for the pollution monitoring system

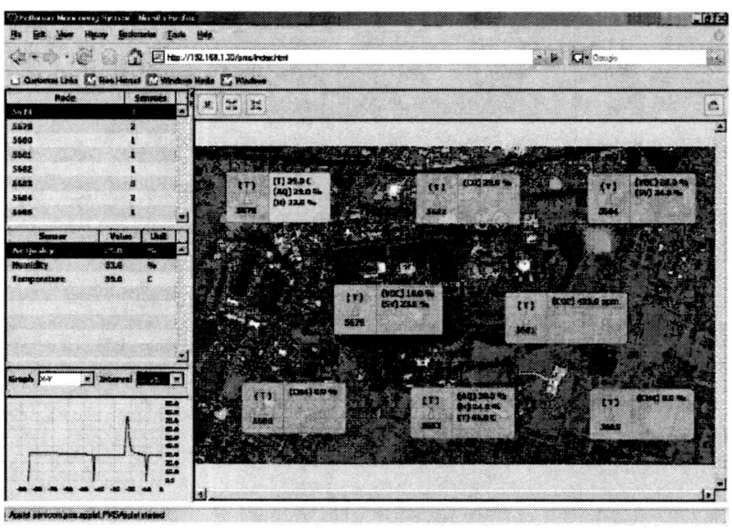

and implemented in this article, gives the power of abstracting the underlying architecture at the gateway. For example, the sensor network can be an Ethernet based IP system or a host of WiFi Hotspots. The underlying structure and nature of the network does not affect the architectural framework. However, this abstraction requires a formalization of the services that must be offered at the sensor network and gateway interface. This set of services can only be determined when the entire end to end service requirements are ascertained. We, thus, would like to develop a structured approach and framework for identifying and designing the services at the interfaces.

REFERENCES

Akyildiz, I. F., & Wang, X. (2005). A survey on wireless mesh networks. *IEEE Communications Magazine*, *43*(9), S23–S30. doi:10.1109/MCOM.2005.1509968

Al-Karaki, J. N., & Ahmed, K. E. (2004). Routing techniques in wireless sensor networks: A survey. *IEEE Wireless Communications*, *11*(6), 6–28. doi:10.1109/MWC.2004.1368893

Cuomo, F., Luna, S. D., Monaco, U., & Melodia, T. (2007, June 24-28). Routing in ZigBee: Benefits from exploiting the IEEE 802.15.4 association tree. In *Proceedings of the IEEE International Conference on Communications 2007*, Glasgow, Scotland (pp. 3271-3276).

Hammons, M. J., & Chisholm, G. (2006). Enabling total asset visibility. *Defense Transportation Journal*, *62*(4).

Ho-In, J., & Yeonsoo, K. (2007, February 12-14). Efficient, real-time short address allocations for USN devices using LAA (last address assigned) algorithm. In *Proceedings of the 9th International Conference on Advanced Communication Theory*, Gangwon-Do, South Korea (pp. 689-693). IEEE.

IEEE. (2003). *Wireless medium access control (MAC) and physical layer (PHY) specifications for low rate wireless personal area networks*. Washington, DC: Author.

International Telecommunications Union. (2005). *The Internet of things*. Retrieved February 17, 2009, from http://www.itu.int/internetofthings

Jung, Y. J., Lee, Y. K., Lee, D. G., Ryu, K. H., & Nittel, S. (2008, July 7-11). Air pollution monitoring system based on geosensor network. In *Proceedings of the IEEE International Geoscience & Remote Sensing Symposium*, Boston (pp. III-1370-III-1373). IEEE.

Ma, Y., Richards, M., Ghanem, M., Guo, Y., & Hassard, J. (2008). Air pollution monitoring and mining based on sensor grid in london. *Sensors Journal: Special Issue on Urban Environmental Monitoring*, *8*, 3601–3623.

Peng, R., Mao-heng, S., & You-min, Z. (2006, December). ZigBee routing selection strategy based on data services and energy-balanced ZigBee routing. In *Proceedings of the 2006 IEEE Asia-Pacific conference on Services Computing*, Guangzhou, China (pp. 400-404). IEEE.

Perkins, C., & Royer, E. (1999, February 25-26). Ad-hoc on demand distance vector routing. In *Proceedings of the 2nd IEEE Workshop on Mobile Computing Systems and Applications*, New Orleans, LA (pp. 90-100). IEEE.

Pujolle, G. (2006, October 3-6). An autonomic-oriented architecture for the Internet of Things. In *Proceedings of the IEEE John Vincent Atanasoff 2006 International Symposium on Modern Computing*, Sofia, Bulgaria (pp. 163-168). IEEE.

Schmid, T., Dubois-Ferrière, H., & Vetterli, M. (2005, June 20-21). *SensorScope: Experiences with a wireless building monitoring sensor network*. Paper presented at the 2005 Workshop on Real-World Wireless Sensor Networks (REAL-WSN 2005), Stockholm, Sweden.

Werner-Aller, G., Swieskowski, P., & Welsh, M. (2005, April). MoteLab: A wireless sensor network testbed. In *Proceedings of the 4th International Conference on Information Processing in Sensor Networks (IPSN)*, Los Angeles (pp. 483-488). IEEE.

Wheeler, A. (2007). Commercial applications of wireless sensor networks using ZigBee. *IEEE Communications Magazine*, *45*(4), 70–77. doi:10.1109/MCOM.2007.343615

ZigBee Alliance. (2004). Retrieved February 17, 2009, from http://www.zigbee.org

This work was previously published in International Journal of Business Data Communications and Networking Volume 5, Issue 4, edited by V. Sridhar and D. Saha, pp. 16-32, copyright 2009 by IGI Publishing (an imprint of IGI Global).

Chapter 17
A GPS Based Deterministic Channel Allocation for Cellular Network in Mobile Computing

Lutfi Mohammed Omer Khanbary
Aden University, Yemen

Deo Prakash Vidyarthi
Jawaharlal Nehru University, New Delhi, India

ABSTRACT

The scarcity of the radio channel is the main bottleneck toward maintaining the quality of service (QoS) in a mobile cellular network. As channel allocation schemes become more complex and computationally demanding, alternative computational models that include knowledge-based algorithms and provide the means for faster processing are becoming a topic of research interest. An efficient deterministic technique, capable of handling channel allocation problems, is introduced as an alternative. The proposed model utilizes the Global Positioning System (GPS) data for tracing the hosts' likely movements within and across the cells and allocates the channels to the mobile devices accordingly. The allocation of the channels to the mobile hosts is deterministic in the sense that the decision of the channel allocation is based on the realistic data received from the GPS about the hosts' movements. The performance of the proposed technique has been evaluated by conducting the simulation experiments for the two parameters—call blocking and handoff failures. Also, a comparison of the proposed model with an earlier model has been carried out to estimate the effectiveness of the proposed technique. Experimental results reveal that the proposed technique performs better and is more realistic as well.

DOI: 10.4018/978-1-60960-589-6.ch017

INTRODUCTION

With the emergence of the miniature portable computing devices and efficient wireless technology, mobile computing discipline came into the existence. The technical challenges that the mobile computing discipline must surmount to achieve the potential are hardly trivial. The issues stem from three essential properties of mobile computing; communication, mobility, and portability. The communication in wireless network is done by modulating radio waves or pulsing infrared light. Wireless networks are linked to the wired network infrastructure through stationary transceivers. The area covered by an individual transceiver's signal is known as a *cell*. Cell radii vary from tens of meters in buildings to hundreds of meters in cities and tens of kilometers in the countryside. Cellular systems use small cells due to frequency reuse, less transmission power, and thin interference (Maha Patra, Roy, Banerjee, & Vidyarthi, 2006).

Each cell has a base station (*BS*) responsible for communication with mobile users in that cell. All the base stations are connected with a wired network under the control of a mobile switching center (*MSC*). MSC is the interface between wireless communication network and the other wireless or wired communication networks (Hac & Chen, 1999).

Radio spectrum, used in cellular network, is a scarce resource and thus the efficient channel allocation is important to the reliability and performance of the system. In order to divide a given radio spectrum into radio channels, many techniques are popular. Three most widely used techniques are frequency division multiple access (*FDMA*), time division multiple access (*TDMA*), and code division multiple access (*CDMA*). Further, a channel cannot be used by the two cells concurrently if the geographical distance between them is less than a threshold distance, called minimum channel reuse distance (D_{min}) (Yang, Jiang, Manivannan, & Singhal, 2005). It is because they may interfere with each other resulting in *co-channel interference* (Baiocchi, Priscoli, Grilli, & Sestini, 1995; Prakash, Shivaratri, & Singhal, 1995; Prakash, Shivaratri, & Singhal, 1999).

Handoffs technique is significant in any cellular system because it is undesirable for a mobile host to break its call while moving from one cell to another. It is more so in mobile computing as it may sometimes result in starting the computation from scratch. Preferable is to block a new initiated call than to break an existing one. Several techniques are used to implement handoffs in ways that cope well with the traffic variation while maintaining a high level of utilization. Handoff is an important issue particularly in microcellular systems where the cell radius is small (Lauro & Aghvami, 2003; Murthy & Manoj, 2004). Handoff is quite a common phenomenon that arises due to mobility (Maha Patra et al., 2006; Schiller, 2003) and is to be taken care of with utmost priority and great sincerity as any handoff drop may wash off the whole computation in mobile computing.

In addition to the limited number of radio channels, radio interference is another constraint in communications within cellular networks. It may significantly deteriorate the quality of transmission if the channel allocation algorithm does not take this into consideration. Extensive research has been done in the area of channel allocation in wireless networks with these issues (see Boukerche, Tingxue, & Kauther, 2005; Khanbary & Vidyarthi, 2008; Tipper & Teresa, 2002; Yang et al., 2005; Zhang & Yum, 1991).

The objective of channel allocation is to assign a required number of channels to each cell so that the need of the cells toward the mobile devices are satisfied and at the same time efficient frequency spectrum utilization is provided with minimum interference effects. Channel assignment (allocation) is therefore an important activity of resource management and its efficient implementation increases the fidelity, capacity, and the quality of service in cellular systems. The search for better channel allocation methods continues to be one of the major challenges in wireless communication network design (Hac & Chen, 1999).

The Global Positioning System (GPS) is a satellite-based precise positioning system available worldwide, where a constellation of satellites continuously transmits coded information that makes it possible to precisely identify locations of an object on earth by measuring its distance from the satellites. GPS has become one of the most popular positioning/tracking systems for its precision, continuous measurement, and global availability. GPS has revolutionized the field of navigation and made precise navigation available to a wide array of users in recent years. The basic function of the GPS system is to provide three dimensional coordinates of a GPS antenna. Since GPS communicates via radio waves which travel at a known speed (the speed of light), it is also a means of determining a user's position in time. From any position on the surface of the earth, there are always sufficient satellites in view to provide a timing and location service (Kwon, Kim, & Dong, 2001).

The present work proposes a channel allocation scheme based on GPS-derived information on the mobile devices' movement. It assures to allocate a channel whenever a mobile device is intended to initiate a call or continues a call while roaming across the cells.

The rest of the paper is organized as follows. The next section is a short discussion on the GPS. The proposed GPS based channel allocation model is then described. The performance of the proposed model is also evaluated and compared with the FTCA model of Khanbary and Vidyarthi (2008). Observations, out of the experiment on the model, and the comparative study with the concluding remarks appear last.

ROLL OF GPS IN CHANNEL ALLOCATION

GPS provides specially coded satellite signals that can be processed in a GPS receiver, enabling the receiver to compute position in three dimensions (*X*, *Y*, and *Z*) and time (*T*). There are 24 satellites in the GPS constellation. GPS satellites are constantly moving. The satellites are arranged in six orbital planes and take approximately 12 hours to orbit the earth (Moore & Crossley, 1999). Measuring the travel time of microwave electromagnetic waves from a number of GPS satellites, all at known locations, a trilateration technique is used to estimate the devices position to the few meter level. By augmenting this procedure with data from other GPS receivers or GPS correction providers, position estimates as precise as a few millimeters can be achieved under constrained conditions (Bisnath, Wells, Howden, & Stone, 2003; Zheng, Wang, & Nihan, 2005).

Due to the market requirements for location based services, the mobile wireless communication and GPS positioning have been forced to merge. Assisted GPS integrates mobile phone and GPS receiver to ensure that it works normally in certain bad environments by aiding information provided via wireless mobile phone. Users can benefit more from assisted GPS than from conventional GPS (Wang & Green, 2000). The location of mobile users can enhance a wide range of mobile services and location-based navigation. These services are forecasted to substantially boost the market for value added mobile services. The combination of wireless communication and GPS positioning will provide opportunities for service innovation and revenue generation. More and more people will benefit from it, both in driving and walking.

Mobile location has played a vital role in wireless communication from its inception. Cellular radio systems have an inherent granular positioning capability by virtue of the fact that a mobile typically connects to the closest serving site. Recently, mobile location has become a hot topic for the purpose of satisfying the location requirement and for driving Location Information Applications in many countries. With the introduction of new positioning technologies and services, this capability will eventually become razor sharp for new wireless location services and wireless exploring (Wang & Green, 2000).

GPS is a key utility in much of the world's infrastructure. It is being used not only as a primary resource for accurate position and velocity information of an object, but also as an important precise timing. GPS is being employed to help its users perform tasks faster, cheaper, and more accurately than before. Currently, the precision of the GPS positioning is dropped to sub centimeter for the static station and sub decimeter for the kinematics object. This accuracy is achieved by the Differential GPS (DGPS) technique. In DGPS, common system errors on the satellite and receiver are cancelled or reduced by making difference observables of the code pseudo range and phase signals from GPS. Thus the performance of the basic GPS system has been augmented by the Differential GPS technique for military as well as civilian uses (Gupta, Low, & Tan, 1998).

Future mobile devices will increasingly have multiple sources of location information associated with them, such as GPS, Blue tooth or 802.11 wireless LAN. In fact, cellular phones with GPS receivers and 802.11 wireless LAN are already getting available (Paiidya, Jain, & Lupu, 2003). New generations of GPS receivers are small, light, cheap, fairly accurate and easy to use. GPS has great potential for many applications (Zheng et al., 2005).

GPS has been efficiently used by the researchers to improve the overall performance of the applied areas. Researchers have developed several location estimation algorithms to improve the precision of the GPS and hence reduce the location errors (Paiidya, Jain, & Lupu, 2003). In the proposed model GPS has been used for channel allocation problem.

THE PROPOSED MODEL

The present work uses GPS information to track movement of mobile devices across the cells for the purpose of channel allocation. Global Positioning System (GPS) can be very useful for tracking the mobile devices roaming across the cells. It has been used for locating the position of the mobile users in case of the emergency calls (Wang & Green, 2000) and could further be explored for allocating the channels in a particular cell.

The GPS provides support to our model for the mobile host likely movement within the cells as well as for the crossing of the mobile hosts across the cells by providing and updating online information. Thus, it is possible to assign the channels to the mobile hosts inside the cells and to the other cells that a group of hosts are likely to move. With the help of the GPS's online data, the cells will be provided a priori information about the required number of channels to deal with the blocked hosts and handoff problems. This will avoid the blocking of the new initiated calls and the disconnections of the ongoing (handoff) calls reducing the blocked calls and the handoff failures in the system. Hence, the efficiency of the whole system will improve. The model also uses the reserved channels technique and the channels reuse technique for handling the handoff calls (Khanbary & Vidyarthi, 2008).

The proposed model, with the support of the GPS, minimizes the number of blocked hosts and the handoff failures (as a part of the blocked hosts) in the mobile computing network system. Model is deterministic in the sense that the allocation is based on the actual movement of the mobile hosts, information for which is obtained from the GPS.

ASSUMPTIONS

Certain assumptions, made in the model, are listed below.

- As in most of the cellular models, cells are hexagonal of equal sizes.
- Base stations are at the center of the cells.
- The mobile users have GPS-equipped devices (i.e., all the mobile devices have built-in GPS receivers).

- DGPS is employed for better precision.
- Errors due to locating the actual position of the mobile hosts in the close areas are negligible.
- The data about positions of the hosts inside the cells is observed and recorded through the GPS.
- The data regarding the host's movements within the cells is observed and recorded by the GPS.
- The recorded GPS data is provided and updated online to the base stations to utilize it.

THE MODEL

In the model, the total available channels is divided into two subsets; the first subset (*primary channels*) is distributed within the cells of the network for handling the new calls inside the cells, and the second subset (*reserved channels*) is kept for handling the handoff mobiles crossing the cells. The initial distribution of the channels (*primary* and *reserved*), within the cells, is according to the initial demand of the cells obtained by the GPS. Further, the channels are assigned dynamically to the cells on demand according to the GPS data on the hosts' movements. All neighboring cells have equal probability of being the mobile user's destination cell. The constraint for co-channel interference is taken into consideration (see co-channel interference section).

The model works under resource planning scheme (Prakash et al., 1999). Borrowing of the channels between the cells is possible only from the first subset, that is, from the set of primary channels. The borrowed channel must be returned to the owner cell as soon as the call is over.

To handle the handoff, the policy adopted is as follows. If there are free channels in a cell (no blocked hosts), these channels are utilized to handle the handoff hosts moving to that cell. If the free channels are less than the number of handoff hosts then the remaining required channels are provided from the reserved pool. If enough channels are not there in the reserved pool to handle these crossing over hosts then a request is sent to the neighboring cells to borrow the channels provided it results in no co-channel interference. If the entire scheme fails, handoff failure occurs in the cell. The handoff channels are supported to the hosts on the first come first serve (*FCFS*) basis.

A new call, in any cell, is supported by the channel from the primary set of channels of that cell. If no free channels are available then the cell checks the neighboring cells for the free channels provided it results in no co-channel interference. Otherwise the call fails (blocked).

Co-Channel Interference Constraint

A cell can lend the same channel to any of its neighbors, for using it concurrently, provided the two borrowing cells C_i and C_j satisfy.

- $C_i - C_j \neq 1$.
- $C_i - C_j \neq$ number of cells in a row.
- $C_i - C_j \neq$ number of neighbor cells in cell pattern.

It is assumed that cells have been numbered in increasing order of enumeration, as shown in Figure 1. The above step will insure that same channel is not being used by the two neighbors in which the co-channel interference may occur.

Figure 1. Celluar network

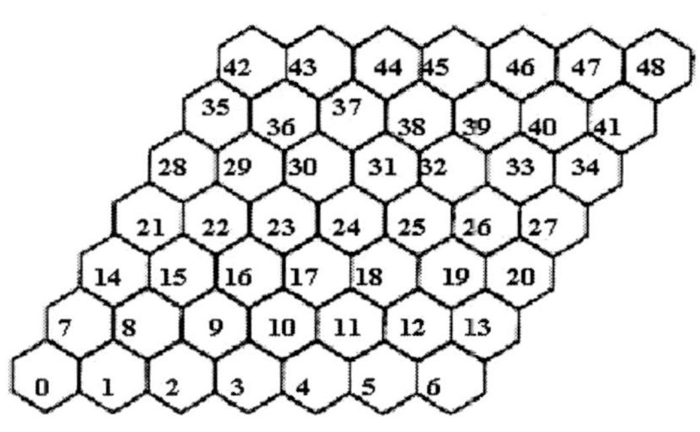

PERFORMANCE ANALYSIS

The efficiency of the model is improved by providing the host movement's data, using the GPS, a priori to the base station. The time by which the mobile host crosses a cell boundary is a critical factor and should be taken into consideration. With the help of the GPS, time is calculated and utilized to improve the performance of the model.

The following parameters are used to address the problem.

L is the distance between two neighboring base stations.

$L/2$ is the distance between a base station and its cell boundary.

d_i is the distance of a mobile host (MH_i) from its base station at any instance of time.

V is the velocity of the mobile host (movement) obtained and updated by the GPS.

L_i is the remaining distance of a mobile host MH_i to cross a cell boundary.

t_{ri} is the remaining time for a mobile host MH_i to cross a cell boundary.

For a mobile host MH_i in cell C_i, the remaining distance L_i to cross the boundary of the neighboring cell C_j, (figure 2) is obtained by

Hence for a mobile host MH_i, that belongs to the cell C_i and is moving with the velocity V_i, the remaining time t_{ri} to cross the boundary of a cell C_j is given by

For a group of mobile hosts MH_1, MH_2, MH_3, …, MH_n moving to the boundary of the cell C_j, (Figure 3) the remaining time t_{ri} for each mobile host to cross the boundary of C_j, at any time t_1, is

The average handoff time (T_H) for n mobile hosts to cross the boundary of the neighboring cell C_j is given by

The average handoff time (T_H) is a critical factor since the base station of the cell, to which the mobile hosts is tending, will search for the free channels. More so the neighboring cells should respond to the channel request within specific time and the response time (T_{RS}) should be less than the average handoff time, that is, $T_{RS} < T_H$.

The value T_H should be taken periodically at t_1, t_2, t_3, ….. This will provide enough time to search channels for handoff hosts before the mobile hosts crosses the cell boundary and thus reducing the handoff failures. If a cell needs a channel for a new call (not handoff) then T_{RS} is only a threshold timer T_{tSh} that has a certain value.

Figure 2. Hosts movements within the cells

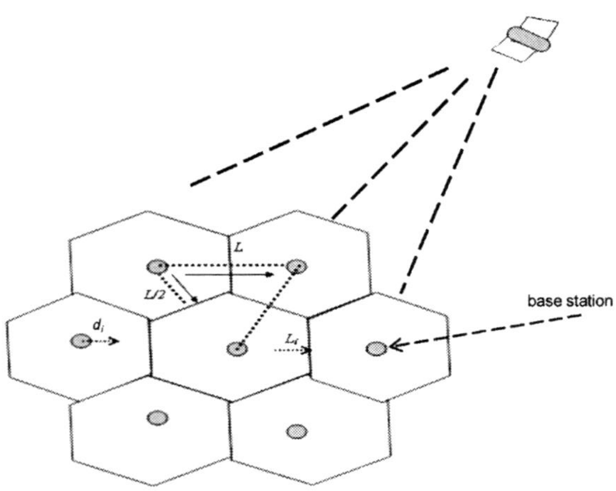

Figure 3. Mobile hosts crossing the boundary of cell C_j

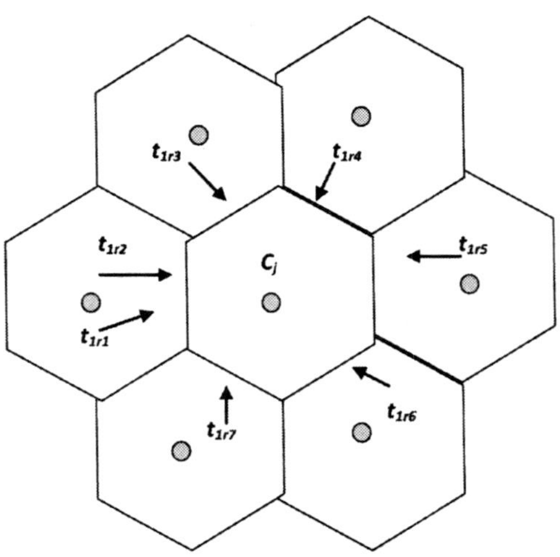

The Algorithm

This section proposes the channel allocation algorithm based on the above discussed model to optimize the number of channels in the cellular network. The algorithm is as follows.

Initialization Step: *blocked_call =0, handoff_failure =0.*

Input Step: *total number of channels and mobile hosts.*

Preprocessing Step: *Divide the total number of channels into primary and reserved set.*

Assign the channels (primary and reserved) to each cell according to the initial demand. Also, distribute the hosts among the cells in proportion to each cell capacity.

Processing Step: *When a cell, C_r, needs a channel to support a call then // two possible // scenarios available*

```
A.  If the call is a new call then
          if
then pick a channel  prim_i>0  //
prim is the primary channels

        else  if  prim_i=prim_i-1  then
while r∈prim_i search and borrow a
prim from neighbor cell C_j

          if    prim_i=0 then pick a
channel T_RS<T_tsh     //  provided co-
channel    // interference constraint
satisfies
            borrow_i>0

              if    the call ends re-
turn the borrowed channel
                 r∈borrow_i  // incre-
ment primary channels number in the
lender cell C_j

                if  borrow_i=borrow_i-1
then drop the call          //  call
blocked occurs
                    total_blocked_
call =blocked_call+1.

  B.  If the call is a handoff
call then          // based on
their a priori location information
// the number of crossing over hosts
is calculated //and queued

          if      prim_j=prim_j+1 then
pick a channel borrow_i=0 reserve_i>0

          if    r∈reserve_i then
                if
reserve_i=reserve_i-1 then   pick a
channel  reserve_i=0 prim_i>0

         if  r∈prim_i & prin_i=prim_i-
1then
              calculate reserve_i=0
// based on equation (3)

                   while prim_i=0 search and
borrow a prim from neighbor cell C_j

                        if  T_H  then
pick a channel  T_RS<T_H // pro-
vided co-channel         // in-
terference constraint satisfies
//  the required no. of channels  is
borrowed
                   borrow_i>0
                     if   the handoff
call ends or the host crossed another
cell boundary
                         return the
borrowed channel
          r∈borrow_i
```

```
// increment lender cell.

                    if  borrow_i=borrow_i-1
then drop the call  //  handoff fail-
ure occurs
                    total_handoff_fail-
ure = handoff_failure +1.
```

Output Step: *total_blocked_call, total_handoff_failure*.

The algorithm starts by initializing the number of blocked calls and handoff failures to zero. The search process for a prime channels from the neighboring cells will start only when $prime_i=0$, so the timer T_{tsh} is initialized. The search process for handoff channels will start only when both $prime_i=0$ and $reserve_i=0$, so the timer T_H is initialized. The channel requests from the handoff hosts are queued up as per the hosts' location information from the target cell based on the GPS data. If a channel becomes available in that cell, and its handoff queue is nonempty, the channel is used to satisfy the handoff request at the head of the queue. Only when the borrowed channel is zero ($borrow_i=0$), then the call is blocked or the handoff failure occurs. Rest of the algorithm is self explanatory. This algorithm is suitable for the channel allocation problem that requires the decisions to be made quickly in the real time.

THE EXPERIMENTS

The performance of the proposed algorithm is evaluated, in this section, by conducting the experiment. The simulated program for the experiment is written in C++. The model is compared with the FTCA model (Khanbary & Vidyarthi, 2008) for both the blocked hosts and the handoff failures.

Simulation is conducted under non-uniform traffic pattern of the hosts which is more realistic.

Simulation Parameters

- The simulated cellular network consists of 20 cells.
- The experiment is conducted for varying number of channels and hosts.
- The reserved channels, for all the experiments, are 30% to the total number of available channels and are distributed amongst the cells in proportion to the distribution of the mobile hosts. For example in the experiments with 200 channels the reserved channels are 60.
- The handoff probability is considered to be 30%, which is in conformity with that of the reserved channels.
- The collected borrowed handoff channels are given to the hosts in the FCFS manner.
- For experimental purpose, handoff mobile hosts are given various speeds and directions randomly.
- For borrowing a channel for a new call, the waiting time for a cell to get reply from the neighboring cells T_{tsh} is 8 milliseconds.
- For borrowing a channel for handoff call the waiting time is T_H.
- The experiments are conducted at various time instances t_1, t_2, t_3, and t_4 and the average is taken.

The input values for the channels and the hosts are:

- Number of channels: 50, 100, 150, 200, 250, 300.
- Number of hosts: 50, 100, 150, 200, 250, 300.

The results are represented in the performance graphs, where the x-axis represents the number of channels for all the experiments. The y-axes represent the blocked hosts for some experiments and the handoff failures for some other experiments.

Blocked Hosts Experiment

In this section we performed the experiment to calculate the number of blocked hosts at different times (time instances) and the average is taken. As the values are being randomly generated, it is more realistic to conduct the experiment at different time instances.

The graph in Figure 4 shows the number of blocked hosts at time instance t_1.

The graph in Figure 5 shows the number of blocked hosts at time instance t_2.

The graph in Figure 6 shows the number of blocked hosts at time instance t_3.

The graph in Figure 7 shows the number of blocked hosts at time instance t_4.

The *average* blocked hosts graph for t_1, t_2, t_3, and t_4 in Figures 4, 5, 6 and 7 is as in Figure 8.

Handoff Failures Experiment

In this section we performed the experiment to calculate the number of handoff failures at different times and the average is taken. The following graph in Figure 9 shows the number of blocked hosts at time instance t_1.

The graph in Figure 10 shows the number of handoff failures at time instance t_2.

The graph in Figure 11 shows the number of handoff failures at time instance t_3.

The graph in Figure 12 shows the number of handoff failures at time instance t_4.

The *average* handoff failures graph for t_1, t_2, t_3, and t_4 in Figures 9, 10, 11 and 12 as bellow in Figure 13.

Figure 4. Results for blocked hosts at time instance t_1

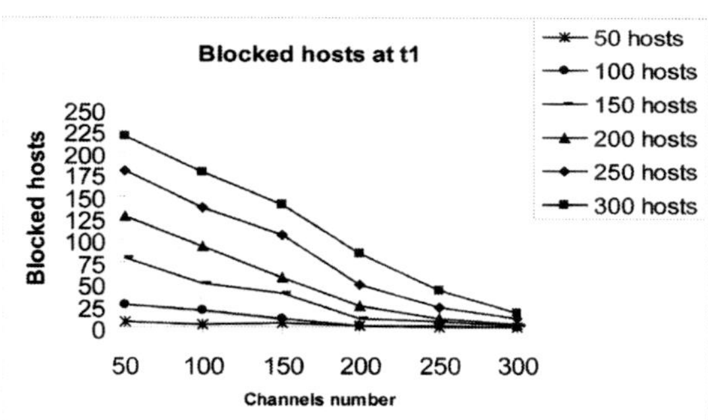

Figure 5. Results for blocked hosts at time instance t_2

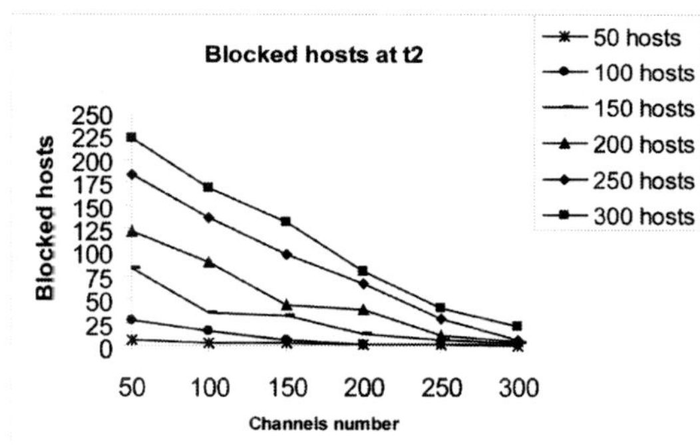

Figure 6. Results for blocked hosts at time instance t_3

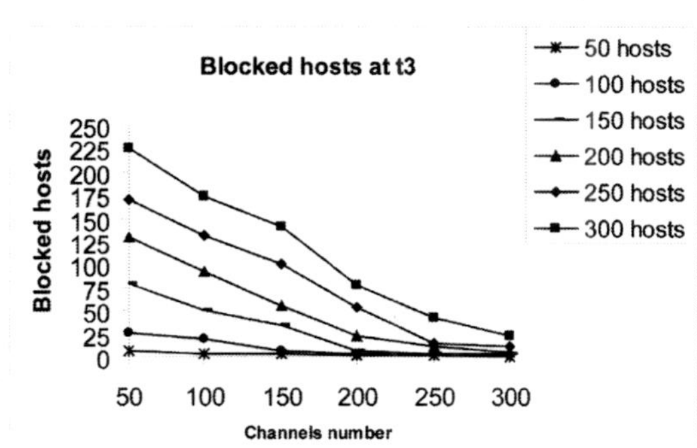

Figure 7. Results for blocked hosts at time instance t_4

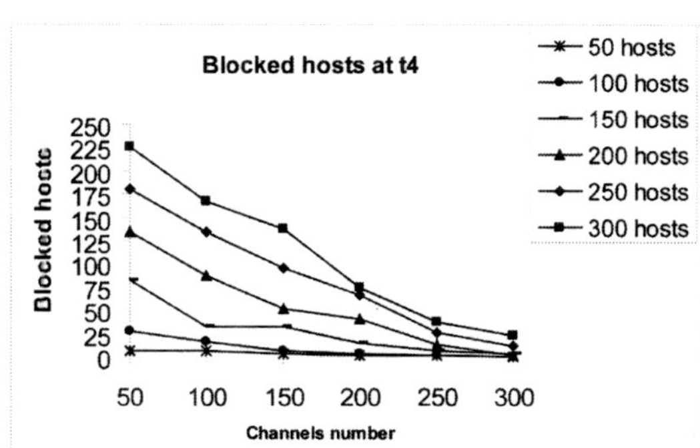

Figure 8. Average blocked hosts of t_1, t_2, t_3, *and* t_4

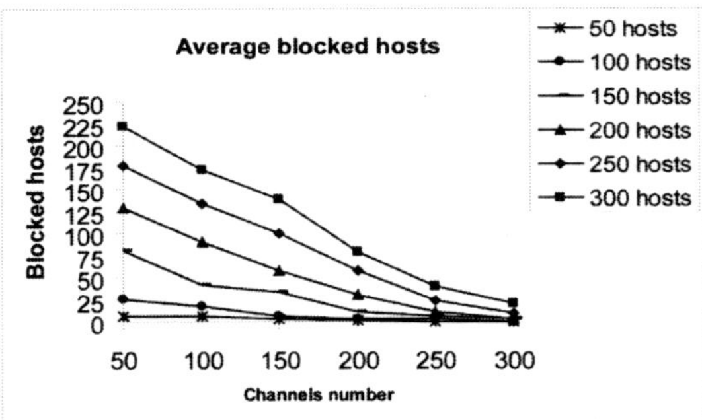

Figure 10. Results of handoff failures at time instance t_2

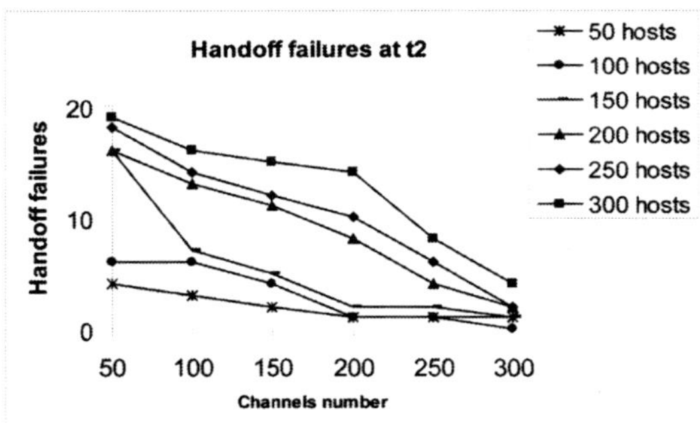

Figure 12. Results for handoff failures at time instance t_4

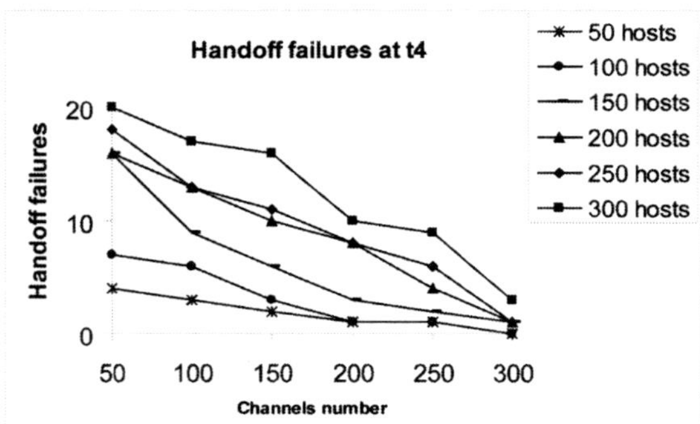

Comparative Study

To observe the efficacy of the proposed GPS model, it has been compared with the Fault-Tolerant Channel Allocation (FTCA) model in (Khanbary & Vidyarthi, 2008). The comparison is made for both the blocked hosts and the handoff failures. Most of the parameters and environment, in this experiment, are kept same to the FTCA model.

Figure 9. Results for handoff failures at time instances t_1

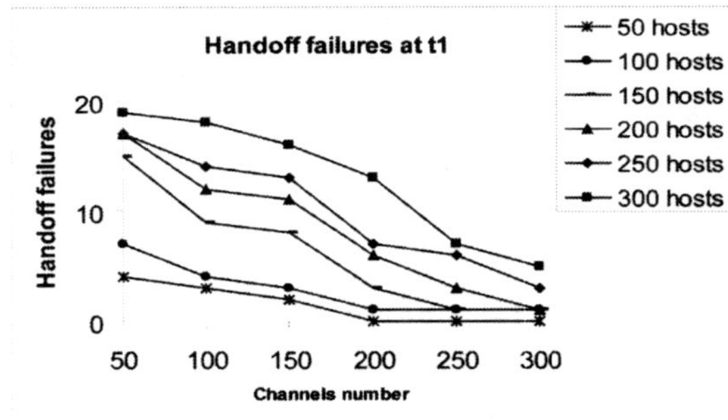

Figure 11. Results for handoff failures at time instance t_3

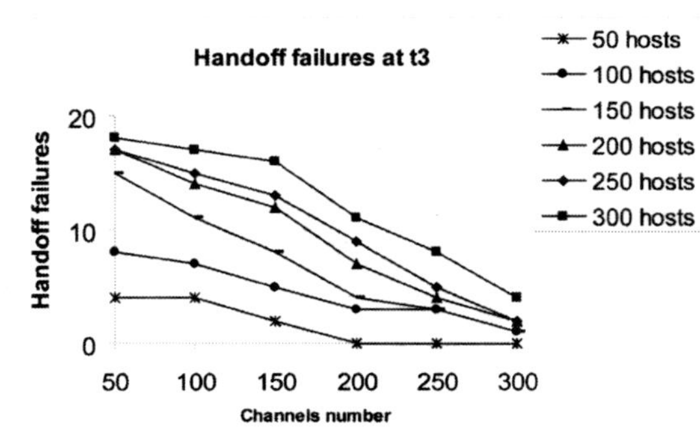

Figure 13. Average handoff failures of t_1, t_2, t_3, *and* t_4

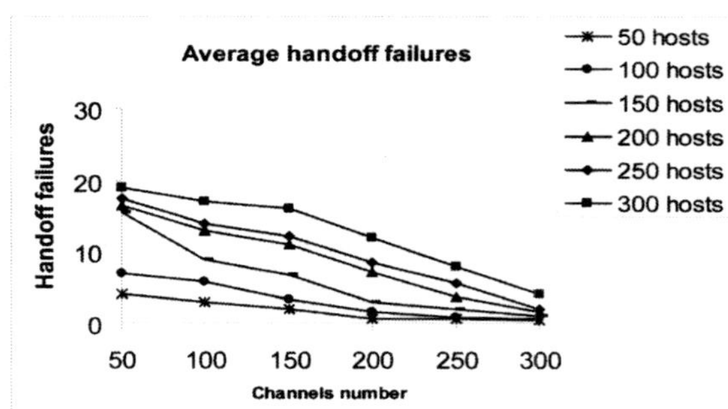

Comparison of the Blocked Hosts

In this section the results of the *blocked hosts* obtained by this model is compared with the results obtained by the FTCA model. The experiment is done up to 20 generation.

The following graph (Figure 14) shows the comparison with blocked hosts of FTCA in 20 generations.

Comparison of Handoff Failures

This section compares the results of the handoff failures of the current model with that of FTCA model. Again the experiment is done up to 20 generations.

The following graph (Figure 15) shows the comparison of the handoff failures of the two models.

OBSERVATIONS AND CONCLUSION

Based on the experiment conducted for the performance evaluation of the model and the comparative study, the observations are derived.

Observations of the Experiments

- From the blocked hosts and the handoff failure graphs, it is evident that by increasing the number of channels, there is a significant decrease in the number of blocked hosts and the handoff failures (Figures 4–13). The reason is obvious as more channels will facilitate more number of hosts.
- In many graphs points, the number of blocked hosts and the handoff failures reduce to zero (Figure 4–13). That is due to the advantage of using the GPS for utmost precision of the channel distribution amongst the cells.
- The average number of blocked hosts and the average handoff failures are less in comparison with the input values of hosts and channels.
- Both the reserved channel and the efficient channel reuse techniques help significantly to reduce the number of blocked hosts and handoff failures even when the number of input channels is less than the input hosts. It also keeps the initial values of the handoff failure very low.
- Although in some cases the number of channels is small compared to the number of hosts, but the model still resulted in less blocked hosts.
- The efficient usage of the GPS location information for distributing the channels to the base stations according to the hosts' likely movements greatly reduces

Figure 14. Comparison of blocked hosts for GPS model and FTCA model

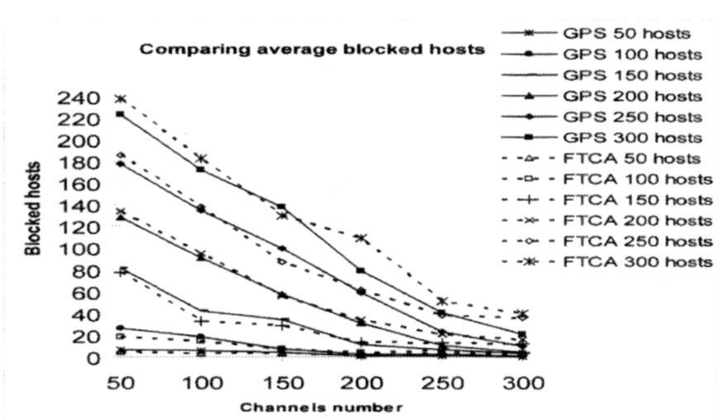

Figure 15. Comparison of handoff failures between GPS model and FTCA model

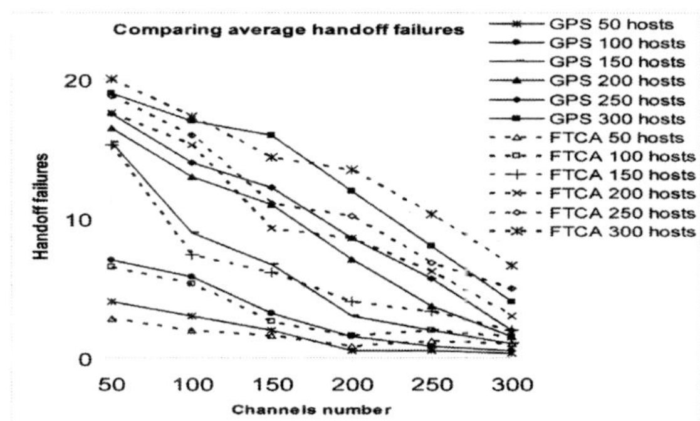

the blocked hosts and the handoff failures. This is as each base station in the cellular network borrows only the required number of channels (not more or less), from the neighboring cells, for handling the hosts' requests resulting in better channel utilization.

- Using a threshold time, calculated from the GPS data, to borrow the required number of channels for the handoff mobile hosts a priori, makes the system more realistic. The crossing over hosts need the handoff channels immediately and the threshold constraint will reduce the chance of acquiring extra number of channels to handle the handoff hosts due to the timeout set. The host crossing a cell boundary can not wait infinitely for a handoff channel and be provided a handoff channel immediately. Thus the use of timing for the channel requests ensures fairness. In the absence of the timeout it will be easier to borrow more channels from neighbors since the searching time is unlimited.

Observations on Comparative Study

By comparing the model with the FTCA model in (Khanbary & Vidyarthi, 2008), the following observations are conspicuous.

- The proposed model shows better results than the FTCA model for both the average blocked hosts and the average handoff failures (Figures 14 & 15).
- The performance of the GPS model is remarkable when the total number of input hosts is 200 and above. Similarly the GPS model shows much better results when the total number of input channels exceeds 150. Hence the model outperforms the FTCA model in resource management when the network size grows (i.e., hosts and channels increase).
- As the distribution of channels in FTCA model is random (Khanbary & Vidyarthi, 2008), while the distribution of channels in our model depends on the GPS, increasing the number of input channels in the GPS model helps in better and more precise management of the channels within the cells.
- The GPS model depends on time as a critical factor for searching and allocating the required number of channels for the new as well as the handoff calls which is not the case in the FTCA model. Thus GPS model is time constrained and it makes borrowing of channels more restricted and at the same time the GPS model is more realistic than the FTCA model.
- The FTCA model performs at least twenty iterations (generations) to obtain better results and in each generation GA operations such as selection, crossover and mutation are involved. Thus, the FTCA model incurs more overhead than the GPS model in which processing is done in just one step.
- The growing population size adds more complexity to the FTCA model in addition to the time to the convergence (Fernando, Goldberg, & Pelikan, 2000; Khanbary & Vidyarthi, 2008). The time taken by the GPS model is less than the FTCA model.
- In the GPS model, keeping the number of hosts constant and increasing the number of channels, there is always a fall in the blocking hosts and the handoff failures. In the FTCA model, there are many cases where the blocking hosts and the handoff failures increase even by increasing the number of input channels, for the same number of input hosts.

CONCLUSION

In this work a GPS based channels allocation model is proposed. The proposed work shows how efficiently the GPS system can be utilized for the effective channels usage in the mobile computing network.

It is observed that the efficient usage of the GPS for distributing the channels to the base stations according to the hosts' movement greatly reduces the number of blocked hosts and the handoff failures. To evaluate the performance of the proposed GPS model, we conducted experiments for observing the number of blocked hosts and handoff failures. It is found that the performance of the model is quite good in serving the channels to the hosts.

It is also shown, by the comparison study that the GPS model performs better than the FTCA model proposed in (Khanbary & Vidyarthi, 2008) for both the average blocked hosts and the average handoff failures. Using the GPS data effectively for calculating the required time for the borrowing channels, in both new initiated as well as the handoff calls, makes the GPS model more realistic than the FTCA model. The model is deterministic as it is based on the actual data rather than using speculation. Further, the time taken to furnish the result using GPS algorithm is better than that of FTCA as the latter is based on GA for which result converges in number of iterations.

The authors would like to accord their sincere thanks to the anonymous reviewers for their nice and valuable suggestions resulting in the quality improvement of the paper.

REFERENCES

Baiocchi, A., Priscoli, F., Grilli, F., & Sestini, F. (1995). The geometric dynamic Channel allocation as a practical strategy in mobile networks with bursty user mobility. *IEEE Transactions on Vehicular Technology*, *44*(1), 14–23. doi:10.1109/25.350265

Bisnath, S., Wells, D., Howden, S., & Stone, G. (2003, September 22-26). The use of a GPS-equipped buoy for water level determination. In *Proceedings of IEEE OCEANS 2003,* San Diego, CA (pp. 1241-1246). IEEE.

Boukerche, A., Tingxue, H., & Kaouther, A. (2005, September). Design and performance evaluation of a QoS-based dynamic channel allocation protocol for wireless and mobile networks. In *Proceedings of 13th IEEE / ACM International Symposium on Modeling, Analysis, and Simulation of Computer and Telecommunication Systems,* Atlanta, Georgia (pp. 445-452). IEEE.

Fernando, G. L., Goldberg, D. E., & Pelikan, M. (2000, July 8-12). Time complexity of genetic algorithms on exponentially scaled problems. *Proceedings of the Genetic and Evolutionary Computing Conference,* Las Vegas, NV (pp. 151-158). Morgan Kaufmann.

Gupta, S. K., Low, M. K., & Tan, C. W. (1998, April 20-23). A new approach to simulate GPS measurements. In *Proceedings of the 1998 IEEE Position Location and Navigation Symposium,* Palm Springs, CA (pp. 236-242). IEEE.

Hac, A., & Chen, Z. (1999, September 19-22). Hybrid channel allocation in wireless networks. In *Proceedings of the IEEE VTS 50th Vehicular Technology Conference*, Houston, TX (pp. 2329-2333). IEEE.

Khanbary, L. M. O., & Vidyarthi, D. P. (2008). A GA based fault-tolerant model for channel allocation in mobile computing. *IEEE Transactions on Vehicular Technology, 57*(3), 1823–1833. doi:10.1109/TVT.2007.907311

Kwon, J., Kim, W., & Dong, L. (2001, July 9-13). Absolute kinematic GPS positioning for remote area. In *Proceedings of the IEEE International Geoscience and Remote Sensing Symposium,* Sydney, Australia (pp. 2067-2069). IEEE.

Lauro, O., & Aghvami, A. H. (2003). *Resource allocation in hierarchical cellular systems.* Norwood, MA: Artech House.

Maha Patra, S. S., Roy, K., Banerjee, S., & Vidyarthi, D. P. (2006). Improved genetic algorithm for channel allocation with channel borrowing in mobile computing. *IEEE Transactions on Mobile Computing, 5*(7), 884–892. doi:10.1109/TMC.2006.99

Moore, P., & Crossley, P. (1999). GPS applications in power systems, part 1. *IEEE Power Engineering Journal, 13*(1), 33–39. doi:10.1049/pe:19990110

Murthy, C. S. R., & Manoj, B. S. (2004). *Ad Hoc wireless networks architectures and protocols.* Indianapolis, IN: Pearson Education.

Paiidya, D., Jain, R., & Lupu, E. (2003, September 7-10). Indoor location estimation using multiple wireless technologies. In *Proceedings of the 14th IEEE International Symposium on Personal, Indoor and Mobile Radio Communication Proceedings,* Beijing, China (pp. 2208-2212). IEEE.

Prakash, R., Shivaratri, N., & Singhal, M. (1995, August 20-23). Distributed dynamic channel allocation for mobile computing. In *Proceedings of 14th ACM Symposium on Principles of Distributed Computing,* Ottawa, Ontario, Canada (pp. 47-56). ACM Publishing.

Prakash, R., Shivaratri, N., & Singhal, M. (1999). Distributed dynamic fault-tolerant channel allocation for cellular networks. *IEEE Transactions on Vehicular Technology, 48*(6), 1874–1888. doi:10.1109/25.806780

Schiller, J. (2003). *Mobile communications.* Indianapolis, IN: Pearson Education.

Tipper, D., & Teresa, D. (2002). Providing fault tolerance in wireless access networks. *IEEE Communications Magazine, 40*(1), 17–30. doi:10.1109/35.978050

Wang, S. S., & Green, M. (2000, September 24-28). Mobile location method for non-line-of-sight situation. In *Proceedings of IEEE 52nd Vehicular Technology Conference,* Boston (pp. 608-612). IEEE.

Yang, J., Jiang, Q., Manivannan, D., & Singhal, M. (2005). A fault-tolerant distributed channel allocation scheme for cellular networks. *IEEE Transactions on Computers, 54*(5), 616–629. doi:10.1109/TC.2005.71

Zhang, M., & Yum, T. (1991). The non-uniform compact pattern allocation algorithm for cellular mobile systems. *IEEE Transactions on Vehicular Technology, 40*(3), 387–391. doi:10.1109/25.289419

Zheng, J., Wang, Y., & Nihan, N. L. (2005, March 19-22). Quantitative evaluation of GPS performance under forest canopies. In *Proceedings of the 2005 IEEE International Conference on Networking, Sensing and Control,* Tucson, AZ (pp. 777-782). IEEE.

This work was previously published in International Journal of Business Data Communications and Networking Volume 5, Issue 4, edited by V. Sridhar and D. Saha, pp. 33-51, copyright 2009 by IGI Publishing (an imprint of IGI Global).

Chapter 18
Energy-Efficient Scalable Self-Organizing Routing for Wireless Mobile Networks

Melody Moh
San Jose State University, USA

Xuquan Lin
Echelon Corporation, USA

Subhankar Dhar
San Jose State University, USA

ABSTRACT

The instant deployment without relying on an existing infrastructure makes the mobile ad hoc networks (MANET) an attractive choice for many dynamic situations. An efficient MANET protocol may be applied to other important emerging wireless technologies such as wireless mesh and sensor networks. This chapter proposes a hierarchical routing scheme that is scalable, energy-efficient, and self-organizing. This chapter presents a new algorithm: the Dynamic Leader Set Generation (DLSG). This algorithm dynamically selects leader nodes based on traffic demand, locality, and residual energy level, and deselects them based on residual energy. Therefore, energy consumption and traffic load are distributed throughout the network. The network also reorganizes itself surrounding the dynamically selected leader nodes. Time, space, and message complexities are formally analyzed; implementation issues are also addressed. Incorporating the IEEE 802.11 medium access control mechanism including the power saving mode, performance evaluation is carried out by simulating DLSG and four existing hierarchical routing algorithms. It shows that DLSG successfully extends network lifetime by 20-50% while achieves a comparable level of network performance.

DOI: 10.4018/978-1-60960-589-6.ch018

INTRODUCTION

A mobile ad hoc network (MANET) is a temporary network formed dynamically by wireless mobile hosts without a pre-setup infrastructure. Unlike in a traditional infrastructure-based wireless network where each host routes packets through an access point or a mobile router, in MANET each host routes the packets and communicates directly with its neighbors. MANET may be viewed as a general network model that is applicable to other emerging wireless network technologies such as wireless mesh networks and sensor networks.

Since MANET offers much flexibility than traditional wireless networks, its demand and potential applications have rapidly increased. It is excellent for highly dynamic situations such as military operations, disaster recovery including its healthcare needs, and ad-hoc teleconferencing. In the near future it may also be applied to vehicular ad-hoc networks and countless forms of social gatherings.

Growing along with the demand is the research challenges in designing efficient MANET protocols. There are several major factors to be considered. Firstly, as its demand is rapidly rising, so is the number of mobile hosts shared a MANET. Secondly, since MANET is designed to work without an existing network infrastructure, it needs to adapt dynamically to the changing network topology. Yet, the third critical factor to be considered is energy; mobile hosts usually rely on limited battery power. Thus, it is vital that a MANET protocol be scalable, energy-efficient, dynamic, and self-organization/self-healing (Banerjee and Misra, 2002; Chen and Gerla, 2002; Pei et al., 1999; Royer, 2004; Sivakumar et al., 1999).

To address the *scalability* issue, this paper focuses on hierarchical routing, as hierarchy makes a routing protocol scalable. We investigate hierarchical routing based on the idea of dominating sets (DS). Minimum connected dominating sets (MCDS) and its variations have been used for hierarchical routing in MANET such that nodes in a DS form a virtual backbone and are responsible for routing packets through the entire network. Both DS and MCDS and their use in MANET routing will be described further in the Related Studies section.

To address the *energy efficiency* issue, we study energy awareness of these dominating-set based routing algorithms. Most existing works have mainly focused on heuristics for finding a small dominating set (since MCDS is NP-complete) and evaluating the size of dominating sets and the message costs. We incorporate IEEE 802.11 medium access control (MAC) schemes, with power saving mode (PSM) (IEEE, 1999), into these routing algorithms and assess them in terms network lifetime, throughput, and delay.

To address the *dynamic* issue, we propose a new algorithm, *Dynamic Leader Set Generation (DLSG),* which dynamically selects leader nodes based on traffic demand, locality, and residual energy level. As a result, the network lifetime is greatly increased.

Finally, to address the *self-organization, self-healing* issue, the new algorithm DLSG not only dynamically selects new leaders; it further follows a threshold mechanism, such that a leader node may de-select itself when its energy level falls below some threshold. The network therefore re-organizes itself around new leaders. Furthermore, a simple self-healing mechanism is also suggested (in Section 3.2.4).

The rest of the paper is organized as follows. Section 2 gives a broad overview of related work on energy-efficient routing. In section 3, we present our proposed algorithm DSLG. Performance evaluation of our algorithm is discussed in section 4. We discuss conclusions on section 5.

BACKGROUND AND RELATED STUDIES

Routing in MANET has attracted numerous research efforts in the last decade. The MANET routing protocols may be broadly classified into the following four main categories: (1) Reactive – such as Ad-hoc On-demand Distance Vector protocol (AODV) (Perkins et al. 2003), (2) Proactive (Table-driven) – such as Optimized Link State Routing protocol (OLSR) (Clausen et al., 2001), (3) hybrid – such as Zone Routing Protocol (ZRP) (Beijar, 2002), and (4) hierarchical – such as Dynamic Address Routing (DART) (Eriksson et al., 2004). Many existing works however does not explicitly consider energy efficiency in their designs. Few of them have integrated a layer-2 medium access control (MAC) protocol with power-awareness in their design or evaluation.

Wireless devices are often powered by batteries that have a finite amount of energy. In some ad hoc networks, it may not be always possible to change a battery once it runs out of energy. As a consequence, the conservation of energy is of foremost concern for those networks (Chen et al., 2001; Singh and Raghavendra, 1998; Stojmenovic and Lin, 2001; Wu et al., 2001). A good routing protocol should therefore be *energy-efficient.*

DLSG may be seen as a hierarchical, hybrid, energy-aware protocol. This is because it relies on a proactive local routing protocol, and a global routing protocol executed by the leader nodes that are selected on-demand (reactive). Furthermore, the leader selection is based on the residual energy, which makes the protocol energy-aware.

In the rest of the section, we first describe the most relevant works of the four reference algorithms i.e., DS-based hierarchical routing algorithms. This is followed by a explanation of power consumption in MANET. Finally, a survey of major energy-efficient routing protocols for MANET is presented.

Dominating-Set Based Hierarchical Routing

Routing based on a connected dominating set is a promising approach, where the search space for a route is reduced to the nodes in the set. A subset *D* of nodes of a graph *G* is a dominating set (DS) if every node not in *D* is adjacent to at least one node in *G*. Furthermore, Minimum Connected Dominating Set (*CDS*) problem is described as follows. Find a smallest subset *D* of nodes, such that the subgraph induced by *D* is connected and *D* forms a DS. Since *CDS* is a NP complete problem, several approximation algorithms have been proposed in this regard (Guha and Khuller, 1998).

Hierarchical routing schemes can be divided into two groups; one that is based on clustering and the other based on *virtual backbone* construction. Both the set of clusters and the set of backbone nodes form a second-level subnetwork. However, in the clustering schemes, the second-level nodes (*i.e.,* the cluster heads) need not be connected - in fact, it is more desirable to have the cluster heads spread out, whereas in the backbone-based routing schemes the second-level nodes need to be connected to form the backbone.

Ad hoc networks are typically represented by a connected graph where all the links are bi-directional. Several researchers have used minimum CDS to induce a virtual backbone in ad hoc networks (Das and Bhargavan, 1997; Heinzelman et al., 2001).

One advantage of clustering-based hierarchy over backbone-based hierarchy is that clustering usually results in a smaller set of second-level nodes and communication path between second-level nodes are not fixed *i.e.* it can take any route, which gives it more flexibility. In recent studies, it has been shown that *CDS* problem can also be used to find cluster-like hierarchical structure by extending the problem of finding minimum *k*-hop connected, *k*-dominating set (*k-CDS*) problem. In this structure (for $k>1$, *k* a positive integer), the dominating nodes, together with their *k*-hop neighbors, form *virtual clusters* instead of a virtual backbone, as the dominating nodes are no longer directly connected (*k*-hop connected).

Hierarchical routing has gained special attention for ad hoc networks for their scalability and flexibility. In order to orchestrate hierarchical routing, various clustering algorithms have been developed for this purpose (Bannerjee and Khuller, 2001). Both deterministic and probabilistic distributed algorithms are used to construct clusters (Heinzelman et al., 2000; Hsu et al., 2005, Iwata, 1999; Liang and Haas, 2000; Ramanathan and Streenstrup, 1998; Tseng et al., 2002). The average communication costs in construction and maintenance of such hierarchies is shown to be logarithmic with the network size (Lov´asz, 1975). Hierarchical cluster-based structures provide efficient routing of messages in large dynamic networks. With this approach, the network is divided into several interconnected clusters of nodes, which might or might not overlap. This approach is good for scalability. However, all these clustering strategies do not guarantee shortest/minimal path routing that is quite desirable in many situations to minimize end-to-end delay.

This paper chooses four algorithms as references for performance comparison: the algorithm proposed by Wu and Li (Wu and Li, 2001), and three variations of an algorithm by Dhar, Rieck, Pai, and Kim (Dhar et al., 2004; Dhar and Rieck, 2007; Kim et al., 2004; Rieck et al., 2005). They are described below.

J. Wu and H. Li have proposed a heuristic for selecting DS based on local neighborhood information and have discussed two rules that eliminate redundant nodes in the DS that result in a very small DS. It is worth mentioning that applying rules 1 and 2 reduces the size of the DS at the cost of incurring more communication.

In order to guarantee shortest path routing, Rieck et al proposed several algorithms based on the notion of a *k*-SPR set, which is defined as follows.

Definition A set S of nodes will be called a *k*-SPR set (*k* is a positive integer) if given any pair of nodes *u* and *v* of *G* such that $\delta(u,v) = k+1$, there exists a $w \in S$ with $\delta(u,w) + \delta(w,v) = k+1$ and $w \neq u$, $w \neq v$.

Here *k*-SPR stands for '*k-shortest path routing*', and under reasonable assumptions, a *k*-SPR set can be shown to be a *k*-hop connected, *k*-dominating set that has the shortest path property (Rieck et al 2005). M. Q. Rieck and a group of researchers proposed three variations of the *k*-SPR sets, which share the same basic properties but differ in how DS nodes are (Dhar et al., 2004; Rieck et al 2005). There three variations of of the *k*-SPR algorithms are summarized in Table 1.

Table 1. Summary of k-SPR Algorithms

k-SPR version	Dominating node selection criterion	Self-Organization: DS re-selection
*k*SPR-I	Highest Identity	None
*k*SPR-C	Highest Covering Number	None
*k*SPR-E	Highest Residual Energy	Reselect the DS once in a while based on the current residual energy

In order to blend the desirable features of cluster-based routing with those of backbone-based routing, one might consider applying the algorithm of Wu and Li to the *k*-hop closure G_k of the original graph *G*, which represents the network. While this approach leads to a virtual backbone involving a small number of nodes, the message routing paths that result can be unnecessarily long, since the backbone will not have the shortest path property in general. By contrast, the *d*CDS algorithm (Rieck et al., 2005) which was later renamed as *k*-SPR-I for clarity produces a larger set of backbone nodes, but one which guarantees shortest path routing. In order to accomplish this, all of the shortest possible paths of length $k+1$ are discovered and the highest IDs of the nodes along such paths are considered. The details are discussed in (Rieck et al., 2005).

The *k*-SPR-C algorithm is a variation of the *k*-SPR-I algorithm. In *k*-SPR-I, the node with the highest ID was selected to be included in the k-SPR set (Dhar and Rieck, 2007). In *k*-SPR-C, the node with the highest priority was selected. The priority of each node is defined to be the ordered pair of numbers (covering number, ID), that are lexicographically ordered. The *k*-SPR-E (``E" for ``energy") algorithm is similar to *k*-SPR-C, except that nodes are elected so as to have the smallest possible costs (highest residual energy), rather than the highest possible covering numbers.

Note that among the above four algorithms (Wu and Li, and three *k*-SPR), *k*-SPRE is the only one that explicitly addressed the issues of self-organization and energy efficiency. Yet, it did not explain clearly the DS reselection rules (such as when or how often it should be re-selected).

Recently, Rieck and Dhar (Rieck and Dhar, 2008) extended the notion of a k-SPR set from previous work in the context of an edge-weighted graph. Under a reasonable assumption, such a set is still *k*-dominating, and k-hop connected. When a decreasing sequence of such sets is used, together with a hybrid route discovery strategy (partly proactive, partly reactive), the result is a highly scalable and efficient routing protocol that turns out to use only minimal cost paths.

Power Saving Mode (PSM)

Energy dissipation in a MANET can be controlled and accounted in two main places: the first is in the MAC layer discussed below. The second is in data transmission and reception, which is usually proportional to (some power of) the distance between two nodes, and will be described more in the Performance Evaluation section (Section 4).

IEEE 802.11 (IEEE 1999), the standard for MAC in wireless local area networks, first consists of the basic protocol: the Distributed Coordination Function (DCF). It is based on the CSMA/CA (Carrier Sense Multiple Access with Collision Avoidance) scheme. In addition, the standard also defines the power saving mode (PSM). In a simple sense, the PSM scheme allows nodes that are neither transmitting nor receiving to go to the sleep mode during the current beacon interval, thus saves energy greatly (IEEE 1999). Further improvements over PSM have also been proposed (Jung and Vaidya, 2005).

Energy-Efficient Routing Protocols

The energy-efficient protocols for ad hoc networks can be classified into the following four major categories: *minimum transmits power protocols*; *transmit power control protocols*, *maximum lifetime protocols*, and *power save protocols*. This classification roughly follows the chronological order of energy-aware protocol development.

Minimum Transmit Power Protocols: Some of the earliest power-aware routing protocols are minimum transmit power protocols, which incorporate power consumption in link cost calculation and thus finding a path that consumes the least power for the end-to-end transmission of a packet. These protocols inherently presume the use of a *variable transmit power system* which is discussed next in more detail; if a common transmit power level is used for every transmission regardless of the distance to the receiver, the energy cost for a route will be proportional to the number of links in the route, in which case the minimum transmit power protocols will degenerate into shortest path routing. One problem often associated with the minimum transmit power protocols is that they tends to use the same route over and over, eventually draining the battery power of those nodes prematurely, which was later addressed by *maximum lifetime protocols*.

Transmit Power Control Protocols: Another approach is to use a power control scheme which suitably varies transmit power to reduce energy consumption, depending on the distance to the receiver. To do this, a node needs to know (or estimate) the distance to each neighbor and calculate the minimum transmit power level required to ensure good reception by the receiver, both of which are not so easy to achieve. Varying transmit power level has other consequences too. For example, virtual collision avoidance using RTS-CTS exchange may not work correctly under variable transmit power scheme. Low power transmissions cannot be sensed by distant nodes, which may then initiate transmissions using sufficient power to disrupt ongoing transmissions (Feeney, 2002). Two papers by Jung and Vaidya (Jung and Vaidya, 2005) are proposed to solve this problem. Also several researchers (Narayansawamy et al., 2002) try to combine power control scheme with clustering where clusters are determined whether they are reachable at a given power level. In addition to providing energy saving, power control can potentially be used to improve spatial reuse of the wireless channel.

Maximum Lifetime Protocols: As pointed out before, the main disadvantage of the minimum transmit power protocol is that it always selects the least-energy cost routes and as a result, nodes along these routes tend to die soon because of the battery energy exhaustion. This is doubly harmful since the nodes that die early are precisely the ones that are needed most to maintain the network connectivity (and hence useful service life). Therefore, it is better to use a higher energy cost route (e.g. a multi-hop route) if it avoids using nodes that have a small amount of remaining battery energy (also called the *residual energy*). This gave rise to another family of power-aware routing protocols, which try to maximize network lifetime.

Power Save Protocols: More recently, a number of papers found that the energy savings achieved by reducing the energy used in transmitting or receiving packets are significantly smaller than those achieved by taking advantage of low power mode (or *sleep* mode) of the physical interface. Power-saving mechanisms allow a node to enter a doze state or sleep state by powering off its wireless network interface when deemed reasonable. Singh and Raghavendra produced some pioneering works on power-saving mechanism in their power-aware multi-access protocol with signaling (PAMAS) (Singh and Raghavendra, 1998). The main challenge of utilizing the power save mode is that if nodes go in and out of the sleep mode without coordination, the throughput of the network drops significantly as each transmitter has to wait until the receiver is awake to receive the data. This is done in the MAC layer as opposed to the network layer.

Span (Chen et al., 2001) is one of several ad hoc networking protocols based on the notion of a dominating set. In Span, "coordinators" — a group of nodes that form a connected dominating set over the network — do not sleep. Non-coordinator nodes follow a synchronized sleep/wake cycle, exchanging traffic using an algorithm based on the beaconing and traffic announcement methods of IEEE 802.11 IBSS power save. The routing protocol is integrated with the coordinator mechanism so that only coordinators forward packets, acting as a low latency routing backbone for network. Span is intended to maximize the amount of time nodes spend in the sleep state, while minimizing the impact of energy management on latency and capacity.

Recent Advances in MANET Routing

In this section we describe some recent works in the routing developments in MANET, including the extension of OSPF (Open Shortest Path First) routing protocol over MANET, and security issues in MENET routing.

Isaacs, et al. (Isaacs, 2010) recently studied the extension of OSPF routing protocol for MANET. This is based on the fact that OSPF is the most widely used intra-domain routing protocol on the Internet. Extending OSPF to handle the unique concerns of MANETs would be ideal as OSPF is well-known and well-tested. This study reviewed several major OSPF extension works over MANET. It also compared these extension protocols, both qualitatively in their protocol designs and quantitatively by summarizing their individual simulation results.

On emerging issue in MANET routing is security. While instant deployment makes MANET an attractive choice for many dynamic situations, such flexibility however comes with a consequence – these networks are much more vulnerable to attacks. Traditional protection mechanisms, including authentication and encryption, are ineffective against attacks such as selfish nodes and malicious packet dropping. Recently, reputation systems have been proposed to enforce cooperation among nodes. These systems have provided useful countermeasures and have been successful in dealing with selfish and malicious nodes. Moh and Li presented a survey of major contributions in this field (Moh, 2010). They also discussed limitations of these approaches and suggested possible solutions and future directions.

Most existing reputation systems in MANET considered only node reputations when selecting routes; in such case, reputation and trust are only ensured within a one-hop distance when routing decisions are made. This often fails to provide the most reliable, trusted route. Li and Moh proposed a system that is based on *path reputation*, which is computed from the reputation and trust values of each and every node in the route (Li, 2009). The approach has significantly enhanced the reliability of the resulting routes.

AODV routing protocol (Perkins et al., 2003) is by far the most widely used protocol for MANET. While there have been numerous security enhancement works over AODV, one recent attempt is by Chang et al. (Chang, 2009). The authors presented a powerful, versatile security suite for AODV. It utilized powerful authentication and user-adjustable encryptions based on digital certificate chaining and popular ciphers such. The security strength can be customized to reach a desired balance with the performance of the network. One attractive characteristic is the highly modular design, which allows additional security strength to be added on top of any existing frameworks.

PROPOSED ALGORITHM: DYNAMIC LEADER SET GENERATION (DLSG)

In this section, we first describe the motivation and the major design principles. The proposed DLSG algorithm is then described in detail. This is followed by a formal analysis of time, space and message complexities and protocol overhead. Finally, several practical implementation issues are discussed.

Motivation and Design Principles

A DS may theoretically be a good choice as a virtual backbone for hierarchical routing. Yet, due to the dynamic nature of MANET and the need for self-organization, DN are not an ideal choice for backbone nodes. This is because DN selection and reselection requires a network-wide flooding and it consumes a lot of energy. This has been observed from a preliminary study that simulated the four referenced algorithms described in the previous section. It has been shown that DN reselection in k-SPRE consumed a great deal of energy. Another observation is that DN usually die far earlier than non-DN since they need to send and receive much more data.

In a power-stringent MANET, it is therefore impractical to always maintain a DS: Nodes generally move randomly, and maintaining a DS (or a k-dominating set) requires frequent DN reselections, which are expensive in terms of energy consumption. Rather, it is better to dynamically select leader nodes (virtual backbone nodes) that are able to quickly route packets to their destinations.

The Dynamic Leader Set Generation (DLSG) algorithm proposed in this paper is based on the above observation. It re-selects *leader* nodes (instead of DN) *dynamically, on demand,* and based on a combination of *residual energy levels* and *close proximity* in the geographical forwarding. DLSG selects a smaller set of leader nodes that has high residual energy, which increases network lifetime. Furthermore, it selects a leader only from L nodes among the k-hop neighbors that are closer to the destination node. This greatly reduces energy consumption since (1) queries of residual energy level are only sent to L nodes (instead of network flooding as in DN reselection), and (2) energy consumed by any node during transmission is proportional to the transmission distance.

DLSG further extends its network lifetime by allowing leaders to de-select themselves based on a *threshold* scheme. Observed that if several global routing paths share a common leader node, then this node would drain out its energy very quickly since it would have transmitted and received many packets. In DLSG, a leader node switches to the sleep mode when its energy level falls below a pre-determined *threshold value*, and a new leader node may later be selected based on traffic demand.

DLSG Algorithm Description

Initialization and Local Routing

Variables, tables, and the local routing table are initialized below:

1. Each node computes its *Threshold Energy Level* (see Sec 3.2.3). All nodes except the source nodes operate in *PSM*.
2. *Leader Status* is initialized to false, and both *Leader Node List* and *Potential Leader Node List* are initially empty, for all nodes.
3. Each node finds its *k*-hop neighbors and thereby knows their (x, y) geographical locations. Each node maintains the local routing table covering its *k*-hop neighbors using a local proactive routing algorithm such as Dijkstra's algorithm or OLSR (Clausen et al., 2001). This is used to send packets to a destination node within *k* hops.

Dynamic Leader Set Generation (DLSG) for Global Routing

DLSG Algorithm described below is used to send a packet to a destination node beyond *k* hops. Note that *L* is generally much smaller than the total number of *k*-hop neighbors a node has.

1. When a node has a packet to be transmitted to a destination beyond *k* hops, it first checks its *Leader Node List*. If a leader node exists in the *list,* then it is selected.
2. Otherwise, the node selects *L* nodes among its *k*-hop neighbors that are closer to the destination using geographical forwarding approach. These selected nodes are stored in its *Potential Leader Node List*.
   a. It sends *energy queries* to every node in the *Potential Leader Node List*.
   b. Upon receiving energy queries, these potential nodes respond with their current energy level.
   c. Upon receipt of energy level replies, the node determines the *highest* energy node and declares it as a *leader node*. It adds this node to its *Leader Node List*. It notifies the selected node about its new status.
   d. Upon receiving leader-node notification, the node sets its *Leader Status* to be true, empties the *Potential Leader Node List*, and then broadcasts its new status to all its *k* hop neighbor nodes.
   e. Upon receiving the broadcast packet of new leader status, all the *k*-hop neighbor nodes update their *Leader Node List.*
3. Once a leader node is selected, the packet is forwarded to the leader node using local routing.
4. Upon reaching the leader node, if the destination node is not within *k*-hops, then the above steps are repeated; otherwise, the packet reaches the destination through local routing.
5. The selected leader node continues to be powered on until the current energy level of the node falls below the pre-set threshold value (see next subsection). Thereafter it enters the sleep mode to conserve energy.

Threshold Mechanism

The following scheme allows a leader to de-select itself when its energy level falls below a threshold value. This value is based on a fixed percentage, and the value is re-computed whenever a node is selected as a leader node, or whenever it becomes active from sleep mode.

1. Once a leader node's current energy level falls below the threshold energy level the leader node sets its *Leader Status* to false.
2. It informs all of its *d*-hop neighbor nodes by broadcasting the non-leader status notification.
3. Upon receiving the broadcast packet all its *k*-hop neighbor nodes remove the leader node from their *Leader Node List.*

Setting the Threshold Value

a. Let p be a pre-determined percentage value, and e_i be the initial energy level when a node has just become a leader node. Then e_{th}, the threshold value for energy level, may be calculated as follows:

$$e_{th} = p \times e_i, \ (0 < p < 1)$$

b. While the current energy level $e(t)$ is larger than e_{th} the node continues to be active; after which it switches to sleep mode.

c. Every time a node is switched from sleep mode to active mode, the node computes a new threshold value based on its current energy level e_c:

$$e_{th} = p \times e_c, \ (0 < p < 1)$$

Handling Node Mobility and Failures

Most mobile nodes have some built-in mechanisms to detect their own movements and the movement of their neighbors (through the beacon packets). This mechanism usually resides in the layer-2 schemes. This information may be passed to its layer-3 routing protocol. Alternatively, many local routing protocols are able to detect the aliveness of its neighbors (through the *HELLO* messages). *DLSG* described above may therefore be readily extended to handle node mobility and failures, as described below.

When a node moves or fails, consider the following cases:

1. If a node detects the movement or failure of some of its neighbors by either its layer-2 mechanism or its local routing protocol, then it updates its local routing table, *Leader Node List*, and *Potential Leader Node List* accordingly.
2. Otherwise (node movement/failure is not detected) a time-out mechanism is employed in *DLSG* for all queries of energy level (Step 2c). Any neighbor nodes fail to reply after the timeout period will be removed from a node's local routing table, *Leader Node List*, and *Potential Leader Node List*.
3. If a node is able to detect its own movement (or even failure) before hand, it informs all of its *k*-hop neighbor nodes (similar to Step 2 of the Threshold mechanism). All the neighbor nodes thus update their routing tables, *Leader Node List*, and *Potential Leader Node List* accordingly.

Complexities and Protocol Overhead

Let D be the maximum degree of the network graph; i.e., the maximum number of 1-hop neighbors any network node has. The maximum number of *k*-hop neighbors a node may have is therefore D^k. Note that k and L are both small, tunable constants; in the simulation $k = 4$ and $L = 3$.

Space Complexity

The algorithm needs a local routing table for k-hop neighbors which takes $O(D^k)$, a *Leader Node List* which takes $O(1)$, and a *Potential Leader Node List* which takes $O(L)$. Thus, the space complexity is no greater than that of a local routing table for k-hop neighbors.

Time Complexity

The dictating steps are Step (2), which takes $O(D^k \log(D^k))$ to sort all the k-hop neighbors according to proximity, and Step (2c), which takes $O(L)$ using insertion sort, or $O(L\log L)$ for a common sorting. Note however that this step does not happen frequently – it takes place only when the Leader Node List is empty; i.e. the leader de-select itself due to the threshold mechanism (next subsection). The worst-case time complexity is $O(D^k \log(D^k))$; the same as a proactive local routing (such as Dijkstra's) when the routing table needs to be updated.

Message Complexity

Based on traffic demand, a source node queries L of its k-hop neighbors about their energy levels in choosing leader nodes (Step 2a), which takes $O(L)$ messages. After selecting a leader node, it notifies the selected leader node, which then broadcast to all its k-hop neighbors (Step 2d) of $O(D^k)$. Finally, there is another broadcast to k-hop neighbors in Step 2 of the threshold mechanism, of $O(D^k)$. Thus, the message complexity is $O(L) + O(D^k) + O(D^k)$, which is $O(D^k)$; the same as a simple k-hop proactive local routing.

Protocol Overhead

Protocol overhead usually refers to the extra control messages involved in the protocol, but it may also refer to extra time and space needed. The protocol overhead of *DLSG* has been discussed above in terms of time, space, and message complexities, which are the same order as those of a proactive local routing protocol.

Finally, the extra energy needed for *DLSG*, including its leader election and routing, is represented by message complexity (control messages sent/received for leader election and routing). The actual energy usage may be calculated according to Table 1. Detailed results are demonstrated in Section 4.

Implementation Issues

Initialization and Local Routing

The proposed algorithm is distributed in nature. Initially, it is assumed that a node communicates and learns its k-hop neighbors, including their geographical locations, and is able to build a local routing table. A proactive routing protocol, such as some wireless, mobile version of the Dijkstra's algorithm could be used as a local routing protocol. Since k is a tunable constant, by keeping it small ($k = 4$ in our simulation), there would be no scalability problem. Alternatively, if scalability is a concern, OLSR (Clausen et al., 2001) may be applied. These routing protocols usually use some kind of *HELLO* messages by which the neighboring nodes exchange keep-alive and link state information. Energy level and geographical location information may therefore be included in these *HELLO* messages.

DLSG and Threshold Mechanism

All the steps in the DLSG algorithm and the threshold mechanism require a node to either communicate with its k-hop neighbors or make some local decision. Thus, they may be readily implemented as an enhancement to a local routing protocol (that offers communication within d-hop neighbors).

Performance Evaluation

We developed our own simulation tool using Java. The implementation was run on a single machine while simulating the distributed nature of the algorithm. Each node gathers the information it needs from its neighboring nodes and declares its results. In the following, we first present simulation setting, including the energy model and simulation parameters. The next three subsections illustrate results on network lifetime, throughput, and average end-to-end delay. This section is concluded by a summary of simulation results.

Simulation Setting

Energy Model and Power Saving Mode

In order to accurately model energy consumption, the first-order energy model is used, as shown in Table 2 (Heinzelman et al., (2000), which is also used in some earlier works in sensor networks (Culpepper et al., 2004, Moh et al.,2006). Each node consumes E_{elec} = 50 nJ/bit to receive and to transmit one bit, $\in_{amp}$ = 100pJ/bit/m^2 is consumed to transmit a bit within one square meter, s is the distance in meter between two nodes, and b is the message size in bits.

In addition, two cases of PSM are considered:

Case 1 (PSM): All the nodes follow IEEE 802.11 DCF and PSM so that energy is saved when these nodes are idle. Note that DN in the four referenced algorithms do not follow PSM; they can route the packets anytime.

Case 2 (NPSM or Non-PSM): All the nodes follow IEEE 802.11 DCF but not PSM.

Routing

The five algorithms, the proposed DLSG and the four references, are used to create a set of nodes serving as the virtual backbone for hierarchical routing. For local routing with their k-hop neighbors, nodes use Dijkstra's algorithm (thus all paths within the local routing are shortest paths), but any proactive routing algorithm (such as OLSR (Clausen et al., 2001)) may be used. For global routing, leader nodes or DN form the backbone and use the geographic forwarding.

Table 2. Energy Model

Operation	Energy Dissipation
Transmission	$E_{tx}(b,d) = E_{elec} * b + \in_{amp} * b * s * s$
Reception	$E_{rx}(b) = E_{elec} * b$

Simulation Parameters

The major simulation parameters are summarized in Table 1, k = 4 (used in k-SPRx algorithms), and L = 3 (used in DLSG). Initially, each node is assigned to a random, unique geography location with the coordinate (x, y). Each node considers all the nodes within its transmission range as neighbors. The maximum degree of the network graph representing is 4. Ten different source and destination pairs are randomly picked to generate ten data flows. Note that each DS-based algorithm creates a different set of DN, and thus different data paths

Network Lifetime

Figures 1 and 2 show the results on network lifetime; one for the four existing algorithms and the new proposed *DLSG* algorithm (with no threshold, thus p=0), and the other for the *DLSG* algorithm with various threshold values. Clearly from Figure 1, for PSM case, *DLSG* has the longest network lifetime - even longer than Wu-Li and k-SPRE algorithms.

Figure 2 evaluates *DLSG* with various energy threshold values. When the threshold value is very small ($p < 0.6$), network lifetime is also small since a leader node does not go to sleep until it has only a small percentage of energy left, and so its energy drained out more quickly. When the

Table 3. Simulation Parameters

Parameter	Value	Parameter	Value
Simulation area	1000×1000 m^2	Simulation duration	Till 1st node dies
Network size	100 nodes	Sifs	0.028 ms
Transmission range	250 m	Difs	0.128 ms
Packet size	320 bytes	Beacon period	200 ms
Traffic load	5.12 Mbps	ATIM window	10 ms
Initial energy for each node	300 J	*k*-SPRE reselection interval	1 minute

Figure 1. Network Lifetime (min)

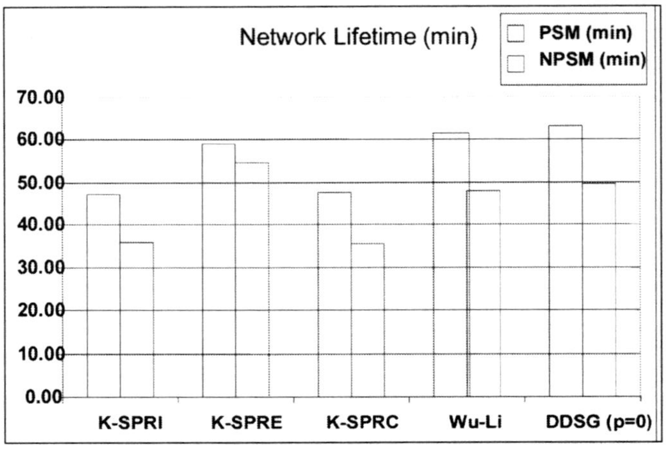

Figure 2. Network Lifetime (sec.) vs. Threshold

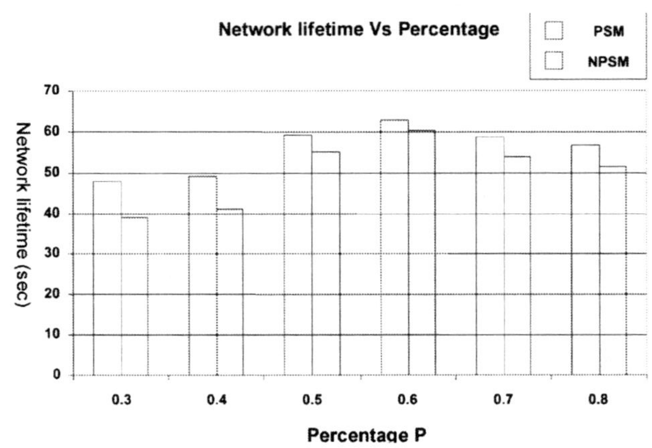

threshold value increases, so does energy saving and therefore network lifetime. However, when the threshold value increases further ($p > 0.6$), energy saving reduces. This is due to frequent changes of leader nodes, which requires frequent exchanges of energy level and status notifications. Overall, with a good threshold value ($p = 0.6$), DLSG improves network lifetime by more than

Figure 3. Throughput (Mbps)

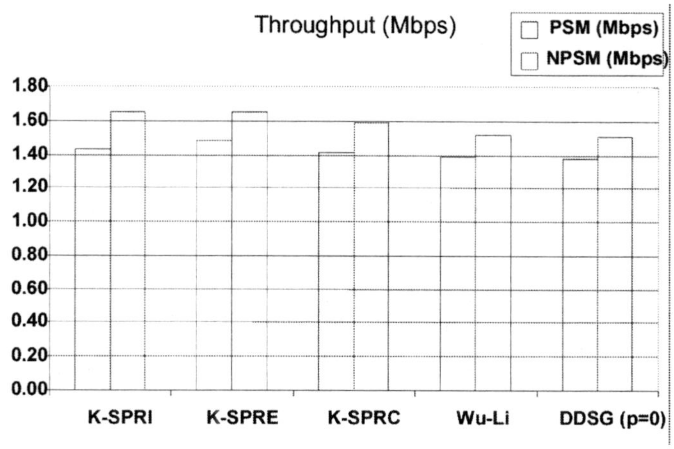

Figure 4. Throughput (Mbps) vs. Threshold

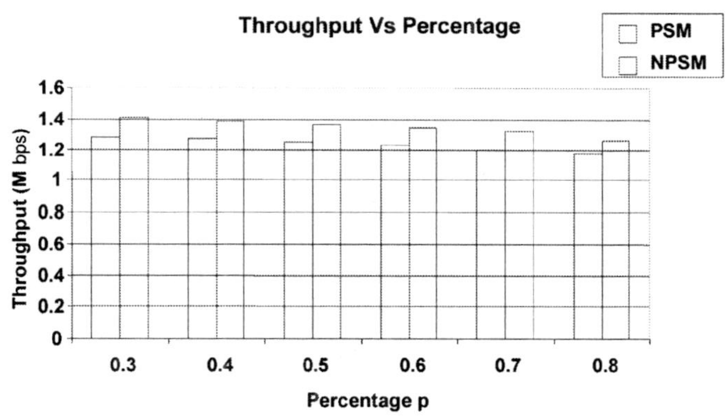

20% than when no threshold is used (*DLSG* in Figure 1), and by more than 50% comparing with some of the *k*-SPR algorithms.

Throughput

Figures 3 and 4 show the results on throughput. From Figure 3, the newly proposed *DLSG* maintains an equally high throughput as Wu-Li (note that it also has a longer network lifetime in Figure 1); its throughput is only very slightly lower than the three *k*-SPR algorithms.

From Figure 4, as threshold value p increases, throughput faintly decreases due to more frequent changes of the leader nodes. Yet, the decrease is very small. With a good threshold value ($p = 0.6$), throughput is only slightly smaller than the largest throughput when $p = 0.3$, yet it is still higher than when $p = 0.8$. This shows that with a good choice of threshold value, energy is well-preserved (showing the longest network lifetime), while throughput is still maintained in a stably high level.

End-to-End Delay

Figures 5 and 6 show the results of the average end-to-end delay. In Figure 5, *DLSG* with PSM has a longer delay (about 30%) due to all the nodes (except source nodes) operating in PSM.

Figure 5. Average end to end delay (ms)

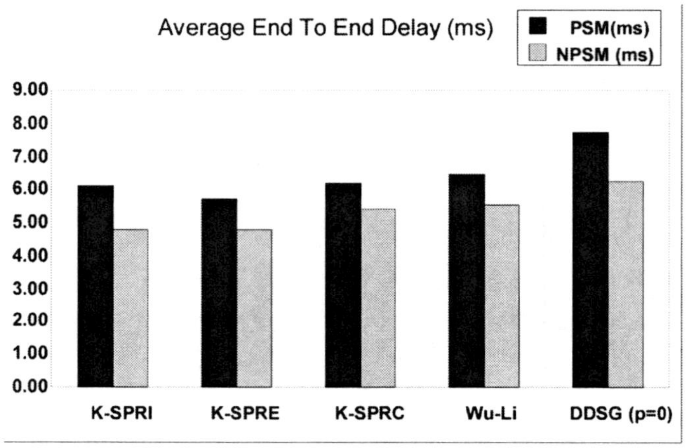

Figure 6. Average End-to-end Delay (msec) vs. Threshold

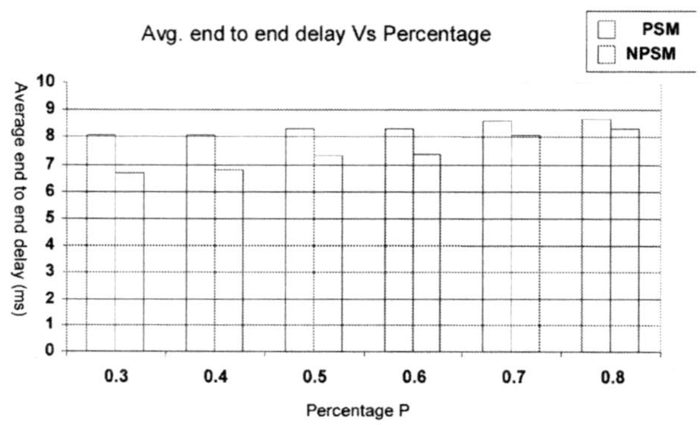

When applying the threshold mechanism, the good value ($p = 0.6$) still has a shorter delay than do the higher threshold values ($p = 0.7$ and 0.8), while providing a longer network lifetime.

SUMMARY OF SIMULATION RESULTS

Above simulation experiments clearly show that *DLSG* is most effective in extending network lifetime (Figure 1). With the initial energy of 300J, it can achieve a network lifetime of 60-70 minutes which is remarkable for many MANET models. DLSG also preserves a high throughput (Figure 3) along with some slight increase in delay (Figure 5).

When the threshold mechanism is applied, with an appropriate choice of threshold value ($p = 0.6$), network lifetime may be further extended (Figure 2), while maintaining a reasonably high throughput (Figure 4) without a much longer delay (Figure 6). Note that while a threshold value of 0.6 gives the best results in the simulation, choosing an optimal threshold value remains to be further investigated.

CONCLUSION

Addressing the challenge of achieving scalable, energy-efficient, self-organizing routing in the highly dynamic MANET, this paper has proposed a new hierarchical routing algorithm, *Dynamic Leader Set Generation* (*DLSG*). It dynamically selects the leader nodes based on traffic demand, residual energy level, and proximity. The leader nodes also de-select themselves based on a threshold mechanism to avoid early energy drained-out. The new algorithm may be readily implemented as an enhancement of a proactive local routing protocol without increasing its time, space, and message complexities. Simulation evaluation has shown that, with a suitable choice of threshold value, it achieves up 20-50% increase of network lifetime, with a similar level of network throughput, and a slight increase in delay. Future works may include (1) an extension of simulation with node movements and node failures, (2) a mathematical analysis of the optimal threshold values, and (3) a cross-layer design incorporating *DLSG* with a multi-hop medium access control scheme (Hsu et al., 2005; Lin et al., 2008).

ACKNOWLEDGMENT

M. Moh was supported in part by Fujitsu Labs of America and S. Dhar was supported by a Lucas Fellowship. This work is extended from an earlier publication (Moh, 2009).

REFERENCES

Banerjee, S., & Khuller, S. (2001). *A Clustering Scheme for Hierarchical Control in Multi-hop Wireless Networks*. Proceedings of IEEE International Conference on Computer Communications (INFOCOM), held in Anchorage, Alaska.

Banerjee, S., & Misra, A. (2002). *Minimum energy paths for reliable communication in multi-hop wireless networks*. Proceedings of Annual Workshop on Mobile Ad Hoc Networking & Computing (MobiHOC 2002), 146–156.

Beijar, N. (2002). *Zone Routing Protocol (ZRP). Licentiate course on Telecommunications Technology*. Ad Hoc Networking.

Belding-Royer, E. M. (2004). Routing Approaches in Mobile Ad Hoc Networks. In S. Bagagni et al.(Eds). Mobile Ad Hoc Networking, 275-300. Washington, DC: IEEE Press.

Chang, R., Gundala, S., Moh, T.-S., & Moh, M. (2009). *VESS: A Versatile Extensible Security Suite for MANET.* (Invited paper) Proc. IEEE Pacific Rim Conference on Communications, Computers and Signal Processing (PACRIM'09), 944 – 950, held in Victorian, Canada, Aug 2009.

Chen, B., Jamieson, K., Balakrishnan, H., & Morris, R. (2001). *Span: An Energy-efficient Coordination Algorithm for Topology Maintenance in Ad Hoc Wireless Networks*. Proc. ACM Annual International Conference on Mobile Computing and Networking (Mobicom), 85-96.

Clausen, T., Jacquet, P., Laouiti, A., Muhlethaler, P., Qayyum, A., & Viennot, L. (2001). *Optimized Link State Routing Protocol for Ad Hoc Networks*. IEEE International Multitopic Conference (INMIC), held in Pakistan.

B. J. Culpepper, L. Dung, & M. Moh. (2004). *Design and analysis of Hybrid Indirect Transmissions (HIT) for data gathering in wireless micro sensor networks*. ACM Mobile Computing and Communications Review (MC^2R), 61-83, Jan/Feb 2004.

Das, B., & Bharghavan, V. (1997). *Routing in Ad-Hoc Networks Using Minimum Connected Dominating Sets*. Proc. IEEE Int'l Conf. Communications, 376-380

Dhar, S., & Rieck, M. Q. (2007). An Energy-Efficient Heterogeneous Dual Routing Scheme for Mobile Ad Hoc and Sensor Networks. *International Journal on Mobile Network Design and Innovation*, *2*(2), 105–115. doi:10.1504/IJMNDI.2007.015060

Dhar, S., Rieck, M. Q., Pai, S., & Kim, E. J. (2004). Distributed Routing Schemes for Ad Hoc Networks Using dSPR sets. *Journal of Microprocessors and Microsystems. Special Issue on Resource Management in Wireless and Ad Hoc Mobile Networks*, *28*(8), 427–437.

Eriksson, J., Faloutsos, M., & Krishnamurthy, S. (2004). *The Case for Dynamic Addressing.* Proceedings of IEEE International Conference on Computer Communications (INFOCOM).

Feeney, L. M. (2002). *A QoS Aware Power Save Protocol for Wireless Ad Hoc Networks.* Proc. IFIP Mediterranean Ad Hoc Networking Workshop (Med-Hoc-Net), held in Sardegna, Italy.

S. Guha & S. Khuller. (1998). *Approximation Algorithms for Connected Dominating Sets.* Algorithmica, 20.

Heinzelman, W., Chandrakasan, A., & Balakrishnan, H. (2000). *Energy-Efficient Communication Protocol for Wireless Microsensor Networks.* IEEE Proc. Hawaii International Conf. on System Sciences.

Hsu, C., & Tseng, Y. C. Hsu, & Y. Tseng. (2005). *Cluster-Based Semi-Asynchronous Power-Saving Protocols for Multi-hop Ad Hoc Networks.* Proc. IEEE International Conference on Communications (ICC).

IEEE. (1999). *Wireless LAN Medium Access Control and Physical Layer specifications.* IEEE 802.11 Standard.

Isaacs, K., Hsieh, J., & Moh, M. (2010). Extending OSPF for MANET Routing. In Khan, S., Lloret, J., Ortiz, J. H., & Loo, J. (Eds.), *Mobile Ad hoc Networks: Current Status and Future Trends*. New York: Taylor & Francis Publications.

A. Iwata, C-C. Chiang, G. Pei, M. Gerla, & T.-W. Chen. (1999). Scalable Routing Strategies for Ad Hoc Wireless Networks. *IEEE Journal on Selected Areas in Communications, Special Issue on Ad-Hoc Networks*, 1369-79.

E. Jung & N. Vaidya. (2005). *Improving IEEE 802.11 Power Saving Mechanism.* IEEE Trans. on Automatic Control. July 2005.

Kim, E. J., Dhar, S., & Rieck, M. Q. (2004). *Coordinator-Oriented Forwarding for Energy Efficiency in Large Scale Wireless Ad hoc Networks.* Proc.s of the International Conference on Information Systems (ISNG), 44-49, held in Las Vegas.

Li, J., Moh, T.-S., & Moh, M. (2009). *Path-Based Reputation System for MANET Routing.* Proc. of the 7th International Conference on Wired/Wireless Internet Communications, 48 – 60, held in Enschede, the Netherlands.

Liang, B., & Haas, Z. J. (2000). *Virtual Backbone Generation and Maintenance in Ad Hoc Network Mobility Management.* Proc. IEEE International Conference on Computer Communications (INFOCOM), 1293-1302, held in Tel Aviv.

Lin, X., Moh, M., & Moh, T.-S. (2008). *On Energy-Efficient Semi-Asynchronous MAC Protocols for Multi-hop Ad-hoc Networks.* Proc. of 13th IEEE Asia-Pacific Computer System Architecture Conference (ACSAC), held in Hsin-Chu, Taiwan.

Lov'asz, L. (1975). On the Ratio of Optimal Integral and Fractional Covers. *Discrete Mathematics*, *13*, 383–390. doi:10.1016/0012-365X(75)90058-8

Moh, M., Kim, E.-J., & Moh, T.-S. (2006). Design and Analysis of Distributed Power Scheduling for Data Aggregation in Wireless Sensor Networks*Int.* [IJSNet]. *Journal of Sensor Networks*, *1*(3-4), 143–155. doi:10.1504/IJSNET.2006.012030

Moh, M., Kukanur, R., Lin, X., & Dhar, S. (2009). Energy-efficient, self-organizing routing for wireless mobile networks. [Hershey, PA: IGI Global Publishing]. *International Journal of Business Data Communications and Networking*, *5*(4), 52–69. doi:10.4018/jbdcn.2009091704

Moh, M., & Li, J. (2010). A survey of reputation and trust systems for mobile ad-hoc network routing. In Stavroulakis, P., & Stamp, M. (Eds.), *Handbook of Information and Communication Security*. New York: Springer.

Pei, G., Gerla, M., Hong, X., & Chiang, C.-C. (1999). *A Wireless Hierarchical Routing Protocol with Group Mobility*. Proceedings of IEEE International Conference on Computer Communications and Networks (ICCCN), held in Boston, MA.

Perkins, C. E., Royer, E. M., & Das, S. R. (2003). *Ad Hoc On-demand Distance Vector Routing. RFC 3561*. IETF.

Ramanathan, R., & Streenstrup, M. (1998). Hierarchically organized, multi-hop mobile wireless networks for quality-of-service support. *Mobile Networks and Applications*, *3*, 101–119. doi:10.1023/A:1019148009641

Rieck, M. Q., & Dhar, S. (2008). Hierarchical Routing in Ad Hoc Networks using *k*-dominating Sets. *ACM Mobile Computing and Communications Review*, *12*(3), 45–57. doi:10.1145/1462141.1462147

Rieck, M. Q., Pai, S., & Dhar, S. (2005). Distributed Routing Algorithms for Multi-hop Ad Hoc Networks Using *d*-hop Connected *d*-Dominating Sets. [COMNET]. *Computer Networks Journal*, *47*(6), 785–799. doi:10.1016/j.comnet.2004.09.005

Singh, S., Woo, M., & Raghavendra, C. S. (1998). *Power-Aware Routing in Mobile Ad Hoc Networks*. Proc. ACM Annual International Conference on Mobile Computing and Networking (Mobicom).

Singh, S. R., & Raghavendra, C. S. (1998). PAMAS - Power Aware Multi-access Protocol with Signaling for Ad-Hoc Networks. *ACM Computer Communication Review*, *28*, 5–26. doi:10.1145/293927.293928

Sivakumar, R., Sinha, P., & Bharghavan, V. (1999). CEDAR: a Core-Extraction Distributed Ad Hoc Routing Algorithm. *IEEE Journal on Selected Areas in Communications*, *17*(8). doi:10.1109/49.779926

Stojmenovic, I., & Lin, X. (2001). Power-aware Localized Routing in Wireless Networks. *IEEE Transactions on Parallel and Distributed Systems*, *12*(11), 1122–1133. doi:10.1109/71.969123

Tseng, Y. C., Hsu, C. S., & Hsieh, T. Y. (2002). *Power-Saving Protocols for IEEE 802.11- Based Multi-hop Ad Hoc Networks*. Proceedings of IEEE International Conference on Computer Communications (INFOCOM), held in New York.

Tsu-Wei Chen & Mario Gerla. (1998).*Global State Routing: A New Routing Scheme for Ad-hoc Wireless Networks*. Proc. IEEE International Conference on Communications (ICC).

Wu, F. Dai, M. Gao, & I. Stojmenovic. (2002). On Calculating Power-aware Connected Dominating Sets for Efficient Routing in Ad Hoc Wireless Networks. *IEEE/KICS J. Communication Networks*, *4*(1), 59–70.

Wu, J., & Li, H. (2001). A Dominating-Set-Based Routing Scheme in Ad Hoc Wireless Networks. *Telecommunication Systems*, *18*(1-3), 13–36. doi:10.1023/A:1016783217662

Compilation of References

A. Iwata, C-C. Chiang, G. Pei, M. Gerla, & T.-W. Chen. (1999). Scalable Routing Strategies for Ad Hoc Wireless Networks. *IEEE Journal on Selected Areas in Communications, Special Issue on Ad-Hoc Networks*, 1369-79.

Aa, W. d., & Elfring, T. (2002). Realizing innovation in services. *Scandinavian Journal of Management*, *18*, 155–171. doi:10.1016/S0956-5221(00)00040-3

Agrawal, P. (2008). IP Multimedia Subsystems in 3GPP and 3GPP2: Overview and scalability issues. *IEEE Communications Magazine*, *46*(1), 138–145. doi:10.1109/MCOM.2008.4427242

Agüero, R., García-Arranz, M., & Muñoz, L. (2010). Accurate Simulation of 802.11 Indoor Links: A "Bursty" Channel Model Based on Real Measurements. *EURASIP Journal on Wireless Communications and Networking*, 2010.

Ahn, C. W., Kang, C. G., & Cho, Y. Z. (2003). Soft Reservation Multiple Access with Priority Assignment (SRMA/PA): A Distributed MAC Protocol QoS-guaranteed Integrated Services in Mobile Ad-hoc Networks. *IEICE Transactions on Communications. E (Norwalk, Conn.)*, *86-B*(1), 50–59.

Akimaru, H., & Kawashima, K. (1999). *Teletraffic, theory and applications*. Berlin, Germany: Springer-Verlag.

Akyildiz, I. F., & Wang, X. (2005). A survey on wireless mesh networks. *IEEE Communications Magazine*, *43*(9), S23–S30. doi:10.1109/MCOM.2005.1509968

Alderman, J. (2003). IPv6: Asia's Agent of Change. www.japaninc.com*; J@pan Inc* (pp. 24-27).

Aleksy, M., Butter, T., & Schader, M. (2008). Architecture for the development of context-sensitive mobile applications. *Mobile Information Systems*, *4*(2), 105–117.

Al-Karaki, J. N., & Ahmed, K. E. (2004). Routing techniques in wireless sensor networks: A survey. *IEEE Wireless Communications*, *11*(6), 6–28. doi:10.1109/MWC.2004.1368893

Al-Karaki, J. N., & Chang, J. M. (2004). A simple distributed access control scheme for supporting QoS in IEEE 802.11 wireless LANs. *Wireless Communications and Networking Conference* (Vol. 1) (pp. 213–218). *IEEE.*

Amichai-Hamburger, Y. (2002). Internet and personality. *Computers in Human Behavior*, *18*, 1–10. doi:10.1016/S0747-5632(01)00034-6

Annamalai, V., Gupta, S. K. S., & Schwiebert, L. (2003). On Tree-Based Convergecasting in Wireless Sensor Network. *Proceedings of the IEEE Wireless Communications and Networking (WCNC 2003)*, New Orleans, LA, USA.

ARCEP (2005). Annual Report.

AT&T. (2008). *Fixed-Mobile Convergence Designed to Lift Enterprises to a Higher Level of Performance*, White Paper. Retrieved January 2009, from http://www.business.att.com/enterprise/resource_item/Portfolio/enterprise-mobility-enterprise/Whitepaper/fixed_mobile_convergence_pov/

Atuahene-Gima, K. (1996). Market orientation and innovation. *Journal of Business Research*, *35*, 93. doi:10.1016/0148-2963(95)00051-8

B. J. Culpepper, L. Dung, & M. Moh. (2004). *Design and analysis of Hybrid Indirect Transmissions (HIT) for data gathering in wireless micro sensor networks*. ACM Mobile Computing and Communications Review (MC^{2}R), 61-83, Jan/Feb 2004.

Bader, A.-M., Nidal, N., Abu, A. N., & Ho, H. (2008). Efficient bandwidth management in Broadband Wireless Access Systems using CAC-based dynamic pricing. *33rd IEEE Conference on Local Computer Networks, (pp 484 – 491).*

Baiocchi, A., Priscoli, F., Grilli, F., & Sestini, F. (1995). The geometric dynamic Channel allocation as a practical strategy in mobile networks with bursty user mobility. *IEEE Transactions on Vehicular Technology*, *44*(1), 14–23. doi:10.1109/25.350265

Banerjee, S., & Khuller, S. (2001). *A Clustering Scheme for Hierarchical Control in Multi-hop Wireless Networks*. Proceedings of IEEE International Conference on Computer Communications (INFOCOM), held in Anchorage, Alaska.

Banerjee, S., & Misra, A. (2002). *Minimum energy paths for reliable communication in multi-hop wireless networks*. Proceedings of Annual Workshop on Mobile Ad Hoc Networking & Computing (MobiHOC 2002), 146–156.

Barceló, F. (2004). Performance analysis of handoff resource allocation strategies through the state-dependent rejection scheme. *IEEE Transactions on Wireless Communications*, *3*(3), 900–909.

Barceló, F., & Jordán, J. (2000). Channel holding time distribution in public telephony systems (PAMR and PCS). *IEEE Transactions on Vehicular Technology*, *49*(5), 1615–1625.

Barceló, F., Paradells, J., & Setaki, F. (2003). Design and modelling of Internode: A provider provisioned mobile VPN. *Mobile Networks and Applications*, *8*(1), 51–60.

Barceló, F., & Ramón, A. (1997, December). PRMA performance evaluation based on queueing theory. In *Proceedings of the IEEE International Conference on Personal Wireless Communications-ICPWC'97,* Mumbai, India (pp. 455- 459). IEEE.

Barceló, F., & Sánchez, J. I. (1999, June). Probability distribution of the inter-arrival time to cellular telephony channels. In *Proceedings of the 49th IEEE Vehicular Technology Conference (VTC'99),* Houston, TX (pp. 762-766). IEEE.

Bayer, M. (2007). *SQLAlchemy 0.3 Documentation*. http://www.sqlalchemy.org/docs/

Beijar, N. (2002). *Zone Routing Protocol (ZRP). Licentiate course on Telecommunications Technology*. Ad Hoc Networking.

Belding-Royer, E. M. (2004). Routing Approaches in Mobile Ad Hoc Networks. In S. Bagagni et al.(Eds). Mobile Ad Hoc Networking, 275-300. Washington, DC: IEEE Press.

Bicknell, D. (2007). *Preparing your business for IPv6* (pp. 26–28). Computer Weekly.

Bisnath, S., Wells, D., Howden, S., & Stone, G. (2003, September 22-26). The use of a GPS-equipped buoy for water level determination. In *Proceedings of IEEE OCEANS 2003,* San Diego, CA (pp. 1241-1246). IEEE.

Blefari-Melazzi, N., Sorte, D. D., & Reali, G. (2003). Accounting and pricing: a forecast of the scenario of the next generation Internet. *Computer Communications*, *26*, 2037–2051. doi:10.1016/S0140-3664(03)00146-4

Bohl, O., Manouchehri, S., & Winand, U. (2007). Mobile information systems for the private everyday life. *Mobile Information Systems*, *3*(3-4), 135–152.

Bohm, W. (2003). *Policy based architecture for the umts multimedia domain. Proceedings second ieee international symposium on network computing and applications* (pp. 275–285). IEEE Comput. Soc.

Boldrini, C., Conti, M., Iacopini, I., & Passarella, A. (2007). HiBOp: a history based routing protocol for opportunistic networks. *Proc. IEEE WoWMoM*.

Boleng, J., Navidi, W., & Camp, T. (2002). *Metrics to enable adaptive protocols for mobile ad hoc networks*. In Proc. of the International Conference on Wireless Networks (ICWN'02), pp. 293–298.

Bolotin, V. A. (1994, June). Telephone circuit holding time distribution. In *Proceedings of the 14th International Teletraffic Congress,* Antibes Juan-les-Pins, France (pp. 125-134). Elsevier Science.

Boudour, G., Teyssié, C., & Mammeri, Z. (2008). Reservation Clash Handling to Optimize Bandwidth Utilization in MANETs. *In proceedings of the International Conference on Communication Theory, Reliability, and Quality of Service (CTRQ'2008),* Bucharest, Romania, (pp. 77-82).

Boukerche, A., Tingxue, H., & Kaouther, A. (2005, September). Design and performance evaluation of a QoS-based dynamic channel allocation protocol for wireless and mobile networks. In *Proceedings of 13th IEEE/ACM International Symposium on Modeling, Analysis, and Simulation of Computer and Telecommunication Systems,* Atlanta, Georgia (pp. 445-452). IEEE.

Bracewell, R. N. (2000). *1* (3rd ed.). New York: McGraw-Hill.

Bregni, S., Cioffi, R., & Dècina, M. (2006, November 27-December 1). *An empirical study on statistical properties of GSM telephone call arrivals.* Paper presented at the IEEE Global Telecommunications Conference (Globecom'06), San Francisco.

Bryan, K. L., Ren, T., DiPippo, L., Henry, T., & Fay-Wolfe, V. (2007). *Towards Optimal TDMA Frame Size in Wireless Sensor Networks (Technical Report, TR-xxx).* University of Rhode Island.

Bundesnetzagentur (2009). *Tätigkeitsbericht 2008/2009* (in German)

Burger, E. (2006). *RFC 4483 A Mechanism for Content Indirection in Session Initiation Protocol (SIP) Messages.* IETF.

Burgess, J., Gallagher, B., Jensen, D., & Levine, B. N. (2006). MaxProp: routing for vehicle-based disruption tolerant networks. *Proc. IEEE INFOCOM.*

Burns, B., Brock, O., & Levine, B.N. (2005). MV routing and capacity building in disruption tolerant networks. *Proc. IEEE Infocom.*

Cabral, J. E. D. O. (1998). Survey on technological innovative behavior in the brazilian food industry. *Scientometrics, 42*, 129–169. doi:10.1007/BF02458353

Camarillo, G., & Garcia-Martin, M. (2006). *The 3G IP Multimedia Subsystem: Merging the Internet and the Cellular worlds.* Hoboken, NJ: J. Wiley & Sons.

Camp, T., Boleng, J., & Davies, V. (2002). A survey of mobility models for ad hoc network research. *Wireless Communication and Mobile Computing, 2*(5), 483–502. doi:10.1002/wcm.72

Canalys (2006). *Research release 2006/081.* Retrieved January 2009, from http://www.canalys.com/pr/2006/r2006081.pdf

Canalys (2008). *Global Mobile Navigation Device Shipments Hit 39 Million in 2007.* Retrieved January 2009, from http://www.canalys.com/pr/2008/r2008031.htm

Candan, I., & Salamah, M. (2007). Analytical modeling of a time-threshold based bandwidth allocation scheme for cellular networks. *Computer Communications, 30*, 1036–1043.

CapGemini (2005). Growing Mobile Data Revenues, Opportunities in Infotainment. *Telecom & Media Insights,* (11), November 2005

Carpenter, B. E. (2003). Future applications and middleware, and their impact on the infrastructure. *Future Generation Computer Systems, 19*, 191–197. doi:10.1016/S0167-739X(02)00145-0

Cavin, D., Sasson, Y., & Schiper, A. (2002). *On the accuracy of MANET simulators.* In Proc. of the second ACM international workshop on Principles of mobile computing table of contents'. Toulouse, France.

Cellular-news. (2009). *Bouygues Telecom to Trial Alcatel-Lucent's IMS Platform.* Published 28 May 2009 in http://www.cellular-news.com/story/37708.php

Chaintreau, A., Mtibaa, A., Massoulie, L., & Diot, C. (2007). The diameter of opportunistic mobile networks. *Proc. ACM CoNEXT Conference,* (pp. 1-12).

Chang, J.-Y., & Chen, H.-L. (2003). Dynamic-Grouping bandwidth reservation scheme for multimedia wireless networks. *IEEE Journal on Selected Areas in Communications, 21*(10), 1566–1574. doi:10.1109/JSAC.2003.814863

Chang, R., Gundala, S., Moh, T.-S., & Moh, M. (2009). *VESS: A Versatile Extensible Security Suite for MANET.* (Invited paper) Proc. IEEE Pacific Rim Conference on Communications, Computers and Signal Processing (PACRIM'09), 944 – 950, held in Victorian, Canada, Aug 2009.

Charging architecture and principles (2007). *Stage 2.* TS 32.240, 3GPP.

Charging management: IP Multimedia Subsystem (IMS) charging (2007). TS 32.260, 3GPP.

Charging management; Online Charging System (OCS): Applications and interfaces (2006). TS 32.296, 3GPP.

Chen, F., Dietrich, I., German, R., & Dressler, F. (2009). An Energy Model for Simulation Studies of Wireless Sensor Networks using OMNeT++. [PIK]. *Praxis der Informationsverarbeitung und Kommunikation, 32*(2), 133–138. doi:10.1515/piko.2009.0023

Chen, J.-L., Chao, H.-C., & Kuo, S.-Y. (2006). IPv6: More than protocol for next generation Internet. *Computer Communications, 29*, 3011–3012. doi:10.1016/j.comcom.2005.11.001

Chen, B., Jamieson, K., Balakrishnan, H., & Morris, R. (2001). *Span: An Energy-efficient Coordination Algorithm for Topology Maintenance in Ad Hoc Wireless Networks*. Proc. ACM Annual International Conference on Mobile Computing and Networking (Mobicom), 85-96.

Chen, X., & Fang, Y. (2003). An Adaptive Bandwidth Reservation Scheme in Multimedia Wireless Networks. *Global Telecommunications Conference, GLOBECOM '03. IEEE, 5*, 2830–2834.

Chlamtac, I., Ganz, A., & Karmi, G. (1992). Lightpath communications: an approach to high bandwidth optical WAN's. *IEEE Transactions on Communications, 40*(7), 1171–1182. doi:10.1109/26.153361

Chlamtac, I., & Kutten, S. (1987). Tree-based Broadcasting in Multihop Radio Networks. *IEEE Transactions on Computers, C-36*(10). doi:10.1109/TC.1987.1676861

Chlebus, E. (1997, June). Empirical validation of call holding time distribution in cellular communications systems. In *Proceedings of the 15th International Teletraffic Congress,* Washington, DC (pp. 1179-1188). Elsevier Science.

Cho, H. S., & Sung, D. K. (1998). Channel holding time distribution in cross- and cigar-shaped urban microcells. *IEICE Transactions on Communications. E (Norwalk, Conn.), 81-B*(6), 1275–1279.

Christensen, T. K., Nielsen, B. F., & Iversen, V. B. (2001, December). Distribution of channel holding time in cellular communication networks. In *Proceedings of the 17th International Teletraffic Congress,* Salvador de Bahia, Brazil (pp. 471-480). Elsevier Science

Chung, Y.-L., & Tsa, Z. (2006). Modeling and Analysis of Dynamic Spectrum Allocation of Two Wireless Communication Systems. *IEEE 17th International Symposium on Personal, Indoor and Mobile Radio Communications* (pp. 1–5).

Clausen, T., Jacquet, P., Laouiti, A., Muhlethaler, P., Qayyum, A., & Viennot, L. (2001). *Optimized Link State Routing Protocol for Ad Hoc Networks*. IEEE International Multitopic Conference (INMIC), held in Pakistan.

Clyman, J. (2007). The Insider's Guide to Windows Vista. (cover story). *PC Magazine, 26*, 74–83.

Cooper, D. E., Ezhilchelvan, P., & Mitrani, I. (2004). High coverage broadcasting for mobile ad-hoc networks. *Lecture Notes in Computer Science, 3042*, 100–111. doi:10.1007/978-3-540-24693-0_9

Cooper, D. E., Ezhilchelvan, P., Mitrani, I., & Vollset, E. (2005). Optimization of encounter gossip propagation in mobile ad-hoc networks. *Proc. 13th MASCOTS*, (pp. 529-532).

Cormen, T. H., Leiserson, C. E., & Rivest, R. L. (2000). *Introduction to Algorithms*. The MIT Press.

CRAWDAD. (2008). *CRAWDAD – A community resource for archiving wireless data at Dartmouth*. Retrieved from http://www.crawdad.org

Cronbach, L. J. (1951). Coefficient alpha and the internal structure of tests. *Psychometrika, 16*, 297–334. doi:10.1007/BF02310555

Cuomo, F., Luna, S. D., Monaco, U., & Melodia, T. (2007, June 24-28). Routing in ZigBee: Benefits from exploiting the IEEE 802.15.4 association tree. In *Proceedings of the IEEE International Conference on Communications 2007,* Glasgow, Scotland (pp. 3271-3276).

Current Analysis. (2008). *IMS Status Report: A Protracted Adoption – Advisory Report*. Published 20 June 2008 in http://www.currentanalysis.com/m/ericsson/CurrentAnalysis-IMS.pdf

Current Analysis. (2010). *DT Offers Fresh Approach to Converged Fixed, Mobile and IT.* Published 8 Mar 2010 in http://www.currentanalysis.com/h/2010/DT-DeutschlandLAN.asp

Dahlman, E., Parkvall, S., Sköld, J., & Beming, P. (2007). *3G Evolution: HSPA and LTE for Mobile Broadband* (1st ed.). New York: Elsevier Ltd.

Das, B., & Bharghavan, V. (1997). *Routing in Ad-Hoc Networks Using Minimum Connected Dominating Sets.* Proc. IEEE Int'l Conf. Communications, 376-380

DasBit, S., & Mitra, S. (2003). Challenges of computing in mobile cellular environment - a survey. *Computer Communications, 26,* 2090–2105. doi:10.1016/S0140-3664(03)00117-8

Davis, F. D. (1989). Perceived Usefulness, Perceived Ease Of Use, And User Accep. *Management Information Systems Quarterly, 13,* 319. doi:10.2307/249008

Davis, F. D., Bagozzi, R. P., & Warshaw, P. R. (1989). User Acceptance Of Computer Technology: A Comparison Of Two. *Management Science, 35,* 982. doi:10.1287/mnsc.35.8.982

Davis, J. A., Fagg, A. H., & Levine, B. N. (2001). Wearable computers as packet transport mechanisms in highly partitioned ad-hoc networks. *Proc. 5th IEEE Intl. Symp. on Wearable Computers.*

de Waal, C., & Gerharz, M. (2008). *BonnMotion – A mobility scenario generation and analysis tool.* Retrieved from http://web.informatik.uni-bonn.de/IV/Mitarbeiter/dewaal/BonnMotion/.

Dell, P., Kwong, C., & Ying, L. (2008). Some reflections on IPv6 adoption in Australia. *Info, 10,* 3–9. doi:10.1108/14636690810874034

Deutsche Telekom. (2009). *Invitation to the Extraordinary Shareholders' Meeting.* Bonn, October 2009

Dhar, S., & Rieck, M. Q. (2007). An Energy-Efficient Heterogeneous Dual Routing Scheme for Mobile Ad Hoc and Sensor Networks. *International Journal on Mobile Network Design and Innovation, 2*(2), 105–115. doi:10.1504/IJMNDI.2007.015060

Dhar, S., Rieck, M. Q., Pai, S., & Kim, E. J. (2004). Distributed Routing Schemes for Ad Hoc Networks Using dSPR sets. *Journal of Microprocessors and Microsystems. Special Issue on Resource Management in Wireless and Ad Hoc Mobile Networks, 28*(8), 427–437.

Diameter applications: 3GPP specific codes and identifiers (2008). TS 29.230, 3GPP.

DTN Research Group. (2008). *Delay Tolerant Networking Research Group.* Retrieved from http://www.dtnrg.org/wiki

Dubois-Ferriere, H., Grossglauser, M., & Vetterli, M. (2003). Age matters: efficient route discovery in mobile ad hoc networks using encounter ages. *Proc. 4th ACM Intl. Symp. MobiHoc,* (pp. 257-266).

Duncan, O. D. (1966). Path analysis: Sociological example. *American Journal of Sociology, 72,* 1–16. doi:10.1086/224256

Durham, D., Boyle, J., Cohen, R., Herzog, S., Rajan, R., & Sastry, A. (2000). *RFC 2748 The COPS (Common Open Policy Service) Protocol.* IETF.

Durrett, R. (2001). *Essentials of Stochastic Processes.* Harrisonburg, VA: Springer.

E. Jung & N. Vaidya. (2005). *Improving IEEE 802.11 Power Saving Mechanism.* IEEE Trans. on Automatic Control. July 2005.

ECOSYS. (2004-2007). *ECOSYS project website.* Retrieved January 2009, from http://www.optcomm.di.uoa.gr/ecosys/deliverableslist.html

Editorial. (2003, July 21). IPv6It's time. *eweek,* 48.

Elmasri, R., & Navathe, S. B. (1994). *Fundamentals of Database Systems* (2nd ed.). Benjamin/Cummings.

Emory, C. W., & Cooper, D. R. (1991). *Business research method* (4th ed.). Homewood, IL: Irwin.

End-to-end Quality of Service concept and architecture (2007). TS 23.207, 3GPP.

ERG (European Regulators Group). (2009). *Report on Fixed-Mobile Convergence: Implications on Competition and Regulatory Aspects.* Retrieved November 2010 from http://www.erg.eu.int/doc/publications/2009/erg_09_06_report_on_fixed_mobile_convergence.pdf

Eriksson, J., Faloutsos, M., & Krishnamurthy, S. (2004). *The Case for Dynamic Addressing*. Proceedings of IEEE International Conference on Computer Communications (INFOCOM).

ETSI, *Universal Mobile Telecommunications System (UMTS)* (1998). *Selection procedures for the choice of radio transmission technologies of the UMTS (UMTS 30.03 version 3.2.0).*

European Communities. (2007). *Europe in Figures: Eurostat Yearbook 2006-07*. Luxembourg

Evangelista, R., Perani, G., Rapiti, F., & Archibugi, D. (1997). Nature and impact of innovation in manufacturing industry: Some evidence from the Italian innovation survey. *Research Policy, 26*, 521–536. doi:10.1016/S0048-7333(97)00028-0

Everett, C. (2008). An IP address for every machine. *Wireless Asia, 11*, 28–28.

Exane BNP Paribas & Arthur D Little. (2006). *Facing Off on Convergence*. Retrieved January 2009, from http://www.cellular-news.com/story/16742.php

Fall, K., & Varadhan, K. (1999). *ns Notes and Documentation*. The VINT Project, UC Berkeley, LBL, USC/ISI, and Xerox PARC. Available http://www-mash.cs.erkeley.EDU/ns/.

Fang, Y., Chlamtac, I., & Lin, Y. B. (1998). Channel occupancy times and handoff rate for mobile computing and PCS networks. *IEEE Transactions on Computers, 47*(6), 679–692.

Fazio, M., Villari, M., & Puliafito, A. (2006). IP Address Autoconfiguration in Ad Hoc Networks: Design, implementation and Measurements. *Computer Networks Journal Elsevier Science, 50*(7), 898–920. doi:10.1016/j.comnet.2005.04.017

Feeney, L. M. (2002). *A QoS Aware Power Save Protocol for Wireless Ad Hoc Networks*. Proc. IFIP Mediterranean Ad Hoc Networking Workshop (Med-Hoc-Net), held in Sardegna, Italy.

Fehnker, A. Fruth, M. McIver, A. K. (2007). *Graphical modelling for simulation and formal analysis of wireless network protocols*. In Proc. Workshop on Methods, Models and Tools for Fault-Tolerance (MeMoT'07).

Ferguson, P., & Huston, G. (1998). *Quality of Service: Delivering QoS on the Internet and in Corporate Networks* (1st ed.). New York: John Wiley & Sons, Inc.

Fernando, G. L., Goldberg, D. E., & Pelikan, M. (2000, July 8-12). Time complexity of genetic algorithms on exponentially scaled problems. *Proceedings of the Genetic and Evolutionary Computing Conference,* Las Vegas, NV (pp. 151-158). Morgan Kaufmann.

FICORA. (2009). *Market overview 2/2009 by FICORA (in Finnish).* Retrieved from http://www.ficora.fi/attachments/suomimq/5jdCGapBJ/Markkinakatsaus_2_2009.pdf

Forum Rep, U. M. T. S. 33 (2003). *3G Offered Traffic Characteristics*.London: UMTS Forum.

Forum Rep, U. M. T. S. 37 (2005). *Magic Mobile Future 2010-2020*, London: UMTS Forum.

Forum White Paper, U. M. T. S. (2009). *Mobile Broadband Evolution: the roadmap from HSPA to LTE,* London.

Forum White Paper, U. M. T. S. (2010). *Recognising the Promise of Mobile Broadband*. London

Francois, J. M., & Leduc, G. (2005). Mobility prediction's influence on QoS in wireless networks: a study on a call admission algorithm. *Third International Symposium on Modeling and Optimization in Mobile, Ad Hoc, and Wireless Networks* (pp. 238-247).

Galeana, H., Ferrus, R., & Olmos, J. (2007, March). Transport Capacity Estimations for Over-Provisioned UTRAN IP-Based Networks. *Wireless Communications and Networking Conference (WCNC)* (pp. 4295-4300). Washington, DC:IEEE

Galstad, E. (2007). *Nagios*. http://www.nagios.org/

Gamberini, G., Giudici, F., Pagani, E., & Rossi, G. P. (2008). Impact of history on epidemic broadcast in DTNs. *Proc. 1st IFIP Wireless Days Conference*.

Gandham, S., Zhang, Y., & Huang, Q. (2008). Distributed Time-optimal Scheduling for Convergecast in Wireless Sensor Networks. [Elsevier.]. *Computer Networks, 52*(3), 610–629.

Garcia, R., & Calantone, R. (2002). A critical look at technological innovation typology and innovativeness terminology: A literature review. *Journal of Product Innovation Management*, *19*, 110–132. doi:10.1016/S0737-6782(01)00132-1

Gartner (2006). *Forecast: GPS in Mobile Devices, Worldwide, 2004-2010*. Retrieved January, 2009, from http://gartner.com/DisplayDocument?id=498745

Georgantas, N. V., Papadoukakis, N. K., Liveris, A. D., & Sykas, E. D. (1999). Modelling of a bandwidth management scheme for DECT wireless access to a local area network. *Conference on Local Computer Networks* (pp. 43–48).

Getz, M., Siegfried, J. J., & Anderson, K. H. (1997). Adoption of innovations in higher education. *The Quarterly Review of Economics and Finance*, *37*, 605–631. doi:10.1016/S1062-9769(97)90013-2

Ghassemian, M., Friderikos, V., & Aghvami, A. H. (2005). *On mobility metrics applied for ad hoc network protocol evaluation.* In Proc. of the 7th IFIP International Conference on Mobile and Wireless Communications Networks (MWCN 2005), Marrakech, Morocco.

Giordano, S. (2002). Mobile ad-hoc networks. I. Stojmenovic (Ed.), *Handbook of Wireless Networks and Mobile Computing* (pp. 325–343, 371–391.) New York: Wiley.

Giudici, F., Pagani, E., & Rossi, G. P. (2009). Self-adaptive and stateless broadcast in delay and disruption-tolerant networks. *Proc. Italian Networking Workshop.*

Global Insight, I. H. S. (2009). *Vodafone Spain Launches DSL/3G Bundle*. Published 12 Jun 09 in http://www.ihsglobalinsight.com/SDA/SDADetail17039.htm

GloMoSim. (n.d.). *Global Mobile Information Systems Simulation Library*. Retrieved from http://pcl.cs.ucla.edu/projects/glomosim/.

GPP system to Wireless Local Area Network interworking (2007). TS 23.234, 3GPP.

GPP TR 23.806 (2007). Voice Call Continuity between CS and IMS Study (Release 7)

GPP TS 23.228 (2008). IP Multimedia Subsystem; Stage 2 (Release 8)

GPP TS 26.236 (2008). Packet Switched Conversational Multimedia Applications; Transport protocols (Release 7)

GPP TS 43.318 (2009). *Generic Access Network (GAN)*; Stage 2 (Rel. 9)

Grafakos, L. (2003). *Classical and Modern Fourier Analysis*. Upper Saddle River, NJ: Prentice Hall.

Grewal, J., & DeDourek, J. (2004). Provision of QoS in wireless networks. *Second Annual Conference on Communication Networks and Services Research (pp. 337–340).*

Grillo, D., Skoog, R. A., Chia, S., & Leung, K. K. (1998). Teletraffic engineering for mobile personal communications in ITU-T work: The need to match practice and theory. *IEEE Personal Communications*, *5*(6), 38–58.

Guerin, R. A. (1987). Channel occupancy time distribution in a cellular radio system. *IEEE Transactions on Vehicular Technology*, *35*(3), 89–99.

Guice, J. (1998). Looking backward and forward at the Internet. *The Information Society*, *14*, 201–211. doi:10.1080/019722498128827

Gulliver, S. R., Ghinea, G., Patel, M., & Serif, T. (2007). A context-aware Tour Guide: User implications. *Mobile Information Systems*, *3*(2), 71–88.

Guo, H.-B., & Kuo, G.-S. (2005). A dynamic and adaptive bandwidth management scheme for QoS support in wireless multimedia networks. *IEEE 61st Vehicular Technology Conference* (Vol. 3) (pp. 2081–2085).

Gupta, S. K., Low, M. K., & Tan, C. W. (1998, April 20-23). A new approach to simulate GPS measurements. In *Proceedings of the 1998 IEEE Position Location and Navigation Symposium,* Palm Springs, CA (pp. 236-242). IEEE.

Haas, Z. J., Deng, J., Liang, B., Padadimitratos, P., & Sajama, S. (2002). Wireless Ad Hoc Networks. In Proakis (Ed.) *Encyclopedia of Telecommunications*.New York: John Wiley.

Hac, A. (1991). Bandwidth management in the switch with various traffic burstiness. *Third IEE Conference on Telecommunications* (pp. 343–347).

Hac, A., & Chen, Z. (1999, September 19-22). Hybrid channel allocation in wireless networks. In *Proceedings of the IEEE VTS 50th Vehicular Technology Conference*, Houston, TX (pp. 2329-2333). IEEE.

Hakala, H., Mattila, L., Koskinen, J.-P., Stura, M., & Loughney, J. (2005). *RFC 4006 Diameter Credit-Control Application*. IETF.

Hammons, M. J., & Chisholm, G. (2006). Enabling total asset visibility. *Defense Transportation Journal, 62*(4).

Harras, K. A., Almeroth, K. C., & Belding-Royer, E. M. (2005). Delay tolerant mobile networks (DMTNs): controlled flooding in sparse mobile networks. *Proc. IFIP Networking*.

Heidemann, J., Bulusu, N., Elson, J., & Intanagonwiwat, C. Lan, Ya Xu, K.-C-, Ye, W., Estrin, D., & Govindan, R.(2001). *Effects of Detail in Wireless Network Simulation*. In Proceedings of the SCS Multiconference on Distributed Simulation, pp. 3-11. Phoenix, Arizona, USA.

Heinzelman, W., Chandrakasan, A., & Balakrishnan, H. (2000). *Energy-Efficient Communication Protocol for Wireless Microsensor Networks*. IEEE Proc. Hawaii International Conf. on System Sciences.

Higgins, J. R. (1996). *Sampling Theory in Fourier and Signal Analysis*. New York: Oxford Science Publications.

Hjalager, A.-M. (1997). Innovation patterns in sustainable tourism an analytical typology. *Tourism Management, 18*, 35–41. doi:10.1016/S0261-5177(96)00096-9

Ho-In, J., & Yeonsoo, K. (2007, February 12-14). Efficient, real-time short address allocations for USN devices using LAA (last address assigned) algorithm. In *Proceedings of the 9th International Conference on Advanced Communication Theory*, Gangwon-Do, South Korea (pp. 689-693). IEEE.

Hovav, A., Patnayakuni, R., & Schuff, D. (2004). A model of Internet standards adoption: the case of IPv6. *Information Systems Journal, 14*, 265–294. doi:10.1111/j.1365-2575.2004.00170.x

Howell, K. B. (2001). *Principles of Fourier analysis*. Boca Raton, FL: Chapman and Hall/CRC.

Hsieh, H. Y., & Sivakumar, R. (2002). IEEE 802.11 over Multi-hop Wireless Networks: Problems and New Perspectives. *In proceedings of the IEEE Vehicular technology Conference 2002 Fall: Vol(3).* (748-752).

Hsu, C., & Tseng, Y. C. Hsu, & Y. Tseng. (2005). *Cluster-Based Semi-Asynchronous Power-Saving Protocols for Multi-hop Ad Hoc Networks*. Proc. IEEE International Conference on Communications (ICC).

Hsu, C.-S., Sheu, J.-P., & Tung, S.-C. (2006). An On-Demand Bandwidth Reservation QoS Routing Protocol for Mobile Ad Hoc Networks. *IEEE International Conference on Sensor Networks, Ubiquitous, and Trustworthy Computing* (Vol. 2) (pp. 198–207).

Huang, C.-J., Chuang, Y.-T., Lai, W. K., Sun, Y.-H., & Guan, C.-T. (2007). Adaptive rescurce reservation schemes for proportional DiffServ enabled fourth-generation mobile communications system. *Computer Communications Journal, 1*(7), 1613–1623. doi:10.1016/j.comcom.2007.01.015

Hui, P., Chaintreau, A., Scott, J., Gass, R., Crowcroft, J., & Diot, C. (2005). Pocket switched networks and human mobility in conference environments. *Proc. ACM SIGCOMM Workshop on Delay Tolerant Networking*.

Hwang, W.-S., & Tseng, P.-C. (2005). A QoS-aware residential gateway with bandwidth management. *IEEE Transactions on Consumer Electronics, 51*(3), 840–848. doi:10.1109/TCE.2005.1510493

IEEE 802.11 WG. (2005). Drat supplement to standard for telecommunications and information exchange between systems-LAN/MAN specific requirements-part11: MAC enhancements for Quality of service (QoS). *IEEE 802.11e standard Draft/D13.0,* January 2005.

IEEE Std. 802.11. (1999), Part11: Wireless LAN Medium Access Control (MAC) and Physical Layer (PHY) Specifications, Reference number ISO/IEC 8802-11:1999(E), *IEEE Std 802.11,* 1999 edition.

IEEE. (1999). *Wireless LAN Medium Access Control and Physical Layer specifications*. IEEE 802.11 Standard.

IEEE. (2003). *Wireless medium access control (MAC) and physical layer (PHY) specifications for low rate wireless personal area networks.* Washington, DC: Author.

Infocom (2010). *Residential Fixed Mobile Substitution (FMS) and Convergence (FMC) – Showing Caution After the Hype*. Retrieved November 2010 from http://www.infocom-de.com/pressarchives/press_051010.html

Infrastrutture Software per Reti Ad-hoc Orientate ad Ambienti Difficili (2002-2007). *Consiglio Nazionale delle Ricerche (CNR) with the Strategic IS-MANET Project.* http://zeus.elet.polimi.it/is-manet/.

Intellinet (2009). *Mobile Data Offload for 3G Networks – A Whitepaper*. Retrieved November 2010 from http://www.intellinet-tech.com/Media/PagePDF/Data%20 Offload.pdf

International Game Developers Association. (2005). *Mobile Games White Paper.*

International Telecommunications Union. (2005). *The Internet of things*. Retrieved February 17, 2009, from http://www.itu.int/internetofthings

Internet Telephony. (2010). *Revisiting the Femtocell: Carriers Move To Commercial Deployments, Equipment Prices Drop*. Published 01 October 2010 in http://www.tmcnet.com/voip/features/articles/114440-revisiting-femtocell-carriers-move-commercial-deployments-equipment-prices.htm

Inwhee, J. (2004). Qos-Aware MAC With Reservation For Mobile Ad-Hoc Networks. *In proceedings of the IEEE Vehicular Technology Conference 2004 Fall: Vol(2).* (1108-1112).

IP Multimedia Subsystem (2007). *Stage 2*. TS 23.228, 3GPP.

Isaacs, K., Hsieh, J., & Moh, M. (2010). Extending OSPF for MANET Routing. In Khan, S., Lloret, J., Ortiz, J. H., & Loo, J. (Eds.), *Mobile Ad hoc Networks: Current Status and Future Trends*. New York: Taylor & Francis Publications.

ITU-T Recommendation G.992.2 (1998). *High bit rate Digital Subscriber Line (HDSL) transceivers*.

ITU-T Recommendation G.992.1 (1999). *Asymmetric digital subscriber line (ADSL) transceivers,* G.992.2 (1999). *Splitterless asymmetric digital subscriber line (ADSL) transceivers,* G.992.2 (2002). *Asymmetric digital subscriber line transceivers 2 (ADSL2),* G.992.2 (2003). *Asymmetric Digital Subscriber Line (ADSL) transceivers - Extended bandwidth ADSL2 (ADSL2plus)*.

ITU-T Recommendation Q.2630.2. (2000). *AAL Type 2 Signaling Protocol (Capability Set 2)*.

Jayaputera, J., & Taniar, D. (2005). Data retrieval for location-dependent queries in a multi-cell wireless environment. *Mobile Information Systems*, *1*(2), 91–108.

Jedrzycki, C., & Leung, V. C. M. (1996, April 28-May 1). Probability distribution of channel holding time in cellular telephony systems. In *Proceedings of the IEEE 46th Vehicular Technology Conference*, Vancouver, British Columbia, Canada (pp. 247-251). IEEE.

Johne, A. (1999). Using market vision to steer innovation. *Technovation*, *19*, 203–207. doi:10.1016/S0166-4972(98)00112-6

Jones, E. P. C., Li, L., & Ward, P. A. S. (2005). Practical routing in delay tolerant networks. *Proc. ACM SIGCOMM Workshop on Delay Tolerant Networking*, (pp. 237-243).

Juang, P., Oki, H., Wang, Y., Martonosi, M., Peh, L., & Rubenstein, D. (2002). Energy-efficient computing for wildlife tracking: design tradeoffs and early experiences with Zebranet. *Proc. ASPLOS.*

Jung, S. H., Hong, J. W., Chang, W., & Lie, C. H. (2009). Quality of service estimation for soft handoff region ratio and call admission control in CDMA cellular systems. *Computers & Industrial Engineering*, *56*, 1429–1441.

Jung, Y. J., Lee, Y. K., Lee, D. G., Ryu, K. H., & Nittel, S. (2008, July 7-11). Air pollution monitoring system based on geosensor network. In *Proceedings of the IEEE International Geoscience & Remote Sensing Symposium,* Boston (pp. III-1370-III-1373). IEEE.

Juniper (2006). *Mobile Navigation Finds its Way*. Retrieved January, 2009, from http://www.wirelessweek.com/article.aspx?id=110108

Kano, S. (2000). Technical innovations, standardization and regional comparison - A case study in mobile communications. *Telecommunications Policy*, *24*, 305–322. doi:10.1016/S0308-5961(00)00020-3

Katz, J. E., & Aspden, P. (1998). Theories, Data, and Potential Impacts of Mobile Communications - A Longitudinal Analysis of U.S. National Surveys. *Technological Forecasting and Social Change*, *57*, 133–156. doi:10.1016/S0040-1625(97)00089-9

Keegan, A., & Turner, J. R. (2002). The management of innovation in project-based firms. *Long Range Planning*, *35*, 367–388. doi:10.1016/S0024-6301(02)00069-9

Kenny, D. A. (1979). *Correlation and causality*. New York: Wiley.

Khanbary, L. M. O., & Vidyarthi, D. P. (2008). A GA based fault-tolerant model for channel allocation in mobile computing. *IEEE Transactions on Vehicular Technology*, *57*(3), 1823–1833. doi:10.1109/TVT.2007.907311

Kim, E. J., Dhar, S., & Rieck, M. Q. (2004). *Coordinator-Oriented Forwarding for Energy Efficiency in Large Scale Wireless Ad hoc Networks*. Proc.s of the International Conference on Information Systems (ISNG), 44-49, held in Las Vegas.

Kimura, T. (1994). Approximations for multi-server queues: System interpolations. *Queueing Systems*, *17*, 347–382.

Kuckartz, M. (2001). Innovation market - the economic exploitation of property rights in high-quality inventions. *World Patent Information*, *23*, 67–70. doi:10.1016/S0172-2190(00)00098-3

Kuhl, C. (2007). 'Vista' worries far off on cable's horizon: New OS supports IPv6, viewed as potential upgrade for online services. *CED*, *33*, 42–46.

Kuhne, R. (2007). Charging in the ip multimedia subsystem: A tutorial. *IEEE Communications Magazine*, *45*(7), 92–99. doi:10.1109/MCOM.2007.382666

Kulkarni, S. S., & Arumugam, M. U. (2005). *SS-TDMA: A Self-Stabilizing MAC for Sensor Networks. Sensor Network Operations*. IEEE Press.

Kumar, S., & Zahn, C. (2003). Mobile communications: Evolution and impact on business operations. *Technovation*, *23*, 515–520. doi:10.1016/S0166-4972(02)00120-7

Kuo, G.-S., Ko, P.-C., & Kuo, M.-L. (2000). A probabilistic resource estimation and semi-reservation scheme for flow-oriented multimedia wireless networks. *IEEE Wireless Communications and Networking Conference, (WCNC 2000)* (Vol. 3) (pp. 1046 – 1051).

Kurkowski, S., Camp, T., & Colagrosso, M. (2005). Manet simulation studies: the incredibles. *ACM SIGMOBILE Mobile Computing and Communications Review*, *9*(4), 50–61. doi:10.1145/1096166.1096174

Kwon, J., Kim, W., & Dong, L. (2001, July 9-13). Absolute kinematic GPS positioning for remote area. In *Proceedings of the IEEE International Geoscience and Remote Sensing Symposium*, Sydney, Australia (pp. 2067-2069). IEEE.

Land, K. C. (1969). Principles of path analysis. *American Sociological Association 1*.

LaRose, R., & Hoag, A. (1996). Organizational adoptions of the internet and the clustering of innovations. *Telematics and Informatics*, *13*, 49–61. doi:10.1016/0736-5853(96)00003-2

Lauro, O., & Aghvami, A. H. (2003). *Resource allocation in hierarchical cellular systems*. Norwood, MA: Artech House.

Lederer, A. L., Maupin, D. J., Sena, M. P., & Zhuang, Y. (2000). The technology acceptance model and the World Wide Web. *Decision Support Systems*, *29*, 269–282. doi:10.1016/S0167-9236(00)00076-2

Lee, D. K., Zhu, M., & Hu, H. (2005). When location-based services meet databases. *Mobile Information Systems*, *1*(2), 81–90.

Lee, U., Magistretti, E., Zhou, B., Gerla, M., Bellavista, P., & Corradi, A. (2006). MobEyes: smart mobs for urban monitoring with a vehicular sensor network. *IEEE Wireless Communications*, *13*(5).

Lenders, V., Karlsson, G., & May, M. (2007). Wireless ad hoc podcasting. *Proc. 4th IEEE Conf. SECON*, (pp. 273-283).

Lewis, P. A. W., & Orav, E. J. (1988). *Simulation Methodology for Statisticians. Operations Analysts and Engineers* (*Vol. I*). New York: Wadsworth & Brooks/Cole Advanced Books & Software.

Li, E. (1999). *Report on Taiwan's network usage for 1998* (pp. 38–41). Advertising Magazine.

Li, C. C. (1975). Path analysis: A primer. Paci®c Grove

Li, J., Moh, T.-S., & Moh, M. (2009). *Path-Based Reputation System for MANET Routing*. Proc. of the 7th International Conference on Wired/Wireless Internet Communications, 48 – 60, held in Enschede, the Netherlands.

Li, X., Zeng, Y. Z., Kracker, B., Schelb, R., Görg, C., & Timm-Giel, A. (2008). Carrier Ethernet for Transport in UMTS Radio Access Network: Ethernet Backhaul Evolution. *Vehicular Technology Conference, 2008. VTC Spring 2008* (pp. 2537 – 2541). Washington, DC:IEEE

Liang, B., & Haas, Z. J. (2000). *Virtual Backbone Generation and Maintenance in Ad Hoc Network Mobility Management.* Proc. IEEE International Conference on Computer Communications (INFOCOM), 1293-1302, held in Tel Aviv.

Lim, S., Cao, G., & Das, C. R. (2001). *A Differential Bandwidth Reservation Policy for Multimedia Wireless Networks* (pp. 447–452). Parallel Processing Workshops.

Lim, S. Y., Taniar, D., & Srinivasan, B. (2005). On-Mobile Query Processing Incorporating Multiple Non-Collaborative Servers. *Ingénierie des Systèmes d'Information, 10*(5), 9–38. doi:10.3166/isi.10.5.9-38

Lin, J. C.-C., & Lu, H. (2000). Towards an understanding of the behavioural intention to use a web site. *International Journal of Information Management, 20*, 197–208. doi:10.1016/S0268-4012(00)00005-0

Lin, C. R., & Gerla, M. (1997). Asynchronous Multimedia Multi-hop Wireless Networks. *In proceedings of the IEEE Conference on Computer Communications (INFOCOM'97),* Kobe, Japan: *Vol(1).* (118-125).

Lin, X., Moh, M., & Moh, T.-S. (2008). *On Energy-Efficient Semi-Asynchronous MAC Protocols for Multihop Ad-hoc Networks.* Proc. of 13th IEEE Asia-Pacific Computer System Architecture Conference (ACSAC), held in Hsin-Chu, Taiwan.

Liu, K., Petr, D. W., Frost, V. S., Zhu, H., Braun, C., & Edwards, W. L. (1997). Design and analysis of a bandwidth management framework for ATM-based broadband ISDN. *Communications Magazine, IEEE, 35*(5), 138–145. doi:10.1109/35.592108

Lorenz, P. (2006). QoS in Wireless Networks. *The IET Workshop on Practical Applications for Wireless Technology* (pp. 1–22).

Lov'asz, L. (1975). On the Ratio of Optimal Integral and Fractional Covers. *Discrete Mathematics, 13*, 383–390. doi:10.1016/0012-365X(75)90058-8

Lu, G., & Krishnamachari, B. (2007). Minimum latency joint scheduling and routing in wireless sensor networks. [Elsevier.]. *Ad Hoc Networks, 5*(6), 832–843. doi:10.1016/j.adhoc.2007.03.002

Lu, H.-P., & Yeh, D.-C. (1998). Enterprises' perceptions on business process re-engineering: A path analytic model. *Omega, 26*, 17–27. doi:10.1016/S0305-0483(97)00029-7

Lu, G., Krishnamachari, B., & Raghavendra, C. S. (2004). An Adaptive Energy-Efficient and Low-Latency MAC for Data Gathering in Wireless Sensor Networks. *Proceedings of the 18th International Parallel and Distributed Processing Symposium (IPDPS 2004)*, Santa Fe, NM, USA.

Ma, H., Alotaibi, E., & Roy, S. (2006). *IEEE 802.11 Mesh Networks, OPNETWORK 2006.* Washington, D.C.: Analysis and Simulation Model of Physical Carrier Sensing.

Ma, Y., Richards, M., Ghanem, M., Guo, Y., & Hassard, J. (2008). Air pollution monitoring and mining based on sensor grid in london. *Sensors Journal: Special Issue on Urban Environmental Monitoring, 8*, 3601–3623.

Mackay, C. E., Dunmore, M., Chown, T., & Carvalho, G. (2003). A Scenario-Based Review of IPv6 Transition Tools. *IEEE Internet Computing, 7*, 27–35. doi:10.1109/MIC.2003.1200298

Madeley, S. (2007). *Fixed-mobile Convergence - Who Owns The Customer? Connect-World, Europe II.* Retrieved November 2010 from http://www.connect-world. com/index.php/magazine/europe/item/1194-fixed-mobile-convergence-who-owns-the-customer

Maha Patra, S. S., Roy, K., Banerjee, S., & Vidyarthi, D. P. (2006). Improved genetic algorithm for channel allocation with channel borrowing in mobile computing. *IEEE Transactions on Mobile Computing*, *5*(7), 884–892. doi:10.1109/TMC.2006.99

Manoj, B. S., & Siva, R. M. C. (2002). *Real*-time traffic support for ad hoc wireless networks. *In proceedings of the* IEEE *International Conference on Networks (ICON 2002),* Singapore. (pp. 335-340).

Manu, F. A., & Sriram, V. (1996). Innovation, marketing strategy, environment, and performance. *Journal of Business Research*, *35*, 79. doi:10.1016/0148-2963(95)00056-9

Manvi, S. S., & Venkataram, P. (2002). Adaptive bandwidth reservation scheme for multimedia traffic using mobile agents. *High Speed Networks and Multimedia Communications 5th IEEE International Conference ISBN: 0-7803-7600-5* (pp. 370- 374).

Mao, J., Wu, Z., & Wu, X. (2007). A TDMA scheduling scheme for many-to-one communications in wireless sensor networks. [Elsevier.]. *Computer Communications*, *30*(4), 863–872. doi:10.1016/j.comcom.2006.10.006

Mayer, G., Khartabil, H., Niemi, A., & Poikselka, M. (2006). *The IMS: IP Multimedia Concepts and Services*. Chichester, England; Hoboken, NJ: J. Wiley.

Mazzarella, NJ. (2006). Advantages of harmonized IMS-based charging architecture in different access technologies. *Bell Labs technical journal, 10*(4), 109-116.

McLoughlin, G. J. (1999). Next generation Internet and related initiatives. *Journal of Academic Librarianship*, *25*, 226–229. doi:10.1016/S0099-1333(99)80204-3

Mehta, S., & Ullah, N. Kabir Md. H., Sultana Mst. N., Kwak K. S. (2009). *A case study of networks simulation tools for wireless networks.* Third Asia International Conference on Modelling & Simulation (AMS'09). OMNeT++ Discrete Event Simulation System (n.d.). Retrieved from http://www.omnetpp.org/.

Metz, C. (2003). Moving Toward an IPv6 Future. *IEEE Internet Computing*, *7*, 25–26. doi:10.1109/MIC.2003.1200297

Modeler, O. P. N. E. T. (n.d.). Retrieved from http://www.opnet.com/products/modeler/home.html.

Moh, M., Kim, E.-J., & Moh, T.-S. (2006). Design and Analysis of Distributed Power Scheduling for Data Aggregation in Wireless Sensor Networks*Int.* [IJSNet]. *Journal of Sensor Networks*, *1*(3-4), 143–155. doi:10.1504/IJSNET.2006.012030

Moh, M., Kukanur, R., Lin, X., & Dhar, S. (2009). Energy-efficient, self-organizing routing for wireless mobile networks. [Hershey, PA: IGI Global Publishing]. *International Journal of Business Data Communications and Networking*, *5*(4), 52–69. doi:10.4018/jbdcn.2009091704

Moh, M., & Li, J. (2010). A survey of reputation and trust systems for mobile ad-hoc network routing. In Stavroulakis, P., & Stamp, M. (Eds.), *Handbook of Information and Communication Security*. New York: Springer.

Mohamed, M. Z. (1995). Innovation implementations in Malaysian firms: Process, problems, critical success factors and working climate. *Technovation*, *15*, 375. doi:10.1016/0166-4972(95)96598-N

Monteiro, E. (1998). Scaling information infrastructure: The case of next-generation IP in the Internet. *The Information Society*, *14*, 229–245. doi:10.1080/019722498128845

Montresor, A., Jelasity, M., & Babaoglu, O. (2005). Gossip-based aggregation in large dynamic networks. *ACM Transactions on Computer Systems*, *23*(3), 219–252. doi:10.1145/1082469.1082470

Moon, J.-W., & Kim, Y.-G. (2001). Extending the TAM for a World-Wide-Web context. *Information & Management*, *38*, 217–230. doi:10.1016/S0378-7206(00)00061-6

Moore, P., & Crossley, P. (1999). GPS applications in power systems, part 1. *IEEE Power Engineering Journal*, *13*(1), 33–39. doi:10.1049/pe:19990110

Murthy, C. S. R., & Manoj, B. S. (2004). *Ad Hoc wireless networks architectures and protocols*. Indianapolis, IN: Pearson Education.

Nagarajanvelmurugan, K. S., & Dananjayan, P. (2007). Resource Allocation Management for Multimedia QoS in Wireless Networks. *International Conference on Signal Processing, Communications and Networking* (pp. 122–127).

Nedjah, N., & de Macedo Mourelle, L. (2006). *Swarm Intelligent Systems*. Berlin, Heidelberg: Springer-Verlag. doi:10.1007/978-3-540-33869-7

Network architecture (2007). *Stage 1.* TS 23.002, 3GPP.

Nunnally, J. C. (1978). *Psychometric theory* (2nd ed.). New York: McGraw-Hill.

OECD (2007). *Fixed-Mobile Convergence: Market Developments and Policy Issues*. March 2007

Ofcom (2006). *The Communications Market.*

Open Source Wireless Network Simulator. (openWNS) (n.d.). Retrieved from http://www.openwns.org/

OPNET technologies Inc.(n.d.). *OPNET Modeler Accelerating Networks R&D.* Retrieved 12 4, 2008, from http://www.opnet.com

Orlik, P. V., & Rappaport, S. S. (1998). A model for teletraffic performance and channel holding time characterization in wireless cellular communications with general session and dwell time distributions. *IEEE Journal on Selected Areas in Communications*, *16*(5), 788–803.

Ozakar, B., Morvan, F., & Hameurlain, A. (2005). Mobile join operators for restricted sources. *Mobile Information Systems*, *1*(3), 167–184.

Padilla, J. J., Paradells, J., & Rodriguez, A. (2006). Supporting QoS over IPv6 wireless networks with IntServ6. *IEEE 17th International Symposium on Personal, Indoor and Mobile Radio Communications* (pp. 1–6).

Paiidya, D., Jain, R., & Lupu, E. (2003, September 7-10). Indoor location estimation using multiple wireless technologies. In *Proceedings of the 14th IEEE International Symposium on Personal, Indoor and Mobile Radio Communication Proceedings,* Beijing, China (pp. 2208-2212). IEEE.

Pan, M. S., & Tseng, Y.-C. (2008). Quick convergecast in ZigBee beacon-enabled tree-basewireless sensor networks. [Elsevier.]. *Computer Communications*, *31*(5), 999–1011. doi:10.1016/j.comcom.2007.12.015

Pedersini, F., Grossi, G., Gaito, S., & Rossi, G. P. (2008). Experimental validation of a 2-level social mobility model in opportunistic networks. *Proc. 1st IFIP Wireless Days Conference.*

Pei, G., Gerla, M., Hong, X., & Chiang, C.-C. (1999). *A Wireless Hierarchical Routing Protocol with Group Mobility*. Proceedings of IEEE International Conference on Computer Communications and Networks (ICCCN), held in Boston, MA.

Peng, R., Mao-heng, S., & You-min, Z. (2006, December). ZigBee routing selection strategy based on data services and energy-balanced ZigBee routing. In *Proceedings of the 2006 IEEE Asia-Pacific conference on Services Computing,* Guangzhou, China (pp. 400-404). IEEE.

Perkins, C. E., Royer, E. M., & Das, S. R. (2003). *Ad Hoc On-demand Distance Vector Routing. RFC 3561*. IETF.

Perkins, C., & Royer, E. (1999, February 25-26). Ad-hoc on demand distance vector routing. In *Proceedings of the 2nd IEEE Workshop on Mobile Computing Systems and Applications,* New Orleans, LA (pp. 90-100). IEEE.

Polastre, J., Hill, J., & Culler, D. (2004). Versatile Low Power Media Access for Wireless Sensor Networks. *Proceedings of the 2nd ACM SenSys Conference*, (pp. 95-107), Baltimore, MD, USA.

Policy and charging control architecture (2008). *Stage 2.* TS 23.203, 3GPP.

Policy and Charging Control over Gx reference point (2008). *Stage 3.* TS 29.212, 3GPP.

Policy and Charging Control signaling flows and QoS parameter mapping (2008). TS 29.213, 3GPP.

Pospischil, R. (1998). Fast Internet. *Telecommunications Policy*, *22*, 745–755. doi:10.1016/S0308-5961(98)00053-6

Prakash, R., Shivaratri, N., & Singhal, M. (1999). Distributed dynamic fault-tolerant channel allocation for cellular networks. *IEEE Transactions on Vehicular Technology*, *48*(6), 1874–1888. doi:10.1109/25.806780

Prakash, R., Shivaratri, N., & Singhal, M. (1995, August 20-23). Distributed dynamic channel allocation for mobile computing. In *Proceedings of 14th ACM Symposium on Principles of Distributed Computing,* Ottawa, Ontario, Canada (pp. 47-56). ACM Publishing.

Prowler: Probabilistic Wireless Network Simulator (n.d.). Retrieved from http://www.isis.vanderbilt.edu/Projects/nest/prowler/

Pujolle, G. (2006, October 3-6). An autonomic-oriented architecture for the Internet of Things. In *Proceedings of the IEEE John Vincent Atanasoff 2006 International Symposium on Modern Computing,* Sofia, Bulgaria (pp. 163-168). IEEE.

Rajaratnam, M., & Takawira, F. (2000). Nonclassical traffic modeling and performance analysis of cellular mobile networks with and without channel reservation. *IEEE Transactions on Vehicular Technology*, *49*(3), 817–834.

Ramanathan, S. (1999). A unified Framework and Algorithm for Channel Assignment in Wireless Ad Hoc Networks. *Wireless Networks*, *5*(2), 81–94. doi:10.1023/A:1019126406181

Ramanathan, R., & Streenstrup, M. (1998). Hierarchically organized, multi-hop mobile wireless networks for quality-of-service support. *Mobile Networks and Applications*, *3*, 101–119. doi:10.1023/A:1019148009641

Rappaport, T. (2002). *Wireless Communications: Principles and Practice* (2nd ed.). Prentice Hall.

RealNetworks (2005). Annual Report

RFC 3916: *Requirements for Pseudo-Wire Emulation Edge-to-Edge (PWE3)*. (2004).:IETF

RFC3985: *Pseudo Wire Emulation Edge-to-Edge (PWE3) Architecture*. (2005).: IETF

RFC 4717: *Encapsulation Methods for Transport of Asynchronous Transfer Mode (ATM) over MPLS Networks* (2006): IETF

Rhee, I. Warrier, A., Min, J., & Xu, L. (2006). DRAND: Distributed Randomized TDMA Scheduling for Wireless Ad-hoc Networks. *Proceedings the 7th ACM International Symposium on Mobile Ad Hoc Networking and Computing (MobiHoc'2006)*, Florence, Italy.

Rieck, M. Q., & Dhar, S. (2008). Hierarchical Routing in Ad Hoc Networks using *k*-dominating Sets. *ACM Mobile Computing and Communications Review*, *12*(3), 45–57. doi:10.1145/1462141.1462147

Rieck, M. Q., Pai, S., & Dhar, S. (2005). Distributed Routing Algorithms for Multi-hop Ad Hoc Networks Using *d*-hop Connected *d*-Dominating Sets. [COMNET]. *Computer Networks Journal*, *47*(6), 785–799. doi:10.1016/j.comnet.2004.09.005

Roger, E. M. (1983). *Diffusion of Innovation* (3rd ed.). New York: Free Press.

Romdhami, L., Ni, Q., & Turletti, T. (2003). Adaptive EDCF: Enhanced Service differentiation for IEEE 802.11 Wireless Ad-Hoc Networks. *In proceedings of the IEEE Wireless Communications and Networking Conference (WCNC 2003),* New Orleans, Louisiana, USA: *Vol (2).* (1373-1378).

Ronald Beaubrun, S. P. (2001). Technological developments and socio-economic issues of wireless mobile communications. *Telematics and Informatics*, *18*, 143–158. doi:10.1016/S0736-5853(00)00026-5

Rosenberg, J., Schulzrinne, H., Camarillo, G., Johnston, A., Peterson, J., & Sparks, R. (2002). *RFC 3261 Session Initiation Protocol*. IETF.

Rothenberg, C. E. (2008). A review of policy-based resource and admission control functions in evolving access and next generation networks. *Journal of Network and Systems Management*, *16*(1), 14–45. doi:10.1007/s10922-007-9096-3

Ruggieri, M., Graziosi, F., & Santucci, F. (1998). Modeling of the handover dwell time in cellular mobile communications systems. *IEEE Transactions on Vehicular Technology*, *47*(2), 489–498.

Ruiz, G., Doumi, T. L., & Gardiner, J. (1998). Teletraffic analysis and simulation for nongeostationary mobile satellite systems. *IEEE Transactions on Vehicular Technology*, *47*(1), 311–320.

Ryu, J., Lee, J., Lee, S.-J., & Kwon, T. (2008). *Revamping the IEEE 802.11a PHY Simulation Models*. In Proc. of the 11th ACM/IEEE Intl. Symposium on Modeling, Analysis and Simulation of Wireless and Mobile Systems (MSWIM'08), Vancouver, British Columbia, Canada, October.

S. Guha & S. Khuller. (1998). *Approximation Algorithms for Connected Dominating Sets*.

Samad, M., & Herman, S. H. (2005). Quality of service for mobile IP services in wireless network. *Asia-Pacific Conference on Applied Electromagnetic* (p. 4).

Saraireh, M., Saatchi, R., Al-khayatt, S., & Strachan, R. (2006). Development and Evaluation of a Fuzzy Inference Engine System to Incorporate Quality of Service in IEEE 802.11 Medium Access Protocol. *Conference on Wireless and Mobile Communications International* (pp. 29–29).

Schiller, J. (2003). *Mobile communications*. Indianapolis, IN: Pearson Education.

Schmid, T., Dubois-Ferrière, H., & Vetterli, M. (2005, June 20-21). *SensorScope: Experiences with a wireless building monitoring sensor network*. Paper presented at the 2005 Workshop on Real-World Wireless Sensor Networks (REALWSN 2005), Stockholm, Sweden.

Schulzrinne, H., Tschofenig, H., Morris, J., Cuellar, J., Polk, J., & Rosenberg, J. (2007). *RFC 4745 Common Policy: A Document Format for Expressing Privacy Preferences*. IETF.

Seeling, P., Fitzek, F. H. P., & Reisslein, M. (2007). *Video Traces for network Performance Evaluation, A Comprehensive Overview and Guide on Video Traces and Their Utilization in Networking Research*. Netherlands: Springer.

Seo, J.-H., Im, C.-H., Heo, C.-G., Kim, J.-K., Jung, H.-K., & Lee, C.-G. (2006). IEEE Transactions on Multimodal function optimization based on particle swarm optimization. *Magnetics*, *42*(4), 1095–1098. doi:10.1109/TMAG.2006.871568

Service requirements for the Internet Protocol (IP) multimedia core network subsystem (2007). Stage 1. TS 22.228, 3GPP.

Sesay, S., Yang, Z., & He, J. (2004). A Survey on Mobile Ad Hoc Wireless Network. *Information Technology Journal*, *3*(2), 168–175. doi:10.3923/itj.2004.168.175

Shan, T., & Yang, O. W. W. (2007). Bandwidth Management for Supporting Differentiated Service Aware Traffic Engineering. *IEEE Transactions on Parallel and Distributed Systems*, *18*(9), 1320–1331. doi:10.1109/TPDS.2007.1052

Sharafeddine, S., Riedl, A., Glasmann, J., & Totzke, J. (2003). On Traffic Characteristics and Bandwidth Requirements of Voice over IP Applications. *In proceedings of the IEEE International Symposium on Computers and Communication (ISCC'03),* Antalya, Turkey: *Vol (2).* (1324-1330).

Shengming, J., Jianqiang, R., Dajiang, H., Xinhua, L., & Chi, C. K. (2002). A Simple Distributed PRMA for MANETs. *IEEE Transactions on Vehicular Technology*, *51*(2), 293–305. doi:10.1109/25.994807

Shiau, W.-L., Chao, H.-C., & Hsu, P.-Y. (2005). Obstacles of enterprises moving towards the next generation internet. [IJIPT]. *International Journal of Internet Protocol Technology*, *1*, 34–39. doi:10.1504/IJIPT.2005.007557

Shiau, W.-L., Li, Y.-F., Chao, H.-C., & Hsu, P.-Y. (2006). Evaluating IPv6 on a large-scale network. *Computer Communications*, *29*, 3113–3121. doi:10.1016/j.comcom.2005.11.011

Silveira, G. D. (2001). Innovation diffusion: Research agenda for developing economies. *Technovation*, *21*, 767–773. doi:10.1016/S0166-4972(01)00007-4

Singh, S. R., & Raghavendra, C. S. (1998). PAMAS - Power Aware Multi-access Protocol with Signaling for Ad-Hoc Networks. *ACM Computer Communication Review*, *28*, 5–26. doi:10.1145/293927.293928

Singh, S., Woo, M., & Raghavendra, C. S. (1998). *Power-Aware Routing in Mobile Ad Hoc Networks*. Proc. ACM Annual International Conference on Mobile Computing and Networking (Mobicom).

Sisaye, S. (1999). An organizational approach for the study of the diffusion of process innovation strategies in internal auditing and control systems. *International Journal of Applied Quality Management*, *2*, 279–293. doi:10.1016/S1096-4738(99)80095-5

Sivakumar, R., Sinha, P., & Bharghavan, V. (1999). CEDAR: a Core-Extraction Distributed Ad Hoc Routing Algorithm. *IEEE Journal on Selected Areas in Communications*, *17*(8). doi:10.1109/49.779926

Sonera (2010). Prices retrieved 9 November 2010, from http://www.sonera.fi/puhelin+ja+liittyma/iphone+hinnat

Spedalieri, A., Martin-Escalona, I., & Barceló, F. (2005, March 13-17). Simulation of teletraffic variables in UMTS networks: Impact of lognormal distributed call duration. In *Proceedings of the IEEE Wireless Communications and Networking Conference (WCNC),* New Orleans, LA (Vol. 4, pp. 2381-2386). IEEE.

Spyropoulos, T., Psounis, K., & Raghavendra, C. S. (2004). Single-copy routing in intermittently connected mobile networks. *Proc. 1st IEEE SECON*, (pp. 235-244).

Stallings, W. (1998). *SNMP, SNMPv2, SNMPv3, and RMON 1 and 2 (3 ed.).* Boston, MA, USA: Addison-Wesley Longman Publishing Co., Inc.

Stewart, B. (2005). *How Many Voice Callers Fit on the Head of an Access Point?* from http://www.oreillynet.com/etel/blog/2005/12/how_many_voice_callers_fit_on.html

Stock, G. N., Greis, N. P., & Fischer, W. A. (2002). Firm size and dynamic technological innovation. *Technovation, 22*, 537–549. doi:10.1016/S0166-4972(01)00061-X

Stojmenovic, I., & Lin, X. (2001). Power-aware Localized Routing in Wireless Networks. *IEEE Transactions on Parallel and Distributed Systems, 12*(11), 1122–1133. doi:10.1109/71.969123

Strategy Analytics (2006). Mobile Video Telephony: Growth Projection Upgrade and Key Success Factors

Su, J., Chin, A., Popinova, A., Goely, A., & de Lara, E. (2004). User mobility for opportunistic ad-hoc networking. *Proc. 6th IEEE Workshop on Mobile Computing Systems and Applications.*

Sundararajan, D. (2001). *The Discrete Fourier Transform-Theory, Algorithms and Applications.* New York: World Scientific Publishing Company. doi:10.1142/9789812810298

Tang, Z., & Garcia-Luna-Aceves, J. J. (1999). A Protocol for Topology-Dependent Transmission Scheduling in Wireless Networks. *In proceedings of the IEEE Wireless Communications and Networking Conference (WCNC 1999): Vol(3).* (1333-1337).

Taniar, D., & Rahayu, J. W. (2004). Global parallel index for multi-processors database systems. *Inf. Sci., 165*(1-2), 103–127. doi:10.1016/j.ins.2003.09.019

Technologies, S. N. (n.d.). *Qualnet family of products.* Retrieved from http://www.scalable-networks.com/products/qualnet.php.

Telecompaper (2009). Retrieved 4 September 2009, from http://www.telecompaper.com/news/article.aspx?cid=689518

Thajchayapong, S., & Peha, J. M. (2006). Mobility patterns in microcellular wireless networks. *IEEE Transactions on Mobile Computing, 5*(1), 52–63.

The 3GPP (2001). *Physical layer aspects of UTRA High Speed Downlink Packet Access* (3GPP TR 25.848 version 4.0.0).: 3GPP Technical Specification Group RAN.

The 3GPP (2002). *IP Transport in UTRAN Work Task Technical Report.* (3GPP TR 25.933 version 2.0.0).: 3GPP Technical Specification Group RAN

The 3GPP (2006). *FDD Enhanced Uplink; Overall Description; Stage 2* (3GPP TS 25.309 version 6.6.0).: 3GPP Technical Specification Group RAN.

The 3GPP (2007). *Enhanced uplink; Overall description; Stage 2.* (3GPP TR 25.319 version 7.2.0).: 3GPP Technical Specification Group RAN

The 3GPP(2001). *High Speed Downlink Packet Access (HSDPA); Overall UTRAN description.* (3GPP TR 25.855 version 5.0.0).: 3GPP Technical Specification Group RAN.

The, A. T. M. Forum (2000, July). *Addendum to Traffic Management V4.1 for an Optional Minimum Desired Cell Rate Indication for UBR.* Retrieved January 8, 2009, from http://www.ipmplsforum.org/ftp/pub/approved-specs/af-tm-0150.000.pdf

Tipper, D., & Teresa, D. (2002). Providing fault tolerance in wireless access networks. *IEEE Communications Magazine, 40*(1), 17–30. doi:10.1109/35.978050

Toh, C. (2002). *Ad hoc Mobile Wireless Networks Protocols and Systems.* Upper Saddle River, NJ: Prentice Hall, Inc.

Tseng, Y. C., Hsu, C. S., & Hsieh, T. Y. (2002). *Power-Saving Protocols for IEEE 802.11- Based Multi-hop Ad Hoc Networks.* Proceedings of IEEE International Conference on Computer Communications (INFOCOM), held in New York.

Tsumochi, J., Masayama, K., Uehara, H., & Yokoyama, M. (2003). *Impact of mobility metric on routing protocols for mobile ad hoc networks.* In Proc. IEEE Pacific Rim Conference on Communications, Computers and Signal Processing(PACRIM03)'.

Tsu-Wei Chen & Mario Gerla. (1998).*Global State Routing: A New Routing Scheme for Ad-hoc Wireless Networks.* Proc. IEEE International Conference on Communications (ICC).

Türke, U., Winter, T., Perera, R., Lamers, E., Meijerink, E., Fledderus, E. R., et al. (2002). *Comparison of different simulation approaches for cell performance evaluation.* Retrieved January 5, 2009, from http://momentum.zib.de/paper/momentum-d22.pdf

UCLA Parallel Computing Laboratory. (2008). *GloMoSim – Global Mobile Information Systems Simulation Library.* Retrieved from http://pcl.cs.ucla.edu/projects/glomosim

Upadhyayula, S., Annamalai, V., & Gupta, S. K. S. (2003). A low-latency and energy-efficient algorithm for convergecast in wireless sensor networks. *Proceedings of IEEE Global Telecommunications Conference (GLOBECOM '03)*, San Francisco, CA, USA.

Uzun, A. (2001). Technological innovation activities in Turkey: The case of manufacturing industry, 1995-1997. *Technovation, 21*, 189–196. doi:10.1016/S0166-4972(00)00033-X

Vahdat, A., & Becker, D. (2000). Epidemic routing for partially connected ad-hoc networks. Technical Report CS-200006, Duke University. This work was previously published in International Journal of Business Data Communications and Networking (IJBDCN) Volume 5, Issue 2, edited by Varadharajan Sridhar, and Debashis Saha, pp. 1-15, copyright 2009 by IGI Publishing (an imprint of IGI Global).

Valkó, A. G., Javier Gomez, K. K., & Campbell, A. T. (1999). On the Analysis of Cellular IP Access Networks. *Proceedings of the IFIP TC6 WG6.1 & WG6.4 / IEEE ComSoc TC on Gigabit Networking Sixth International Workshop on Protocols for High Speed Networks VI, 158,* 205–224.

Verkasalo, H. (2007). *A Cross-Country Comparison of Mobile Service and Handset Usage*, Licentiate's Thesis Series, TKK Helsinki University of Technology

Vodafone (2010). *Vodafone One Net.*Retrieved from http://www.vodafone.com/start/about_vodafone/what_we_do/business_solutions/products/integrated_communications/one_net.html

Vonortas, N. S., & Xue, L. (1997). Process innovation in small firms: Case studies on CNC machine tools. *Technovation, 17*, 427–438. doi:10.1016/S0166-4972(97)00016-3

W3C (2003). *SOAP Version 1.2 Part 0: Primer*. http://www.w3.org/TR/soap12-part0/

Waluyo, A. B., Srinivasan, B., & Taniar, D. (2008). Indexing schemes for multichannel data broadcasting in mobile databases. *International Journal of Wireless and Mobile Computing, 3*(1-2), 79–90. doi:10.1504/IJWMC.2008.019720

Waluyo, A. B., Zhu, F., Taniar, D., Srinivasan, B., & Rahayu, J. W. (2007). Multiple Entity Types Wireless Broadcast Database System. *Proceedings of the 21st International Conference on Advanced Information Networking and Applications (AINA 2007)* (pp. 179-186).

Waluyo, A.B., Srinivasan, B., & Taniar, D. (2005d). Research on location-dependent queries in mobile databases. *Comput. Syst. Sci. Eng., 20*(2).

Wang, S. S., & Green, M. (2000, September 24-28). Mobile location method for non-line-of-sight situation. In *Proceedings of IEEE 52nd Vehicular Technology Conference,* Boston (pp. 608-612). IEEE.

Weerawardane, T. L., Li, X., Timm-Giel, A., & Görg, C. (2006). *Modeling and Simulation of UMTS HSDPA in OPNET*. Paper presented at OPNETWORK 2006, Washington DC.

Weerawardane, T. L., Timm-Giel, A., Görg, C., & Reim, T. (2006). *Impact of the Transport Network Layer Flow Control for HSDPA Performance*. Paper presented at IEE conference 2006, Sri Lanka.

Weerawardane, T. L., Timm-Giel, A., Malafronte, G. C., Gianluca, D., Hauth, S., & Görg, C. (2008). Preventive and Reactive Based TNL Congestion Control Impact on the HSDPA Performance. *Vehicular Technology Conference, 2008. VTC Spring 2008* (pp. 2296 - 2300).:IEEE

Weiser, M. (2001). Whatever happened to the next-generation Internet? *Association for Computing Machinery. Communications of the ACM*, *44*, 61–68. doi:10.1145/383694.383708

Werner-Aller, G., Swieskowski, P., & Welsh, M. (2005, April). MoteLab: A wireless sensor network testbed. In *Proceedings of the 4th International Conference on Information Processing in Sensor Networks (IPSN)*, Los Angeles (pp. 483-488). IEEE.

West, D. B. (2001). *Introduction to Graph Theory* (2nd ed.). Prentice Hall.

Wheeler, A. (2007). Commercial applications of wireless sensor networks using ZigBee. *IEEE Communications Magazine*, *45*(4), 70–77. doi:10.1109/MCOM.2007.343615

Wiethölter, S., & Hoene, C. (2003). *Design and Verification of an IEEE 802.11e EDCF Simulation Model in ns-2.26.* (Tech. Rep. No. TKN-03-19). Berlin, Germany: Technische Universität Berlin.

Wiethölter, S., Hoene, C., & Wolisz, A. (2004). *Perceptual Quality of Internet Telephony over IEEE 802.11e Supporting Enhanced DCF and Contention Free Bursting*. (Tech. Rep. No. TKN–04-11). Berlin, Germany: Technische Universität Berlin.

Winer, D. (1999). *XML-RPC Specification.* http://www.xmlrpc.com/spec

Workplan, G. P. P. (2010). *Overview of 3GPP Release 11.* Retrieved November 2010 from http://www.3gpp.org/ftp/Information/WORK_PLAN/Description_Releases/Rel-11_description_20100924.zip

Wright, M. (2007). Unique IP addresses will prove main IPv6 benefit. *EDN*, *52*, 14–14.

Wu, F. Dai, M. Gao, & I. Stojmenovic. (2002). On Calculating Power-aware Connected Dominating Sets for Efficient Routing in Ad Hoc Wireless Networks. *IEEE/KICS J. Communication Networks*, *4*(1), 59–70.

Wu, J., & Li, H. (2001). A Dominating-Set-Based Routing Scheme in Ad Hoc Wireless Networks. *Telecommunication Systems*, *18*(1-3), 13–36. doi:10.1023/A:1016783217662

Xhafa, A. E., & Tonguz, O. K. (2008). Handover performance of priority schemes in cellular networks. *IEEE Transactions on Vehicular Technology*, *57*(1), 565–577.

Xu, S., & Saadawi, T. (2001). Does the IEEE 802.11 MAC Protocol Work Well in Multihop Wireless Ad Hoc Networks? *IEEE Communications Magazine*, *39*(6), 130–137. doi:10.1109/35.925681

Xuan, K., Zhao, G., Taniar, D., & Srinivasan, B. (2008). Continuous Range Search Query Processing in Mobile Navigation. In *Proceedings of the 14th International Conference on Parallel and Distributed Systems (ICPADS 2008)* (pp. 361-368).

Yamaguchi, H. Konishi, K.& Higashino, T. (2007). *Efficient Parallel Simulation of Mobile Wireless Networks by Run-time Prediction of Multi-hop Propagation Delay*. In Proc. of TridentCom 2007.

Yang, J., Jiang, Q., Manivannan, D., & Singhal, M. (2005). A fault-tolerant distributed channel allocation scheme for cellular networks. *IEEE Transactions on Computers*, *54*(5), 616–629. doi:10.1109/TC.2005.71

Yavatkar, R., Pendarakis, D., & Guerin, R. (2000). *RFC 2753 A Framework for Policy Based Admission Control.* IETF.

Zaki, Y., Weerawardane, T. L., Li, X., Timm-Giel, A., Malafronte, G. C., & Görg, C. (2008). Effect of the RLC and TNL Congestion Control on the HSUPA Network Performance. *Mosharaka International Conference on Communications, Computers and Applications, MIC-CCA 2008* (pp. 1-7).

Zander, R., & Karlsson, J. M. (2005). Combining Bandwidth Borrowing and Reservation in Cellular Networks. *International Journal of Wireless Information Networks*, *12*(3), 187–201. doi:10.1007/s10776-005-0004-z

Zhang, M., & Yum, T. (1991). The non-uniform compact pattern allocation algorithm for cellular mobile systems. *IEEE Transactions on Vehicular Technology*, *40*(3), 387–391. doi:10.1109/25.289419

Zhao, G., Xuan, K., Taniar, D., & Srinivasan, B. (2008). Incremental k-Nearest-Neighbor Search on Road Networks. *Journal of Interconnection Networks*, *9*(4), 455–470. doi:10.1142/S0219265908002382

Zheng, J., Wang, Y., & Nihan, N. L. (2005, March 19-22). Quantitative evaluation of GPS performance under forest canopies. In *Proceedings of the 2005 IEEE International Conference on Networking, Sensing and Control,* Tucson, AZ (pp. 777-782). IEEE.

Zhu, C., Corson, & M. S. *A* five-phase reservation protocol (FPRP) for mobile ad hoc networks. *In proceedings of the IEEE INFOCOM'98*, San Francisco: *Vol(1).* 322-331.

Zhu, H., Cao, G., Yener, A., & Mathias, A. D. EDCF-DM: A Novel Enhanced Distributed coordination Function for Wireless Ad Hoc Networks. *In proceedings of the IEEE International Conference on Communications (ICC2004),* Paris: *Vol(7).* 3886- 3890.

ZigBee Alliance. (2004). Retrieved February 17, 2009, from http://www.zigbee.org

Zonoozi, M. M., Dassanayake, P., & Faulkner, M. (1995, November). Mobility modelling and channel holding time distribution in cellular mobile communication systems. In *Proceedings of the IEEE Globecom,* Singapore (pp. 12-16, 1995). IEEE.

About the Contributors

Varadharajan Sridhar is a Research Fellow at Sasken Communication Technologies (Bangalore, India). He received his BE from the University of Madras (India), Post Graduate Diploma in industrial engineering from the National Institute for Training in Industrial Engineering (Mumbai, India), and PhD in MIS from the University of Iowa (USA). He had taught at Ohio University and American University in the US; University of Auckland in New Zealand, Indian Institute of Management (Lucknow, India) and Management Development Institute (India). Dr. Sridhar's primary research interests are in the area of telecommunication management and policy and global software development. He has published many research articles, business cases, and chapters in edited books in his area of research. Dr. Sridhar is a member of the Committee on Allocation and Pricing of Spectrum for Access Services set-up by the Indian Government. He was the recipient of the Nokia Visiting Fellowship awarded by the Nokia Research Foundation. He is on the editorial board of the Journal of Global Information Management and is a member of ACM and AIS. Dr. Sridhar's book titled *The Telecom Revolution in India: Technology, Regulation and Policy* will be published by the Oxford University Press India in 2011.

Debashis Saha is a professor with the MIS Group, Indian Institute of Management (IIM) Calcutta. Previously, he was with CSE Department at Jadavpur University (Kolkata, India). He received his BE (Hons) degree from Jadavpur University (Kolkata, India), and the MTech and PhD degrees from the Indian Institute of Technology (Kharagpur) all in electronics and telecommunication engineering. His research interests include pervasive communication and computing, network operations, management and security, wireless networking and mobile computing, ICT for development, and network economics. He has supervised thirteen doctoral theses, published about 200 research papers in various conferences and journals, and directed four funded research projects on networking. He has co-authored several book chapters, a monograph, and five books including Networking Infrastructure for Pervasive Computing: Enabling Technologies and Systems (Norwell, MA: Kluwer, 2002) and Location Management and Routing in Mobile Wireless Networks (Boston, MA: Artech House, 2003). Dr. Saha is the recipient of the prestigious career award for Young Teachers from AICTE, Government of India, and is a SERC Visiting Fellow with the Department of Science and Technology (DST), Government of India. He is a Fellow of West Bengal Academy of Science and Technology (WAST), Senior Life Member of Computer Society of India, Senior Member of IEEE, member pf ACM, and a member of the International Federation of Information Processing Working Group's 6.8 and 6.10. He is the founding Chair of Calcutta Chapter of IEEE Communications Society.

***

Atef Abdelkefi received the engineer diploma in computer science in 2008 from the National School of Computer Science of Tunisia (ENSI). He is now a M.S. student at LAAS-CNRS. His master thesis work involves Transport protocol for QoS support in mobile Ad-hoc networks.

Mohammad Anbar received his B.Tech. in Electronics Engineering from Tishreen University, Lattakia, Syria, in the year 2003, and M.Tech. in Computer Science from Jawaharlal Nehru University, New Delhi, India in the year 2007. Currently, Mohammad Anbar, is Ph.D. student at the School of Computer & Systems Sciences, Jawaharlal Nehru University, New Delhi. His research interest includes Wireless Communication, Mobile computing, Particle Swarm Optimization.

François Armando received the Engineering degree and the M.S. degree (DEA/MASTER) in Network and Telecommunication from the INSA of Toulouse in 2005. He is currently Ph.D student at LAAS-CNRS. His thesis work deals with self adaptive Transport architecture for QoS within MEO-like group communication activities.

Francisco Barcelo-Arroyo earned a degree in telecommunications engineering and a Ph.D. from the Universitat Politécnica de Catalunya (UPC), in 1986 and 1997 respectively. In 1987, he joined the Faculty of Telecommunications Engineering of Barcelona at UPC. After graduation, he did research in the areas of digital network synchronization and switching. Since 1997, he has been an associate professor at the Department of Telematic Engineering at UPC. Dr. Barcelo-Arroyo participated and led a number of research projects on performance modeling of wireless systems and networks and is currently involved with projects supported by the Spanish Government and the European Commission (IST Internode, IST Emily). Recently he chaired a working group on mobility issues in the framework of project COST290 on wireless multimedia teletraffic and was the coordinator of R&D activities for the IST Liaison project. His current research interests lie in the study of the evaluation of the capacity and teletraffic performance of wired and wireless networks and location based services in wireless networks with or without infrastructure.

Somprakash Bandyopadhyay is a Professor in Management Information Systems Group of Indian Institute of Management, Calcutta. He is a PhD in Computer Science from Jadavpur University Calcutta, India and B.Tech in Electronics and Electrical Communication Engineering from Indian Institute of Technology, Kharagpur, India. He has over two decades of international and domestic experience in R&D of emerging technologies and their commercial applications. He has more than 60 research publications and has coauthored three books. He has initiated Ad Hoc Network Research and Application Group in 1998 to provide an environment for research and application development necessary to support next generation wireless and mobile communication systems. He is on the Board of Directors of PervCom Consulting Pvt Ltd., Kolkata, India.

Christophe Chassot received the Engineering degree and the M.S. degree (DEA) in Computer Science from the National Polytechnic Institute of Toulouse (INPT) in 1992, and the Ph.D. degree in Computer Science from INPT, in 1995. He is now Professor at the National Institute of Applied Sciences of Toulouse (INSA). He is also associate Researcher at the Laboratory of Analysis and Architectures of Systems (LAAS), a laboratory of the National Center of the Scientific Research (CNRS). His main fields of interest include end-to-end communication architectures and signaling for self-adaptive management of QoS in heterogenous networks environements.

Chen-Yao Chung received his Master's degree from the department of Information Management at the Da-Yeh University, Taiwan in 1996. He is currently working toward his Ph.D. degree in the Department of Business Administration, National Central University in Chung-Li, Taiwan. He was a computer system engineer at Industrial Technology Research Institute, Taiwan. He was a certified instructor of ERP and has trained innumerable certified engineers at Chinese Enterprise Resource Planning Society. His current research interests include ERP, Data Mining, Data Warehousing, and Networking.

Subhankar Dhar is an Associate Professor in the Department of Information Systems at the Lucas Graduate School of Business at San José State University. He is also an affiliate faculty member of the Silicon Valley Center for Entrepreneurship. Dr. Dhar's research interests are in the areas of distributed, mobile and pervasive computing. In addition, he is also interested in information technology outsourcing. He teaches a variety of courses including telecommunications, data communications and networks, and distributed information systems. He serves as a member of the editorial board of International Journal of Business Data Communications and Networking. He also served as a member of the organizing committee of various international conferences including International Conference on Broadband Networks (BroadNets) and International Workshop on Distributed Computing (IWDC). Dr. Dhar has several years of industrial experience in software development, enterprise resource planning, consulting for Fortune 500 and high-tech industries including product planning, design, and information systems management.

Khalil Drira received the Engineering and M.S. (DEA) degrees in Computer Science from ENSEEIHT (INP Toulouse), in June and September 1988 respectively. He obtained the Ph.D. and HDR degrees in Computer Science from UPS, University Paul Sabatier Toulouse, in October 1992, and January 2005 respectively. He is since 1992, Chargé de Recherche, a full-time research position at the French National Center for Scientific Research (CNRS). Khalil DRIRA's research interests include formal design, implementation, testing and provisioning of distributed communicating systems and cooperative networked services. His research activity addressed and addresses different topics in this field focusing on model-based analysis and design of correctness properties including testability, robustness, adaptability and reconfiguration.

Anurag D is a doctoral student at Indian Institute of Management Calcutta. His research interests are in the networking aspects of wireless ad-hoc networks specifically focusing on the low power and low computational capability of IEEE 802.15.4. He has numerous publications on both theoretical and practical work on wireless sensor network and has been awarded a Doctoral Scholarship by Infosys Technologies. Prior to joining IIM Calcutta, he worked with Lucent Technologies, India. He holds a bachelors degree in Electronics and Communications Engineering. He is on the Board of Directors of PervCom Consulting Pvt Ltd., Kolkata, India.

Israel Martin-Escalona earned the degree in Telecommunications Engineering from the Universitat Politècnica de Catalunya (UPC) in 2001. He began his Ph.D. in Telematics Engineering in 2002 in the Department of Telematics Engineering of the Technical University of Catalonia (UPC). In 2003 he joined the School of Telecommunications Engineering at UPC, where he as been teaching networking fundamentals, teletraffic and network simulation. He is currently applying for the degree in Computer Science in the Computer Science Faculty of Barcelona at UPC. He has been involved with several research projects supported by the Spanish Government and the European Commission (IST Emily and IST Liaison among them). His research has focused on mobile networks and particularly on the development of location systems in cellular environments with and without network infrastructure.

Maria Fazio received the degree in Electronic Engineering in 2002 and the PhD in 2006 at the University of Messina (Italy). She has been exchange visitor at the Department of Computer Science of the University of California in Los Angeles in 2005. She received a post-doc fellowship in 2006. She has collaborated with many abroad research groups at the University of California in Los Angeles, Indian Institute of Technology in Guwahati, National Technical University in Athene, Amirkabir University of Technology and Iran University of Science and Technology in Tehran. She is currently Assistant Researcher at the University of Messina (Italy). Current research activities include distributed systems and mobile networks, especially with regard to wireless multi-hop networks, and Cloud computing, with particular attention on the integration of different communication technologies, federation and services provisioning.

Boudour Ghalem obtained his Engineering degree in Computer Science in 2004 from the University of Science and Technology of Oran and his MSc degree in Computer Science in 2006 from the Paul Sabatier University – Toulouse. He is currently a PhD student at the IRIT laboratory in Paul Sabatier University. His main research interests include: Wireless Sensor Networks, MAC protocols for wireless and mobile networks, and QoS Support in ad-hoc networks.

Andreas Timm-Giel, (Dipl.-Ing, 1994, Dr.-Ing., 1999, Member IEEE and VDE/ITG) From 1994–1999 he was group leader at the University of Bremen in the area of mobile and satellite communications. From 2000 to 2002 he was with MediaMobil GmbH and M2SAT Ltd. as Technical Project Leader. In December 2002 he joined the Communication Networks at the University of Bremen as senior researcher and lecturer. He is leading several industry, national and EC funded research projects. Since 2006 he is additionally directing the interdisciplinary activity "Adaptive Communications" of TZI. His research interests are adaptive mobile and wireless communications and sensor networks.

Francesco Giudici received his Master degree in Computer Science from the University of Milan, Italy, in 2005. He is currently a Ph.D. student at the Information Science and Communication Department of the University of Milan. He is currently involved in the CARTOON national project, on the design of communication protocols over Opportunistic Networks. His research interests include routing protocols and architectures for mobile wireless networks, especially vehicular and delay tolerant networks.

António Grilo received the Information Technology Engineering degree in 1996, the M. Sc. degree in Electrotechnical and Computer Engineering in 1998 and the PhD in Electrotechnical and Computer Engineering in 2004, all from the IST, Technical University of Lisboa, Portugal. He is Assistant Professor at IST, where he lectures subjects related with computer networks and telecommunications, in under-graduate and graduate courses. In 1995 he joined INESC, Lisboa, where he is currently a senior researcher. His current research areas are Wireless and Mobile Networks and Network Centric Military Communications. From 1996 until now he has been working in several European projects, namely ACTS projects ATHOC and AROMA, and IST projects MOICANE, OLYMPIC, AIRNET and UbiSeq&Sens. He is currently working in IST project WSN4CIP.

Carmelita Görg's professional stations were the University of Karlsruhe (diploma), Aachen University of Technology (PhD, lecturer) and the University of Bremen (professor) where she established the Communication Networks Group (tzi / Center for Communication and Information Technology, Mobile Research Center). C. Görg has published over 100 scientific papers. She is a member of the board of the ITG (German Information Technology Society). Research interests: Performance Analysis of (Wireless/ Mobile) Communication Networks and their applications.

Karim Guennoun received the M.S. degree (DEA) in Computer Science from Paul Sabatier University (UPS) of Toulouse in 2002, and the Ph.D. degree in Computer Science from UPS, in 2006. Since 2008, he is assistant professor in the Hassania Engineering School of Casablanca. He is also associate Researcher at the Engineering Sciences Laboratory of Casablanca. His main fields of interest include QoS provisioning and dynamic software architecture formal description, management and verification.

Jarmo Harno received his MSc degree from the University of Helsinki in 1983. After working in SW industry he joined Nokia in 1987, and has worked as systems analyst and manager in R&D, Quality Assurance and Product Management. Recently he has worked as senior research scientist with Nokia Research Center and Nokia Siemens Networks on techno-economics, and completed his PhD in Aalto University School of Science and Technology in 2010. As doing research on the future telecom technologies and services, he has participated EU IST project TONIC (2001–2002) and EUREKA CELTIC project ECOSYS (2004–2007) leading the Mobile Work Package. He is an author of several journal articles and conference presentations relating to techno-economics and technology strategy.

Geert Heijenk (1965) received his M.Sc. in Computer Science from University of Twente, the Netherlands, in 1988. He has worked as a research staff member at the same university and received his Ph.D. in Telecommunications in 1995. He has also held a part-time position as researcher at KPN research, the Netherlands, from 1989 until 1991. From 1995 until 2003, he was with Ericsson EuroLab Netherlands, first as a senior strategic engineer, and from 1999 as a research department manager. From 1998 until 2003 he was also a part-time senior researcher at the University of Twente. Currently, he is a full-time associate professor at the same university. Geert Heijenk has been a visiting researcher at University of Pennsylvania, Philadelphia and a visiting associate professor at University of California, Irvine. He is a senior member of the IEEE. His research interests include mobile-, wireless-, and ad-hoc networks, with a focus on sensor and vehicular networks.

Mikko Heikkinen, M.Sc. (Tech.), is Research Scientist at the Department of Communications and Networking, Aalto University School of Science and Technology. His current research interests are related to techno-economics of mobile peer-to-peer services and technologies, energy efficiency of mobile devices and services, and mobile usage measurements. He did his Master's Thesis in the ECOSYS project on Fixed-Mobile Convergence. In 2010, he was visiting the Massachusetts Institute of Technology.

Dirk von Hugo received his Diploma and Doctorate degree in Physics from Technical University Darmstadt (Germany) in 1984 and 1989, respectively. Since 1990 he has been active in various areas of radio communications research within Deutsche Telekom Group, including contribution to standardization organizations and cooperation in nationally and internationally funded R&D projects. Beside CELTIC project Ecosys he has participated recently in EU funded ICT projects E3 (End to End Efficiency) and CARMEN (CARrier grade MEsh Networks). At DTAG Laboratories he is working in the area of mobile and wireless seamless communication. His current research interests cover heterogeneous mesh networking and efficient intelligent radio system concepts. He has (co-)authored some patents and numerous publications for international scientific journals and conferences in the fields of radio engineering and wireless communication.

Ping-Yu Hsu graduated from the CSIE department of National Taiwan University in 1987, received the master's degree from the Computer Science Department of New York University in 1991, and the PhD degree from the Computer Science Department of UCLA in 1995. He is a professor in the Business Administration department of National Central University in Chung-Li, Taiwan, and the director of the ERP center in the University. He also works as the secretary-in-chief of the Chinese ERP association. His research interest focuses on business data applications, including data modeling, data warehousing, data mining, and ERP applications in business domains. His papers have been published in IEEE transactions on Software Engineering, Information Systems, Information Sciences and various other journals.

Lutfi Mohammed Omer Khanbary received his B.E.E. degree in 1995 from Aden University, Aden, Yemen, and the M.Tech. degree in 2006 from Jawaharlal Nehru University (JNU), New Delhi, India, where he is currently pursuing his Ph.D. degree in Computer Science with the School of Computer and Systems Sciences, on study leave from the Department of Computer Science and Engineering, Aden University, Yamen. His research interests include mobile computing, network management, and genetic algorithms.

Mario Kind received a Dipl.-Inf (FH) degree in communications and information technology at Deutsche Telekom Hochschule für Telekommunikation Leipzig (FH), University of Applied Sciences (HfTL). He is employed as expert in the area of broadband access networks. His main working area is the economic evaluation of business, technology and service trends in the internet, telecommunication and broadcast industry. Mario is author or co-author of several papers which has been published in international telecommunications conferences and journals.

K.R.Renjish Kumar is a PhD student at Aalto University School of Science and Technology. He is also Senior Consultant, Global Media & Telecom, at Wipro Technologies, India with focus on communications service provider business. Prior to joining Wipro, he was involved in techno-economic analysis, thought leadership and technology design in the telecom and IT sector during his stint with Capgemini consulting - India, Helsinki University of Technology - Finland, Siemens ICM - Singapore and Cognizant Technology Solutions - India. Renjish has a Masters (Computer science) from National University of Singapore (NUS) and B.E. (Electronics and communications) from National Institute of Technology (N.I.T) Surathkal, India. He has participated in CELTIC/EUREKA project ECOSYS, involving key players of the European telecom industry as team member and work package leader for Fixed-Mobile

Convergence. Renjish has published over 20 papers in international conferences and journals on topics covering technology as well as business issues in telecom and Internet.

Xi Li received her bachelor degree (B.Sc.) in Electrical Engineering at Sun Yat-sen University (Zhongshan University), China in 1999 and her master degree (M.Sc.) in Electronics & Telecommunication engineering at the Dresden University of Technology, Germany in 2002. In 2003, she joined the Communication Networks Group (TZI / Center for Communication and Information Technology, Mobile Research Center) of the University of Bremen as a scientist researcher and PhD candidate. She has worked in the European IST-Xmotion project and since 2003 started working in the industrial research project funded by Nokia Siemens Networks (NSN) on network dimensioning for 3G UMTS radio access networks including Rel99, HSPA and at present LTE. Xi Li has published many scientific papers in the field of dimensioning of UMTS networks and is a member of IEEE. Her research interests are performance analysis of UMTS networks and analytical dimensioning methods.

Say Ying Lim received Bachelor of Business Systems, Master of Business Systems, and PhD in Computer Science all from Monash University, Australia, where she worked on mobile query processing under the supervision of Prof B. Srinivasan and A/Prof D. Taniar. She is currently an academic staff at the School of Business, Sunway Campus, Monash Malaysia. Her research is to adopt mobile technologies in E-Commerce.

Xuquan Lin was born in China. He came to USA for study all by himself after he finished his high school. He obtained both his bachelor degree and master degree in computer science at San Jose State University (SJSU). All the soft skills and technical skills he has learned during his 6-year study abroad have provided him a strong foundation for his career. During his study at SJSU, he worked as research intern at Fujitsu labs of America; and worked as software intern at HP. Before he finished his master degree, he joined the Networked-Energy-Service-System group at Echelon to provide advanced metering solution. Also, he has previously published two refereed papers based on his master thesis. In his spare time, he is interested in research on energy-saving protocols and applications.

Fei Liu (1980) received her B.Sc. degree in Computer Science from Jiangsu Institute of Petrochemical Techonolgy, China, in 2002, and her M.Sc. degree in Telematics from University of Twente, the Netherlands, in 2004. From 2004 till now, she is a Ph.D candidate at the same university. Her research interests include mobile-, wireless-, and ad-hoc networks, especially in modeling and performance evaluation.

Mário Macedo is Assistant Professor at Universidade Nova de Lisboa (UNL), which joined in the year of 1986, and researcher at Instituto de Engenharia de Sistemas e Computadores (INESC-ID) at Lisbon, since 1995. He graduated in Electrical Engineering (1983) in the Instituto Superior Técnico of Lisbon (IST), has a M.S. degree in Computer Science from IST (1993), and a Ph.D. degree in Telecommunications from UNL (2000). He has been involved on several research projects with INESC in the area of local area networks (LANs), namely Hybrid Fibre Coax/IEEE802.14, IEEE802.11, ad hoc networks, and sensor networks. His main contributions are in the fields of MAC protocols and scheduling, and applications of control methods. More recently, he has been interested in slot assignment algorithms for TDMA-based sensor networks.

Gennaro Ciro Malafronte received master degree in Electronic Engineering in 1999 and PhD in Electrical Engineering on Electrical and Electronic Measurements in 2003 at the University of Naples "Federico II". In 2001, he joined Siemens COM S.p.A., now NSN S.p.A.. Among his activities he was responsible for the definition of HSDPA/HSUPA Iub/Iur congestion control algorithms and he contributed to the 3GPP for the definition of standardized procedures for Iub/Iur congestion control. Dr. Gennaro Ciro Malafronte holds a number of patents in the field of Iub/Iur HSDPA/HSUPA congestion control. His current field of interest is Transport Optimization for LTE systems.

Zoubir Mammeri received his M.S, Ph.D and HDR in computer science from National Polytechnic Institute of Lorraine (France). Since 1998, he is full professor at Paul Sabatier University of Toulouse. His research interests include: quality of service, QoS-based routing, packet scheduling, mobile ad hoc networks, sensor networks, security in sensor and vehicular networks, real-time systems, scheduling, real-time distributed systems. In the last 15 years, he served as program committee member of more than one hundred international conferences and workshops. He published several papers and books on real-time systems and communication networks. He is a member of IEEE and IFIP.

Melody Moh was born in Taiwan and had been in Singapore during high school years. She obtained her BSEE from National Taiwan University. Her MS and Ph.D., both in Computer Science, were from University of California - Davis. Dr. Moh joined the Department of Mathematics and Computer Science, San Jose State University (SJSU) in 1993, and has been a Professor in the Department of Computer Science since Aug 2003. Prior to joining SJSU, she had two-year industry experience in the area of distributed computing in the Silicon Valley. In addition, she was with 3Com Technology Development Center while on sabbatical leave in 1999/2000., and with Fujitsu Laboratories of America in 2006/07. Dr. Moh's research interests include mobile, wireless networks, sensor networks and RFID for biomedical applications, and mobile Internet. She has published over 100 refereed technical papers in international journals and conferences, has supervised over 50 graduate students on their theses, and has consulted for various companies.

Thomas Monath received the Dipl.-Ing. degree at the University in Rostock in communication engineering in 1997. He is specialized in techno-economic analyses of telecommunication networks and project leader of strategic access network evolution projects within Deutsche Telekom Laboratories. He has been involved in several European projects that focussed on broadband access network evolution. He is author or co-author of several papers that have been published in international telecommunication conference proceedings and journals.

Nidal Nasser received his B.Sc. and M.Sc. degrees with Honors in Computer Engineering from Kuwait University, State of Kuwait, in 1996 and 1999, respectively. He completed his Ph.D. in the School of Computing at Queen's University, Kingston, Ontario, Canada, in 2004. He is currently an Associate Professor in the Department of Computing and Information Science at University of Guelph, Guelph, Ontario, Canada. He is the founding and director of Wireless Networking and Mobile Computing Research Laboratory at Guelph (WiNG: http://www.cis.uoguelph.ca/~nasser/index.php/Research). He has authored several journal publications, refereed conference publications and book chapters. He has also given tutorials in major international conferences. He is an Associate Editor of the Journal of Computer

Systems, Networks, and Communications, Wiley's International Journal of Wireless Communications and Mobile Computing and Wiley's Security and Communication Networks Journal. He has been a member of the technical program and organizing committees of several international IEEE conferences and workshops. Dr. Nasser is a member of several IEEE technical committees. His current research interests include, multimedia wireless cellular networks, wireless sensor networks and heterogeneous wireless data networks interconnection, with special emphasis on the following topics, radio resource management techniques, performance modeling and analysis and provisioning QoS. Dr. Nasser received Fund for Scholarly and Professional Development Award in 2004 from Queen's University. He received the Best Research Paper Award at the ACS/IEEE International Conference on Computer Systems and Applications (AICCSA'08).

Mário Nunes graduated with the Electronics Engineer degree in 1975, Ph.D. degree in Electronics Engineer and Computers in 1987, and the Aggregation degree in the same area in 2006, all from the Instituto Superior Técnico, Technical University of Lisbon, Portugal. He is now Associated Professor at Instituto Superior Técnico, where he teaches in telecommunications and networking areas in graduate and postgraduate courses. He has been responsible for the INESC participation in several european projects, namely RACE, ACTS and IST programs in the areas of fixed and wireless networks. Since 2001 he is Director of INESC Inovação, where he is coordinator of the Telecom Area. He is author of two books and submitted 10 patents. He is a Senior Member of IEEE.

Elena Pagani received her Master degree in Computer Science from the Università degli Studi di Milano, Italy, in 1992. From 1992 to 1993 she had a grant of the Italian National Research Council for the Telecommunication Research Project. In 1999 she received her Ph.D. degree in Computer Science from the Università degli Studi di Milano. From March 2006 she is associate professor, at the Information Science and Communication Department of the Università degli Studi di Milano. Her research interests concern networks protocols and architectures, wireless technologies, distributed systems, and performance evaluation.

Antonio Puliafito is a full professor of computer engineering at the University of Messina, Italy. His interests include parallel and distributed systems, networking, wireless and GRID and Cloud computing. He is currently the director of the RFIDLab, a joint research lab with Oracle and Intel on RFID and wireless. He is currently the vice-president of the Consorzio Cometa whose aim is to enhance and exploit high performance computing. He is also the responsible for the University of Messina of two big Grid Projects (TriGrid VL, http://www.trigrid.it, and PI2S2, http://www.pi2s2.it) funded by the Sicilian Regional Government and by the Ministry of University and Research, respectively. He is currently a member of the general assembly and of the technical committee of the Reservoir project, an IP project funded from the European Commission under the seventh FP to explore the deployment and management of IT services across different administrative domains, IT platforms and geographies. He is the scientific director of Inquadro s.r.l., a spin-off company of the University of Messina whose main business is RFID and its application both in public and private sectors.

Gian Paolo Rossi received the degree in Computer Science from the University of Milano in 1976. After working for the Joint Research Centre of the EC in Ispra, from 1975 to 1980, he joined the Computer

Science Department at the University of Milano where he is now Professor. His research interests cover distributed computing and computer networks, wired and wireless. He has designed and experimented several network protocols and algorithms under MIUR, CNR and EC grants. Currently he is leading the CARTOON national project, on the design of communication protocols over Opportunistic Networks.

Siuli Roy is a PhD in Computer Science from Jadavpur University, India with deep domain knowledge in RFID and Ad-hoc wireless networking. She is actively involved in R&D in the area of evolving wireless technologies for last eight years and is working as a Principal Researcher in Ad Hoc Networks and Application Group at IIM Calcutta. She has also worked as a Visiting Researcher in ATR, ACR Lab, Japan. Her research interests includes wireless sensor networks and ad hoc wireless networks. Siuli has significant work experience in startup environment. She is on the Board of Directors of PervCom Consulting, a startup company in India, developing devices and system solutions with sensors and active RFIDs for tracking, monitoring and management of objects and environment.

Ming Shang received his M.Sc. degree in the Department of Computing and Information Science at the University of Guelph, Ontario, Canada, in 2008. Before that, he worked for Nortel Networks mainly on the Soft Switching and Voice over IP development. He is now working for Research In Motion Limited focusing on Testing and Test Control Notation (TTCN). His research interests include policy control, Service Delivery Environment (SDE), IP Multimedia Subsystem (IMS), Next Generation Networks (NGN) and TTCN.

Wen-Lung Shiau received the MS in Computer Science from Polytechnic University, New York, U.S. in 1995 and the PhD degree from the Department of Business Administration, National Central University in Chung-Li, Taiwan in 2006. He is an assistant professor in the Information Management department of Ming Chuan University in Taoyuan County, Taiwan, He was a certified instructor of Cisco (CCSI), Novell (CNI), and Microsoft (MCT) and trained innumerable certified engineers. He wrote more than 50 books and published the first IPv6 book in traditional Chinese. His current research interests include ERP, Data Mining, Networking, and Telecommunication. His papers have been published or accepted in Journal of E-Business, International Journal of Internet Protocol Technology, Journal of Internet Technology, Computer Communications, JEIM and Expert Systems with Applications.

Bala Srinivasan is a professor of information technology and head of school of the Clayton School of Information Technology at the Faculty of Information Technology, Monash University, Australia. He was formerly an academic staff member of the Department of Computer Science and Information Systems at the National University of Singapore and the Indian Institute of Technology, Kanpur, India. He has authored and jointly edited 6 technical books and authored and co-authored more than 150 international refereed publications in journals and conferences in the areas of multimedia databases, data communications, data mining and distributed systems. He is a founding chairman of the Australasian database conference. He was awarded the Monash Vice-Chancellor medal for post-graduate supervision. He holds a Bachelor of Engineering Honours degree in electronics and communication engineering, a masterÕs and a PhD degree, both in computer science.

David Taniar holds Bachelor, Master, and PhD degrees - all in Computer Science, with a particular specialty in Databases. His current research is applying data management techniques to various domains, including mobile and geography information systems, parallel and grid computing, web engineering, and data mining. Every year he publishes extensively, including his recent co-authored book: High Performance Parallel Database Processing and Grid Databases (John Wiley & Sons, 2008). His list of publications can be viewed at the DBLP server (http://www.informatik.uni-trier.de/~ley/db/indices/a-tree/t/Taniar:David.html). He is a founding editor-in-chief of Mobile Information Systems, IOS Press, The Netherlands. He is currently an Associate Professor at the Faculty of Information Technology, Monash University, Australia. He can be contacted at David.Taniar@infotech.monash.edu.au.

Massimo Villari received his PhD in 2003 Computer Science School of Engineering and the Laurea degree (bachelor's degree + masters) in 1999 in Electronic Engineering, University of Messina, Italy. Since 2006 he is an Assistant Professor at University of Messina. He is actively working as IT Security and Distributed Systems Analyst in cloud computing, virtualization and Storage for the European Projects "RESERVOIR" and "VISION". Previously, he was an academic advisor of STMicroelectronics, help an internship in Cisco Systems and worked on the MPEG4IP and NEMO projects. He investigated issues related with user mobility and security, in wireless and ad hoc and sensor networks. He is IEEE member. Currently he is strongly involved on EU Future Internet initiatives, specifically Cloud Computing and Security in Distributed Systems. His main research interests include virtualization, migration, security, federation, and autonomic systems.

Nicolas Van Wambeke received both the Engineering and Master of Research degrees in Network and Telecommunications from the National Institute of Applied Science (INSA) of Toulouse in 2006. He is currently pursuing doctoral studies at INSA and performs his research activities at the Laboratory for Analysis and Architecture of Systems (LAAS), a part of the french National Center of Scientific Research (CNRS). His main research interests deal with next generation internet architectures for seamless, context-aware, optimized communications.

Deo Prakash Vidyarthi, received Master Degree in Computer Application from MMM Engineering College Gorakhpur, India in the year 1991 and PhD in Computer Science from Jabalpur University (work done at Banaras Hindu University, Varanasi) India, in the year 2002. He taught UG and PG students in the Department of Computer Science of Banaras Hindu University, Varanasi, India. Currently, Dr. Vidyarthi is an Associate Professor at the School of Computer & Systems Sciences, Jawaharlal Nehru University, New Delhi He has 17 years of teaching and research experience. He has more than 45 publications in many International Journals and Peer reviewed conferences of repute. His research interest includes Parallel and Distributed System, Grid Computing, Mobile Computing. Dr. Vidyarthi is the member of International Society of Research in Science and Technology (ISRST).

Yasir Zaki received his bachelor (B.Sc.) degree in Electronics and Communication at the University of Baghdad, Iraq in 2004 and master degree (M.Sc.) in Communication and Information Technology at University of Bremen, Germany in 2007. After completing his bachelor degree he was awarded a DAAD scholarship to finish his master studies in Germany and after completing his master he joined the Center for Computer Science and Information Technology (TZI) of the University of Bremen in the Communication Networks group as a scientist researcher and a PhD candidate in 2007. He worked in the industrial research project funded by Nokia Siemens Networks on performance optimization of UMTS/HSPA networks. Starting from 2008 he started working in the 4WARD European project that is focusing on the Future Internet.

Index

Symbols

A

B

C

D

E

F

G

H

O

P

Q

R

S

T

U

V

W

Z

CPSIA information can be obtained at www.ICGtesting.com
262421BV00008B/1-54/P